Secrets of Nature

Transformations: Studies in the History of Science and Technology
Jed Buchwald, general editor

Sungook Hong, *Wireless: From Marconi's Black-Box to the Audion*

Myles Jackson, *Spectrum of Belief: Joseph von Fraunhofer and the Craft of Precision Optics*

William R. Newman and Anthony Grafton, editors, *Secrets of Nature: Astrology and Alchemy in Early Modern Europe*

Alan J. Rocke, *Nationalizing Science: Adolphe Wurtz and the Battle for French Chemistry*

Secrets of Nature
Astrology and Alchemy in Early Modern Europe

edited by William R. Newman and Anthony Grafton

The MIT Press
Cambridge, Massachusetts
London, England

First MIT Press paperback edition, 2006

© 2001 Massachusetts Institute of Technology
All rights reserved. No part of this book may be reproduced in any form by any electronic or mechanical means (including photocopying, recording, or information storage and retrieval) without permission in writing from the publisher.

MIT Press books may be purchased at special quantity discounts for business or sales promotional use. For information, please e-mail special_sales@mitpress.mit.edu or write to Special Sales Department, The MIT Press, 55 Hayward Street, Cambridge, MA 02142.

This book was set in Sabon by Graphic Composition, Inc. and was printed and bound in the United States of America.

Library of Congress Cataloging-in-Publication Data
Secrets of nature: astrology and alchemy in early modern Europe / edited by William R. Newman and Anthony Grafton.
    p. cm.— (Transformations)
  Includes bibliographical references and index.
  ISBN 0-262-14075-6 (alk. paper); 0-262-64062-7 (pbk : alk. paper)
    1. Astrology—Europe—History. 2. Alchemy—Europe—History.
I. Newman, William R. II. Grafton, Anthony. III. Transformations (M.I.T. Press)

BF1676 .S43 2001
133'.094—dc21

2001030602

10 9 8 7 6 5 4 3 2

Contents

1 Introduction: The Problematic Status of Astrology and Alchemy in Premodern Europe   1
William R. Newman and Anthony Grafton

2 "Veritatis amor dulcissimus": Aspects of Cardano's Astrology   39
Germana Ernst

3 Between the Election and My Hopes: Girolamo Cardano and Medical Astrology   69
Anthony Grafton and Nancy Siraisi

4 Celestial Offerings: Astrological Motifs in the Dedicatory Letters of Kepler's *Astronomia Nova* and Galileo's *Sidereus Nuncius*   133
H. Darrel Rutkin

5 *Astronomia inferior:* Legacies of Johannes Trithemius and John Dee   173
N. H. Clulee

6 The Rosicrucian Hoax in France (1623–24)   235
Didier Kahn

7 "The Food of Angels": Simon Forman's Alchemical Medicine   345
Lauren Kassell

8 Some Problems with the Historiography of Alchemy   385
Lawrence M. Principe and William R. Newman

Contributors   433
Index   435

Frontispiece: Chart of alchemical elections from Thomas Norton, *Ordinal of Alchemy*, found in British Library MS. Add. 10302, late fifteenth century.

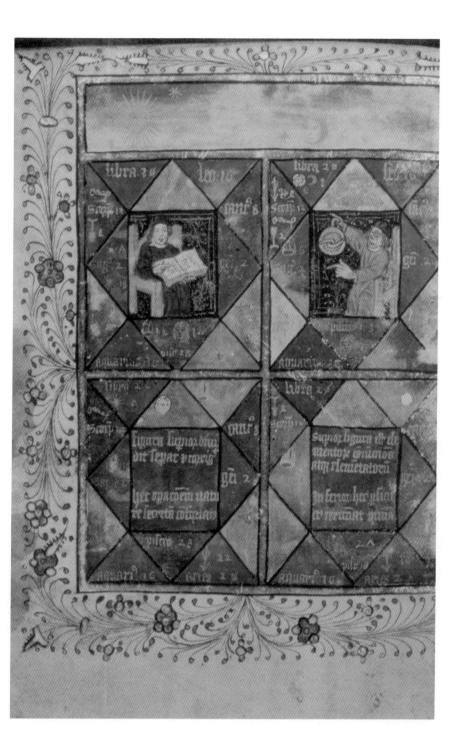

1

Introduction: The Problematic Status of Astrology and Alchemy in Premodern Europe

William R. Newman and Anthony Grafton

The Position of Astrology

One night in 1631, a young Jesuit lay sleeping in his order's college at Würzburg. He slept the sleep of the just, not only because he had found a scholarly vocation, but even more because the Holy Roman Empire had reached an uneasy state of truce. The emperor had conquered his Protestant enemies; no one, the Jesuit later recalled, could even imagine that heresy would revive. Suddenly a bright light filled the room. Waking, he leapt out of bed and ran to the window. He saw the open square before the college full of armed men and horses. Hurrying from room to room, he found that everyone else was still deeply asleep and decided that he must have been dreaming. So he ran to the window, where he saw the same terrifying vision. But when he woke someone to serve as a witness, it had vanished. In the next few days, he became a prey to fear and depression and ran about, as he later recalled, "like a fanatic," predicting disaster. The others made fun of him—until, with satisfying rapidity, invaders materialized and the city fell. Suddenly, the prophet was treated with respect in his own country. Since he taught, among other subjects, mathematics, his friends inferred that he must have used one of his technical skills to forecast the invasion. Surely, they argued, he had used the art of astrology to make his prediction. Nothing else could explain his ability to foresee so unexpected a turn of events.[1]

The young Jesuit, Athanasius Kircher, had actually foreseen the future through direct divine inspiration, a fact he carefully concealed. What matters, from our point of view, is the reaction of his friends. As late as the 1630s, the most highly educated young men in south Germany still found it rational to believe that astrology could enable Kircher to predict vital political and military events. Evidently astrology still enjoyed a level of credibility that now seems hard to fathom, and that in a highly educated and deeply Christian milieu. Kircher, who became not only a brilliant archaeologist and Oriental scholar, but also a practitioner of natural science so adept that his public demonstrations won him at least one charge of being a magician, evidently agreed with his friends' belief in the ancient art of predicting the future through the stars, even though he had not had recourse to it in this case.[2] Astrology was not classified as occult or dismissed as superstitious: it was, in fact, a recognized, publicly practiced art.

This story illustrates the principle, as vital as it is easily ignored, that the past is another country. In educated circles in the United States and Europe, astrology seems merely risible now. No member of the elite wants to be caught with an astrologer. The revelation that former U.S. President Ronald Reagan and his wife, Nancy, regularly consulted the astrologer Joan Quigley was trumpeted by their liberal critics and ignored by their conservative allies. And when the *Economist* noted, a few years ago, that a Brazilian stock advice service that relied on the stars had made enormous profits for its customers, it covered the phenomenon only in order to heap ridicule on all concerned, although the service had scored multiple successes. Astrology has, and can have, no currency in our skeptical, myth-shredding intellectual economy. Even the most astute scholars share these views. Theodor Wiesengrund Adorno, for example, argued that only credulous fools with authoritarian personalities would resign psychic control over their lives to the stars. That explained, in his view, the fact that the art flourished in Los Angeles. Only in the German preface to his essay, originally written in English, on the *Los Angeles Times'* astrology column was Adorno honest enough to admit that the Germans of the 1920s and 1930s had exhibited similar tendencies themselves.[3]

No one who starts from presuppositions like this can hope to understand the pull of astrology in the present, to grasp and explain the fact that between 20 and 50 percent of the population of the world's developed countries, in western Europe, North America, and Asia, believe mildly or strongly in astrology, right now.[4] It is all the more necessary, then, to adopt a different attitude when we turn toward the nature and role of astrology and related disciplines in the past. What E. P. Thompson magisterially condemned as "the enormous condescension of posterity" can only hinder us from understanding the beliefs and practices of practitioners of astrology and those who used their services.

Renaissance astrologers, for example, drew up luxurious, custom-made manuscript genitures for the rulers of Renaissance Europe: not only for their spouses and children, but also, of course, for their enemies. They tabulated the births and fates of men, women, and monsters in collections of genitures, first in manuscript and then in print.[5] On request, they investigated what the planets had to foretell at a particular moment about specific marriages, journeys, and investments or about their clients' physical and mental health.[6] Often they stalked city streets and squares, hawking almanacs: pamphlets, usually of eight or sixteen pages, in which they explained why planetary conjunctions or eclipses foreshadowed disaster.[7] Astrological doctrines inspired some of the most spectacular works of Renaissance art, from the frescoes of the Palazzo Schifanoia in Ferrara to Albrecht Dürer's *Melencolia I*.[8] Astrological practices influenced and sometimes ruled behavior in the most modern, forward-looking sectors of Renaissance culture. At the court of the Estensi in the 1440s, the brilliant young Marquis Leonello changed his clothes according to an astrologically determined rhythm, choosing each day a color that would draw down the favorable influence of a particular planet. In Republican Florence, in the same years, the mercenary captains who led the city's armies received their batons of command at an astrologically determined time.[9] In much of early modern Europe, in other words, astrology was a prominent feature of the practice of everyday life. A good many of the most eminent protagonists of the Scientific Revolution, finally, joined in the production of genitures and almanacs, and a number, including Kepler, worked hard to reform the art in the light of philosophical criticisms and new scientific data.[10]

Yet few historians are willing, even now, to give astrology its due. Tradition weighs heavily against doing so. Even before Friedrich von Bezold, Aby Warburg, and other pioneers began to study the subject systematically in the last years of the nineteenth century, Jacob Burckhardt had described the humanists who revived the art as so many sorcerers' apprentices. They wished to find in ancient culture ways to express their new, objective understanding of the world around them and their new, subjective understanding of their own individuality. Sadly, they made the error of believing that astrology was one of these. In fact, the revival of this fetid, authoritarian superstition led them both to misread the cosmos as a whole and to subjugate the individual to universal laws. Astrology, in other words, hampered the rise of a new culture—until Pico della Mirandola gave the superstition its death blow with his brilliant dialogues.[11]

Warburg knew far more than Burckhardt about astrology; he knew, for example, that it permeated European society at every level long after Pico's death. Yet he inherited not only the great Swiss historian's method, but also his attitude to superstition. Warburg found the primitive elements in advanced systems of thought as terrifying as they were fascinating. He regarded astrology as a threat to reason, one against which philosophers and theologians had had to struggle, in antiquity and in the Renaissance. He himself, during the mental crisis caused by World War I, wandered the streets of Hamburg looking for dark-faced, "Saturnian" children to whom he would give chocolates in the hope of warding off the threat posed by the most malevolent of planets. Warburg saw astrology as incoherent and debased and its practitioners as credulous. For all the fascination with which he studied ancient images of stellar demons, the *Table Talk* of Martin Luther, who denounced astrologers as incompetent, inspired the warmest enthusiasm in him. Luther mocked the genitures that Italian astrologers had put into circulation and that connected his birthday with celestial portents like the great conjunction of 1484. After all, he pointed out, the date of his own birth was uncertain even to him. Luther showed himself even more intolerant when a conjunction in the sign of Pisces, which took place in 1524, led many astrologers to predict that a second universal flood would take place, but none of them foresaw the Peasants' Revolt of 1525. Astrology was related to reason, Warburg thought, only because it provoked the exercise of that all too rare faculty. It belonged to that "Jerusalem" of Eastern superstitions that, over the centuries, had required all the efforts of "Athenian" critical reason to dispel them.[12] Similar prejudices recur regularly in some of the best modern studies.

Thus, the Bologna historian Ottavia Niccoli has brilliantly explicated the large role that astrology played in that curious, enormous literature of threat and promise, the dozens of pamphlets that bearded, ragged itinerants hawked and preached through the streets and squares of Italian cities in the years around 1500. But her heart also clearly lies with the representatives of reason: sophisticated urbanites, in this case, many of them poor as well as skeptical, who mocked the flood when it did not happen on time.

She eloquently evokes the public derision they directed toward astrologers and their clients in the form, for example, of carnival plays that made fun of the credulity of the clerics who had ordered penitential processions and of the ordinary citizens who had fled their homes for high ground. These rituals, Niccoli holds, deprived the figure of the prophet, in Italy, of much of its cultural prominence: after the 1530s, no deluge. Once again, skepticism and realism accompany the attack on astrology; credulity and superstition explain its continued practice. But in fact, prophets inspired by astrology flourished in Italy throughout the sixteenth century.[13] Astrological portents helped Tommaso Campanella decide that the time was ripe for the Calabrian conspiracy he led in the years just before 1600, and astrological eugenics and medicine played a central role in his blueprint for a new and just society, the *City of the Sun*.[14]

Astrology formed more than a set of abstract theories and beliefs. It was also a coherent body of practices, strongly supported by institutions. Modern economists retain their value to employers even when events overtake or refute their concrete forecasts about currency, interest rates, and stock markets. Similarly, the Renaissance astrologer retained his perceived authority and utility even when his individual predictions failed.

Two brief case studies, one drawn from Italy, the other from the Holy Roman Empire, may suggest a new way of looking at early modern astrology: not as a fatty blockage of the intellectual arteries, a bit of philosophical detritus inadvertently fished up from the past along with Aristotelian and Stoic theories about matter and the cosmos, but as one of the many highly practical sets of intellectual tools that Renaissance thinkers forged and honed for dealing with the same problems that they also attacked with what now seem the shinier tools of social and political analysis.

No Renaissance text offers a richer or more unexpected peep into the astrologer's atelier than Leon Battista Alberti's dialogues *On the Family*. In book 4 of this famous description of the life and beliefs of a great Florentine clan, Piero Alberti explains how he won favor at the court of Giangaleazzo Visconti, duke of Milan and inveterate enemy of the Florentines. Piero had used every technique known to Renaissance analysts of the ways of courts—or to modern sociologists specializing in network theory—to haul himself up the greasy rope of court favor. Unexpectedly, his deep knowledge of Italian literature gave him his main chance. By reciting poems, he attracted the attention of someone who already stood high at the

Visconti court, Francesco Barbavara. This was no easy feat. Like Saint-Simon 250 years later, Piero noted that the courtier must never leave the presence of power. Accordingly, he had spent whole days without food, "pretending to have other concerns," just waiting "to encounter and greet" his patron, even though Barbavara was already on the lookout for talented men to support, since his own high position at court rested on his ability to dispense help from his favor bank to those who clung to lower rungs on the ladder.[15] Only precisely calibrated, pragmatic social tactics like these could have enabled Piero, a poor man, to become the friend of Barbavara, who, in turn, brought him to Giangaleazzo's attention.

Piero, himself an exile from Florence, found the duke gracious and eager to befriend him. But he did not have to depend on his family name or his knowledge of the sonnet form to ingratiate himself. "At that time," he recalled, "the learned astronomers were anxiously expecting some sort of great trouble, for the sky showed them clear indications of upheaval, particularly of the overthrow of republics, governments and persons in high command. It was their almost unanimous opinion that the comet which shone brightly in the middle of the sky and was for months visible even in the daytime would not be shining so long if it did not portend, as comets usually do, the end and death of some famous, powerful prince, like the duke himself."[16] Giangaleazzo met this prediction with an impressive, even princely mixture of pride and resignation: "The heavenly intelligences' concern to give him a rare and marvelous omen and sign, he said, surely proved to the world that the divine and immortal spirits in the skies were interested in his life and death."[17]

Piero, however, divined that Giangaleazzo's bravado masked "some considerable inward anxiety."[18] Fortunately, he had information at his disposal that enabled him to allay it. The Alberti firm, which had played a prominent role for more than a century in the medieval Mediterranean trading system, had offices throughout the world, from England, Flanders, and France in the north to Catalonia, Rhodes, Syria, and Barbary in the south and east. These branches, like those of all Florentine firms, were normally headed by members of the family and remained in constant communication, using their own highly efficient private message services. They kept Piero "well informed of the revolts, mobilization of ships and men, shipwrecks, or whatever was going on in those regions worth knowing."[19] In this case, the Rhodes office called Piero's attention "on the instant" to the death of Timur, the ruler of the great city of Samarkand. Piero was able to show Giangaleazzo that the comet had betokened the death of another, even greater prince. His deft rereading of the portent kept the duke "benevolent" to his servant.[20] Skillful interpretation of the astrologers' predictions, combined with skillful sifting of the news, made it possible to climb the court ladder.

The story cannot be taken literally. Timur actually died in 1405, three years after Giangaleazzo. Piero, or the author, may have confused the death of Timur with the capture of Bayazid, which took place at the battle of Angora in July 1402, a few months before Giangaleazzo died. Or Alberti may simply have invented the anecdote he needed.[21] But the form in which Piero's story appears matters more to the cultural historian than its factual correctness. When Alberti wrote book 4 of *On the Family*, he was already embarked on a dazzling career as a court adviser on matters scientific and architectural. He would become a favorite of Leonello d'Este, Federigo da Montefeltro, and Ludovico Gonzaga, among other rulers with advanced artistic and literary tastes.[22]

Like his friend Lapo da Castiglionchio, who wrote his bitter dialogues on *The Advantages of Life in the Papal Curia* in 1438, Alberti brought to bear on his subject all the bitter realism of the outsider.[23] A special paranoid, microscopic attentiveness to the unspoken rules of social success enables such observers to grasp and describe the norms of a hierarchical society more vividly and fully than more successful insiders can. Unlike Lapo, moreover, this outsider became an insider and thus benefited from direct, close-up observation of the phenomena he hoped to understand.

Leon Battista made, as he explained in his autobiography, a conscious decision to transform his life into a performance. He concentrated, he recalled, on performing public acts in a way that would win the respect of all beholders and transmuted his ways of walking, riding and conversing into dazzling works of art.[24] His ability to frame and apply the arcane precepts that could turn ordinary conduct into something of aesthetic value has been emblematic for a century and a third—ever since Burckhardt—of the new objectivity that characterized the period, the new ability to observe manners and mores that formed an essential part of "the discovery of the world and of man."

Burckhardt, as we have seen, regarded astrology and objectivity or realism as radically opposed. Astrology subordinated individuals and nations to larger, impersonal forces embodied in the revolutions of the planets. It rested on traditional, unverified beliefs rather than empirical evidence. And it treated its subjects not as individuals, to be understood through the new tools of introspection, but as types. True Renaissance intellectuals, like Pico, fought against astrology's influence on behalf of the real and the individual.

For Alberti, however, or at least for his imaginary character, astrological investigations evidently did not conflict with hard-nosed realism. Piero did not mock Giangaleazzo for his credulity; he argued that the duke and his advisers had misinterpreted a true portent. And elsewhere in Alberti's dialogue, another character made the implications of the author's view of astrology even clearer. Lionardo, a young, learned member of the family, asks his older relative Adovardo to explain the nature of friendship. Drawing on the ancient historians and philosophers, Lionardo cites numerous classical models for making, retaining, and ending friendships. And he suggests that general precepts could be drawn from these. Adovardo demurs:

Ah, Lionardo, what a mass of further information would be needed before one could really discuss this matter in its breadth and extent. It is as though some student had heard from the astronomers that Mars disposes the force of armies and the outcome of battles, Mercury establishes the various branches of knowledge and governs the subtlety of minds and marvelous skills, Jove controls ceremonies and the souls of religious men, the sun reigns over worldly offices and principalities, the moon precipitates journeys and the fluctuations of spirit among women and mobs, Saturn weighs down and slows our mental processes and undertakings—and so he would know the character and power of each. But if he did not know how to evaluate their effect according to their place in the sky and their elevation, and what favorable or unfavorable effect their rays have on each other, and how their conjunctions are able to produce good or ill fortune, surely that student would be no astrologer. The mere recognition of those bare principles is indispensable to any understanding of the art, but even with them you have only just entered the domain of other, almost innumerable laws necessary if you would foresee and understand the things which they sky tends to produce. Similarly, these very useful and numerous examples and sayings, which you say are so amply provided by the best authors, do not give us all the help we need.[25]

Alberti depicts astrology here as the model for a rigorous art of social relations. It rests on clear general principles, but it also explains in detail how these interact with one another in everyday life. Instead of naming single influences, the astrologer must trace and evaluate the whole complex web of influences spun by the planets and the stars. And only an analysis of friendship like this—one based on general principles, but as applied to minutely particular cases—could be really useful.

Alberti believed these sentiments as firmly as the characters into whose mouths he set them. He exchanged astrological letters with the well-known Florentine medical man and astronomer Paolo Toscanelli, a serious observer of comets. In these letters Alberti predicted the immediate future of the Roman church. He dated and timed the points when he began and completed a number of his written works, not just to the day but to the hour, probably, so Riccardo Fubini has suggested, in order to connect them to the configurations of the planets that accompanied his work.[26] In *On the Art of Building* he described how ancient founders had started to build the walls of cities at astrologically propitious times, a rule actually followed for some of the most prominent building projects of the fifteenth and sixteenth centuries.[27] Alberti prided himself, as readers of *On the Family* and *On Painting* know, on his realism, his ability to portray the human body and human society as they really were. Evidently, he saw the astrological techniques that he had mastered and continued to apply throughout his life as neither a delusion or a diversion, but a natural continuation of his efforts to master the laws that formed cities, clans, and individuals.

Similar considerations ensured that astrologers like Cardano and Gaurico would find a welcome at many courts, where aspiring courtiers eagerly read their work. Cardano, in fact, not only drew up genitures for important men and inserted them in his commentary on Ptolemy's textbook, but also composed, at the end of his life, a brilliant and much-discussed manual for courtiers. Here he told them, in language like that of Gracián, how to build a hard, featureless shell around themselves to compete effectively with others while giving nothing at all away.[28] Astrology did not represent a failure of objectivity but exemplified it in action.

A second scene from astrological life took place in historical time rather than the literary imagination. To be more precise, it was enacted in the imperial free city of Augsburg, where Hieronymus Wolf sat down, in 1564, to write an autobiography.[29] Wolf is no longer a household name, even in learned households. In his own time, however, he enjoyed a considerable reputation as a popularizer if not as a scholar. A pupil of the great humanist and astrologer Joachim Camerarius, he attracted the qualified admiration of Philipp Melanchthon. Wolf met with misfortunes of many kinds, from hazing in his schoolboy days to abuse from scholarly colleagues, and he gave detailed accounts of his woes in his picaresque Latin account of his life. But he also achieved a great deal. He translated the Attic orators Demosthenes and Isocrates into Latin, taught rhetoric and poetics, and produced an indispensable—if very faulty—edition of early astrological texts in 1559.[30]

Wolf did more than edit Greek commentaries on Ptolemy's *Tetrabiblos*, however. For all his good Protestant's confidence in salvation, he also believed in the uses of astrology. In fact, Wolf had been asked by friends to write his autobiography even before he started to do so in 1564, so that he could rebut his detractors and preserve the memory of his life. But he waited to begin until he reached the age of forty-eight, only one year short of the year that astrologers had defined as the dangerous climacteric, forty-nine, when he feared he might die.[31] He started his work by considering the positions of the stars at his birth. When the position of the horoscope, the point on the zodiac that was actually rising at his birth, did not seem to him to match the actual course of his life, he decided that the clock had been wrong and considered whether the standard procedures astrologers regularly applied to adjust times when they drew up genitures might yield better results.[32] Wolf was not certain that the many cruel and violent events he experienced really reflected the malevolent influence of Mars or that the many vicious attacks he underwent really resulted from the equally malevolent influence of Saturn. After all, he admitted, divine providence or human magic accounts for many of the events predicted by genitures.[33] But he clearly began his self-examination by poring over the stars that had presided over his birth.

Wolf's reading of his own geniture cannot be fully reconstructed, for reasons that are revealing in themselves. The eighteenth-century scholar Johann Jacob Reiske, who published Wolf's autobiography, omitted the rest of this passage, which he described as "a long passage full of astronomical and horoscopic rubbish," so as not to bore his readers. The twentieth-century Byzantinist Hans-Georg Beck, who translated the original Latin text into German, left out even the truncated horoscope analysis that Reiske had seen fit to print.[34]

For all Wolf's own doubts and the retrospective censorship carried out by his modern students, however, the main point emerges clearly. At some point Wolf, who was himself no astrologer, had a professional draw up a full-scale analysis of his own geniture. Dozens of these documents survive, in print and manuscript, in libraries and archives across Europe; they formed a principal source of income for astrologers like the historical Nostradamus, who, as Jean Dupèbe and Pierre Brind'Amour have shown, created an efficient little boutique where genitures for clients across Europe were drawn up and interpreted.[35]

Genitures normally laid out the positions of the planets at the moment of the client's birth. They explained what consequences these would have for his health, his wealth, his travels, his marriage, his fortune, and his death. And they often included "revolutions": analyses of the positions of the planets at the anniversary of his birth, year by year for fifty or sixty years. Genitures in effect amounted to graphical representations of the client's future bodily and mental health, travels, and career. The client could compare them, detail by detail, to his subjective sense of his own experiences.

In fact, the subjects of such genitures often scrutinized them carefully. Machiavelli's friend Francesco Guicciardini, to cite an example recently studied in detail, showed acid skepticism toward astrology in his *Ricordi*, where he insisted that the art survived only because of the human tendency toward confirmation bias: both the astrologers and their clients remembered only the successful predictions and forgot the more numerous false ones. But he had his geniture drawn up by an astrologer, Ramberto Malatesta, a small-scale feudal lord who fell back on this profession after murdering his wife and being driven from his estate. And he annotated it, expanding the astrological signs into full names of planets and signs—surely a sign of interest.[36]

Others did far more. Cardano brooded so systematically on his geniture that he wrote three full-scale commentaries on it, one after another, all dedicated to showing how precisely the events of his life bore out what the stars had foretold. Ultimately he produced a kind of astrological autobiography: a full-scale account of his own life that began from his geniture and was organized, like a horoscope, topically rather than chronologically.[37] Sir Thomas Smith, Queen Elizabeth's ambassador to France and the author of a famous Aristotelian study of the English constitution, wrote a very similar self-analysis.[38] These astrological autobiographies rank among the frankest works of introspection, the most richly vivid confessional texts, written in the fifteenth and sixteenth centuries. Their authors freely discussed their physical ailments, including venereal diseases and torrential flows of urine; their personal defeats, including the derision they had experienced from social superiors and intellectual enemies; and their character traits, including shyness, awkward conduct, and a tendency to alienate others. Astrology did not, as Burckhardt thought it should, hinder self-scrutiny. On the contrary, it promoted introspection, providing the astrologer's clients with a template of questions to ask themselves about their personalities, their everyday lives, and the larger trajectories of their careers.

Throughout his text, Wolf drew correlations between his ailments and misfortunes and the stars in their courses. But he also made clear that he—and his father before him—consulted astrologers regularly for advice on matters large and small. Wolf's father, for example, said one day that he would have to make way for his son, who had already grown so large that he could wear the shoes he would inherit: this perhaps showed, Wolf thought, that he had consulted an astrologer about his life chances.[39] Wolf himself described more than one strategically placed conversation with an astrologer. In fact, astrologers became his principal authority figures. When he found himself besieged by magically caused noises and misfortunes, he recalled that he had run into the astrologer Georg Joachim Rheticus (better known as the first popularizer of Copernicus) in the street. Rheticus had read his palm and told him to expect trouble from a woman.[40] When he finished writing his most ambitious commentary, the astrologer Cyprian Leowitz (better known as the author of a famous treatise on the great conjunctions of Jupiter and Saturn) told him that his work would make him rich.[41] Others, whom Wolf did not name, advised him about the astrological causes of the case of gonorrhea that he contracted though still a virgin, since Venus resented the fact that he neglected her in favor of Mercury.[42]

In each case the practices Wolf described were widely disseminated, even standard. As Keith Thomas and Michael MacDonald showed long ago in classic works, the default occupation of the Renaissance astrologer was not drawing up full-scale genitures but providing small-scale, short-term advice in the form of "elections" and "interrogations": precise counsel, based on the position of the heavens at a given moment, about the likely outcome of a particular enterprise. Such advice giving could be risky: Cardano, for example, counseled the astrologer against trying to use the stars to tell a client if a given child was his own or someone else's, since violence could ensue. But it was also highly reasonable, in a world that lacked statistics, tables, and insurance policies, to try to use the best mathematical techniques available to foresee, and thus control, the future. Cardano, for example, though he felt considerable distaste for interrogations, still saw them as a reasonable way to determine the gender of a fetus in the womb. The only alternative to reading the heavens, after all, was reading the pregnant woman's body to find the angle at which her fetus hung, a difficult proposition at the best of times, and one whose results were no easier to interpret than the planets.[43]

Though Wolf earned his bread translating and teaching the classics and regretted the time and eyesight he had to waste on barbarous Byzantine Greek, he was not a cloistered, credulous figure. His library, on which he spent such money as he had, became an encyclopedic collection, stuffed with up-to-date theology and philosophy as well as costly Aldine Greek texts. When he tried to diagnose his ailments, he used the modern medical theories of Paracelsus as well as the ancient ones of Hippocrates and Galen. When he continually checked his bodily and mental health against the movements of the planets, he followed the advice of Marsilio Ficino, widely disseminated in the Holy Roman Empire. According to Ficino, the scholar's body was vulnerable, even labile: planetary influences washed constantly through it, like the tides, causing health and illness, inspiring exaltation and depression. Only an astrologically regulated regime of self-medication could keep these influences—from which Ficino himself suffered grievously—under control.[44] In Wolf's own circle, Philipp Melanchthon taught astrology, Joachim Camerarius edited astrological classics, and Caspar Peucer argued that even the devils and angels who sat on the shoulders of every Christian used astrology to unlock the secrets of their characters and lead them more effectively into salvation or damnation.[45]

Wolf's example, like Alberti's, shows astrology in an unfamiliar light. In each case, astrology emerges not as the object of contempt or ridicule but as a rational art of living, comparable to and compatible with the more obviously modern arts of pragmatic politics and courtiership on the one hand and Paracelsian medicine on the other. In the one case astrology helped one to see the world as it really was; in the other, to see inside the darkest recesses of the psyche. These testimonies deserve to be heard. Astrology in the early modern world was not a pathological, but a normal, piece of intellectual equipment. One could use it to find one's way through the dangerous labyrinths of public and private life, to form one's character, and even to devise therapies for one's illnesses. Literati and natural philosophers worked with and tried to improve the tools that astrology afforded them, believing implicitly, in many cases, that human lives were tied to the movements of the planets. Trying to understand the society and culture of early modern Europe without taking astrology into account is exactly as plausible as trying to understand modern society without examining the influence of economics and psychoanalysis.

Astrology and Alchemy as Celestial and Tellurian Twins

In 1564, the same year that Hieronymus Wolf began his autobiography, the English astrologer and polymath John Dee was seeing his *Monas hieroglyphica* through its printing in Antwerp. The *Monas* contains a long and detailed apology, written in the form of a dedication to Maximilian of Habsburg, soon to become Emperor Maximilian II. In this introduction, Dee claims that the astrological sign for Mercury could form the basis of a new scientific language in which alchemy too would figure prominently. Here Dee was able to combine his own ideas about an astrology reformed along lines supplied by natural philosophy and optics with an alchemy that had undergone recent refurbishing at the hands of Marsilio Ficino, Agrippa von Nettesheim, and Paracelsus. Just as early modern astrology was not the pallid caricature that we find in modern newspapers, so Renaissance alchemy was a very distant cousin of the burlesque parody that modern culture has inherited from the scornful dismissals of the Enlightenment *philosophes*. It was not the Art without art, whose beginning is lying, whose middle is labor, and whose end is beggary, of a contemptuous Nicolas Lemery, writing at the end of the seventeenth century.[46] We cannot, like Lemery and his colleagues at the Académie des sciences, create a special dustbin for alchemy and astrology as equally egregious examples of arrant nonsense.

At the same time, early modern alchemy was not a contemplative discipline focusing on internal spiritual development, an idea that would be popularized by nineteenth-century occultists and their later followers. The Mary Anne Atwoods and Eliphas Lévis of the occult revival were quite content to see both astrology and alchemy as encoded forms of wisdom whose real goal was the rechanneling of an internal "Mesmeric fluid," expressed in the form of planetary and metallurgical symbols. And yet it is clear that Renaissance figures such as Dee did see some reason for relating the two fields. But what precisely was that reason? Was it simply the obvious fact that the all-pervasive realm of astrology could be used to find favorable times to begin alchemical operations, in the same way that it could provide the best times for purging a patient, building a building, or starting a war? Or did Dee and other alchemists have something else in mind when they spoke of alchemy as *astronomia inferior* and referred to the science of the stars as a sort of celestial alchemy? Did the disciplines of alchemy and astrology have a privileged and integral relationship with one another that distinguished them from other fields? The quest for an answer supplies the problematic of this book.[47]

Let us clarify the issues as follows. Astrology was a form of divination along with oneiromancy, arithmology, and a host of other techniques for auguring and at times altering the future, whereas alchemy was an artisanal pursuit concerned with the technologies of minerals and metals. The fundamental practices of the two fields were vastly different. Furthermore, if we withdraw our minds from the modern cultural stereotype of "the occult sciences," it is not immediately obvious that the two fields shared a closely related theoretical framework. Already in the second century of our era, the Alexandrian mathematician Claudius Ptolemy observed that astrology was a natural part of mathematical astronomy. Whereas astronomy predicted the positions of the planets, astrology predicted their effects on the earth: Both sciences were therefore part of a larger endeavor concerned with celestial prognostication.[48] Alchemy, on the other hand, had close ties to Aristotelian and Stoic theories of matter, and its early practitioners were enamored of religious themes drawn from what John Dillon has called "the underworld of Platonism."[49] In the Middle Ages, alchemy was not usually considered a mathematical science at all but found itself subordinated to the study of natural philosophy and often compared to the science of medicine.[50]

And yet, if we turn to early modern writers on alchemy and astrology, we shall see that John Dee was not the only figure to view the two fields as part of an overarching discipline in a way that suggests something more than a casual overlap. In his *Theatrum chemicum britannicum* of 1652, the antiquarian and founding member of the Royal Society Elias Ashmole had the following to say about the relationship of alchemy and astrology: "In the *operative* part of this Science [i.e., alchemy] the *Rules* of *Astronomie* and *Astrologie* (as elsewhere I have said) are to be consulted with. . . . So that *Elections*, (whose *Calculatory part* belongs to *Astronomie*, but the *Judiciary* to *Astrologie*) are very necessary to begin this work with."[51]

Ashmole supported himself with a neoplatonic view of the cosmos inherited in large part from the magical writer Agrippa von Nettesheim, according to which the universe is divided into supercelestial, celestial, and natural realms and the sublunary world is the final recipient of divine ideas transmitted by the planets and stars.[52] The secret virtues of material things are infused and stirred to action by the celestial bodies, which are the proximal agents of mundane generation and corruption. Hence it seems perfectly natural that Ashmole would invoke the rules of catarchic astrology in a discussion of alchemy: Since the ingredients of the philosophers' stone are subject to the influences of the heavens, one should time one's alchemical endeavors so that they fall under propitious celestial configurations. Indeed, Ashmole continues, the astrology of elections should be employed in virtually all terrestrial pursuits, including "*Dyet, Building, Dwelling, Apparell,*" the planting of crops, and of course, the manufacture of talismans.[53]

Yet despite the seductive logic of a view that subordinates all sublunary activities to astrological planning, it was not a conclusion to which alchemists as a whole subscribed. The two disciplines had led separate lives too long for their marriage to be an easy one. Indeed, a few years before Ashmole was burning his fingers in casting talismans of caterpillars, moles, and rats,[54] his contemporary and fellow Royalist Thomas Vaughan wrote a critique of such practices: "*The common Astrologer, he takes a stone, or some peece of Metall, figures it with ridiculous Characters, and then exposeth it to the Planets, not in an Alkemusi, but as he dreams himself, he knows not how. . . . It is just thus with the common Astrologer, he exposeth to the Planets a perfect compounded Body, and by this means thinks to performe the Magician's Gamaea, and marry the Inferior and Superior Worlds. It must be a Body reduc'd into Sperm, that the Heavenly Feminine moysture which receives and retains the Impresse of the Astrall Agent, may be at Liberty, and immediately expos'd to the Masculine Fire of Nature.*"[55] Even though Vaughan, like Ashmole, was an apostle of

Agrippa, he denied the efficacy of normal talismanic magic and by implication traditional astrology as a whole. Metals, even when molten, were "perfect, complete bodies," which could not receive the stellar influx. Vaughan thus argued that fully formed matter cannot be imprinted by celestial influences: it must first be reduced to a "nonspecificated" form approaching the Aristotelian first matter.[56] The means of attaining this mysterious substance, which was also the initial ingredient of the philosophers' stone, was alchemy.

The historian is presented here with a peculiar dilemma. On the one hand we have Ashmole claiming that alchemy, along with most other physical pursuits, depends on astrologically determined times. On the other hand we have Vaughan asserting that such propitious moments are without significance unless the alchemist has already produced an inchoate "sperm" for the stars to work upon. In the one case alchemy presupposes astrology, in the other, the contrary. The two followers of Agrippa have arrived at exactly opposite conclusions. What are we to make of this perplexing situation?

The problem is not without significance for the modern understanding of the "occult sciences." It is commonly supposed that the view held by Ashmole was uniformly represented in astrology, alchemy, and natural magic over the *longue durée*. The final chapters of Keith Thomas's deservedly famous *Religion and the Decline of Magic*, for example, are built on the assumption that astrology served as the basis and justification of the other "occult sciences."[57] Hence when astrology declined, it was only natural that belief in alchemy should crumble as well. Brian Vickers takes a more ambitious position, arguing that alchemy and astrology, as well as the other occult sciences, shared a common "mentality" characterized by such traits as a tendency to heap up symbols "paratactically" rather than employing critical thinking, an inability to distinguish object from signifier, and a stagnation of ideas over time. Like Thomas, Vickers adopts astrology as his model, arguing that alchemy and the other occult sciences formed a "unified system" of belief.[58] The same assumption, that alchemy and astrology formed part of a seamless garment, often seems to underlie the research of Frances Yates as well. Although Yates employed her customary modesty when approaching the technical details of the disciplines, it is clear that her notion of a "Hermetic-Cabalist tradition" included alchemy and astrology as sister sciences.[59]

The "Unity of the Occult Sciences" Reexamined

Given the position of such prominent scholars, it may come as a surprise to learn that Vaughan's dismissal of traditional astrology was not a minority view among alchemists. From the time of its entrance into the Latin West of the twelfth century, alchemy as a discipline was notably cool to astrology. It is true that the Latin alchemists acquired the Arabic (and ultimately Greek) practice of substituting the planetary names for the metals, so that gold became *sol*, silver *luna*, copper *venus*, iron *mars*, tin *jupiter*, lead *saturn*, and quicksilver *mercury*. Yet this simple substitution code was only one element in a complex and variant set of *Decknamen* or "cover names" alchemists used. In the rich alchemical glossaries of the Middle Ages, quicksilver's planetary designation had to compete with such names as "the fleeting," "the runaway," "the fugitive slave," "the cloud," "the lightning," "the heavy water," "the spirit," "the fluid," and "water of life," to name but a few.[60] The same was true of the other metals.

The use of planetary *Decknamen* did not necessarily signal the dependence of alchemy upon astrology. It is true that alchemy was sometimes called *astronomia inferior* or *astronomia terrestris* in the Middle Ages, in reference to the planetary *Decknamen*, but this implied only a very superficial relationship.[61] Let us consider an influential Latin text of the early thirteenth century, the *De perfecto magisterio* of pseudo-Aristotle. The author first announces that alchemy should be called "inferior astronomy" as a point of comparison to "superior astronomy" because alchemy deals with stones that are "fixed" in the fire (i.e. nonvolatile) just as astronomy deals with stars that are "fixed in the fiery firmament." Similarly, both disciplines deal with planets, which are erratic, and are borne in a direction contrary to the firmament.[62] He continues: "The stones that are called stars, are sol, luna, mars, saturn, jupiter, venus, niter, calx, carbuncle, emerald, and the other stones that do not flee the fire; but the stones that are called planets are quicksilver, sulfur, arsenic, sal ammoniac, tutia, magnesia, and marchasite. For these do not withstand the fire, but gradually flee upwards and escape."[63]

Despite his elaboration of this trope, the author then presents copious alchemical recipes without any further appeal to the heavens. Here he displays considerable mineralogical skill, giving workable recipes for refining the precious metals and purifying salts, but is entirely unconcerned with employing elections. This is in fact the usual case with medieval alchemy.

If we consider the rather idiosyncratic *Liber secretorum alchimie* of Constantine of Pisa, apparently written in the mid-thirteenth century, a still greater divergence between astrological theory and alchemical practice appears. In its modern edition, the author expends some five pages in describing the need for the alchemist to observe lunations in congealing mercury, but in the twenty-five pages of recipes that follow, no further mention is made of any astrological theme.[64] Evidently the author felt no need to link his allegiance to astrological theory to actual alchemical practice. Another illuminating case may be found in the *Ars alchemie* attributed to Michael Scot, the famous astrologer and philosopher of Frederick II von Hohenstaufen. The text is of early date and may contain elements going back to the genuine Michael Scot. It is all the more surprising, then, that only one recipe out of the thirty or so contained therein has any reference to astrology.[65] This recipe advises that "if you wish to make luna from mercury, then put the mercury in a furnace on the day of luna (i.e., Monday) in the hour of luna, and do this in the augmentation of luna."[66] The recipe contains some further elaborations, but the astrological import, if it can be called that, is clear: To make silver from quicksilver, one must begin the process on the hour and day of the moon when the moon is waxing. If this is astrology, it is of a type that bears little relation to Elias Ashmole's complex nativities and elections. The directions require neither astronomical tables nor any technical knowledge to be carried out. One need only know the day of the week, the time of day in the system of planetary hours, and the approximate phase of the moon. This has more relationship to the age-old common sense of the farmer than to the sophisticated astrological knowledge of an Albumasar or an Alchabitius.

Let us consider one more instance of an alchemist making use of astrology, this time on the threshold of the Renaissance. The late-fifteenth-century *Ordinall of Alchimy* by the Bristol alchemist Thomas Norton (?1433–1513 or 1514) contains a striking illustration of four alchemical elections for beginning different operations in the "great work," which is reproduced in Ashmole's *Theatrum*. There is strong evidence that Ashmole's illustration is based on a presentation copy prepared by Norton himself, and a similar presentation manuscript is still extant in MS Add. 10302 in the British Library in London: Hence we can assume that Norton oversaw the making of the manuscript illuminations.[67] The illustration is followed in Ashmole's text by these comments of Norton's: "Wherof

Concord most kindly and convenient / Is a direct and firie *Ascendent,* / Being signe common for this Operation, / For the multitude of their Iteration: / Fortune your *Ascendent* with his *Lord* also, / Keeping th'aspect of *Shrewes* them fro; / And if thei must let, or needly infect, / Cause them to looke with a *Trine* aspect. / For the *White warke* make fortunate the *Moone,* / For the Lord of the *Fourthe house* likewise be it done; / For that is the *Thesaurum absconditum* of olde Clerks; / Soe of the *Sixt house* for *Servants* of the Werks; / Save them well from greate impediments, As it is in Picture, or like the same intents."[68] All of this sounds at first like a serious commitment to catarchic astrology on Norton's part, but a close inspection of the illustration to which he refers reveals a different story altogether. All four elective schemes are constructed for a latitude of about 52°N, a good approximation for Bristol, and all four employ the Alchabitius house system.[69] The four schemes also agree in placing the ascendent in Sagittarius, which is indeed a member of the fiery triplicity, as Norton advises. In all four schemes, likewise, Jupiter will be the Lord of the Ascendent, since Sagittarius is one of his two domiciles. Since Jupiter is found "bodily" in Sagittarius in two of the four schemes (the second and third, reading clockwise from the top left), and in his other domicile, Pisces, in the first, it is clear that Norton was trying to put him in a propitious spot. It is not so easy, then, to understand why the second and possibly the third scheme have put the maleficent Saturn in conjunction with Jupiter or why the fourth scheme has Jupiter situated in Leo, in the sixth house, and possibly in conjunction with Saturn as well.[70] In the second instance, and perhaps the third and fourth as well, the Lord of the Ascendent, Jupiter, has surely been put into aspect with a "shrewe."

But these violations of Norton's rules raise trivial inconveniences compared to the insurmountable fact that three of the four figures are obviously impossible from an astronomical point of view. In the first scheme, Venus and the sun have an angular separation of 114°, in the second they are separated by 176°, and in the third by either 113° or 143° (depending on whether one interprets the figure to mean that Venus is in Aries or Pisces). As Ashmole admits in his commentary, these schemes therefore exceed the actual maximum separation of Venus from the sun, for which he gives the figure 48°.[71] Ashmole was also forced to point out that even if one disregarded the problems posed by Venus, the planetary positions Norton gave for the superior planets and the sun did not correspond to any period within the time when Norton could have written his book.⁷² As Ashmole put it, "Withall, the *Planets* as they stand here placed in *Signes* and *Houses* are not so that these *Figures* were the *Elected* times for the *Authors* owne *Operations* (or any others in that *Faculty*) but are rather *fained* and invented, onely to bring them within the compasse of his *Rules*. And to satisfie my selfe herein, I have taken some paines to *Calculate* the places of the *Planets* for severall years about the *Authors* time, but cannot find the three *Superiors* and place of the [sun] to be in those *Signes* wherein he has posited them."⁷³ Despite Ashmole's acumen, he was not able to bring himself to the evident conclusion that Norton was an astrological incompetent. Commenting on the unorthodox planetary symbols employed by Norton, Ashmole used their "hieroglyphic" character as evidence that Norton was "a learned *Astrologian*" who would not divulge his secrets to the vulgar but instead employed "*Vailes* and *Shadows,* as in other parts of the *Mistery*." The very presence of seeming mistakes in Norton's elections could be used as evidence that the Bristol alchemist was not a "vulgar *Plumet*," but a Hermetic sage. Such an attitude does not accord well with historical scholarship.

In the case of Norton, then, what seems at first a serious commitment to catarchic astrology turns out to be mere window dressing, for Norton's elections are manifestly unworkable. All the evidence considered up to now points to the conclusion that alchemy and astrology were two quite distinct disciplines in the Middle Ages, although on some occasions they overlapped, as indeed astrology overlapped with medicine, architecture, and a host of other pursuits. This impression is heightened by the fact that alchemists did not merely ignore astrology as a rule: In some famous instances they openly expressed their disregard for or even hostility to it. One trend-setting instance of trivializing astrology can be found in the *Summa perfectionis* of pseudo-Geber, written around the end of the thirteenth century by an occidental. The text was one of the most influential works of alchemy produced in the Middle Ages and was still widely cited in the seventeenth century.⁷⁴ The *Summa* contains a long scholastic introduction in which various objections to alchemy are first raised, then systematically rebutted. One of these objections is the following:

And similarly, being and perfection are given by the stars, as it were by the first things perfecting and moving the matter of generation and corruption to the being and not-being of species. This, moreover, happens in an instant, when one or many stars from its own motions arrives at a determinate position in the firmament, by which is bestowed perfect being, because everything acquires being for itself from a certain position of the stars in a moment. And there is not one position only but many mutually diverse ones just as their effects are diverse. And we cannot know thoroughly their diversity and distinction, for they are unknown and infinite. How therefore will you correct the defect of your work from your ignorance of the diversities among the stellar positions due to their motion?[75]

This attack on alchemy argues that the unknowable and indeed infinite crowd of celestial configurations makes exact prediction of their effects on matter impossible. Such an argument presupposes that astrology itself is invalid, for what is true of stellar effects on alchemical ingredients must be true a fortiori of stellar effects in general. Hence we might expect the *Summa* to reply with a defense of astrological prognostication. If so, we must be prepared for a rude shock, for "Geber's" response is quite the contrary of our expectations:

And if they should say that the perfection of the metals is derived from the position of one or more stars, which perfection we do not know, we say it is not necessary for us to know this position, since there is no species of generables and corruptibles in which the generation and corruption of each of its individuals fails to occur every day, whence it is manifest that the position of the stars is every day perfective and corruptive of whatever species of individuals. It is not therefore necessary for us to wait for this position of the stars, even if it should be useful. But it is sufficient merely to arrange the matter for nature so that she, herself wise, in turn coordinate it with the suitable positions of the mobile bodies. . . . For we see, when we want to lead a worm into being from a dog, or other putrescible animal, [that] we do not consider immediately the position of the stars, but rather the disposition of the ambient air, and other causes of putrefaction in that.[76]

The gist of "Geber's" argument is that nature herself induces perpetual generation and corruption on earth, in every sort of mutable being, without waiting for specific celestial aspects.[77] His reference to the artificial spontaneous generation of worms without regard to astrological elections shows that the significant factor in generation is the ambient air and other proximal causes—not the celestial bodies themselves. Possibly the author thinks that the celestial rays are collected in the ambient and then absorbed by a given type of matter in accordance with its particular characteristics. At any rate, he is quite willing to sacrifice the election of times by admitting that it is at best otiose and at worst impossible. Instead of defending catarchic astrology, he rejects it.

The presence of such anti-astrological arguments in the *Summa* is of the highest significance because of the text's gargantuan influence. One author who seems to have taken these clues to heart and developed them further is Bernard of Trier, whose *Epistola ad Thomam de Bononia* is one of the lesser classics of alchemical literature. Bernard's letter, which sometimes bears the date 1385, is addressed to Thomas of Bologna, physician to Charles V and VI of France and the father of the well-known writer Christine de Pizan.[78] Thomas, who had fallen under suspicion for sending a dubious medicine to the French king and to the dukes of Burgundy and Berry, wrote to Bernard in the apparent hope of gaining support for his recipe. In the course of his letter, Thomas refers to the generation of metals beneath the surface of the earth. This offers Bernard the opportunity of launching into a veritable tirade on the subject of astral causation.

Bernard begins his attack by affirming the traditional alchemical theory that metals are generated out of sulfur and mercury. He vehemently denies the notion of "some" that the main agency in congealing mercury is the sun rather than sulfur. Indeed, Bernard states, the form of gold is not perfected by the heat of the sun in mines as some say, but rather by the power of the motion of the sun's orb. It is not even a heat from the sun's own sphere that perfects gold, but rather from all the celestial spheres together (*universaliter totius coeli*).[79] The sun's rays per se do not penetrate the earth at all, nor does any other influence, although the motion of the orbs is indeed the cause of heat. Consequently, there is no connection between gold and the sun except that the sun is the hottest of the planets and gold the hottest of the metals. The *Deckname* "sol" has been given to gold in recognition of this fact, which has led fools to think that "each of the seven planets generates one metallic species by its own proper influence to which species it agrees in property and nature."[80]

Bernard's argument, which he develops into an interesting discussion of the reflection of the stellar rays by the elementary spheres, draws on a traditional argument that the sun heats by the motion of its sphere, rather than by means of its rays.[81] What is of interest to us, however, is the fact that Bernard uses this argument in a fashion reminiscent of his antiastrological contemporary Nicole Oresme to deny the significance of causation by celestial influences. Although Bernard restricts himself to a rejection of planetary rays in the formation of metals, by implication the alchemist who is intent on reproducing the natural generation of metals by artificial means can also ignore elections.

We have seen that the alchemists of the Middle Ages often disregarded, and in some instances even attacked, astrology. Alchemy and astrology were widely recognized to be distinct disciplines with their own respective methods and goals. How then could Elias Ashmole state in the 1650s that "*Iudiciall Astrologie* is the *Key* of *Naturall Magick,* and *Naturall Magick* the *Doore* that leads to this *Blessed Stone* [i.e., the philosophers' stone]"?[82] The answer, of course, lies in the conceptual universe that Ashmole inhabited, a world vastly different from that of the medieval alchemists. Although ostensibly commenting on the *Ordinall* of Thomas Norton, Ashmole had in fact imported the neoplatonic magic of Marsilio Ficino and his acolytes, above all Agrippa von Nettesheim.

Several major changes in the standard view of alchemy can be traced directly to Ficino and Agrippa. First, in his *De vita coelitus comparanda,* Ficino explicitly linked the vital spirit of the cosmos with the alchemical quintessence, a physical substance that could be extracted by means of distillation and other techniques. As Ficino himself says,

between the tangible and partly transient body of the world and its very soul, whose nature is very far from its body, there exists everywhere a spirit, just as there is between the soul and body in us, assuming that life everywhere is always communicated by a soul to a grosser body. . . . When this spirit is rightly separated and, once separated, is conserved, it is able like the power of seed to generate a thing like itself, if only it is employed on a material of the same kind. Diligent natural philosophers, when they separate this sort of spirit of gold by sublimation over fire, will employ it on any of the metals and will make it gold. This spirit rightly drawn from gold or something else and preserved, the Arab astrologers call Elixir. But let us return to the spirit of the world. The world generates everything through it (since, indeed, all things generate through their own spirit); and we can call it both "the heavens" and "quintessence."[83]

Ficino's brief reference to alchemy in *De vita* had an influence out of all proportion to its length.[84] The association that the prominent neoplatonist drew between the alchemical quintessence and the spirit of the world gave alchemy a cosmic character that it had lacked in the Middle Ages, when it was seen primarily as a pursuit devoted to metals, minerals, and items of chemical technology. The claim that alchemy could isolate the vital principle of the world was something new.[85]

Ficino's follower, Agrippa von Nettesheim, appropriated and expanded Ficino's linkage of the alchemical quintessence to the *spiritus mundi,*[86] adding another important feature to the new understanding of alchemy in the form of an alchemically colored treatment of the four elements. Relying on his teacher Johannes Trithemius of Sponheim, Agrippa argued that each of the four elements was actually "threefold" and contained its purer and simpler cognates within itself. Agrippa stated clearly that without direct knowledge of these simpler elements, one could not obtain success in natural magic.[87] He further asserted the cosmic significance of alchemy in the second book of *De occulta philosophia,* where he supplied a list of correspondences to the number one, or *monas.* In the elementary world, Agrippa argued that the number one is represented by "the philosophers' stone—the one subject and instrument of all natural and trans-natural virtues."[88] The accompanying text supports this statement further: "There is one thing created by God, the subject of all the wonderfulness on earth or in the heavens: this thing is itself animal, vegetable, and mineral *in actu,* it is found everywhere yet known to very few, mentioned by none under its proper name but veiled in innumerable *figurae* and enigmas, without which neither alchemy, nor natural magic can attain its complete goal."[89] Agrippa's description of the "one thing," replete with language traditionally used for the philosophers' stone, reinforced the new status that Ficino had imparted to alchemy. It also made it possible to interpret one of the hallowed texts of alchemy, the *Tabula smaragdina* or *Emerald Tablet* of "Hermes Trismegistus," in a new light, albeit one inherited from Trithemius.

The *Emerald Tablet* is first found in the *Kitāb sirr al-khalīqa* or *Book of the Secret of Creation* attributed to "Balinas" (pseudo-Apollonius of Tyana, ca. 8th c.). Hermes there says that "that which is above is the same as that which is below" and follows this cryptic utterance with still more obscure material about the conversion of the "one thing" into earth by means of fire and a descent from heaven into earth. Medieval authors usually saw an encoded alchemical recipe in these lines, but in the sixteenth century the *Emerald Table* served as one basis for the comprehensive unification of alchemy and a neoplatonizing cosmology under discussion. Trithemius, whom Agrippa follows, took it literally as a cosmological statement concerning the soul of the world.

Elias Ashmole presented this neoplatonic view of alchemy as a sister science to natural magic in his *Theatrum chemicum britannicum.* The alchemist and the catarchic astrologer were no longer the representatives of distinct fields who might on occasion interact. They were now the same figure, the Hermetic sage who held the key to occult wisdom in its entirety.

Ashmole himself expresses nicely, in his commentary on Norton, the ideal of the universal *archimagus* who, like Prospero, controls the whole of nature: "*Wisemen* conceive it no way *Irrationall* that it should be possible for us to ascend by the same degrees through *each* world, to the very *Originall world* it selfe, the Maker of all things and first *Cause*. But how to conjoyne the *Inferiours* with the *vertue* of the *Superiours* (which is marrying *Elmes* to *Vines*) or how to call out of the hidden places into open light, the dispersed and seminated *Vertues,* (i.e. *Virtutes in centro centri latentes,*) is the work of the *Magi,* or *Hermetick Philosophers* onely; and depends upon the aforesaid *Harmony*."[90]

The Renaissance integration of the occult sciences that Ashmole represents should not make us forget, however, that that other follower of Agrippa, Thomas Vaughan, disparaged the practices of contemporary astrologers. For it was possible, of course, to interpret Agrippa's comments to mean that the science of celestial influence was invalid unless one had already acquired the first matter of the philosophers' stone. This was the conclusion that Vaughan drew, and on the basis of Agrippa's text alone, it was perfectly legitimate. Even during the heyday of Renaissance neoplatonism, astrology and alchemy lived independent lives, despite the vast inkwells devoted to the rhetorical embellishment of occult philosophy.

A particularly enlightening example of the alchemists' view of astrology in this period can be seen in a Munchausen-like story told by one "Edwardus Generosus," who claimed to have created two philosophers' stones, one for the sun and the other for the moon. The stone of the sun was intensely hot, whereas the lunar stone concentrated the light and frigor of the moon. Hence it could be used to freeze small animals, and the author describes at length the great sport he had in this pursuit. First, like a small boy playing pranks with a flashlight under his blanket, Edward employed the lunar stone to freeze fleas beneath the sheets of his bed. Having had such success on minute vermin, Edward then tells us how he graduated to larger animals. He expresses this experiment in astrological terms: "Then I bethought me of a new device. Now I will make such a strange conjunction as neither Haly, Guido, Bonatus or their Mr Ptolomaeus ever saw ye like in all ye conjunctions of ye stars. I first took a quick live mouse, an Humble Bee, & a nimble frogg, & all these I put into a fine bowl glass together, having that care that none should either creep leap or fly away wthout my leave. Then I held my glass wth my nimble quick ^creatures^

[water *deleted*] in it directly under yᵉ beam of my fair beautiful [luna] till she had benummed all their vitall spirits to death, making them ^all^ as cold as any lead quickly."[91] Despite the farcical dispatching of Edward's nimble creatures, the passage says something significant about the relationship of alchemy and astrology in the early modern period. Edward's jocular "conjunction" is not one of stars or planets but of the small animals that he elects to torment. Although few alchemists followed Edward in turning astrology into a burlesque, their astrological allusions were often of a similarly analogical nature rather than a practical one. Even the widely revered description of alchemy as terrestrial astronomy was a tropological association, comparing one discipline to the other rather than using the tools of the former in the operation of the latter.

The Chapters in This Volume

As we have urged in the foregoing, the association between alchemy and astrology is problematic rather than transparent. The two disciplines diverge from one another in at least as many instances as they have points of intersection. For this reason, the remaining seven chapters in this book do not uniformly concern both alchemy and astrology, but for the most part treat one or the other pursuit independently. The subject of the first two papers, Girolamo Cardano (1501–76), was one of the most famous astrologers of the Renaissance. Germana Ernst's contribution in chapter 2 gives a comprehensive overview of Cardano's career as an astrologer, emphasizing his attempts to purify the discipline by returning it to its Ptolemaic roots. Writing after the devastating attack on astrology by Giovanni Pico della Mirandola and dissatisfied with the current status of the discipline as a realm for unlearned and opportunistic diviners, Cardano wanted to restore the Ptolemaic linkage of astrology to natural philosophy. Yet Cardano was not merely a natural philosopher, but a trained practitioner of medicine as well. Hence chapter 3, by Anthony Grafton and Nancy Siraisi, examines the links between Cardano's medicine and his astrology. Astrology had been taught as an adjunct to academic medicine since the High Middle Ages, so there was a long-standing tradition of connecting the two at Cardano's disposal. And yet, as Grafton and Siraisi show, the Italian polymath made surprisingly little of astrological medicine. For the most part, he seems to have viewed the fields of astrology and medicine as separate disciplines with their own theoretical (and practical) bases. It is worth adding here that the same bifurcation applied a fortiori when Cardano considered alchemy. Despite his deep allegiance to the "occult" science of astrology, he had little sympathy for alchemical practice, lumping it together with such nefarious arts as poisoning and the invocation of demons.[92]

Chapter 4, by Darrel Rutkin, is a study of Johann Kepler's dedication to the *Astronomia Nova* of 1609. Although considerable scrutiny has focused recently on the astrological motifs Galileo employed in his own dedicatory letter to Cosimo II in the *Siderius nuncius* of 1610, Rutkin shows that Kepler had already employed similar astrological conceits in his dedication to the Holy Roman Emperor, Rudolph II. This raises the interesting possibility, which Rutkin explores in some depth, that Galileo borrowed the dedicatory use of a ruler's natal chart from Kepler himself. Although the evidence of such influence remains inconclusive at present, the very fact that two such eminent astronomers framed their scientific discoveries within the borders of natal astrology gives further testimony to the infiltration of genethlialogy into every aspect of Renaissance culture. It should not surprise us that the same astronomer who predicted Count Wallenstein's victories on the battlefield should also seek success in the Martial horoscope of his ruler.

Despite their profound awareness of astrological matters, neither Kepler nor Galileo seems to have found much fascination in alchemy. This cannot be said of the subject of the chapters by Nicholas Clulee and Didier Kahn, namely the anonymous Rosicrucians who burst upon the European scene in the second decade of the seventeenth century. Clulee's contribution in chapter 5 considers the claims Frances Yates made for the Elizabethan magus John Dee in setting the stage for the Rosicrucian movement. Basing herself in part on a pseudonymous document attributed to one "Philippus à Gabella," Yates maintained that Dee's visit to the Continent in the 1580s had planted the seeds for a scientific and religious upheaval that would emerge with the publication of the Rosicrucian manifestos in 1614–16. Clulee argues that the treatise attributed to Gabella, although published with the second of the Rosicrucian treatises, was tangential to the movement and extremely derivative. It does not provide evidence of Dee as the hidden patriarch of Rosicrucianism, a negative inference certified both by the eclecticism of its borrowings and its apparent ignorance of the major themes of Rosicrucianism. Despite Gabella's shortcomings as an interpreter of Dee, however, he does provide compelling evidence for the dispersion of Dee's ideas. In particular, Gabella transmits the notion of alchemy as inferior astronomy, which we have discussed above. Clulee shows at length how Dee appropriated this idea from sources such as Trithemius and Agrippa and passed it on to writers who were more concerned with practical alchemy, such as Gabella.

Kahn's contribution in chapter 6 also concerns the Rosicrucian movement, but within the specific context of France during the 1620s. To be exact, Kahn discusses the origin and effects of the sensational posting of Rosicrucian broadsheets in Paris in 1623. In magisterial detail, Kahn shows that the originator of this Rosicrucian hoax was a teenage prankster named Étienne Chaume. Trying to beguile the contemporary Lullists and other aficionados of arcane knowledge, Chaume managed to create an incendiary situation in which charges of libertinism, atheism, and maleficium came to be intertwined. The astonishing ramifications of Chaume's boyhood prank have carried on even into the twentieth century, for a multitude of sober scholars have accepted on scanty evidence that such luminaries as Descartes and Gassendi were embroiled in the Rosicrucian scare. Here and elsewhere Kahn throws new light on the dense thicket of intellectual and religious controversy that surrounded the Rosicrucian movement in early modern France.

As Kahn points out in his chapter, the Rosicrucian literature was deeply pervaded by alchemical ruminations but more concerned with millenarianism and Biblical prophecy than with the details of technical astrology. Such a relative disregard for mathematical astrology cannot be imputed to the subject of Lauren Kassell's chapter 7, namely Simon Forman (1552–1611). Forman's fusion of alchemy, astrology, and cabala paralleled the interests of his older contemporary John Dee as well as those of the younger aspirant to secret wisdom, Elias Ashmole. The integrated vision of the occult sciences propounded by Agrippa von Nettesheim thoroughly conditioned all three authors. Unlike Dee and Ashmole, however, Forman was an autodidact, having risen from an impoverished background to become a famous purveyor of the occult to those who could afford his services. Kassel provides a densely documented record of Forman's reading practices, showing how he fused practical alchemy, heterodox Biblical interpretation, medicine, and astrology into a characteristic brew that would impress his clients and feed his own appetite for secret wisdom.

The final chapter, by Lawrence Principe and William Newman, provides an extended criticism of the existing historiography of alchemy. As Principe and Newman argue, much of the contemporary historical writing on alchemy has unwittingly absorbed themes drawn from nineteenth-century occultism. Anachronistic promoters of so-called spiritual alchemy in that period, such as Mary Anne Atwood and Ethan Allen Hitchcock, avoided the embarrassing fact that many alchemical recipes do not make obvious chemical sense by arguing that such recipes were really not about chemistry at all, but veiled prescriptions for perfecting the alchemist's soul. Twentieth-century apologists of the mysteries, such as Carl Gustav Jung and Mircea Eliade, adopted this viewpoint and put it into the language of psychology and anthropology. Historians of alchemy have in turn employed the views of Jung and Eliade as interpretive tools without realizing their dubious origins. The result is a remarkable incursion of occultist beliefs into the very framework of historiography, a situation possibly unparalleled in other fields of research.

In conclusion, the surprising new material provided by many of the chapters in this volume should help to dispel the myopic stereotypes that have come to dominate the historical study of the occult sciences since its revival in the 1960s. We can no longer accept the pansophic optimism of an Elias Ashmole as reflecting the common situation of alchemical and astrological practitioners in his own time, not to mention the *longue durée*. Only by dropping the blinders acquired from studying Agrippa and Dee, influential though they were, will we be able to understand the diversity of interests displayed by single-minded aficionados of one or another branch of the occult sciences. The Ashmolean image of the universal mage is as absent from such astrological experts as Kepler and Galileo as it is from alchemical mavens of the stamp of a Boyle or Newton. In short, the chapters in this volume present a much-needed new perspective on the historical study of the so-called occult sciences.

Notes

1. Athanasius Kircher, *Vita a semetipso conscripta,* Vienna, Österreichische Nationalbibliothek, MS 13752, 18–20, at 19–20: "Vnde multi secreto examinantes, qua ratione Vrbis invasionem tam constanter praedixissem, putabant Astrologiae id arte factum fuisse, sed, uti ad visionem aperiendam non obligabar, ita alto silentio rem repressi, relinquendo unicuique potestatem de praedictione judicandi quod vellet."

2. On Kircher, see, e.g., T. Leinkauf, *Mundus combinatus* (Berlin: Akademie, 1993); P. Findlen, *Possessing Nature* (Berkeley: University of California Press, 1994).

3. T. W. Adorno, *The Stars down to Earth and Other Essays on the Irrational in Culture*, ed. S. Crook (London and New York: Routledge, 1994).

4. H. Wiesendanger, *Zwischen Wissenschaft und Aberglaube* (Frankfurt: Fischer, 1989).

5. See, e.g., A. Geneva, *Astrology and the Seventeenth-Century Mind* (Manchester: Manchester University Press, 1995); A. Grafton, *Cardano's Cosmos* (Cambridge, Mass.: Harvard University Press, 1999).

6. See, e.g., K. Thomas, *Religion and the Decline of Magic* (London: Scribner, 1971); M. MacDonald, *Mystical Bedlam* (Cambridge: Cambridge University Press, 1981).

7. See, e.g., P. Zambelli, ed., *"Astrologi hallucinati": Stars and the End of the World in Luther's Time* (Berlin and New York: de Gruyter, 1986); O. Niccoli, *Prophecy and People in Renaissance Italy,* trans. L. G. Cochrane (Princeton: Princeton University Press, 1990); R. Barnes, *Prophecy and Gnosis* (Stanford: Stanford University Press, 1988); R. Westman, "Copernicus and the Prognosticators: The Bologna Period, 1496–1500," *Universitas,* (December 1993): 1–5.

8. See esp. the classic essays of A. Warburg, *The Renewal of Pagan Antiquity,* ed. K. W. Forster, trans. D. Britt (Los Angeles: Getty Research Institute, 1999).

9. E. Casanova, "L'astrologia e la consegna del bastone al capitano generale della repubblica fiorentina," *Archivio storico italiano,* 5th ser., 7 (1891): 134–44.

10. M. E. Bowden, "The Scientific Revolution in Astrology" (Ph.D. diss., Yale University, 1974); J. V. Field, "A Lutheran Astrologer: Johannes Kepler," *Archive for History of Exact Sciences,* 31 (1984): 225–68; B. Bauer, "Die Rolle des Hofastrologen und Hofmathematicus als fürstlicher Berater," in *Höfischer Humanismus,* ed. A. Buck (Weinheim: VCH, 1989), 93–117; G. Ernst, *Religione, ragione e natura* (Milan: Franco Angeli, 1991).

11. J. Burckhardt, *Die Kultur der Renaissance in Italien: Ein Versuch* (Darmstadt: Wissenschaftliche Buchgesellschaft, 1955).

12. See esp. Warburg's great study *Heidnisch-antike Weissagung in Wort und Bild zu Luthers Zeiten,* Sitzungsberichte der Heidelberger Akademie der Wissenschaften, 1919 (Heidelberg: Winter, 1920); trans. in Warburg, *The Renewal of Pagan Antiquity,* 597–697.

13. See Niccoli, *Prophecy and People,* and the important review by J.-M. Sallmann, *Annales: Economies, Sociétés, Civilisations,* 47 (1992): 144–46.

14. See Ernst, *Religione, ragione e natura,* and the older study by G. Bock, *Thomas Campanella* (Tübingen: Niemeyer, 1974).

15. L. B. Alberti, *The Society of Renaissance Florence,* trans. R. Neu Watkins (Columbia, S.C.: University of South Carolina Press, 1969), 256; *I libri della famiglia,* in *Opere volgari,* ed. C. Grayson (Bari: Laterza, 1960–73), I, 274.

16. Alberti, *Society,* 256; *Opere volgari,* I, 274: "Erano in que'tempo gli animi de' dotti astronomi solliciti e pieni di varia espettazione, quanto el cielo porgea loro

manifesti indizii di permutazioni ed eversioni di republiche, stati e summi magistrati; e quasi comune sentenza, statuivano non poter lungi essere che quella stella crinita, quale a mezzo il cielo splendidissima e diurna continuati i di appariva in que' mesi, per sua notata consuetudine predicesse fine e morte di qualche simile al Duco famosissimo e supremo principe."

17. Alberti, *Society,* 256; *Opere volgari,* I, 274–75: "E già era chi di questa promulgata opinione forse fatto avea al Duca certo; a cui, magnifica risposta, dicono, e degna di principe, rispose el Duca: sé non acerbo cadere dai mortali, ove così resti persuaso sé essere stato al cielo tanto a cura, e parerli morte gloriosa questa, ove doppo a sé poi viva diuturna fama; ché quelle intelligenze celeste così per sé esposero raro e maraviglioso segno e indizio, onde manifesto ciascuno compreenda che que' lassù divini animi immortali di sua vita e morte stati erano curiosi."

18. Alberti, *Society,* 256–57; *Opere volgari,* I, 275: "Ma pur credo per questo tenea qualche ad altri poco manifesta, ma dentro in sé non piccola agitazion d'animo . . ."

19. Alberti, *Society,* 256; *Opere volgari,* I, 274: "Questi nostri Alberti d'Inghilterra, di Fiandra, di Spagna, di Francia, di Catalogna, da Rodi, di Soria, di Barberia, e di que' tutti luoghi ove oggidì ancora reggono e adirizzano mercantia, quanto gli avea per mie lettere pregati, così o tumulti, armate, esserciti o legge nuove, affinità fra principi, publice amicizie, armi o incendii, naufragi, o qualunque cosa accadesse per le provincie nuova e degna di memoria, subito me ne faceano certo."

20. Alberti, *Society,* 257; *Opere volgari,* I, 275.

21. See Alberti, *Society,* 257, translator's note.

22. See A. Grafton, *Leon Battista Alberti* (New York: Hill & Wang, 2000), chap. 6.

23. For an edition, translation, and study of Lapo's text, see C. S. Celenza, *Renaissance Humanism and the Papal Curia* (Ann Arbor: University of Michigan Press, 1999).

24. See R. Fubini and A. N. Gallorini, "L'autobiografia di Leon Battista Alberti: Studio e edizione," *Rinascimento,* ser. 2, 12 (1972):21–78; and R. Neu Watkins, "L. B. Alberti in the Mirror: An Interpretation of the *Vita* with a New Translation," *Italian Quarterly,* 30 (1989): 5–30.

25. Alberti, *Society,* 272; *Opere volgari,* I, 291–92: "Eh! Quanti precetti qui necessari mancherebbono, Lionardo, a chi volesse lato e diffuso disputarne! come se chi forse avesse dagli astronomi udito che Marte disponga impeto di esserciti e furore d'arme, Mercurio instituisca varie scienze e suttilità d'ingegno e maravigliose arte, Giove moderi le ceremonie e animi religiosi, el Sole conceda degnità e principati, la Luna conciti viaggi e movimenti feminili e plebei, Saturno aggravi e ritardi nostri pensieri e incetti; e tenesse di tutti così loro natura e forza, dove nolli fusse noto in qual parte del cielo e in quanta elevazione ciascuno per sé molto o meno vaglia, e con che razzi l'uno all'altro porga amicizia o inimicizia, e quanto coniunti possano in buona o mala fortuna, certo sarebbe non costui astrologo. Ma quella semplice cognizione di que' nudi principii, a volere bene in quella arte venire

erudito, sarà tale che senza esse nulla potrà; con esse non però arà che introito ad aprendere l'altre quasi infinite ragioni a prevedere e discernere le cose, a quale il cielo tende per produrle. Così qui ora que' tutti essempli e sentenze, quali affermo sono apresso gli ottimi scrittori utilissimi e copiosissimi, non però prestano quanto aiuto ci bisogna."

26. See Fubini and Gallorini, "L'autobiografia."

27. L. B. Alberti, *De re aedificatoria*, 4.3, ed. G. Orlandi, trans. P. Portoghesi, I, 288–97.

28. Grafton, *Cardano's Cosmos*, chaps. 7, 9.

29. H. Wolf, "Ad Io. Oporinum commentariolus de vitae suae ratione ac potius fortuna," in *Oratores Graeci*, ed. J. J. Reiske, vol. 8 (Leipzig: Sommer, 1773), 772–876; German translation in H.-G. Beck, *Der Vater der deutschen Byzantinistik* (Munich: Institut für Byzantinistik, 1984), 1–144.

30. See Beck, "Nachwort," in *Der Vater der Byzantinistik*; H. R. Velten, *Das selbst geschriebene Leben* (Heidelberg: Winter, 1995), 94–101.

31. Wolf, "Commentariolus," 773: "Id ego tum, cum neque recepissem, neque recusassem, in hoc usque tempus, *klimakêtrôn enestôtôn, hôs fasin hoi peri ta astra deinoi*, prorogavi."

32. Ibid., 781: "Quod si horologium fefellit (ut saepe fit, ac potius plerumque) potuit meus natalis paulo esse maturior, ut ex trutina Hermetis, quam vocant, Aquarii 14 horoscopet, vel in septimam incidere, ut initium Geminorum ex antegressa oppositione secundum Ptolemaicam correctionem tenuerit orientem. Caeterum utramque hanc rationem non contemnendis argumentis evertit Ant. Montulmius." On these procedures, see J. North, *Horoscopes and History* (London: Warburg Institute, 1986).

33. Wolf, "Commentariolus," 781: "Iam quod ad investigationem puncti horoscopantis ex eventibus attinet, ei quoque rei haud scio quantum sit tribuendum. Nam tametsi martialia pleraque subita, violenta et cruenta, Saturnia vero insidiosa, virulenta, luctuosa, diuturna perhibentur, tamen et eundem casum ob alias causas non sine rationibus accommodari video, et multis seu divinitus seu *magikôs* evenerunt, quorum significationes in genesi apparent."

34. Ibid., 781–82 n.: "Hic longum locum, astrologicarum et horoscopicarum nugarum plenum omisi, ne taedio lectoribus et derisui essem, neu bonum Wolfium fannis exponerem"; Beck, *Der Vater der Byzantinistik*.

35. See M. de Nostradamus, *Lettres inédites*, ed. J. Dupèbe (Geneva: Droz, 1983); P. Brind'Amour, *Nostradamus astrophile* (Ottawa and Paris: University of Ottawa Press, 1993).

36. R. Castagnola, ed., *I Guicciardini e le scienze occulte*, (Florence: Olschki, 1990).

37. Grafton, *Cardano's Cosmos*, chap. 10.

38. London, British Library MS Sloane 325; J. G. Nichols, "Some Additions to the Biographies of Sir John Cheke and Sir Thomas Smith," *Archaeologia*, 38 (1859): 98–127 at 103–12, 116–20.

39. Wolf, "Commentariolus," 808: "Interdum etiam iocabatur in staturam meam proceriorem, quum diceret, dandum mihi esse locum, cui sui calcei iam essent apti (fortassis enim ex Astrologo audierat, cum eo usque vixisset, receptui cavendum fore) . . ."

40. Ibid., 831: "Nec multo ante, cum in platea in Georgium Ioach. Rhaeticum incidissem, isque inter alia etiam Chiromantiae facta mentione, manum meam inspexisset: brevi, inquit, magnum tibi a muliere periculum impendet."

41. Ibid., 860: "Multo autem ante Cyprianus Leovitius inspecta conversione annalium (quos iam tum confectos habebam ad annum aetatis sexagesimum usque) subita mihi et opum et dignitatis incrementa promitti affirmabat."

42. Ibid., 797: "nec prius tamen voti compos fio, quam in morbum incidissem, indignatione Veneris, ut opinor (neque enim accusandi sunt, quos criminis convincere non possumus) immissum, quae se prae Mercurio a me negligi videret."

43. See A. Grafton and N. Siraisi, "'Between the Election and My Hopes,'" chap. 3 in this volume.

44. See, e.g., M. Ficino, *Three Books on Life,* ed. and trans. J. V. Clark and C. Kaske (Binghamton, N.Y.: Medieval and Renaissance Texts and Studies, 1989). D. P. Walker, *Spiritual and Demonic Magic from Ficino to Campanella* (London: Warburg Institute, 1958); W.-D. Müller-Jahncke, *Astrologisch-magische Theorie und Praxis in der Heilkunde der frühen Neuzeit, Sudhoffs Archiv,* Supplement (Stuttgart: Franz Steiner, 1985).

45. See Barnes, *Prophecy and Gnosis;* S. Kusukawa, *The Transformation of Natural Philosophy* (Cambridge: Cambridge University Press, 1995); and the forthcoming Habilitationsschrift by B. Bauer,

46. William R. Newman and Lawrence Principe, "Alchemy versus Chemistry: The Etymological Origins of a Historiographic Mistake," *Early Science and Medicine,* 3 (1998): 32–65, cf. 61.

47. Although the literature devoted to alchemy and astrology hardly touches upon the problematic interaction between these two subjects, a first effort has been made by Joachim Telle, "Astrologie und Alchemie im 16. Jahrhundert," in *Die okkulten Wissenschaften in der Renaissance,* ed. August Buck (Wiesbaden: Harrasowitz, 1992), 227–53. See also Jacques Halbronn, "Les resurgences du savoir astrologique au sein des textes alchimiques dans la France du XVIIe siècle," in *Aspects de la tradition alchimique au XVII siècle,* ed. Frank Greiner (Paris: S.E.H.A., 1998), 193–205.

48. Claudius Ptolemy, *Tetrabiblos,* trans. and ed. F. E. Robbins (Cambridge, Mass.: Harvard University Press, 1940), 2–3. Ptolemy does not, of course, employ Greek terms equivalent to our "astronomy" and "astrology." Both practices are part of *astronomia,* and their primary difference lies in the two realms that make up their respective purviews: the superlunary and sublunary regions.

49. John Dillon, *The Middle Platonists: A Study of Platonism (80 B.C. to A.D. 220)* (London: Duckworth, 1977), 384.

50. See Petrus Bonus, *Margarita pretiosa,* in Lazarus Zetzner, ed., *Theatrum chemicum* (Strasbourg: Eberhard Zetzner, 1660), 5:511–713, cf. 508, 513, 528.

See also Dominicus Gundissalinus, *De divisione philosophiae*, ed. Ludwig Baur (Muenster: Aschendorff, 1903), 20–24, and Thomas Aquinas, *In librum Boethii de trinitate*, in *S. Thomae Aquinatis opera omnia*, ed. Robert Busa (Stuttgart-Bad Canstatt: Frommann-Holzboog, 1980), 4:533.

51. Elias Ashmole, *Theatrum chemicum britannicum*, ed. A. G. Debus (New York: Johnson Reprint Corporation, 1967; facsimile of London, 1652), 450.

52. Ashmole, *Theatrum*, 446.

53. Ibid., 451. For Ashmole's talismans of vermin, see C. H. Josten, *Elias Ashmole* (Oxford: Clarendon Press, 1966), 2:537, 619, and 4:1697.

54. Josten, *Elias Ashmole*, 4:1679.

55. Thomas Vaughan, *Lumen de lumine* (London: H. Blunden, 1651), 19–20.

56. William R. Newman, *Gehennical Fire: The Lives of George Starkey, an American Alchemist in the Scientific Revolution* (Cambridge: Harvard University Press), 213–22.

57. Keith Thomas, *Religion and the Decline of Magic* (New York: Scribner's 1971), 631–32.

58. Brian Vickers, "On the Function of Analogy in the Occult," in A. G. Debus and Ingrid Merkels, eds., *Hermeticism in the Renaissance* (Washington: Folger Books, 1988), 265–92. For the claim that astrology's demise sounded the death knell for the other occult sciences, see 286.

59. Frances Yates, *The Rosicrucian Enlightenment* (London: Routledge and Kegan Paul, 1972), xi–xii, 17, and passim.

60. Julius Ruska, "Al-Razi's Buch Geheimnis der Geheimnisse," *Quellen und Studien zur Geschichte der Naturwissenschaften und der Medizin*, 6 (1937): 1–246, cf. 15–16. See also G. Carbonelli, *Sulle fonti storiche della chimica e dell'alchimia in Italia* (Rome: 1925), 192–200, where a number of *glosarii* are reproduced.

61. For the expressions *astronomia inferior* and *astronomia terrestris*, see Julius Ruska, *Turba philosophorum: Ein Beitrag zur Geschichte der Alchemie*, in *Quellen und Studien zur Geschichte der Naturwissenschaften und der Medizin*, 1 (1931): 80.

62. pseudo-Aristotle, *De perfecto magisterio*, in *Bibliotheca chemica curiosa*, ed. J. J. Manget (Geneva: Chouet, G. de Tournes, Cramer, Perachon, Ritter, & S. de Tournes, 1702), 1:638.

63. Ibid. In at least one fourteenth-century manuscript of the *De perfecto magisterio*, the normal Latin names for the seven metals are given, instead of the planetary *Decknamen* (Yale MSS, New Haven, Conn., Mellon 2, fol. 1r, ll. 30–31).

64. Constantine of Pisa, *The Book of the Secrets of Alchemy*, ed. Barbara Obrist (Leiden: Brill, 1990), pp. 75–79, 106–31. Obrist suggests on p. 35 that Constantine's advice that an amalgam be buried in dung when the sun is in Gemini or Leo should be considered astrological. It seems more likely that the recipe is simply calling for the process to be carried out in the hot weather of summer.

65. S. Harrison Thomson, "The Texts of Michael Scot's *Ars Alchemie*," *Osiris*, 5 (1938): 523–59: The first six of the thirty-six numbered headings do not consist of

recipes per se, although they do contain interesting directions for identifying "salts" and testing them for purity.

66. Thomson, "Scot's *Ars Alchemie*," p. 556.

67. John Reidy, *Thomas Norton's "Ordinal of Alchemy"* (Oxford: Oxford University Press, 1975), xiv.

68. Ashmole, *Theatrum*, 100. The critical text supplied by Reidy in his excellent edition of Norton is substantially the same, although it employs a more archaic spelling. See Reidy, *"Ordinal,"* 91.

69. John North, *Chaucer's Universe* (Oxford: Clarendon Press, 1988), 79–85, where North shows how to arrive at the Alchabitius houses by using the unequal hour lines of an astrolabe. North gives another exposition of the Alchabitius or "standard" system in his *Horoscopes and History* (London: Warburg Institute, 1986), 3–6.

70. Although the printed illustration in Ashmole gives no degrees for Jupiter in the second and third scheme, British Library MS Add. 10302 in the second scheme gives 6° Sagittarius for Jupiter, and Saturn is accompanied by the number "2," evidently meaning 2° Sagittarius, despite the fact that Saturn has been placed in the twelfth house.

71. Ashmole, *Theatrum*, 452. It is not clear why Ashmole states that *all four* of the schemes err in the maximum elongation of Venus, however, for the fourth has Venus and the sun separated by only 31°.

72. Reidy, *"Ordinal,"* lii, shows that 1477, the date that Norton gives at the end of the text as the year he began its composition, is very likely correct. If we follow Ashmole's procedure over the period of Norton's life (?1433–1513 or 1514) and employ Bryant Tuckerman's *Tables*, Ashmole's statement will be found to be correct.

73. Ashmole, *Theatrum*, 452.

74. William R. Newman, "L'influence de la *Summa perfectionis* du pseudo-Geber," in *Alchimie et philosophie à la renaissance*, ed. Jean-Claude Margolin and Sylvain Matton (Paris: Vrin, 1993), 65–77; "The Alchemical Sources of Robert Boyle's Corpuscular Philosophy," *Annals of Science*, 53 (1996): 567–85.

75. William R. Newman, *The "Summa perfectionis" of pseudo-Geber* (Leiden: Brill, 1991), 643–44.

76. Ibid., 649.

77. The *Summa* does not of course deny the role of the celestial bodies in inducing sublunary generation and corruption. But it is one thing to assert that the superlunary regions are responsible for this and quite another thing to claim that man can make precise predictions about terrestrial events on the basis of that knowledge. The *Summa* admits in effect that man cannot make such predictions but then asserts that it is unnecessary to do so at any rate, since nature will join the appropriate matter to the appropriate rays.

78. Lynn Thorndike, *A History of Magic and Experimental Science* (New York: Columbia University Press, 1934), 3:611–27. For a recent treatment of Christine

de Pizan, see Joan Cadden, "Charles V, Nicole Oresme, and Christine de Pizan: Unities and Uses of Knowledge in Fourteenth-Century France," in *Texts and Contexts in Ancient and Medieval Science,* ed. Edith Sylla and Michael McVaugh (Leiden: Brill, 1997), 208–44.

79. *Bernardi Trevirensis ad Thomam de Bononia Medicum Caroli octavi Francorum Regis Responsio,* in Manget, *Bibliotheca,* 2:399–408, cf. 403. In MS Paris Bibliotheque nationale 11201, fol. 26r, of the Bibliothèque Nationale in Paris, the passage reads substantially the same.

80. Bernard, *Epistola,* 403. Bibliothèque Nationale, MS 11201, fol. 26r–v: "ex quo surgit error stultorum. credunt enim quod quilibet septem planetarum generat specialiter per suam propriam influentiam unam speciem metallicam cui specie[?] in proprietate convenit et in natura assimilatur."

81. Edward Grant, *Planets, Stars, and Orbs: The Medieval Cosmos, 1200–1687* (Cambridge: Cambridge University Press, 1994), 611–15.

82. Ashmole, *Theatrum,* 443.

83. Marsilio Ficino, *Three Books on Life,* ed. and trans. Carol V. Kaske and John R. Clark (Binghamton, N.Y.: Medieval and Renaissance Texts and Studies, 1989), 255–57.

84. Sylvain Matton, "Marsile Ficin et l'alchimie: sa position, son influence," in Margolin and Matton, *Alchimie et philosophie,* 123–92.

85. Matton makes a compelling case for this in "Marsile Ficin et l'alchimie"; see 164–90. At the same time, Roger Bacon was a prominent medieval forerunner of Ficino's position. See William Newman, "The Philosophers' Egg: Theory and Practice in the Alchemy of Roger Bacon," *Micrologus,* 3 (1995): 75–101.

86. Cornelius Agrippa, *De occulta philosophia libri tres,* ed. V. Perrone Compagni (Leiden: Brill, 1992), 113–14.

87. Ibid., 91. For a treatment of this passage, see Newman, *Gehennical Fire,* 213–21.

88. Agrippa, *De occulta philosophia,* 257.

89. Ibid., 256: "una res est a Deo creata, subjectum omnis mirabilitatis, quae in terris et in coelis est: ipsa est actu animalis, vegetalis et mineralis, ubique reperta, a paucissimis cognita, a nullis suo proprio nomine expressa, sed innumeris figuris et aenigmatibus velata, sine qua neque alchymia neque naturalis magia suum completum possunt attingere finem."

90. Ashmole, *Theatrum,* 446.

91. Cambridge, Cambridge University, Kings College MS 22, p. 13.

92. Thorndike, *A History of Magic and Experimental Science,* 5:564, 571.

2

"Veritatis amor dulcissimus": Aspects of Cardano's Astrology

Germana Ernst

Toward Scotland

The most finished product of Girolamo Cardano's work on astrology was his commentary on the *Quadripartitum* of Ptolemy. He was led to undertake this by a combination of curious and unexpected circumstances in which, as in other cases, chance and an obscure providential design somehow intersected. In April 1552, Cardano was in Lyons, about to depart for Paris, and from there for far-off Scotland, where Archbishop John Hamilton, who suffered from a particularly stubborn form of asthma, anxiously awaited him. Cardano had become acquainted with a schoolmaster who, after asking for a medical opinion, insisted on bringing him to his house, where he promised to show him a boy who could see demons in a vessel.[1] The visions turned out to be nonsense, but at the man's house Cardano encountered Antonio Gogava's translation of the two last—and very difficult—books of Ptolemy's *Quadripartitum*.[2] Seized by a sudden feeling of desire and inspiration, Cardano asked for and obtained a copy of the text. On the ship that took him up the Loire to Paris, he began to work with feverish intensity on a project that he found absolutely urgent.

The work was published at Basel in 1554, with a dedication to the Scottish archbishop. It took the form of a splendid folio volume, with which, Cardano declared, he was entirely satisfied.[3] A Lyon edition followed, very similar in content but more modest in form. The last edition appeared posthumously at Basel in 1578 and was reprinted in volume 5 of Cardano's *Opera omnia*. It departed considerably from the other two. Cardano enlarged and reworked a good many passages but he also mutilated or deleted many others. The most famous of these—but not the only ones—were those that discussed the horoscope of Christ.[4] These final changes in the text may well have been the result of a late effort at self-censorship that Cardano was forced to undertake. If so, they were inspired by the same considerations that led to the disappearance of Ptolemy's name from Cardano's curious "classification" of the greatest geniuses of all times in *De subtilitate,* where Ptolemy had previously occupied an honorable second position, after Archimedes and before Aristotle.[5]

The Twelve Horoscopes

The "great voyage" that Cardano began on February 23, 1552, and that kept him away from home for about ten months gave him the opportunity not only to visit strange places and cities, but also to meet illustrious individuals. Many of these individuals presented him with gifts and indications of their esteem, which to some extent made up for the incomprehension and persecution he had experienced at home.

But honors and money were not the only benefits that Cardano received on his voyage. He also had the opportunity to frequent courts and other high social circles, to gain an intimate knowledge of the life and habits of important people, and to gather a rich harvest of astrological data. Cardano appended to the *Commentary* on Ptolemy twelve genitures to serve as examples of a particularly rich and meaningful kind. They were meant to embody and to confirm the theoretical principles stated in the text. Cardano described them as, among other things, "worthy of admiration, made famous by events, and calculated with the utmost precision." The first five belonged to persons he had met for the first time on his voyage.[6]

The series began with the controversial geniture of Edward VI. When Cardano encountered the young king, whose precocious erudition filled him with astonishment and admiration, in London, he uttered a series of reassuring predictions. But Edward's premature death in July 1553—officially from phthisis, though Cardano suspected poison—refuted these in a most embarrassing way. Cardano found himself compelled to reexamine the astral data and their interpretation so that he could account for these events. He also added a page in which he explained and justified the extreme reticence he himself had shown on this occasion. By revealing his own fears, he might have provoked public disturbances or encountered personal danger, like earlier astrologers who had imprudently ventured to predict unhappy ends for their sovereigns. Cardano had left the English court as rapidly as possible, disgusted and frightened because he had perceived, through political acumen rather than astrological expertise, the dangers posed by the insidious traps of power.[7]

Next came the genitures of Hamilton; of Claude de Lavalle, the French ambassador in London; and of the Hellenist and royal tutor John Cheke, who, after various vicissitudes, suffered disaster on the accession of Mary Tudor. His horoscope included disquieting references to the dangers that threatened him. In the fifth place appeared another geniture which would cause Cardano no little embarrassment: that of Aimar de Ranconnet, a prominent member of the Parlement of Paris who was strongly linked to him by esteem and friendship. The two redactions of Ranconnet's horoscope, that of 1554 and that of 1578, were very different.[8] In the first, Cardano showed the highest admiration for someone who combined the gravity of public office with unusual gifts of humanity and learning. He expressed his deep gratitude for the warm greeting that he had received from Ranconnet, even though he had appeared before him poorly dressed and had spoken to him very simply. The incident showed that Ranconnet was endowed with a rare capacity to see past appearances: he could actually read minds.

But once again, as a result of Ranconnet's tragic death in 1559, Cardano was forced to revise his text. He had to inform the reader that his illustrious friend had been imprisoned on the infamous charge of incest and had died mysteriously in prison. Some said that he had committed suicide out of shame, others that he had been strangled, and still others that he had been burned at the stake, at night and under a false name—astrologically, the most plausible hypothesis. Despite Cardano's warm feelings for his friend, his respect for the truth of astrology compelled him to record this dishonorable end, one entirely unworthy of such a person.

The other seven horoscopes came from Cardano's earlier works. One of them belonged to the soldier of fortune Giovan Giacomo Medici, whose portrait, as a bold man of arms, was the precise image of a disillusioned Machiavellian hero. His brilliant career began with a crime, which he later entirely justified by the outstanding virtues that he showed. Anyhow, as Cardano observed, no one had ever risen to power unless some crime opened the way: "Who ever moved from the rank of private citizen to that of a prince without committing a crime? . . . Only crime ever enabled anyone to attain the highest honors, by opening the way: if virtue and fortune then stand by him, the way to power is clear."[9] In the eighth place appeared the very detailed horoscope of the author himself, to which we will return, between those of two of his closest friends, the doctor Guglielmo Casanate and Cardinal Francesco Sfondrato. The latter passed, in the space of a few years, from the rank of private citizen to that of bishop, then to that of cardinal, only to miss, by a hair, being elected pope on the death of Paul III. He then died, a few months later, from uncertain causes ("Is this not a marvellous and unusual story?" commented Cardano).[10]

The tenth and eleventh horoscopes were closely connected. They belonged to Paul III and his natural son Pier Luigi Farnese, prince of Parma and Piacenza, who could not avoid dying, skewered by the daggers of conspirators, even though insistent rumors had predicted the event. When Cardano drew up the pope's horoscope, he underlined the great man's strong predilection for predictions of every kind, copying out and commenting on a prognostication by Paris Ceresarius. This Mantuan astrologer, with his red hair and beard, tall, handsome, and rich, had passed at a relatively late age from the study of law and the classics to that of astrology, as we are told by Luca Gaurico, who pointed out with some contempt that Cardano had borrowed Paris's imprecise method of laying out the astrological houses. In the prognostication, Paris had predicted, very precisely and many years in advance of the event, that Cardinal Farnese would be elected pope, as well as the precise moment of his death. Cardano added some personal reminiscences that offer us, in passing, a view of a court filled with anxieties and buzzing with divinatory practices of all kinds. Cardano says that he saw with his own eyes a text in which a demon confirmed Ceresarius's predictions. His extraordinary ability at prediction gave rise to the rumor that his mantelpiece held marble heads that alerted him to coming disasters.[11]

Cardano chose this series of twelve genitures to reveal the connections between the stars and remarkable events that happened in the lives of outstanding individuals. They made it possible to provide concrete applications and verifications of Ptolemy's principles, explaining not only cases of extraordinary virtue and the peaks of success and power, but also painful diseases, reversals of fortune, dangers, and the threat of violent death. The series ended in a highly appropriate way with the geniture of Erasmus.

Here Cardano underlined the contrast between the obscurity of his subject's origins and the splendor of his learning, which induced popes and princes to compete to show him their favor.

Cardano's Astrological Works

Cardano's commentary on Ptolemy represented the end of a long journey that had begun when he was very young. The lessons he had learned from his father as a young boy included the elements of astrology, based on the Arabic texts that he later rejected.[12] From the start the young astrologer revealed a sharp interest in analyzing his own personality. He began at a very young age laying out the self-portrait to which he would devote decades of work, making it more and more extensive and precise.[13]

As an adult, Cardano received his strongest inducement to study astrology from Filippo Archinto, the future apostolic protonotary and archbishop of Milan. In his early period of residence at Gallarate, in 1532–33, Cardano planned and wrote, at Archinto's behest, the first sketch of his text *De iudiciis astrorum,* which he would enlarge in later years. Archinto was also responsible for Cardano's obtaining a post in 1534 that required him to teach arithmetic, geometry, and astrology in the Scuole Piattine in Milan, though only on feast days and for a small stipend. Doing so earned Cardano a dubious reputation as an astrologer.[14]

It was no accident that Cardano's first surviving publication was an astrological *Pronostico* that seems, on internal evidence, to have been directed at Pope Paul III and perhaps written at his request.[15] The most striking quality of this text, which refers the reader to a larger, forthcoming Latin *Pronostico,* "which will cover more years and do so more extensively," is the extreme reticence Cardano showed about it. So far as I know, he never cited it explicitly, as if he wished to delete it from the list of his works.[16] Still, within the rather conventional framework of general predictions about important events connected with the five great European powers (the pope, the emperor, the king of France, the king of Venice, and the Turkish sultan), the short text reveals some striking features.

This early work marked the beginning, for example, of Cardano's effort to reclaim the dignity of astrological prediction, when correctly understood and practiced, and of his polemic against the "crazy diviners" who, in their ignorance or in their desire to flatter princes, had corrupted "this noble art of astrology" and "defiled its doctrines." The more ignorant they were, Cardano claimed, the more eagerly "they play at divination, where they have not a leg to stand on, with their light heads, vile arrogance and beastial audacity." But his most important remarks had to do with religion. Cardano evoked the terrible decadence of his age and pointed out the need for a deep spiritual renewal ("Both Sacred Scripture and astrology make it absolutely clear that this insatiable greed of ours must have an end."). But he also pointed out that no one should cherish illusions about the ease with which this could be brought about and maintained that matters would go from bad to worse in the future: "there was little faith in the past, less and almost none now; in future it will be completely destroyed."

Paul III's strong interest in the study of the stars induced Cardano to continue down this road. In 1538 he completed and published at Milan two "libelli" that were presented as the first fragments of his great project, the *De judiciis astrorum*: the *De supplemento Almanach* and the *De restitutione temporum et motuum coelestium*. The last chapter of the second text set out ten genitures, five for princes and five for scholars: the first of a long series. In the two dedicatory letters to Archinto, Cardano distanced himself from his envious critics and from the "criminal incompetence" of those who tried to discredit his work, driven as they were by greed to disseminate their own impostures.[17]

The two works were reprinted, in a revised and corrected form, at Nuremberg in 1543, accompanied by a new dedicatory letter to Archinto. Cardano also added a considerable number of genitures—the total now reached 67—and his *Encomium Astrologiae*. The most interesting aspect of this section of the work lay in the fact that Cardano based his praise of astrology as "the most excellent of the sciences" on the euhemerist and allegorical interpretation of mythical characters and stories. Cardano maintained that those who had studied the stars in antiquity had become the rulers of their communities and were venerated as divinities after their deaths. This befell the Egyptian Hermes Trismegistus, the Chaldean Berosus, and the Greek Orpheus, whose lyre with seven strings, raised to the heavens as a constellation, clearly referred to the seven planets and the harmony of the universe. The myths about Phaeton, Endymion, Atlas, Daedalus and Icarus, and Bellerophon concealed the same meanings. The bisexuality of the most famous ancient prophet, Tiresias, referred to the division of the planets into masculine and feminine; the expedition of the

Argonauts in search of the Golden Fleece—one of the fables most beloved of alchemists, who saw it as adumbrating the most secret operations of their art—had to do with the competition among rulers to establish the exact moment of the spring equinox. Even the quarrels of the gods in Homer and Vergil, their councils, and their choosing to favor one hero or another—stories that would be silly and worthless if taken literally, and as absurd and laughable as a vain chimera, and thus entirely unworthy of elevated poets, if they lacked some such hidden meaning—had in fact to be understood as references to the various celestial influences on the world of men.[18]

The reference to Icarus, "who tumbled into the sea of ignorance because he had not entirely mastered his father's art," suggests that Cardano's interpretations of ancient fables probably also represented a cloaked polemic against Andrea Alciato. This famous jurist and close friend of Cardano's was also the declared enemy of all occult interests, as is clear from the enlightened position he adopted with regard to witchcraft, and a sharp opponent of astrology, as Cardano himself recorded, with some regret, in his comments on Alciato's geniture.[19] This attitude is clearly revealed by two of Alciato's *Emblems*, destined to become extremely famous and to serve as the prototype for an enormously successful literary genre. One of these represented Prometheus, chained and mutilated. It warned men not to indulge themselves in the desire of elevating their minds to inaccessible forms of knowledge, pointing out that "what is above us does not pertain to us." The other, entitled *Against the Astrologers,* showed Icarus falling into the ocean after the wax of his wings had melted. The verses that commented on the image cautioned the astrologers against suffering the same end. As Icarus, "who flew too high," had fallen, "so ruin threatens the wise man who tries to fly to God's lap in the heavens, because he wants to know secrets to which our merits do not rise. The higher the rash man rises, the greater the splash he will make when he falls."[20]

In the new edition of Cardano's *Libelli* that appeared at Nuremberg four years later, in 1547, the collection of works had become even larger: Cardano added *De iudiciis geniturarum* and *De revolutionibus* to his two earlier works. The genitures, now assembled in a separate book, had reached the final and definitive number of 100. The volume also included one of Cardano's most successful astrological works, the *Astrologicorum aphorismorum segmenta septem.* Thanks to their sharp, aphoristic style and their strictly technical content, these came to be considered particularly useful for the concrete practice of astrology. In the *Peroratio* of this work Cardano clearly enunciated the project that he would systematically pursue in his commentary on Ptolemy's *Quadripartitum*: to rescue astrology from the infamy into which it had fallen by organizing its valid elements, correcting errors, and eliminating vain superstitions in such a way as to make clear its full right to be considered part of natural philosophy.[21]

Once again Cardano rewrote his dedicatory letter to Archinto, which took on a special pathos in this version. In the introductory section, he outlined the history of astrology, refuting the popular schema according to which it had originally existed in a state of purity and perfection. At first, he argued, men basically resembled animals; the more difficult and noble any form of knowledge, accordingly, the later they arrived at it, and with more difficulty. From these premises he sharply rejected the views of the critics and detractors of astrology, from Pico onward. He rebutted their chief objections and insisted that no one could deny the natural causal action of the stars. As such, this causal action could be altered or blocked by the interference of other causes. Ptolemy reigned supreme in this discipline; the texts written by others "depart so far from the truth that they rather resemble fables." After having asserted the excellence of the art once more, Cardano warmly praised the generous patronage available in Germany, which had enabled such studies to flourish enormously. He felt compelled to lament the unhappy state of Italy, where no one could distinguish true science from fake and only vulgar imitators found any support. His own experiences, the long series of difficulties and enmities that he had encountered, offered eloquent testimony to the truthfulness of what he said. Cardano admitted that he had had to have recourse to dissimulation to protect himself against the violence and enmity of his enemies: "I had to remind myself how great was the envy of these apes who have rebelled against true honor and erudition."[22]

The Dignity of the Art

In a passage in his commentary that disappeared, not by chance, from the edition of 1578, Cardano pointed out that Ptolemy's *Quadripartitum* was the one canonical text of astrology. Only Ptolemy had had the mastery of astronomy needed to create this body of extremely subtle and difficult principles. If his book had not been written or had not been preserved, astrology itself would not exist.²³

Cardano's dedicatory letter to Hamilton included a similarly warm passage in praise of Ptolemy and his work as well as of his modern commentator, who had rescued this wonderful text from oblivion:

> Ptolemy, thanks to his wonderful art, great diligence, and extraordinary effort, and aided by his good health and very long life, not only described the movements of the fixed stars and planets, their sizes and qualities, but also their decrees and predictions—and did so with such intellectual subtlety that he frightened many away from the art and attracted no one to it. This gave birth to the hosts of vicious charlatans, while the discipline—like Ptolemy's own book—lay buried in darkness and oblivion. Ptolemy understood perfectly well that this would happen to him, but he preferred to write the truth in an obscure way, rather than to write deceptive falsehoods in a clear way: he did so in the hope that someday, someone would come along who was fully equipped to explicate his monuments.²⁴

Cardano, in other words, portrayed himself as the most authoritative representative of a line that maintained that a "return to Ptolemy" was necessary to restore dignity and rigor to a discipline seriously corrupted by the "follies" of the Arabs' manuals, which Cardano criticized for offering rules as manifold and minute as they were unfounded and useless.

The demand for a philological and substantive recovery of the Ptolemaic text had been raised as early as the beginning of the sixteenth century. Albertus Pighius, for example, assigned the guilt for the degradation of astrology to those who composed annual prognostications. Ignorant of mathematics and committed only to the superstitious nonsense of the Arabs, they spread innumerable and unbearable lies throughout the one true kind of astrology. He dedicated his work to Agostino Nifo, whom he urged to requite him with the favor of translating Ptolemy's work.²⁵

In his preface to Gogava's aforementioned translation of Ptolemy, Gemma Frisius also denounced the folly of the moderns, who rejected the ancient texts in order to accept only new ones, as others rejected normal foods in their quest for exotic and extravagant ones. In fact, he argued, none of those who had written after Ptolemy had rivaled him. The *Quadripartitum* remained the only foundation and the indispensable point of reference for any serious student of astrology.

Cardano maintained, in his program for the redefinition of astrology as "the conjectural part of natural philosophy," that astrologers must above all free themselves from all the ballast of the various "Albumasars,

Abenragels, Alchabitiuses, Abubatres, Zaheles, Messahalas, and Bethenes."[26] But he also criticized the classical works of Firmicus Maternus and Guido Bonatti, which claimed to offer astrological predictions of particulars so minute and contingent that no scientific account of them could be given. Even the very popular *Centiloquium*, generally attributed to Ptolemy, was spurious and made astrology "a form of evil magic" by including "interrogations" as well as "genitures."[27]

Following the line laid out in the Ptolemaic text itself, which he paraphrased at length, Cardano set out, in the introduction and conclusion of his *Commentary*, to give a precise account of the status of astrology. He described both the dignity and the limitations of the art and at the same time defended it against the accusations of its attackers.

Cardano felt compelled to admit that the situation was genuinely difficult and that the critics of astrology had it all too easy. The discipline had in fact been discredited and corrupted, and by its own practitioners. Cardano condemned not the *art*, but the *artisans:* they were the ones who failed to bring to its study the attention, effort, and mental profundity a discipline of its nobility and difficulty required. Moved either by greed or by ambition, they claimed to possess knowledge that they did not have and promised to give answers that an astrologer could not provide. They continually invented new expedients, taking shameless advantage of the ambiguous and profitable area of "elections and interrogations." One particularly greedy and ignorant astrologer, for example, had forced Ludovico Sforza to follow minute rules, even making him and his courtiers ride horseback in rain and mud.[28]

Astrology, Cardano admitted, was not an "absolutely precise" form of knowledge, endowed with absolute certainty and rigor. But that did not mean that it was "a superstition, a form of prophecy, magic, vanity, an oracle or a presage." It was a natural, conjectural art that set out to formulate probable judgments about future events. There was no reason to deny the legitimacy of doing so, especially when it was granted to doctors, sailors, farmers, and miners.[29]

The one basic presupposition on which astrology rested was the reality of the influence that the celestial bodies exercised on the sublunary world. These influences, obvious in the case of the sun and the moon, which were the supreme rulers of the life of the universe, undeniably also belonged by extension to the planets and the stars, which had the same basic nature.[30]

Cardano discussed the question of these influences at length. He tried both to prove their existence, with a plethora of examples, and to identify the paths by which they were propagated and the ways in which they affected the sublunary world. Only repeated observations could provide the basis for a body of theory as elaborate as that of Ptolemy, which could be confirmed, enlarged, and corrected in its turn by the observation of further facts.

In antiquity, to be sure, Favorinus had raised a tricky objection, one that even possessed a certain persuasive power: Even if astrology were true, it would be useless, because the prediction of bad events would enhance the subject's fear and that of good ones would diminish his happiness. But Cardano refuted this, since prediction actually helps us to accept both the good and the bad with equal moderation. Moreover, not all events predicted for the future must necessarily come to pass: Some can be changed. Future events do not exist "per se" but "in relation": If I foresee that my sheep may die of thirst because of the great heat, I can avert this end by digging a shelter and a spring for them. Astrology was to natural philosophy as the books of Hippocrates and Galen on prognostication were to medicine. Future events are distinguished from present ones not by species and genus, but by the element of time, which is connected to them as an accident.[31]

More generally, the tripartite Aristotelian division of all goods into those of the soul, the body, and wealth and honors clarified the utility of astrology as an art. Like philosophy, astrology was not "profitable" in itself, like medicine and trade. Nor did it promise glory, unlike military and legal pursuits. Nonetheless, it could provide excellent tools for attaining all these goods. Many philosophers had become so famous as to be immortalized, and in recent times some had become very rich as well. In the same way, many astrologers had used their art, however false it was, to enrich themselves.

If, on the other hand, contemplation is really the highest and most divine human activity, then astrology must take its place at the top of the hierarchy of the sciences. For it studied celestial things and future events, that is, the rarest, noblest, and most desirable objects, "as if one took part in the banquets of the gods." To be sure, the weakness of the human mind sometimes overturned this natural ordering: "astrology is very beautiful, but extremely difficult and demanding."[32]

Large-Scale Events and "Laws"

The second book of Cardano's commentary dealt with large-scale events. Commenting on its relatively modest length—as compared to that of the third and fourth books, which dealt with the genitures of individuals—Cardano explained that this was due to the much smaller amount of knowledge that we possess about the general constitutions of the stars. Accordingly, these had to be discussed in a suitably modest way. Nonetheless, a treatment of this subject naturally preceded that of horoscopic astrology, since "universal causes are more powerful than particular ones." By making this distinction between two levels of causality, one could both avoid many errors and answer the objection of those who claimed that collective disasters, like shipwrecks, wars, and plagues, brought people with totally different genitures together in a common death.

As is well known, Ptolemy emphasized, in his discussion of large-scale events, the importance of eclipses and comets. By contrast, he was far more reticent about other factors: for example, the "great conjunctions" so widely discussed in the Arabic tradition. Cardano evidently oscillated between strict adherence to Ptolemaic orthodoxy and the temptation to make room for other suggestions that could fill out the rather meager theory of eclipses and comets, though he also evinced a rather cautious attitude toward the great conjunctions, which he described as "very famous" but also, in reality, as "of no great importance." In themselves, they could offer only very general indications. For example, when they were found in the watery signs of the zodiac, which formed the trigon dominated by Mars, there would be wars, new mechanical inventions, contagious diseases, and heresies. Muhammad and his law belonged to this trigon. But when they were found in the fiery signs, where the Sun and Jupiter exercised a predominant influence, they would bring about monarchies, peaceful periods, and wise men, as, for example, in the time of the monarchy of the Medes, that of Christ, and that of Charlemagne. He dealt with the airy and watery signs in the same way.[33]

Cardano clearly felt the need, however, to fill out Ptolemy's discussion, as is clear from the fact that he did not hesitate, in his discussion of the rise and fall of different "laws," to draw heavily on the commentary of the Arabic astrologer Haly. Cardano's deep interest in these problems is evident from the fact that he did not quote passages from Haly word for word but reworked those that he found useful "for using the stars to predict the events that bring about laws, changes, and heresies."[34]

Cardano's discussion based itself on the correspondences that Ptolemy had established between peoples, climatic zones, and astral influences. He proposed an elegant and ingenious division of the inhabited world into four quadrants, each subdivided in turn into eight triangles, each of which was subject to particular planetary influences, following precise rules. These endowed each people with particular characteristics related to the triangle it inhabited, but also with shared ones due to the common dominance of the sun and moon: "No people is so barbarous that it can be totally free of the effects and customs of the sun and moon."

According to the doctrines that Cardano claimed to derive from Haly, laws have their origin in the central triangles, which are dominated by Mercury, and then spread into the peripheral ones. Mercury is necessary for every law, since these require "much speech and argument and changes in the ordering of one's life" (in the 1554 text Cardano wrote, more negatively: "and lies, when they are necessary, and lightness of brain"), but it is not sufficient to bring them into being on its own. Its associations with the planets that rule different triangles produce the different laws. When Mercury comes into conjunction with Saturn, for example, it will produce the Hebrew law, which will correspond to the characteristics of the planet. On the one hand, it will be "impious and extremely shameful, and will allow avarice and divorce and illicit unions [according to the original text, it will also be "full of lies and abominations"] and leprosy and impurity." But it will also be stable and constant. When Mercury is in conjunction with Jupiter, it will bring Christianity into being. "This is the law of purity, of piety, of chastity, of mercy, of honesty, in which there were many kingdoms and a priesthood deserving of every form of honor," since Jupiter presides over the priesthood. When Mercury is in conjunction with Mars, it means the law of Muhammad, and arms, and wars, violence, and cruelty: In conjunction with Venus, it means the law of the idolaters, with all its indulgence in the pleasures of the flesh.[35]

According to Cardano, the equinoctial signs are also connected with the laws. These signify not popular consent, as Haly claimed, but sudden changes of opinion, which are one of the factors most strongly characteristic of the advent of a new law. These erupt into the world of men with the scorching violence of a burning torch, overturning customs, emotions, and well-established social institutions: "It is like a torch lit in the minds of men, which descends from heaven and moves with the greatest imaginable speed. Men greet it with open mouths, despising the fear of death and the favor of princes, their own interests and those of their sons, to such an extent that some, in violation of every sense of humanity, have to be punished by the executioner for love of the laws."[36]

Though Cardano did not underestimate the importance of eclipses and comets, the powerful effects of which he actually illustrated with a large number of examples, he saw a wide variety of astral factors as relevant to understanding and explaining large-scale events. He gave a long list of these. At this point, in his treatment of the "events of the most general kind, which are of the greatest importance," he considered it appropriate to insert, as the most effective possible example to support his point, the "Birth of the Savior," which became one of the most often cited, sharply criticized, and misunderstood passages in his works.

The Horoscope of Christ

Cardano was conscious of the risk that he was taking when he published the geniture of Christ. He admitted that he had drawn it up more than twenty years earlier but had hesitated, out of religious scruples, to publish it.[37] His fears were not without foundation. As early as 1556, Adrien Turnèbe used the preface to his edition of Plutarch's *De defectu oraculorum* to denounce in no uncertain terms the revival of interest in astrology. He showed special distaste for the "vile and criminal madness" of those who, reaching levels of impiety previously undreamt-of, dared to set out the geniture of the Savior himself, and those who subjected to the stars the One who had created them.[38]

The accusation was regularly repeated, with little variation. Among the criticisms that Francisco Sanches leveled at Cardano in his *De divinatione per somnum* was that he had made the stars superior to "our Savior, the Lord of all things in heaven, on earth and in the underworld." Joseph Scaliger recalled the audacity, at once impious and foolish, of that *"cymbalum astrologorum"* who had published the horoscope of Christ, deducing all the events of his life from the position of the stars. De Thou reproved the "extreme madness" and "impious audacity" with which Cardano had subjected the Creator to the stars.[39]

The argument became a commonplace in the critical literature on astrology, one designed to show whether the author was an atheist or superstitious, and the fact that Naudé discussed the question at some length in his biography of Cardano reinforced this tendency. To restore the scandal—and the supposed originality and dubious fame of the author of the pages in question—to their true dimensions, Naudé pointed out to other critics that the topic was hardly new. Albertus Magnus, who cited Albumasar in his turn, had already dealt with it, as had Cardinal Pierre d'Ailly. So had, more recently, an unprejudiced pupil of Agostino Nifo, the Calabrian Tiberio Russiliano Sesto, who had discussed the question at such length as to make it surprising that Cardano had found anything to add. Naudé concluded that the clever Milanese astrologer had pretended not to be familiar with these precedents, more or less famous as they were, since he would rather be accused of impiety than risk losing the fame that his pages on the horoscope had won him.[40]

In fact, Cardano did put forward, in a number of passages, precise remarks and distinctions that clarified his intentions and offered a defense against potential charges of impiety. In the fourteenth geniture, for example, he discussed one point quite explicitly. Since Spica Virginis came up for discussion, Cardano made clear that the one who created the stars had no need of them, as he had had no need of fasting and prayer. Nonetheless, he had taken in what was best in them, without making any change whatever in the natural order of causality that he had established. Jesus assumed, that is, the temperament and appearance that they produced, but not because they were necessary to his "bodily balance," as was maintained by a certain "complete madman," who claimed that Christ's particular physical constitution had enabled him to walk on water. Cardano confessed that the impurity of his times forced him to make this digression and these distinctions, since some were trying to force astrology to yield unacceptable naturalistic theories, and others took even what was human as divine: "Some make the powers of natural bodies so great that they make astronomy produce wild theories. Others claim that what is human is also divine, since they confuse the prerogatives of man with those of God."[41]

The whole issue of relations between the human and the divine and their delicate equilibrium came into question here, since the problem of the horoscope of the man/God seemed to shatter it. But as Cardano pointed out, acutely, in another passage, although it was doubtless heretical to deny the divinity of Christ, it was equally so to try to make him completely exempt from mortality and all other links to mortality, as others did in their desire to exalt his divinity.[42] The human part of the Savior was clearly subject to the influence of the stars, like every other created being. This did not mean that the stars had made Christ divine, produced his miracles, or brought about the promulgation of his law. But it did mean that God had designed the positions of the stars in advance in such a way that they proved appropriate for this geniture, in which ten very rare and unusual factors came together. The bare diagram of the geniture was a sort of icon, which represented, in advance, miracles that had been determined from eternity: "And those are the ten very rare and unusual factors in this geniture, whose almost divine conjunction provides a kind of advance image of the miraculous works that had been predetermined from eternity."[43]

The astrologer's job was simply to decode the geniture, to make manifest what was implicit in it, and to reveal the precise correspondence between the aspects of the stars and the life of the man Christ on earth. A close analysis would yield a great many forms of confirmation and explanation, on the natural plane, for the law that Christ promulgated, which was "by nature" the law "of piety, justice, faith, simplicity, charity, and was established in perfect form, and would not come to an end, at least until the ecliptics come back together and the universe enters a new state." It would also provide these for the personal qualities and experiences of the man himself: not only for his "natural" ability to know the future, his eloquence, his precocious wisdom, and his brilliant intellect, but also for his melancholy character, his freckled complexion, his poverty, the plots against him, the risks he ran, and his violent death.

The text certainly seems audacious, despite the protests that the author made and the passionate defense he offered in his preface to the reader.[44] For it is, in the end, quite simply a horoscope: a horoscope in which Cardano quietly follows out the thread of the events of Christ's life, continually noting that Saturn was retrograde, that the Sun was in opposition to Mars, that Saturn and Jupiter came into conjunction in Aries. The perfect "congruency" of stars and events, Cardano argued, provided yet one more confirmation of Ptolemy's basic veracity. He concluded, accordingly, with a cry of triumph—and of defiance of the enemies of the art: "let those who deny the truth of the art see if I have changed the times, or miscalculated the positions of the stars, or departed in any way from the teachings of Ptolemy in my explication of what they portended."

Cardano was perfectly aware that the theologians were hostile and knew the risks he would run. From the time of Pico onward, sharp attacks on astrology had tried not only to undermine its claims to possess a "scientific" foundation, by insisting repeatedly that its principles lacked any foundation or consistency, but also to denounce and condemn any effort to reconcile astrology with theology. From Pico's point of view, the worst offender—even worse than the Arabs, with their superstitious "fables"—was the very authoritative Cardinal Pierre d'Ailly, for he had tried in a number of different works to prove the harmony of theology with astrology.[45]

The polemic against astrology took place on an increasingly theological plane. Savonarola translated Pico's *Disputationes* into Italian and summarized them to make them accessible to ordinary people.[46] Gianfrancesco Pico took up his uncle's arguments in book 5 of his *De praenotione*, and in his *De veris calamitatum nostri temporis causis* he sharply attacked Nifo, who had held that there were connections between the positions of the stars and calamities on earth. He insisted that all disasters were caused exclusively by divine initiative and providence, which used them to punish mankind for their sins.[47] These polemics were designed above all to keep supernaturally inspired prophecy and other divine gifts absolutely free from any contamination whatever from astrology.

Against these positions, Cardano argued that astrology and religion not only did not conflict with one another, but were in basic agreement. From the very start of his dedicatory letter to Archbishop Hamilton, he not only celebrated the "excellence" of the art, but also defended its piety. Contemplation of the order and harmony of the celestial spheres and the whole great "machinery of the world" would make man conscious that a single, sovereign intelligence existed. No branch of learning was better equipped than astrology to make man recognize the wisdom, power, and love of God.[48]

Naturally Cardano had in mind a deep and intimate form of religion, one that had nothing in common with the false religiosity so prevalent in the world and so often found in the highest circles. Julius II, for example, had spread innumerable quarrels and caused a vast number of wars. One could properly call him a weapon sent by God to punish the sins of men.

He was responsible, in the first instance, for the ruin of the Roman church and of Christianity as a whole—and of the many shepherds who had changed into wolves, so that the members of Christ lay even more bloodied than they had been on the cross: "Are the blood of Christ and the divine law granted for man's benefit destroyed by you, who should guard the flock? And those members of Christ lie there more bloody than they did on the cross. He was nailed to the cross by his own will, for the benefit of a great many: but you keep on tormenting him."[49]

Other Horoscopes

As is clear from the twelve exemplary genitures that accompanied the *Commentary* on Ptolemy, Cardano felt that an intensive effort at "experimentation" and verification had to accompany his exposition of theoretical principles. As early as 1543, in his first prefatory letter to Filippo Archinto, he declared that the sixty-seven genitures he was publishing were meant to provide a body of examples offering significant information about fundamental aspects of horoscopic astrology. He intended to consider different types of birth (twins, monsters, bastards, difficult births) and death (by poison, thunderbolt, water, capital punishment, arms, falling, and illness); the variety of human tendencies and habits, since he would analyze the genitures of men who were timid, rash, stupid, prudent, possessed by demons, deceitful, simple, heretical, thieves, robbers, and adulterers; as well as the whole range of professions and possible outcomes in life. He would provide genitures for men who had killed their wives, had suffered exile or imprisonment, had become apostates, had fallen from the highest honors into the lowest possible position, or vice versa. The collection of genitures included those of certain princes, which posed more serious problems than the genitures of private citizens. In the first place, more supraindividual factors connected with the kingdoms that they governed interfered with these men: Thus it could be hard for a king to avoid flattery or risky situations if evil was predicted. That was why Firmicus Maternus had made rulers exempt from the predictions that held for ordinary mortals. But Cardano disagreed: That position might have been acceptable for pagan rulers, who loved to be believed to be divine and to make themselves appear so, but it was unacceptable for Christians, animated by true piety, who see nothing superstitious in the stars but also natural causes. After all, every man, without exception, was subject to the changes the stars brought about, just as every man was subject to heat, cold, and suffering.

More than once Cardano found himself forced to confess his own satisfaction when genitures and theoretical principles coincided perfectly. He defied the denigrators of astrology to show that its principles were not certain: "from this it is clear that even if Pico came back to life, astrology would not be uncertain"; "even if you don't believe in astronomy, this one geniture will convince you—unless you are an ass—that the art is not empty"; "let the enemies of astrology come and provide answers for human experiences as remarkable as these, unless they claim that I invented these genitures which I drew from public sources, or that I adapted the planetary positions to fit fictitious places. For the life corresponds absolutely perfectly with the predictions that follow from the horoscope."[50]

Astrology is the most sublime of arts. Like that of the jeweler, it is inexhaustibly rich. But it calls for more than knowledge of a complex body of theory. The astrologer must also have special gifts, must be endowed "with a special kind of acumen, with a great deal of experience, with a mind that seeks only the truth."[51] Cardano knew that he himself possessed these qualities. With obvious pride he described how he had met Georg Joachim Rheticus at Milan on March 21, 1546. Rheticus, who had heard of Cardano's predictive abilities, wanted to put them to the test. He showed Cardano the geniture of an unknown individual, inviting him to comment on it. Cardano analyzed the individual elements of the geniture, inferred the personality of the subject, and then deduced, step by step, to the growing amazement of his interlocutor, that the subject in question had been accused and convicted of forgery and then publicly burned. Rheticus, who demanded an explanation for every single statement, became more and more amazed. But he had to confirm that the subject in question had been a counterfeiter, who had been condemned to exactly the punishment predicted by Cardano. For his part, Cardano, almost incredulous about a prediction so accurate that it seemed comparable to those of the ancients, actually consulted the judge to ask for confirmation of the facts.[52]

A comparison with Luca Gaurico makes it clear that scientific questions played the central role in Cardano's research. Paul III made Gaurico a bishop, even though Cardano considered him one of the charlatans who brought the art into discredit (Gaurico himself insisted, in his celebratory geniture of the pope, that he had not asked for this high office).⁵³ Gaurico's *Tractatus astrologicus* contains long series of horoscopes for popes, princes, men of letters, and men who died by violence. The actual technical elements of these are few and rather general, but the kinds of details that would fill a gossipy chronicle are very plentiful. These naturally make the text highly interesting but contrast sharply with the scientific austerity shown by Cardano, who consistently remained intent on understanding and explaining every fact, even the most trivial or unusual ones, in the light of the configurations of the heavens.

In Gaurico's geniture of Pier Luigi Farnese, for example, he mentioned that the tyrant was stabbed while indulging himself with three youths, and then discussed in some detail the ways in which the corpse was disfigured. Equally terrible is his account of the end of his "black soul" Apollonio—"deformed, dark, his hair black and curly like that of an Ethiopian, without his left eye, a traitor"—who, imprisoned and tortured, died, buried alive, in a ditch within the prison.⁵⁴ Also worthy of mention is Gaurico's portrait of a criminal named Raimondo, a Celestine friar, who ended his days at age twenty-six, stabbed and then burned, because he had not hesitated, in his "love for a boy," to engage in blasphemous and disgusting practices. In his cell he kept a wooden statue of Christ, upside down and bound by the feet, which he struck with a whip. After he had consecrated the hosts, he fed some of them to a chicken and fried some in boiling oil.⁵⁵

Cardano's collection also included individuals like this: for example, the Servite friar Ciriaco, who enjoyed the favor of powerful men and was venerated like a saint by the people, though in reality he was a man of terrible character and a great hypocrite. After amassing an immense amount of wealth by every imaginable sort of crime, he put an end to his own life, committing suicide in prison.⁵⁶ But these details are not ends in themselves: They acquire their meaning from the astrological perspective, which enabled Cardano to explain their behavior and habits, analyze their darkest passions, and unmask their fictions.

The pinnacle of Cardano's intensive efforts at exploration took the form of his own geniture, on which he worked continually from his youth until his old age, and which formed the astrological backbone of his autobiography. This autobiography, as Alfonso Ingegno has rightly observed, has the structure of a detailed horoscope.⁵⁷ The geniture itself was continually enlarged, evolving from the first meager versions until it reached its final, full form, which modified and updated all the rest, in the last edition of the twelve genitures.[58]

Cardano justified adding his own geniture to the other exemplary ones partly from the rigor with which he had studied the facts: "this," he wrote, "is the most precisely executed geniture, and the one which I have worked out with the greatest care." But he also cited the variety of experiences that had characterized his strange and contradictory life, one in which good and bad events occurred in the most unexpected circumstances: "Good and bad things have never happened to anyone in so unexpected a way"; "often, a great good has come about from great evils, or, on the contrary, great disasters have befallen me out of great goods."[59] It seems paradoxical that Cardano, who always set out to control events in one way or another, emphasized so strongly the unpredictability of his own life.

Astrological analysis, rooted in the period between life and death, centered on three basic nuclei: close study of the subject's body and temperament; reconstruction of the family structures to which he belonged, from the parents to children and grandchildren; and the recounting of external events (profession, honors, wealth). With regard to the first cluster of questions, which is also the most interesting from our point of view, Cardano admitted that he ran a considerable risk in revealing his most secret impulses, thoughts, and passions. But he did not hesitate to set himself up as the object of scientific verification, since the love of truth and research outweighed any human reservation:

> I could not know the mind, customs, secret deeds of anyone else as well as I knew my own thoughts, appetites, desires, and the movements of my soul. . . . If I set out to praise or criticize myself, will I not seem stupid or insane? If I remain mute, what help can I bring to the students of this discipline? Let the love of truth and the general welfare win out, then. . . . And if I also confess my vices, what evils will result? Am I not a man? And it is more worthy of a man to confess openly than to dissimulate. Things dissimulated become hidden, while those which we acknowledge can be confessed and avoided. Let the sweet love of truth, accordingly, win the day.[60]

This perspective—the need to overcome worldly conventions in order to attain the level of sincerity required to verify the truth of astrology—provides the context for the collection of impulses and habits that Naudé found so disconcerting and inappropriate. As Cardano saw it, he was subject to mixed planetary influences: It took him sixty adjectives just to reflect the variegated tendencies that the mingled influences of Venus,

Mercury, and Saturn conferred on him. Naudé, who discussed only the last set of impulses, the most "Saturnine" ones, remarked that Cardano would have done better to publish only those of his desires that could be reconciled with his reputation, veiling the less edifying ones in a discrete silence.[61] Evidently he did not realize that Cardano was not engaging in some foolish or shameless form of exhibitionism but following scientific necessity. The point at issue was not decorum but truthfulness.

The second nucleus of the horoscope had to do with family structures and the connections between the genitures of different members of the family. Cardano divided up the phases of his own life in accordance with the impact of these genitures on it and found only one of the nine phases that he listed to be genuinely happy. The central line of his analysis connected Cardano's geniture, on one side, to that of his father, whom he certainly esteemed and revered but also found a little bit overwhelming; on the other side, to those of his sons, on whom, as late as 1553, he had reposed the highest imaginable hopes. From his daughter he expected more problems than benefits, but the genitures of his sons promised "many goods and few evils." Female personalities played less significant, or negative, roles: from his mother, "small, fat, and devout," continually afflicted with attacks of hysteria and therefore not very affectionate, to his wife, who remains for the most part in the shadows after her first, luminous premonitory appearance in a dream, to his sterile daughter, down to the "shameless" daughter-in-law who was the main cause of Cardano's family tragedies.

The tragedy of his elder son's condemnation for murder took place between the two Basel editions of Cardano's *Commentary*.[62] As in the cases of Edward VI and Aimar de Ranconnet, Cardano found himself forced to modify and add to the data and their interpretation, tormentedly doing and redoing the calculations, in order to understand and explain what had happened. Detractors of astrology like the Jesuit Alessandro De Angelis would make excellent use of the episode: De Angelis recounted the tragic story in every detail and then addressed himself, with brutal directness, to Cardano: "Why didn't you keep the axe from your son's neck?"[63] Cardano, who had analyzed 100 genitures of princes and kings, who had explored the darkest recesses of nature, had not been able to foresee his own family tragedy, or at least had not been able to prevent it. What better proof could there be that astrology was false and useless?

Cardano played the role of the restorer of a noble discipline, one who hoped to restore it to its ancient dignity, from the shameful and decadent condition to which it had fallen. In truth, even though his harshest critics recognized that he was the greatest astrologer of the sixteenth century, his art was inexorably being marginalized. The ninth rule of the Tridentine Index condemned divinatory practices and theories. In the solemn preamble of his bull *Coeli et terrae* of 1586, Sixtus V proclaimed that knowledge of future events was reserved exclusively for God. All theories that aspired to such knowledge, including astrology, were to be rejected as deceptive.[64] In the same period, the "repentant" astrologer Sixtus ab Hemminga examined thirty famous genitures, showing that they were full of contradictions and imprecise statements. The most respected astrologers of the sixteenth century, with Cardano at their head, came in for sharp criticism.[65] At the end of the century, the Spanish Jesuit Benito Pereyra brought out a very successful attack, *Adversus fallaces et superstitiosas artes,* against magic, dreams, and astrology, areas of learning that were all generated by the same mad desire to know the future and therefore could produce only illusions and deceit.[66]

De Angelis, a professor at the Collegio Romano, showed a lack of generosity in his diligent and well-documented attack on astrology, as when he reproached Cardano because he had failed to predict the death of his son. On the other hand, when he spoke of the impossible labyrinths of judicial astrology, he was simply reflecting the growing intolerance for a discipline that was becoming more and more incomprehensible, a spider's web so complicated and subtle that in the end, by trying to explain too much, it explained nothing.[67] On the eve of the trial of Galileo, in 1631, Urban VIII—who, "though very expert in astrology, forbade others to pursue it"—confirmed in his bull *Inscrutabilis* that the human intellect, "imprisoned in the shadows of the human body," was prohibited from raising itself to the "secrets" of God.[68]

The complex and fragile scaffolding of astrology rapidly turned into a useless relic, a self-enclosed device, individual parts of which could hardly prove of any use. Vanini and others who tried to find in it some inspiration for their own corrosive impiety were generally disappointed. The best they could find in their exploration of Cardano's astrological work was the Arabic theory of the succession of "laws." They found the horoscope of Christ, basically, silly, and the whole technical apparatus seemed to them incomprehensible and abstruse. Once astrology was no longer scientific and not impious enough, it lost any interest.

Cardano's discussions of astrology can reveal a great deal to anyone who wants to become acquainted with an incredibly elusive personality: that of one who confessed that he was the despair of any artist who tried to reproduce his features. Of all the disparate features of his personality mentioned in his interminable list of his own qualities, the last is probably the most valid of all: that he remained unknowable even for those who lived on intimate terms with him. In another passage Cardano reiterated that the harder he tried to fix his own coherent identity, the less he succeeded.[69]

Cardano believed firmly that astrology could provide him with the thread that would orient him in the labyrinth of life. It would give him the tools with which he could interpret and comprehend the obscure and disorderly world of his emotions. Though Cardano had some elements of superstition, they were not especially prominent in his books on astrology, in which his desire to rationalize what was disorderly and unpredictable prevailed. The motive that permeates and unifies his innumerable works—the desire for knowledge, which is at one and the same time the highest activity of man and the most effective way to exorcise suffering—confronts us in his astrological works as well.

Astrology serves as a kind of "link" between heaven and earth, as the border between the agitation and confusion of this world and the "secrets of eternity," where everything is clear and bright. It enables us to have a more distanced and coherent vision of man, to project his brief, restless existence on earth onto the background of a higher world. As we contemplate the heavens "these things will appear to our mind: the memory of eternity, the fragility of our condition, the vanity of ambition, the bitter recollection of our sins. Hence our disdain for so short a life. Even if it should last a hundred years, what is that in comparison with the vast extent of eternity? Is it not as a point to a circle? What is all the happiness of man? If anyone has felt it, even you, is it not all wind, smoke, dreams?"[70]

Notes

1. Cf. G. Cardano, *Liber de libris propriis* (1554) and (1562), in his *Opera omnia* (Lyons, 1663; reprinted, Stuttgart-Bad Canstatt: Frommann, 1967), I, 72, 89–94, 109–10, 136–37; *In Cl. Ptolemaei Pelusiensis IIII de astrorum iudiciis aut* . . .

Quadripartitae constructionis libros commentaria (Basel, 1554) (cited below as *In Quadrip.* [1554]; when cited without a date, the reference is to the text in Cardano, *Opera omnia*, V), dedicatory letter to John Hamilton, A 3r, 126. For a pleasant reconstruction of Cardano's consultation with his illustrious patient, see C. L. Dana, "The Story of a Great Consultation: Jerome Cardan Goes to Edinburgh," *Annals of Medical History*, 3 (1921): 122–35. Recent studies on Cardano appear in E. Kessler, ed., *Girolamo Cardano: Philosoph, Naturforscher, Arzt* (Wiesbaden: Harrassowitz, 1994); M. Baldi and G. Canziani, eds., *Girolamo Cardano: Le opere, le fonti, la vita* (Milan: Franco Angeli, 1999). For a bibliography of the secondary literature on Cardano, including his astrology, see I. Schütze, "Bibliografia degli studi su Girolamo Cardano dal 1850 al 1995," *Bruniana & Campanelliana*, 4 (1998): 2, 449–67.

2. Ptolemy, *Opus quadripartitum, adiectis libris posterioribus*, trans. A. Gogava; with *De sectione conica orthogona, quae parabola dicitur deque speculo ustorio* (Louvain, 1548). The first two books of the Greek text had been available since 1535 in a Latin translation by Joachim Camerarius (Nuremberg, 1535). Before that time, the medieval Latin translations of the Arabic text had been available. On Cardano's commentary—and, more generally, on his astrological theories—see A. Ingegno, *Saggio sulla filosofia di Cardano* (Florence: La Nuova Italia, 1980), esp. 41ff., 272ff.

3. Cardano, *De libris propriis, Opera*, I, 72. The second edition of Cardano's *Commentary* (Lyons, 1555) does not seem markedly different from the first. But the second Basel edition does show clear differences, as the author himself stated it did: "ita castigavi et auxi ut maxime studiosis satisfacere possit" (Cardano, *Opera*, I, 110).

4. Cardano's "Christi nativitas admirabilis" appeared in the first edition, *In Quadrip.* (1554), 163–66. It reappears in the text in the *Opera omnia*. The decision to reinstate it was clearly made at the last moment, since the printer simply inserted a new pair of pages, 221 and 222, to avoid having to interrupt or redo the existing page numbering.

5. J.-C. Margolin, "Cardan interprète d'Aristote," *Platon et Aristote à la Renaissance*, XVIe Colloque International de Tours (Paris: Vrin, 1976), 307–33. For interesting discussions of Cardano's astrology, see also Margolin, "Rationalisme et irrationalisme dans la pensée de Jérôme Cardan," *Revue de l'Université de Bruxelles*, 21 (1969): 89–128, and J. Ochman, "Il determinismo astrologico di Girolamo Cardano," in *Magia, astrologia e religione nel Rinascimento* (Wroclaw: Wydawnicwo Polskiej Akademii nauk, 1974), 123–29.

6. Cardano, *Liber duodecim geniturarum, Opera*, V, 503.

7. Ibid., 508. For the voyage to Scotland, see also G. Aquilecchia, "L'esperienza anglo-scozzese di Cardano e l'Inquisizione," in Baldi and Canziani, *Cardano*, 379–91.

8. These differences have already been pointed out by F. Secret, "Jérôme Cardan en France," *Studi francesi*, 30 (1966): 480–82; see Cardano, *In Quadrip.* (1554), 422–44, and *Opera*, V, 513.

9. This geniture had already appeared at the end of Cardano's *Aphorismorum astrologicorum segmenta septem* (Nuremberg, 1547); cf. *Opera*, V, 514.

10. *Opera*, V, 515.

11. L. Gaurico, *Tractatus astrologicus* (Venice, 1552), 65 v; Cardano, *Opera*, V, 548.

12. Cardano, *De propria vita*, *Opera*, I, 36.

13. Cardano claimed in his *Liber de exemplis centum geniturarum*, xix, *Opera*, V, 469, that he had worked on his own geniture "for more than thirty years." A few lines below, in listing certain astrological data, he mentioned that he was 44 years old. Evidently, then, he began to study his own geniture as an adolescent.

14. Cardano, *Libellus de libris propriis, cui titulus est Ephemerus*, *Opera*, I, 56–57; further details on this work, which he continued to enlarge, in *De libris propriis*, 63. Archinto's geniture appears in the *Liber centum geniturarum*, *Opera*, V, 477–78. On his connections with Cardano, see F. Secret, "Filippo Archinto, Cardano et Guillaume Postel," *Studi francesi*, 39 (1965): 173–76. For Cardano's reputation as an astrologer, see *De libris propriis*, *Opera*, I, 64, 100. And for another redaction of this text, see M. Baldi and G. Canziani, "Una quarta redazione del *De libris propriis*," *Rivista di storia della filosofia*, 53 (1998): 767–98.

15. *Pronostico o vero iudicio generale composto per lo eccelente messer Hieronymo Cardano phisico milanese dal 1534 insino al 1550 con molti capitoli eccellenti* (Venice, 1534). The only known copy of the text is apparently that kept in the Bibliothèque Nationale de France in Paris (Rés. V, 1179); I have now edited it: "Astri e previsioni: Il *Pronostico* di Cardano del 1534," in Baldi and Canziani, *Cardano*, 457–75 at 461ff. In certain passages Cardano addresses the pope directly, as if he had written this prognostication at his request, perhaps as a sort of act of homage responding to his election to the papal throne: "Circa el stato clericale sono apparuti molti segni . . . quali lasso a sua Santità interpretare"; Cardano adds that he could not say anything about the time of his death "per non aver la sua genitura" (464).

16. In addition to the surviving *Pronostico*, another *Pronostico del anno 1535* (Milan, 1535) is mentioned by Ingegno, *Saggio*, 23 (cf. G. W. Panzer, *Annales typographici ab anno MDI ad annum MDXXXVI continuati* [Nuremberg, 1801], IX, 92). Cardano connected his ambiguous reputation as an "astrologer" with his teaching in the Scuole Piattine and with the writing of the *De iudiciis astrorum* (*De libris propriis*, *Opera*, I, 65). It is not clear if the prognostications fell within the scope of this project, though Cardano alluded explicitly to the two short Latin works on astrology that appeared in 1538 as the beginning of the *De iudiciis*. In the *Ephemerus*, *Opera*, I, 57, Cardano stated that he had written two short medical works, *De malo medendi usu* and the *De simplicium medicinarum noxa*, which was connected with it, to defend himself from the charge of taking an excessive interest—or being only interested—in astrology and maintained that these were the first of his works to appear. One might see a very vague reference to the prognostications in *De propria vita*, *Opera*, I, 16, where Cardano admitted that he had survived hard economic times by, among other things, composing ephemerides.

17. In the *Ephemerus*, *Opera*, I, 57, Cardano claimed that he had completed the two works in fifteen days, "intelligens Pontificem astronomia delectari."

18. Cardano, *Encomium astrologiae, Opera,* V, 727–28.

19. Cardano, *De exemplis centum geniturarum,* xiii, *Opera,* V, 466. Alciato insisted, against those who maintained that witches' pact with the devil really enabled them to engage in "flight" by night, that such trips were really hallucinations and imaginary. See his *Parergon iuris,* VIII, 22; *Omnia opera* (Basel, 1582), IV, 499.

20. I cite *Diverse imprese accomodate a diverse moralità con versi che i loro significati dichiarano insieme con molte altre nella lingua italiana non più tradotte tratte da gli Emblemi di Alciato* (Lyons, 1549), 69–70.

21. Cardano, *Peroratio, Opera,* V, 90–92.

22. Cardano, *Libelli quinque* (Nuremberg, 1547), A 2v, A 4v.

23. Cardano, *In Quadrip.* (1554), 126.

24. Ibid., A 2v. This important prefatory letter, like the other introductory letters in this and other works, was omitted from the *Opera.* In my view, one of the most important limitations of this edition lies in the fact that it presents the texts entirely wrenched out of their original contexts.

25. A. Pighius, *Adversus prognosticatorum vulgus qui annuas praedictiones edunt, et se astrologos mentiuntur astrologiae defensio* (Paris, 1518 [1519]). L. Thorndike underlines the importance of this work in the huge debate that blew up a little later about the flood that was predicted for 1524. See his *History of Magic and Experimental Science,* 8 vols. (New York: Columbia University Press, 1923–58), 5:184ff. On the enormous burst of pamphlet literature generated by this event, see the works of Paola Zambelli, esp. "Fine del mondo o inizio della propaganda?" in *Scienze, credenze occulte, livelli di cultura* (Florence: Olschki, 1982), 291–368.

26. Cardano, *In Quadrip.* (1554), A 2v.

27. Cardano, *Opera,* V, 356.

28. Ibid., 104.

29. Cardano, "Prooemium," *In Quadrip.* (1554), A 4v.

30. See, e.g., *Opera,* V, 99.

31. Ibid., 93, 97. The arguments of Favorinus (second century) form the conclusion of the anti-astrological *Dissertatio* recorded by Aulus Gellius (*Noctes Atticae,* 14.1).

32. Cardano, *Opera,* V, 110; *Aphor. astr.,* I, 34; *Opera,* V, 31.

33. Ibid., 73–74. The theory of the "great conjunctions," which was disseminated above all by the works of Albumasar, connected the main events of human history, including the birth and passing of religions, with the entry of the superior planets (Mars, Jupiter, and Saturn) into particular signs. It thus offered an interesting key to the understanding of universal history and a naturalistic interpretation of religion. See E. Garin, *Lo zodiaco della vita* (Bari: Laterza, 1976), chap. 1. On the theories of the great conjunctions in the Middle Ages, see J. D. North, "Astrology and the Fortunes of Churches," *Centaurus,* 24 (1980): 181–211; T. Gregory, "Temps astrologique et temps chrétien," in *Le temps chrétien de la fin de l'Antiquité au Moyen Age* (Paris: Centre National de Recherche Scientifique, 1984), 557–73, reprinted in

Gregory, *"Mundana sapientia": Forme di conscenza nella cultura medievale* (Rome: Edizioni di Storia e Letteratura 1992), 329–46. For Cardano's theory of the *leges* (religions), see J. Ochman, "Les horoscopes des religions établis par Jérôme Cardan (1501–1576)," *Revue de Synthèse,* ser. III, 77–78 (1975): 33–51.

34. Cardano, *Opera,* V, 188. For the commentary of Hali ibn Rodoan (11th century), see, for example, the text of the *Quadripartitum* included in the astrological miscellany printed "Venetiis, sumptibus haeredum Octaviani Scoti, 1519."

35. On the importance of Mercury and the great conjunctions in Hali's theory, see *Quadripartitum* (Venice, 1519), 28v, 32v, 38v. The sections of Cardano's commentary that dealt with changes in the *leges* were among those most attentively studied by all who hoped to find in astrology hints for a sharp critique of religion: e.g., Giulio Cesare Vanini, who cited substantial passages in his *Amphitheatrum aeternae Providentiae* (Lyons, 1616), 53ff. (cf. G. Ernst, *Religione, ragione e natura* [Milan: Franco Angeli, 1991], 233–36); and the anonymous author of that *summa atheistica,* the *Theophrastus redivivus,* ed. G. Canziani and G. Paganini (Florence: La Nuova Italia, 1981–82), 2:398ff., 460ff., where the whole horoscope of Christ is also cited. For the connections between this text and Cardano—and some insightful remarks on astrological points—see G. Canziani, "Une encyclopédie naturaliste de la Renaissance devant la critique du XVIIe siècle: le 'Theophrastus redivivus' lecteur de Cardan," *XVIIe siècle,* 36 (1985): 379–406.

36. Cardano, *Opera,* V, 199. For Hali's opinion on the equinoctial signs, see *Quadripartitum* (Venice, 1519), 36r; cf. Ingegno, *Saggio,* 275.

37. Cardano, *Opera,* V, 221.

38. Plutarch, *De oraculorum defectu* (Paris, 1556), A ii verso.

39. F. Sanches, *Opera philosophica,* ed. J. de Carvalho (Coimbra, 1955), 103. J. Scaliger, *Prolegomena de astrologia veterum Graecorum,* in M. Manilius, *Astronomica,* ed. Scaliger (Leiden, 1599), B 3v (cf. Ernst, *Religione, ragione e natura,* 234). J.-A. de Thou, *Historiae* (Paris, 1620), I, 155, including the story that Cardano starved himself to death to avoid contradicting the astrological prediction of the date of his death, a story that would be repeated many times, by Naudé and others, down to Bayle. J. Brucker spent a considerable portion of his chapter on Cardano discussing the horoscope of Christ, the reception of which he traced in its basic outlines: *Historia critica philosophiae* (Leipzig, 1766), IV, pt. 2, 75–77.

40. G. Naudé, *Vita Cardani,* Cardano, *Opera,* I, [4r–v]. Cf. Albertus Magnus, *Speculum astronomiae,* ed. S. Caroti, M. Pereira, and S. Zamponi under the direction of P. Zambelli (Pisa: Domus Galilaeana, 1977), 36–37; cf. now P. Zambelli, *The "Speculum astronomiae" and Its Enigma* (Dordrecht: Kluwer, 1992). On Pierre d'Ailly, see note 45. On Tiberio Russiliano Sesto and his very rare work *Apologeticus adversus cucullatos,* see P. Zambelli, "Una disputa ereticale proposta nelle Università padane nel 1519," in *Il Rinascimento nelle corti padane* (Bari: De Donato, 1977), 495–528 [and, more recently, Zambelli, *Una reincarnazione di Pico ai tempi di Pomponazzi* (Milan: Il Polifilo, 1994), including an edition of the *Apologeticus*]. For an English translation of the horoscope of Christ with commentary, see W. Shumaker, *Renaissance Curiosa* (Binghamton, N.Y.: Center for Medieval and Renaissance Studies, 1982), 53–90.

41. Cardano, *Liber centum geniturarum*, xvi, *Opera*, V, 466. Cardano drew his allusion to the "complete madman" from Pietro d'Abano, *Conciliator*, differentia xx (Venice, 1565), 32r. Pietro reported the view of the medical man Gentile da Foligno, who held that Elijah and Christ both owed their special abilities to the "temperamentum" of their bodily constitutions. The passage naturally received a critical note from Symphorien Champier (ibid., 274r). Cardano discussed the question of the "temperatura aequalis" in *Contradicentium medicorum libri*, I.vi.9, *Opera*, VI, 411–13. On this point see Ingegno, *Saggio*, 257ff.

42. The originators of this heresy were the Docetists, who drew on Gnostic doctrines and tended to deny that Christ had had a true body of flesh and blood and that he had undergone real sufferings.

43. Cardano, *Opera*, V, 222.

44. Cardano, *In Quadrip.* (1554), A 5r; see his entire *Ad pium lectorem praefatio*.

45. G. Pico della Mirandola, *Disputationes adversus astrologiam divinatricem*, ed. E. Garin, 2 vols. (Florence: Vallecchi, 1946–52), 1:566–84. Cardinal Pierre d'Ailly (1350–1420), who played a major role at the Council of Konstanz, had written important works in which he tried to maintain the "concord" of theology and astrology. He took a special interest in problems of chronology and eschatology, hoping to produce well-founded estimates for the dates of the advent of the Antichrist and the fate of Christianity.

46. G. Savonarola, *Tractato contra gli astrologi* (Florence, 1497). Reprinted under the title *Opera singolare contra l'astrologia divinatrice*, this work was translated into Latin by the Dominican Tommaso Boninsegni: *Opus eximium adversus divinatricem astronomiam* (Florence, 1581). In his prefatory *Apologeticus*, Boninsegni tried to blunt the sharpest points in the anti-astrological polemics of Pico and Savonarola.

47. G. Francesco Pico, *De rerum praenotione*, V: "De superstitiosa praenotione contra astrologiam divinatricem," *Opera omnia* (Basel, 1573; reprinted, Turin: Bottega d'Erasmo, 1972), 504 ff. Pico's little work *De veris calamitatum nostrorum temporum causis* was a polemic directed against Agostino Nifo's *De nostrarum calamitatum causis* (Venice, 1505). In his *Praefatio*, 2r–v, Nifo confessed that he had once believed that all the disasters that had recently befallen Italy—plagues, deaths of rulers, massacres, famines—were the result of divine wrath. After making a careful analysis of various celestial phenomena, however, including eclipses, comets, and conjunctions, he had changed his opinion and decided that these were the general causes of the misfortunes in question. He claimed to follow Ptolemy, "the prince of the mathematicians" as Aristotle was the god of the philosophers, inasmuch as he had been the only one to combine astronomy with natural philosophy. Nifo described the "astrologastri" who departed from Ptolemy's principles as mere "fabulatores circulatoresque."

48. Cardano, *In Quadrip.* (1554), A 2r.

49. Cardano, *Liber centum geniturarum*, xlviii, *Opera*, V, 484.

50. Ibid., 473, 477, 481.

51. Ibid., 471.

52. Cardano, *Aphorism. astr.*, VII, *Opera*, V, 85–86.

53. Gaurico, *Tractatus*, 21r–v.

54. Ibid., 109r–v.

55. Ibid., 99r–v.

56. Cardano, *Aphorism. astr.*, VII, *Opera*, V, 83. In Cardano's eyes, Gaurico apparently was the embodiment of the bad astrologer, venal and entirely without scientific rigor.

57. See A. Ingegno, "Prefazione," in G. Cardano, *Della mia vita* (Milan: Serra e Riva, 1982).

58. For Cardano's geniture see *Liber centum geniturarum*, xix, *Opera*, V, 468–72, and the two versions found in the *Liber duodecim geniturarum: In Quadrip.* (1554), 430–75, and *Opera*, V, 517–41.

59. Ibid., 468–469, 517.

60. Ibid., 523.

61. Naudé, *Vita Cardani*, Cardano, *Opera*, I [1r–v].

62. The two versions show notable divergences. The most striking naturally have to do with the events surrounding the tragic death of Cardano's son, which took place after he drew up the first version, but many passages on Cardano's honors, wealth, and other fortunes are also substantially changed.

63. A. De Angelis, *In astrologos coniectores* (Rome, 1615), 302.

64. *Index librorum prohibitorum cum regulis confectis per Patres a Tridentino Synodo delectos* (Rome, 1596), 32; for the text of the bull, see *Magnum Bullarium Romanum* (Lyons, 1592), II, 515–17.

65. Sixtus ab Hemminga, *Astrologiae ratione et experientia confutatae liber* (Antwerp, 1583).

66. The work, which takes the form of an elegant, though not especially original, collection of the arguments against the occult sciences, was reprinted a number of times in the 1590s; see Ernst, *Religione, ragione e natura*, 265–70.

67. De Angelis, *In astrologos*, 235.

68. G. Gigli, *Diario romano (1608–1670)*, ed. G. Ricciotti (Rome: Tuminelli, 1958), 253. For the bull see *Magnum Bullarium*, V, 173f.

69. Cardano, *De propria vita*, *Opera*, I, 5; ibid., V, 523, 524.

70. Cardano, *In Quadrip.* (1554), A 2r.

3

Between the Election and My Hopes: Girolamo Cardano and Medical Astrology

Anthony Grafton and Nancy Siraisi

Cardano and Medical Interrogations

Early in the seventeenth century, three short texts by Girolamo Cardano caught the attention of Giovanni Antonio Magini, a Paduan professor who was expert in both astronomy and medicine.[1] He included them in a collection of similar analyses:

Third Observation from Cardano
The onset of disease of Giovanni Antonio de Campioni
10 May 1553 8 PM
Giovanni Antonio de Campioni, at the age of around thirty, fell ill after a journey. He seemed mildly ill in the first instance, down to the fourth day, because the moon was in sextile [a benign aspect] to Venus [a benign planet] and Venus received it; for Venus was in her dignities [in Taurus, her mansion].

And because the moon was quite slow in her course, the disease seemed not to become more serious, since Venus, as I said, held it back. The moon reached the twenty-fifth degree of Gemini in around three days and eighteen hours, since its motion was so slow, and therefore the fourth day was drawn out. But then there took place a conjunction of Jupiter and Mars in Leo, and with the humid stars. This produced a great fire and turbulence in his urine, though these seemed, because of the extension of the fourth day, to begin on the fifth. Now in the seventh day, the disease became worse, since the moon had not yet reached a distance of 90° because of the slowness of its motion, but was in a very bad situation at the beginning of Leo, since it did not strike any beneficent star. Indeed, it struck an antiscion [a degree under the influence] of the sun, which was in the sixth house, and the dragon's head [the ascending node of the moon's path]. Similarly, the disease became more serious on the eighth and ninth days, because the moon came into conjunction with Jupiter and Mars, themselves in conjunction, and because they were among the humid stars, he underwent a sweat. For heat, combined with humidity, creates sweat and much urine, which he passed. On the eleventh day he sweated, but it was with great effort, for the moon overcame Saturn, which was in opposition to it, but there followed a conjunction with Venus.

On the twelfth day he seemed to be very ill, because he raved a great deal. But nevertheless because of the conjunction with Venus his urine appeared concocted. On the thirteenth, since the moon was in quartile to [90° away from] Mercury (for he is the enemy of the horoscope) he was no worse, because he was now moved towards health. But he was also no better, because of Mercury. On the fourteenth he had another sweat, and felt better. But he could not find release from the disease, because the moon had covered only 174° 22', and the disease had to be prolonged to the seventeenth day. But in the fourteenth day the moon reached sextile to [60° away from] Jupiter and Mars and quartile to Venus. Therefore he had a sweat. On the seventeenth he was freed from the disease, since the moon had now passed opposition with its [original position] and reached trine to [120° away from] Venus.

Fourth observation, from Cardano
The onset of disease for the same man, who died on the 14th day
23 May 3 PM
From the start he had the moon quartile to Venus. Lack of temperance in food and drink made him ill, and his condition quickly worsened because of the rapid motion of the moon. On the seventh day he felt considerably worse, for the moon was with the dragon's tail [the descending node of its path], and devoid of any aspect with Jupiter, and moving towards opposition to Venus and the sun was afflicted by the square of Saturn. On the eighth he seemed to be relieved by a flow of blood from the nostrils, but his strength declined because of the moon's opposition to Venus. On the ninth he seemed to breathe a little because the moon was trine to [120° away from] the sun. The tenth took the place of the eleventh, since the moon had reached the angle, that is, the beginning of Pisces, in opposition to Jupiter and Mars. On the eleventh day it was reasonable for him to die, since the moon had come into conjunction with Saturn at the tenth hour, and into quartile to [90° from] the sun at the eighteenth hour. He died on June 5, three hours before noon, and it was the beginning of the fourteenth day, and the moon had arrived at the point exactly opposed to its [original] position.

Fifth observation from Cardano
The beginning of an illness from the transfixion of the arm of Battista Cardano, which caused his death
1552 19 December 4:32 PM
This other patient was my relative, a man of sixty when he was wounded. The moon was apart from Mars, and Mars was with the dragon's head [the ascending node of the moon's [path], and the moon with the dragon's tail [the descending node], and it applied to [approached] Saturn and opposition to Jupiter, which was then unfortunate. But he did not immediately suffer, because the moon was going towards sextile to Mercury, and the wound was the transfixion of an arm. On the fourth day he suffered because the moon was quartile to the sun, but he had no fever because none of the malevolent planets attacked him. For that reason he improved greatly up until the tenth day, so much so that he was able to rise. On the eleventh day he suffered at the third hour of the night, when the moon moved towards opposition with the sun. But this was deadly, since it was the lord of the place opposed to the moon. And afterwards the moon moved towards opposition with Mercury. He was therefore laid low by fever and hemorrhage on the fourteenth day,

which was January 2. At the third hour of the day, when the moon was in exact conjunction with Mars, he died.[2]

Couched in the almost-forgotten language of astrology, these texts served as the captions for figures that Cardano erected and Magini reproduced: quick-paced narratives of celestial developments that took place over a given short period, rather than the formal analyses of horoscopes that Cardano collected and published in two of his most influential writings. To Magini, whose training and practice brought him far closer than any modern reader can hope to be to Cardano's intellectual and professional world, they seemed very revealing, even typical of the balance that Cardano tried to hold in his everyday practice between the two predictive arts of medicine and astrology. We hope in this chapter to pose the question of whether he was right.

Decipherment, naturally, must precede discussion and interpretation. Let us begin by supplying at least some of the glosses needed to follow Cardano through his analytical work. Consider, for example, the third and simplest of the figures. Cardano needed to explain why his relative Battista Cardano became ill and eventually died, some days after being wounded in the arm. Laying out the positions of the sun, moon, and planets for the moment when the wound was inflicted, Cardano arranged them in the houses of an astrological figure laid out in the standard square form: as twelve triangles, superimposed on the twelve signs of the zodiac, beginning from the left at nine o'clock. Then he analyzed their relationships and effects, using a rich and well-established technical vocabulary and following with special care the way the moon, moving rapidly along the zodiac, altered these configurations and thus exerted different effects.[3] In this case, Cardano began from a series of astronomical facts singled out without explanation as significant and given astrological meaning:

the moon and Mars are 180° away from one another on the zodiac, the moon with the descending and Mars with the ascending node of the lunar path (the tail and head of the dragon)

the moon, moreover, is coming into conjunction with Saturn, a malevolent planet, and opposition with Jupiter, a benevolent one; and the moon is moving towards sext with Mercury (60° of separation)

Accordingly, he inferred, the prospects were generally bad. The moon, a neutral planet, underwent the influence of Mars or Saturn, both malevolent planets, whose properties it may share in such circumstances. But it also entered into a benign relation with Mercury, which postponed the predicted ill effects and accounted for the anomaly that this sexagenarian did not fall ill until four days after being wounded.

As the moon moved toward quartile to the sun, or 90° of separation from it, a malign aspect, the old man felt worse, but since none of the actively malevolent planets attacked him, he remained free from fever and his condition improved gradually. On the eleventh day, however, as the moon reached another malign aspect with the sun, opposition, he felt ill at night. And when the moon went into opposition, a malign aspect, to Mercury, Battista showed the symptoms of mortal illness, fever and hemorrhage.

The other two analyses closely resemble this one. In each case, Cardano describes the course of an illness, with all the deadpan—or bedpan—detail one might expect from a faithful reader of the *Epidemics* of Hippocrates. In each case, he pays special attention to the rhythms of the patient's suffering, trying to identify the individual days on which he took a distinct turn for the better or the worse. And in each case he traces counterparts to the tossings and turnings of the fevered sufferer in the positions and movements of the stars above. The movement of the moon, in particular, imparts a clear, quantitative order to the qualitative data: Like a modern chart of fever or weight loss, it provides a continuous, measurable armature to which the attending astrologer or medical practitioner can attach other data. Not interrogations in the normal sense—figures erected to give a prognosis—these brief astrological case histories retroject the movements of the stars into the story of a sickness already endured. The only motive for compiling them would have been personal or scientific curiosity: They could serve no immediate practical end.

The form of analysis Cardano employed here seems at once strange and familiar. It seems strange because Cardano unselfconsciously applied a highly sophisticated set of hermeneutical rules to interpret the positions of the planets, without explaining or even explicitly appealing to them. He knew, without explaining why, that each planet, each house of the figure, and each geometrical figure had certain properties. His rapid-fire astral commentary on the course of each illness has, accordingly, something of the gruff impenetrability of speech barked in an unknown language.

In another sense, however, the outlines of Cardano's enterprise are hauntingly familiar to anyone interested in the Renaissance. He clearly hoped to use the astrological conditions obtaining at a given series of moments to explain the course of the diseases his patients actually endured. And this general project, if not its technical details, is exactly what one would expect of a mid-sixteenth-century medical man like Cardano. Historians of science and literature, in fact, have long treated medical astrology as one of the characteristic sciences of Cardano's period. Over and over again, historians have called attention to the large number of medical men who also studied and practiced astrology, to the many textbooks that explained the principles of astrological medicine and how to apply them, and to the widespread polemics that attended such efforts to solve contested problems like that of the origins of syphilis.

Many primary sources support this general picture. "It is accepted," Magini wrote, in an introductory statement that has many earlier and later parallels, "in accordance with the common opinion of all excellent practitioners of the art of astrology, both astrologers and physicians, that one should construct a celestial figure for the onset of each disease, to make it possible to predict its essence, its critical days, the varieties of its accidents, and finally its outcome. For we can use such a celestial figure to work out whether an illness is lethal or will end in health, long-lasting or short."[4] Many modern secondary works echo or expand on these statements, arguing unequivocally that surgeons carried out operations, pharmacists readied prescriptions, and medical practitioners recommended regimens as their celestial informants dictated.

Magini, who began his work with a detailed and precise bibliography of earlier publications, referred to Cardano as a special authority in the field, one who had discussed the astrological determination of critical days and related problems at length in his commentary on the standard ancient astrological work, Ptolemy's *Tetrabiblos,* and elsewhere.[5] The reader would naturally infer that Cardano generally agreed with Magini on the need to use astrological means for treating a vast range of diseases and disabilities, taking into account the patient's geniture, its "revolution" (the astrological configuration of the corresponding day and time) for the year in question, and the immediate astral circumstances of the illness.[6]

Cardano, as Magini did not need to point out, was both one of the most prolific and prominent medical writers and one of the most influential astrologers of the sixteenth century. His works include commentaries on the classics of ancient medicine and astrology, systematic treatises, and short consilia and horoscopes produced for individual patients and clients. It seems reasonable, accordingly, to take Magini at his word, accepting Cardano as an authoritative witness to the principles and practices of medical astrology in the sixteenth century.[7] That is what we propose to do here, but we intend to do it in a highly economical and carefully defined way. We will begin by making as few assumptions as possible. We will use Cardano's works and those of others not to prove that medical astrology was commonly practiced, but to see how one well-known expert practitioner carried out the task, millennia old in his time, of combining medical and astrological data and techniques. And we will try to remain open to the possibility that medical astrology was in fact less ubiquitous in practice than modern historians have tended to believe and less central to Cardano's own work and thought than anyone would expect.

Cardano Interrogates the Interrogations

Literal explication of these texts, obviously, requires patience and some knowledge of classical astrology. But locating their place in Cardano's larger medical and astrological practice is a more complex, even baffling, enterprise. Cardano regularly insisted that astrologers must be dignified figures, remote and learned, not ambulance chasers eager to make a few shillings by predicting the likely outcome of a case of housemaid's knee. True, he devoted a short treatise—the last section of his enormous commentary on the largest ancient manual of astrology, Ptolemy's *Tetrabiblos*, which appeared in 1554—to the uses of interrogations (figures erected to clarify astrological conditions at a given moment). But his attitude toward them was distant, even stern. At the end of the treatise, giving a list of nine commandments for good astrological practice, Cardano made clear that he saw the making of full-scale horoscopes as the astrologer's proper occupation. He counseled the reader interested in practicing to avoid ever erecting a figure for a skeptic or for widespread public consumption (the latter, to be sure, a warning that he himself did not heed very well).[8]

Only after devoting a long digression to the natal charts that accounted for the undying love of Henry II and Catherine de Médicis did Cardano finally manage to discuss the sorts of short-term astrological inquiry that he carried out ex post mortem in the cases from which we began. He admitted that he had condemned the making of astrological "interrogations" in the past and insisted that many predictions could be made only on the basis of a full birth horoscope.[9] He rebuked those—and there were many of them—who believed that an interrogation could reveal whether a theft had taken place, or if the stolen goods could be recovered. The stars, he insisted, "are causes, not signs; they are bodies, they are noble, they are powerful and strong."[10] To tie them to the trivial details of medical practice was clearly an abuse of celestial patience.

Cardano admitted that interrogations had their uses. They could provide information on one's prospects when sowing seed or making a bet, though not on anything that might be determined by other factors, such as one's upbringing and education (thus Cardano excluded marriage from the list of possible interrogations).[11] They determined the best times for administering medicine and carrying out surgery: "It has been discovered by direct experience," wrote Cardano, echoing a classic work of Arabo-Latin astrology, the *Centiloquium* ascribed to Ptolemy in the Islamic world, "that if such operations are carried out when the moon is in possession of the sign that is connected with the bodily limb in question, it cannot take place without harm befalling someone."[12] Interrogations, in other words, really did, in Cardano's estimation, form part of the art of astrology, "which," he asserted, could be practiced "with no less glory and profit than the medical men in our time practice their art of medicine."[13]

At the same time, however, Cardano also indicated that the art of the interrogation was most appropriate to those medical questions for which he could offer no rigorous way of obtaining an answer. Consider, for example, the question—a classic for makers of interrogations—of what sex an unborn child would belong to. "The safest way to determine this," Cardano said, even though he was writing a treatise on astrology,

is from examination of the belly. For Hippocrates says that male babies hang to the right, female to the left. This can also be inferred from the difference in the breasts. But even though this sign, and the last one, are very sure and hardly ever fool you, it is often very ambiguous, since the difference is so small that it can be recognized only by long habitual training. It is like knowledge of jewelry. One must put the woman down on her back, very precisely, and, as in dislocation of the foot, make a meticulously precise comparison of the sides. Here, as I said, expert physicians often go wrong when the feet are out of joint. Hence it is hardly surprising that when the difference is smaller, and experience has been much rarer, men hesitate. Yet it happens—though barely once among twenty pregnant women—that the right side is more swollen and yet she has a girl in her womb, or vice versa. The same holds for the breasts. But this falls outside the art [of astrology]. To the art belong the directions, the progresses, and the entry into masculine signs for male babies, and female ones for female babies, for the mother or the father, and with more security if for both. But if you lack this aid then take the hour of conception, erect a figure, and see which planets dominate the ascendant and the Medium coelum, and how they are affected with regard to the double form of sex. Some think that when siblings are born after brothers or sisters, they are male when the moon is moving toward New Moon and female when it moves towards Full Moon.[14]

It seems unlikely that Cardano (or any other male medical practitioner) had much experience with direct examination of women's breasts and bellies of the sort he describes here. Indeed, his very next sentence gives the game away, since he explains that the desire for information of this kind, however obtained, stemmed not from the existence of scientific means to gain a reliable answer but from the simple desire to gain money by betting. "You had better practise this before you make judgments, or the art will bring you harm, not profit. The merchants of Antwerp and Lyons make a habit of betting large sums of money on this, for the sake of the competition."[15] Cardano, in short, offered the interrogation as a counsel of despair, for the physician who wished to invest in male or female baby futures but could not obtain access to the prospective mother's belly or breasts.

Interrogations, in other words, do not sound very reliable, even in the context of a book dedicated to their study. And the reason may not be far to seek. Cardano warns, early in the treatise, that some believe that the figure of the interrogation itself could have a positive effect on a sufferer. After all, "some figures seem to help somewhat with pains of the kidneys and stone." But the figure, he insisted, "as a product of art, has no power of action." Trying to erect figures in order to influence the heavens was, as Aquinas had said, mere superstition.[16] Here Cardano suggests that many patients—whose concern, of course, was not to preserve the integrity of the astrologer's discipline but to be cured—saw interrogations as counterparts to the figures engraved on magical amulets: not images of the skies as they were but tools for manipulating them to draw down favorable and avert unfavorable influences. One hears the faint, lost echo here of arguments between the astrologer, insisting on the limits of his art, and his desperate clientele. The evidence, in short, seems puzzling, even contradictory. To solve the puzzles with which it presents us, we must examine the Ozymandian monoliths of Cardano's medical and astrological texts. How far do these complement, qualify, or refute Magini's version of Cardano or Cardano's own presentation?

Astrology in Cardano's Medical Writings

In his writings on medical theory and practice, Cardano showed remarkable restraint on the subject of medical astrology. His copious published medical works not only lack examples of horoscopic astrology such as those just analyzed or expositions of techniques of astrological prediction. They also contain relatively little extended discussion of any topics that can be construed as in some broad sense astrological and only occasionally refer to diagnosis, prognosis, or therapy based on astrological presuppositions. For the general term "medical astrology" in fact embraces a number of different concepts and procedures. In the strictest sense it refers to the use of technical astrology, involving the erection of a figure and the systematic use of tables to obtain the positions of the planets for the sake of guidance in diagnosis, prognosis, and therapy. Medical astrology of this type could, in turn, take a number of different forms and be employed for several different purposes besides the use of interrogations like those described above to analyze the cause and progress of an individual's particular episode of illness. Astrology could be used to predict expected health conditions for an entire community or explain the causes of epidemics. For an individual client or patient (who was not necessarily ill at the time), the physician-astrologer could cast a nativity in which the patient's lifelong prospects for health and illness formed a particular focus of attention.[17] In the case of a sick patient, interrogation of a figure erected for the time of the onset of illness could be used as a means of prognosis: in order to ascertain the expected outcome for the disease, the physician, and the patient, as one physician-astrologer put it.[18] Any conscientious and scientific medical astrologer who employed interrogations of this kind was expected to take great care to ascertain the time of onset of disease as accurately as possible and preferably to use these precise data in conjunction with careful study of the patient's nativity.[19] Given the difficulty in many cases of determining the precise moment of the onset of disease, some sixteenth-century experts, like their medieval predecessors, thought it allowable to cast an election or interrogation for the moment at which the practitioner was first consulted or when someone brought him the patient's urine or an object touched by the patient; another view regarded this practice as superstitious nonsense that the scientific (and Christian) medical astrologer should eschew.[20] Renaissance differences of opinion over this issue seem to mirror the process whereby the traditional medical procedure of inspection of urines itself, once the badge of the respected medieval medical practitioner, came over the course of the sixteenth century to be the mark of the quack.[21] Technical medical astrology could of course also be used retroactively—for example, to explain past epidemics or the cause of death of patients (or historical figures).

In a broader sense, medical astrology also encompassed other beliefs and practices that did not involve the casting and interpretation of horoscopes but presumably, at least in principle, required the consultation of astronomical tables. These included the theory that critical days in illness depended on the phases of the moon and or positions of other planets and the idea that choice of times for administration of therapy (usually medication or phlebotomy) required attention to the planetary positions (some held that even the mixing of compound medicines should take place at astrologically propitious times).[22] Also drawn into medical astrology during the Renaissance was the Hippocratic idea that the physician should pay attention to star risings and settings, as these affected climate and, consequently, health.

Other astrological ideas in medicine were considerably more general and less technical. As is well known, the underlying concept of all astrology—namely, that the heavenly bodies exercised influence on bodies in the terrestrial world—was systematized into a network of supposed correspondences between planets, houses, and zodiacal signs on the one hand and parts of the body, temperamental qualities, humors, virtues, phases of pregnancy, diseases, and varieties of medicinal action (purging, strengthening, and so on) on the other. These commonplace and traditional formulae, often expressed in visual images or tables, appeared in simplified medical handbooks, almanacs, and so on throught the later Middle Ages and Renaissance.[23] Finally, at least in the view of the astrologer Symon de Phares, who included Marsilio Ficino in his list of important astrologers, medical astrology extended to the astral magic and doctrines of sympathies and affinities between planets and herbs, talismans, colors, sounds, and odors espoused by that Florentine neoplatonist.[24]

Cardano's published medical writings fill almost five large, double-columned, folio-sized volumes in the seventeenth-century collected edition of his works and include most of the main types of contemporary medical literature: commentaries, treatises on various branches of *theoria* and *practica*, and *consilia* for individual patients. In them the number, extent, and level of detail of allusions to all the aspects of medical astrology just outlined fall well within the normal range for these genres, which was small. Despite the assertions about the importance of astrology for medicine iterated countless times between the thirteenth and the seventeenth centuries, in general medical works the subject of astrology usually appears only in specific, restricted contexts.[25] Where theoretical discussion was concerned, the standard contexts of extended exposition included consideration of the ancillary sciences necessary to medicine or the place of medicine among the arts and sciences, the astrological causes of epidemics, and, above all, the concept of critical days of illness. Disquisitions on some of these topics—notably critical days and the usefulness of astrology for medicine—can, of course, be found not only in the medical writings of Cardano's contemporaries, but also in the works of earlier scholastic physicians going back at least to Pietro d'Abano (d. 1316).[26]

But although the rich heritage of medieval astrology continued to be drawn on in Cardano's day, both the biological and the intellectual environment of medical astrology was now very different from that of the Middle Ages. In general terms, the menace of new or apparently new and terrifying epidemic diseases, especially plague and syphilis, appears to have played a major part in stimulating the widespread and intense interest in astral influences, and especially the role of the heavenly bodies in health and disease, characteristic of the late fourteenth to seventeenth centuries in Europe. In both astrology and medicine, Renaissance editions and translations facilitated a new attention to and reevaluation of ancient sources. Furthermore, although Marsilio Ficino's own view of astrology was in some respects ambivalent, book 3 of his *De vita* constituted an elegant and powerful restatement of fundamental concepts about celestial and planetary influences on human health by an author of great prestige.[27]

More specifically, four aspects of the relation of astrology and medicine acquired a new urgency in the late fifteenth and early sixteenth centuries. In the first place, improved access to the texts of the Hippocratic corpus, which contains a number of statements about the influence of the stars on climate, environment, and health, raised the question of the extent to which Hippocrates had actually known or endorsed astrology.[28] Secondly, vigorous debates about the causation and transmission of epidemic disease became a central topic of medical discussion from the 1490s through the end of the sixteenth century.[29] Thirdly, and perhaps most importantly,

Pico della Mirandola's withering treatment of the claims of astrology called the entire basis of medical astrology into question.[30] Pico's repudiation of astrology touched off several generations of debate about all aspects of the subject, of which the medical consequences were only one small part. Nevertheless, Pico's denunciations were especially provocative where medicine was concerned, because unlike various other medieval and Renaissance critics of astrology who routinely excluded the usefulness of astrology for medicine, along with navigation and agriculture, from their strictures, Pico explicitly and in detail repudiated specific medical doctrines rooted in astrology (notably that of critical days, to which we shall return shortly). Finally, the general tendency of humanist medicine both to seek authentic ancient sources and to confront them with nature seems to have been associated with a new demand for a medical astrology that worked, one that would be based simultaneously on correct astrological techniques and observed consequences for patients. All of these developments affected the discussions of astrology by Cardano's medical contemporaries. One notable result of the new situation—and perhaps also of the opportunities offered by the age of print—was the proliferation of a new generation of specialized treatises on medical astrology written by physicians. But Cardano, surely one of the people best qualified to write such a treatise, was not among these authors.

The limited attention to astrology in Cardano's medical works is in fact quite striking. He certainly perceived parallels and connections between his commitment to the restoration of Ptolemaic astronomy and his attempt to identify himself with a "new" Hippocratic medicine, an endeavor that depended entirely on the full access to the Hippocratic corpus that sixteenth-century editions and translations afforded for the first time.[31] His selection of Hippocratic treatises on which to comment—notably *Airs Waters Places,* the *Epidemics,* and *On the Seven-Month Child*—seems in itself to reflect connections between his medical and his astrological interests.[32] In his most exalted mood he was capable of assuring his readers that the knowledge of a wise physician encompassed the content of other arts and sciences ranging from theology through architecture, natural history, natural magic, meteorology, and cooking, and including astrology.[33] Moreover, he assured his medical readers that Hippocrates had taught that the heavens were divine and that astrology was necessary, not just useful or desirable, for the physician.[34] Cardano, who also commented on *Prognostic,* evidently perceived parallels between Hippocratic medical prognostication and astrology, both true and ancient though difficult and uncertain or conjectural forms of prediction and explanation. On several occasions, he compared the degree of certainty of the procedures available to the physician and the astrologer. Thus he pointed out that the astrologer had to study conjunctions that lasted only a moment but was not expected to have any effect on them, whereas the *medicus* was supposed to influence a body that he had to judge from sense and not from truth—a thoroughly noncommittal comment.[35] Depending on the mood of the moment and perhaps on the audience for which he was writing, he asserted that now one, now the other of the two arts was more certain. In the opening pages of his commentary on *Prognostic* he stated that "medicine alone makes reliable predictions and teaches procedures and times, and brings and shows certain and evident causes of those things." He added that medicine was more certain than natural philosophy, because medical demonstrations were "similar to mathematical ones and from causes."[36] As we shall see, when writing in an astrological context he took a different view. A few pages later on in his discussion of *Prognostic*, he grouped medicine among other predictive arts—not only astrology, but physiognomy and dream interpretation—without identifying any one of them as the most certain.[37] In the preface to his commentary on the Hippocratic *Epidemics*, which he regarded as a work from which one could learn how to prognosticate, he asserted that medical prognostication was connected with divination, one of the branches of which was astrology.[38] In expounding book 4 of *Airs Waters Places*, which is in reality mostly about the influence of weather, he pointed out that this section had much pertinence for astrologers but added that for physicians the usefulness of such knowledge was glory for the physician himself.[39]

But as all these remarks indicate, Cardano also maintained a clear distinction between astrological and medical procedures. Thus despite his admiration for Hippocrates and conviction that he had attributed great importance to astral or celestial influences, he did not ascribe to Hippocrates knowledge of all of astrology. For example, "Hippocrates did not mean" that purges should be prescribed when the moon was in a watery sign but referred only to changes in the weather. Rather, the knowledge that the moon had different qualities in each of its phases and different effects on the therapeutic environment was derived from the astrological teaching of Ptolemy.[40]

Moreover, the actual discussion of any topics relating to the stars, much less predictive astrology strictly defined, in Cardano's commentaries on Hippocratic texts is no more extensive than in other contemporary expositions of the same Hippocratic works. For example, both Pedro Jaime Esteve, a physician of Valencia, in his commentary on *Epidemics* 2, and Adrien L'Alemant, a professor of medicine at Paris, in his commentary on *Airs Waters Places,* used references to the stars in their Hippocratic texts to support the idea that medical astrology should study chiefly the influence of the fixed stars on the weather. Esteve prefaced his commentary with eighteen pages of exhortation and information about the importance of astronomy for the physician. He devoted special attention to precession, since he thought it vital to explain to his medical readers that the positions of the fixed stars in relation to the signs of the zodiac had changed since the time of Ptolemy. As he pointed out, "although very famous men who lived a little before us, such as Regiomontanus, the glory and outstanding ornament of all mathematicians . . . followed through this with accurate diligence, since their writings are not at hand for everyone, I thought it would be a very useful work indeed if I were to set forth what I had sedulously noted with long attention from their writings."[41] A long list, carefully adjusted to the 1570s, enabled readers to predict the risings of the more prominent fixed stars and thus to foresee their effects on weather and health. Cardano himself complained about the material from his own commentary on Ptolemy's *Tetrabiblos* that L'Alemant inserted into his commentary on *Airs Waters Places;* L'Alemant also enriched that work with many pages of *significationes* for the weather of the fixed stars throughout the year, drawn directly and with acknowledgment from Ptolemy's work on the phases of the fixed stars, which had been translated into Latin by Nicolò Leoniceno.[42] In his own commentaries on both *Airs Waters Places* and the *Aphorisms,* Cardano introduced into his discussion of climate, weather, and seasons a few pages of technical explanation (with diagrams) of star risings and settings and of the annual motion of the sun. In this context, perhaps in rebuttal of the views of Esteve and L'Alemant, he also took care to assure the reader that the effect of the heavenly bodies on disease was not merely a consequence of their effect on the weather but rather a result of occult influence, even though this might work by means of changes in heat, cold, moisture, and dryness. In the commentary on the *Aphorisms* he added a jibe at the laughable ignorance about star risings and settings of the celebrated contemporary writer on materia medica Antonio Musa Brasavola of Ferrara.[43]

Nor did Cardano pay much attention in his published medical writings to the topic of the astrological causation of epidemics. He devoted extended discussion to epidemics and their causation and transmission on two occasions: in his treatise on poisons, published in its final version in 1564, but like many of his writings probably long in preparation, and in a work entitled *Liber de providentia ex anni constitutione*, written in 1563. In the former, one short passage in the single chapter devoted to the causes and varieties of poison contracted from air and water gives the influence of the stars as a cause of the corruption of the air responsible for pestilential fevers. Yet he also stated that the stars are only one of several possible causes of pestilence, the others being winds, waters—and chance. In that chapter, Cardano explained that Ptolemy attributed the causes of epidemics to eclipses or unfavorable positions of the luminaries, whereas others thought conjunctions of the planets were responsible, and that both parties spoke the truth. Certainly, the eclipse of the sun that had taken place in June of the year in which he was writing was a bad sign, given the planetary positions at the time. But great conjunctions had been responsible for a number of epidemics. Cardano mentioned the one that had occurred in 1504–5, the one in 1524 that had caused an epidemic "throughout the whole world," so serious indeed that its effects were powerful as late as 1528, "such that the memory of that year will endure for many centuries," and the one in 1544, as well as the outbreak of syphilis (1504, 1524, and 1544 were indeed great conjunction years).[44]

Cardano thus signaled not only his adherence to a reformed, classicizing astrology based on Ptolemy, but also his unwillingness to abandon the theory of influence of conjunctions derived from Arabic astrology, which had been used since the fourteenth century to explain the occurrence of plague and which various early writers on syphilis also espoused.[45] Another brief passage in his commentary on the *Epidemics* shows his adherence to conjunction theory. He noted, a little pretentiously, that Erasmus Reinhold's recent *Prutenic Tables* showed that a conjunction of Saturn and Jupiter would take place in the very year in which he was writing. This was the first of a series of great conjunctions in the fiery trigon (Leo, Aries, Sagittarius), indicating pestilence to come, as the famous conjunction of 1484 had announced the arrival of the "Indian pestilence" (syphilis)—a remark that is one of very few actual astrological predictions to be found in Cardano's medical works.[46] Yet when he appended to his *consilia* a description of an epidemic that occurred in Milan and Pavia in the spring and summer of 1545 and, according to him, attacked only young girls, he made no attempt to assign astrological or any other causes and was content merely to describe the outbreak as *obscurissimus*.[47]

Even more strikingly, Cardano opened his treatise on the constitution of the year with an explicit announcement of his intention to exclude "the dogma of the astrologers" and rely solely on the teaching of Hippocrates about the effect of weather conditions on disease.[48] The title of the work alludes to the "constitutions" in the Hippocratic *Epidemics*. Perhaps Cardano's seventeenth-century editor Charles Spon expected a work with such a title to be analogous to a "judgement of the year" and astrological in content, for he placed the treatise in a volume mainly filled with astrological works.[49] In fact the work matches excerpts from *Airs Waters Places* and the *Epidemics* with descriptions of the respective climatic conditions favorable to epidemics in Rome and other Italian cities (as Cardano said, even though he had never been to Rome, he knew enough to characterize the physical location, water supply, and climatic environment of both the ancient and the contemporary city).[50] The procedure is a good example of the way in which he attempted to use these Hippocratic works as guides to diagnosis and prognosis and, indeed, of his presentation of himself as a truly Hippocratic physician.[51] He illustrated his account of the *fluxiones* (that is, discharges of any kind) characteristic of autumn in Milan with descriptions of individual cases of these complaints. Being Cardano, he could not resist putting his own experience first, in a decidedly un-Hippocratic fashion: He had had a discharge from his right ear so severe that he was afraid he was going to go deaf. More Hippocratic in character are the following accounts of several cases that had led to fatal outcomes. Most of these capsule case histories end with a postmortem dissection in which Cardano took part. His participation doubtless consisted of being present to offer expert analysis, not of doing the actual cutting, which in one case was performed by Gabriel Cuneo. Among the patients were the Milanese nobleman Cesare Brippio, who was found to have much *sanies* between the liver and the ribs; Pietro Casato, a boy of fourteen, whom, as he was very rich, his relatives suspected of having been poisoned, although Cardano pronounced that he had died *ex fluxione;* the noble Alvise Gonzago, who died on the seventh (i.e., a critical) day of his illness, exactly at the time Cardano prognosticated that he would; and a thirteen-year-old girl, the only daughter of a very noble and very wealthy family and a great heiress, whose relatives also wrongly suspected poison. These cases show that Cardano's interest in anatomy had progressed beyond reading anatomical books and advocating anatomical study on the human cadaver to active involvement in dissection even before he moved to Bologna and came into contact with well-known anatomists there. They also reveal social characteristics of Cardano's medical practice in Milan, among them the ready recourse to postmortem dissection of deceased relatives, notably women and children, among upper-class families and the way in which anxieties centered on poisoning.[52] But they show little or no interest in applying astrology to medicine.

The remainder of the treatise includes a discussion of contagion, in which Cardano cited Fracastoro's *De contagione* with approval,[53] but attributed outbreaks of bubonic, as distinct from other forms of pestilence, directly to the will of God.[54] Cardano concluded that there were four kinds of pestilence. The first was "common pestilence," which resulted from corruption of the air or water; this was "impressed by the stars" and was ultimately "from God and from the heavens." But the mutation of this type of illness into the other three kinds required terrestrial causes: exhalation of poisonous vapor, contagion, or the consumption of rotten food. Thus, bubonic plague itself, Cardano explained, "was not caused by the heavens, nor by the air."[55] Modern scholars share Cardano's view that not all sixteenth-century outbreaks his contemporaries described as "pestilence" were of bubonic plague, or bubonic plague alone, but rather encompassed a variety of epidemic diseases.[56] More striking, however, is Cardano's explicit elimination of celestial causation from a major category of contemporary epidemic disease. It is unclear whether Cardano's remarks about the astrological causes of specific epidemics in *De venenis*, a treatise several times referred to in the work now under discussion, mean that he did not think those particular epidemics were outbreaks of bubonic plague or are just another example of his fine indifference to consistency. Whatever the case, he reinforced his determination to exclude astrological explanations from *De providentia ex anni constitutione* with the remark in the concluding section that "[i]t is necessary only to observe those things that have evident qualities and perceptible, strongly established, and firm, long-lasting constitutions, lest a very beautiful discovery be fouled in the manner of astrologers."[57]

Probably the astrological topic to which Cardano recurred most frequently in his medical works was that of critical days. Both Hippocrates and Galen supplied powerful medical authority for the idea that diseases reached a critical turning point for good or ill after a certain, fixed number of days. The Hippocratic corpus provided both examples of events that took place on a specific number of days from the onset of illness recorded in the case histories in the *Epidemics* and, in the aphoristic works, cryptic general rules about the importance of numbering days. One of the most important statements of the general idea occurred in *Aphorisms* 2.24, which asserted that "[t]he fourth of the set of seven is indicative. The eighth is the beginning of another set of seven. But the eleventh is also worthy of consideration; for it is the fourth of the second set of seven. But again the seventeenth will also be considered; it indeed is the fourth from the fourteenth, but the seventh from the eleventh."[58] Whatever the original basis of these and similar statements, the concept of critical days had already received astrological treatment in Galen's *De diebus decretoriis,* book 3 of which links them primarily to the motions of the moon. Galen held that the moon's influence on the atmosphere, and consequently on human health, varied according to its position both in relation to the signs of the zodiac and to the sun, being strong at quadrature: hence the importance of the seventh and fourteenth days.[59] Thus in medical tradition the theory of critical days had a double basis. The actual occurrence of critical days was regarded as empirically determined by "observation" of the sick—one of many instances in which repeated assertions in authoritative texts that something was so were held to constitute empirical evidence—and codified into rules by Hippocrates. Astrology, for its part, provided an explanation of the phenomenon. Hence discussion of the subject might involve technical astrological exposition or controversy but did not necessarily do so. Indeed, no less an authority than Avicenna had advised that it was sufficient for the *medicus* to know that critical days occurred and that their cause lay outside the art of medicine, or that if the cause should be sought the proper approach was a medical one via sense and experience.[60] Yet Avicenna had also mitigated the effect of this excellent advice by including a certain amount of astrological discussion of the subject.

The doctrine of critical days provided endless opportunity for elaboration and controversy, as the existence of numerous treatises on the subject written between the thirteenth and the fifteenth century attests.[61] In the early fourteenth century, Pietro d'Abano had already accused Galen of astronomical ignorance because his formulations were based on the assumption that the motion of the moon was uniform and on an artificial "medicinal month."[62] Other problems included inconsistencies among the actual numbers in different Hippocratic texts, whether some critical days were more indicative than others, why Galen had said crisis occurred on the twentieth rather than the twenty-first day, whether crisis could occur on days adjacent to the supposed critical days, whether the day was a day of twenty-four hours, whether crisis was better at night, and what was the total number of days of illness through which the pattern of critical days could be expected to recur.

In Cardano's lifetime, however, the subject of critical days took on a new and urgent importance among medical men. The earlier literature was not, of course, forgotten. Although writers of treatises published after 1500 paid little attention to the body of scattered and largely anonymous short medieval treatises on the subject, the views expressed in the standard works of such major authorities as Avicenna, Avenezra, and Pietro d'Abano continued to be cited—in praise or blame—with some frequency.[63] Works on astrological medicine by well-known medical astrologers of the fifteenth century were reprinted in the sixteenth.[64] But Pico della Mirandola's rough handling of astrological critical-day theory in his *Disputationes adversus astrologiam divinatricem*, published in 1496, provided a new and more challenging context for all subsequent discussion. Though accepting the occurrence of critical days in illness, Pico entirely repudiated their astrological causation. In the course of a long and penetrating analysis, he pointed out, among many other more recondite objections, the obvious difficulty, with respect to ascribing an astrological cause to critical days, that people did not necessarily fall ill in synchrony with the phases of the moon.[65] Pico's onslaught brought forth a number of efforts to restate the astrological theory of critical days in a way that would place it on an unimpeachably sound footing. One of the first and most important authors to attempt this task was Agostino Nifo, a practicing physician as well as a philosopher (his most famous patient was the Spanish general and governor of Naples Gonsalvo Fernandez, known as the Great Captain).[66]

Nifo's preface lined up revered ancients in favor of the rational analysis of critical days and medieval Arabs and empirics against, thus leaving no humanist reader in doubt as to where sympathies should lie. He described his work as mixed, in that it would treat the subject from the standpoint of both medical observations and astrological reasons.[67] In the medical portion, Nifo described various subcategories of critical days, explained the periods of the circuits ascribed, under separate planetary influences, to each of the humors, and stressed the importance (and difficulty) of determining the time of the onset of illness with sufficient precision for astrological purposes. The astrological section is essentially an elementary survey of the subject. But after repudiating the objections of Pico, *vir gravissimus*, to the theory that the planets governed the motions of the humors as well as to some of the formulations of Galen, Nifo remarked that "Pietro d'Abano and Pico wrote many things against Galen that are frivolous, which are overthrown by my book."[68] In this work, dedicated to a Venetian patrician, Nifo thus signaled a classicizing revision of the theory of critical days that, having discarded the errors of Galen, medieval interpretations, and the objections of astrology's strongest modern critic, would rest securely on the foundations of a purified astrology: namely, as he made clear in another work, that of Ptolemy.[69]

But Pico's views on critical days found supporters as well as opponents among the medical profession. Cesare Ottato, a physician from Naples, insisted on the reality and medical importance of critical days in illness, confirmed both by authority and "the experience of both former and present practitioners," and expatiated at length on subcategories, days of occurrence, and so on.[70] Nor did he repudiate all medical astrology, since he thought that knowledge of a patient's geniture helped the physician to judge a crisis.[71] But he was emphatic that "we do not wish to admit these aspects of hostile stars, not angles, not masculinity or feminity, not coldness, etc., which, with their mathematical precisions, are all vain."[72] As a result, "the sayings of physicians in prognosticating about crisis are arbitrary and vain" and, even more bluntly, "physicians who actually practice in great cities and public hospitals say that they never find any perceptible difference at all in giving medication or phlebotomizing if the moon is full or in conjunction or at any other time of opposition or consummation of the moon etc., or of the other stars."[73] This last comment produced an indignant rebuttal from Federico Grisogono, who remarked that every washerwoman knew enough to avoid washing clothes at the new moon, when they would rot.[74]

Giovanni Mainardi's denunciations of all kinds of medical astrology, disseminated in a number of editions, are likely to have been considerably more influential than Ottato's litle treatise on critical days. Both Leoniceno and his disciple Mainardi, two of the most important medical humanists, had personal contacts with Pico and shared his views on medical astrology. Mainardi had used one of his widely circulated *Epistolae medicinales* to denounce both medical astrology in general and reliance on the scholastic Pietro d'Abano in particular.[75] On the subject of critical days, Mainardi included a telling anecdote about an occasion on which both Francesco Benzi, his own teacher, and the celebrated astrologer Girolamo Manfredo of Bologna were apparently attending the same patient. According to Manfredo, the impending conjunction of the luminaries portended death to the sufferer, but Dr. Benzi prescribed him some medicine and he promptly recovered.[76] Yet another medical author who repudiated astrological explanations of critical days was Girolamo Fracastoro, who offered the opinion that medical men had been "seduced and persuaded" by astrologers. Attributing the origin of the idea that the moon controlled critical days in illness to Egyptian astrologers, Fracastoro remarked, "[E]ven if some Egyptian god, either Anubis or Osiris, tells me these things, I will not easily believe."[77] For the astrological causation of critical days in illness, the actual occurrence of which he did not question, he substituted a theory of his own. According to Fracastoro, critical days in illness were caused by fluctuations in morbid humors, which had their own determined periodic rhythms.[78] Of course, this explanation, just as much as the astrological one, postulated the existence of occult mathematical rhythms in nature.

But wholesale repudiation of astrological explanations seems to have been a minority view among sixteenth-century medical writers on the theory of critical days. Most were concerned with justifying astrological causation while introducing new refinements into the way in which critical days were categorized or astrologically linked. As already noted, Cardano was not among the authors who devoted entire treatises to this enterprise.[79] But he did dedicate six of his *Contradictiones* to the subject, as well as short sections in various other works.[80] Cardano, like Fracastoro, was critical of Galen's account of astrological causation, but unlike Fracastoro he was not prepared to discard astrological causation as such. Instead, he set out to correct Galen's astrology. He took up the by now usual criticisms of Galen's "medicinal month," suggesting that this section of *De diebus decretoriis* was so puerile that Galen could not actually have written it.[81] In general, Cardano's treatments of the subject compare various Hippocratic and Galenic statements about the days to be considered critical and attempt to iron out apparent inconsistencies either among various Hippocratic texts, especially among the descriptions of various cases in the *Epidemics* and the general statements in the *Aphorisms* and *Prognostic*, or between Hippocrates and Galen. As on many other topics, he strove to justify Hippocrates in the light of his own ideas and freely criticized Galen. Thus he maintained that a Hippocratic cycle of 120 days of illness during which critical days were identified corresponded to (approximately) one third of the solar year and hence was based on the motion of the sun; Galen's mistake had been to concentrate excessively on the connection of critical days with the moon.[82] According to Cardano the true doctrine of critical days was Hippocratic and Ptolemaic and would be properly understood only if Galen's errors and confusion were cleared out of the way: "Here therefore is the whole account of critical days—number, order, explanation, and cause—according to truth and the opinion of Hippocrates, which Galen falsely, confusedly, and inconsistently wrapped up in so much obscuration that those who tried to follow it could never find an end." Thus, "I say that no one can resolve this difficulty unless he commands the Hippocratic art of Ptolemy."[83]

If the discussions of medical astrology in Cardano's medical writings are in some respects more limited in scope than those of his contemporaries, they have in common with many of them a primary concern with theory rather than practice. In this body of literature, theoretical discussion seems considerably more abundant than detailed examples of or instruction in the actual practice of astrological medicine. Many works purport to give brief, easy instruction in the art for medical doctors (of whom, according to the author of one such handbook, scarcely one in a hundred actually knew how to make astrological judgments, although they all knew the tag that said "it is no use to medicate without the counsel of the stars").[84] But even some of these give only general outlines of astrology or describe the supposed qualitative attributes and influences of the planets and signs of the zodiac: Practical techniques were presumably left for direct personal instruction.

In Cardano's published medical works, allusions to the actual practice of astrological medicine are even more restrained than his discussions of theory. This restraint cannot be attributed to his status as a professor of theoretical medicine. Throughout his career, Cardano emphasized his skills as a medical practitioner and freely incorporated anecdotes about his medical practice into his work (to a notably greater extent than many of his contemporaries). But among his collection of 100 errors in the practice of modern physicians, just two refer to astrological ineptitude; in them he accused his colleagues of unnecessarily avoiding giving medicine during certain phases of the moon and misjudging which days were critical.[85] On another occasion he ruminated, "But as for whether it is true that the head is threatened when the moon is in Aries, and the neck when it is in Taurus, and the chest when it is in Cancer, and the heart when it is in Leo, and the viscera when it is in Virgo, I really think I have not been able to pay attention to the matter with enough diligence so that I could either affirm or deny it. Because of that it is better not to condemn that opinion, especially since it is very widely held."[86]

Writing in his medical capacity, Cardano was capable not just of ambivalence, but of sweeping skepticism about the practical usefulness for medicine of the theory of critical days. In one of his tirades against Galen, he wrote, "The unskilled will be outraged because I rend their Galen so openly. . . . But whence does he have so much glory? What does this man have to offer except his critical days, whose cause he did not indeed understand (as I have often taught elsewhere). And even if really they were a thousand times true, what usefulness do they have for the physician?"[87] He also expressed the opinion that ill-informed recourse to astrology on the part of physicians did more harm than good, both because it produced mistaken diagnosis and because it brought the art of astrology into discredit. Thus, he took Adrien L'Alemant—a favorite target because he had managed to get his commentary on *Airs Waters Places* published before Cardano's own—to task for claiming that a woman had died because she was given cassia nigra when the moon was in the fifteenth degree of Capricorn: "It would be better to look for the cause in the medicine or in the disease, and not in the stars . . . nor did he know how the power of the stars should be interpreted.... [T]his is a cause absurd and unworthy of such a great man, since a thousand [other] people took medicine made from scammony on that day who were indeed completely uninjured [thereby]. Therefore I regret that that he gave the unskilled a reason to laugh."[88]

Still more notable by its absence from the medical works is Cardano's own practice of astrological medicine. Nothing in his published medical writings remotely resembles the series, collected by Thomas Bodier, of more than fifty horoscopes for patients ranging from a nobleman to a peasant. Bodier seems to have collected these horoscopes, each cast for the moment of the onset of illness, for the purpose of retroactive analysis of the astrological causes for the outcome of his cases; many record the deaths of the patients.[89] By contrast, the series of Cardano's marvelous cures and prognostications, which he published in several different versions, appear to contain only two brief references to astrological factors, one noting that a patient fell ill on one of the critical days and another that bystanders thought a patient might have been affected by the stars. There are no extended examples of astrological diagnosis, prediction, or choice of times for therapy.[90] The few allusions to astrology in Cardano's *consilia* are equally concise and uninformative. Thus, for example, in a *consilium* for a patient suffering from shortness of breath that occupies sixteen double-columned folio pages, Cardano devotes one short paragraph to the appropriate astrological times for purgation. Similarly in the thirty pages of the famous *consilium* for Archbishop Hamilton of St. Andrews—to prescribe for whom he traveled from Milan to Scotland—one sentence mentions the positions of the moon associated with accessions of fever.[91] (Nor, as we shall see, was the horoscope that he also cast for the archbishop very explicit in terms of medical prognostication or advice.) So much, then, for the role of astrology in Cardano's medical writings. Let us now turn to the part played by medicine in his astrology.

The Role of Medicine in Cardano's Astrology

Renaissance astrology was as deeply and controversially involved with medicine as medicine was with astrology, and Cardano's career as a medical man actually began, or so it seems, with the study and practice of astrology. Cardano had apparently begun erecting and interpreting astrological figures for clients by the early 1530s. He had studied natural philosophy and medicine in Padua and Pavia and was living in the small town of Gallarate, outside Milan, trying to make headway as a medical practitioner. But the College of Physicians of Milan had refused his application for membership, causing him both professional difficulties and financial hardship.[92] How Cardano learned the technical methods of astrology we do not know. His father, Fazio, who taught mathematics in Milan, may have instructed him; Nostradamus, another famous medical man and astrologer whose career offers many parallels to his, claimed to have learned special techniques from his father and grandfather, and Cardano lost no occasion to praise his father's predictive gifts.[93] But Cardano may also have mastered astronomy and astrology as a medical student. Unlike Nostradamus, whose critics ridiculed his inability even to use an almanac without making elementary errors, he mastered the basic techniques for finding the positions of the ascendant and the planets and laying out the houses of the horoscope, as well as for interpreting these quantitative data.[94]

Several independent pieces of evidence suggest that Cardano saw and used astrology as a way to solve his professional difficulties. In 1534 he issued a short *Pronostico* in Italian in which he predicted a variety of events for the subsequent two decades, ranging from the weather to the likely fates of the pope, the church, the Holy Roman Emperor, and the major European states.[95] Medical doctors in Bologna and elsewhere, as we have seen, regularly issued pamphlets like this one, and Cardano made his connection to medicine clear by describing himself, on the title page of his work, as a Milanese physician—though he had not yet gained formal admission to the College of Physicians.[96] It seems likely that Cardano hoped to attract the attention of potential patients, and perhaps that of medical colleagues, with his astrological prowess. Complementary evidence suggests that he succeeded. The collection of horoscopes that he issued, in progressively larger versions, in 1538, 1543, and 1547 includes figures evidently erected in the 1530s for a considerable number of patients and other contemporaries: for example, the horoscope of Francis I, on whom the stars inflicted "numerous bodily ills"; that of an infant born in 1534 who died of a wasting disease; and that of a woman who died from poison in 1535, as well as a number of others.[97] At least some of Cardano's clients were prominent figures in Milanese circles: for example, the humanist writer Gualtiero Corbetta, for whom he drew up an elaborate horoscope that he published in 1538, in his first collection, a year after Corbetta's death, and the historian Galeazzo Capella.[98]

Cardano himself tells us that in this early stage of his career he worked closely with older medical men who were expert in astrology. In more than one work he cited horoscopes drawn up or interpreted by members of the Castiglione family, notably that of a child born in 1509, on whose fate "Giovanni Antonio Castiglione, my fellow-citizen, and a royal physician and a man of great excellence" had pronounced.[99] In his treatise *De iudiciis geniturarum* Cardano analyzed the horoscope of a man "born from very humbly-born parents, who was called Niccolò; but when he left his fatherland, he changed his name to Costanzo, and at Milan he was called Costanzo. At Bologna, however, he was called Niccolò, from the de Symis family." This Costanzo, Cardano remarked with unusual enthusiasm, had the good fortune to have Mercury as the lord of the ascendant in his nativity. Accordingly, "though because of poverty he had not studied letters until his twenty-eighth year, he was so brilliant that he gained a modest knowledge of the humanities. He became a geometer, a mathematician, but above all a famous astrologer, so that he taught those arts publicly at Milan for several years."[100]

Costanzo or Niccolò de Symis left a number of astronomical and astrological works, including an unpublished prognostication for the same year as Cardano's first publication, 1534.[101] Cardano dramatically describes a consultation that they held in the next year for a patient of considerable social eminence, Paolo Sforza: "Weakened by loss of blood from his lungs and more or less wasting away, he had consulted Costanzo de Symis of Bologna as to whether the emperor would make him ruler of Milan in place of his brother. When he showed me the figure, I said that he would die in that year. For the moon was among the Pleiades in the sixth house, in quartile to Mars and Jupiter, which were moving through the fixed house of Saturn. And accordingly he died suddenly of suffocation while traveling."[102]

Several points call for comment here. Evidently Paolo asked medical men for astrological advice; to judge by Cardano's testimony, however, he wanted to know not whether he would recover from his pulmonary ailment, but whether he would be made ruler of Milan. When the physician acted as an astrologer, in other words, he might be called on to offer predictions about any of the many aspects of the client's life that horoscopes normally covered: not just health, but wealth, marriage, children, journeys, and much more. Yet Cardano read the figure in question, or claimed to, as offering an urgent and irreversible medical prognosis, which he gave. The astrologer sought to apply the rules of his art properly, whether or not these yielded the answer his client wanted or a cure for the client's disease. Some years after this early time of struggle, Cardano recalled that some Milanese physicians had criticized him for devoting himself entirely to mathematical studies.[103] Taken together with the other evidence, this remark confirms that he became known, in the first instance, as a medical man who was especially expert in astrology and that his astrological pursuits did not always yield medical results.

The few records that survive of the cases that Cardano took on in his astrological capacity support this analysis. Evidently he was called in when someone felt the need to explain or predict events that seemed, for one reason or another, outside the reach of normal medical practice: to identify the cause of death for a woman, which he took as poison, or to offer a prognosis for a child failing to flourish. In the first case, astrology played the same role that autopsy often played in similar cases: It offered a retrospective explanation for a devastating event. In the second case, it offered a prediction based, so both the astrologer and his clients might think, on much richer data than a medical doctor could lay claim to. For a baby, after all, the doctor had no case history to draw on and few signs to interpret; the astrologer, by contrast, could lay out exactly the same planetary and stellar data as for a grown adult. The evidence, however, does not indicate that Cardano saw these predictions, which were explicitly medical, as more particularly his province than ones in which the cause of death was not medical in any identifiable sense: for example, why Galeazzo Capella was run over by a mad horseman. Astrology might win clients and make a reputation for a young physician whose practice was not growing; but just as Cardano did not apply his astrological tools to the bulk of his medical work, so he did not ask medical questions about the bulk of his early astrological subjects.

The course of Cardano's astrological work in the late 1530s and 1540s amply confirms this diagnosis. His major publication in the field took the form of a series of short texts on aspects of astrology and astronomy accompanied by the protean and ever-expanding mass of his horoscope collection. The texts showed more interest in reforming astronomy than in applying it to problems of health and sickness: In one, Cardano devised one of the numerous schemata of his period for astrological history. And even Cardano's analyses of the horoscopes he collected showed no more interest in medicine than in a variety of other subjects. Of the sixty-seven genitures that appeared in his 1543 collection, fewer then twenty discussed medical questions; of the thirty-three new ones that he added in 1547, only ten did so.[104] In many cases, moreover, the medical point at issue was only one of many topics Cardano touched on. His analysis of the horoscope of the magus Henry Cornelius Agrippa is typical in this respect. Cardano found in it clear evidence that Agrippa had a brilliant intellect, that he would die poor, that he would undergo torture and imprisonment, that he was not handsome, that he would die of poison or strangulation, that he could work with his hands like a skilled craftsman, and, at the very end, that he was "incontinent" with regard to women and that he "took more pleasure than was proper in four-footed animals": a ha'pennyworth of medical characterology to offset an intolerable deal of prediction about very different questions.[105]

Cardano's collections, which were printed by the prestigious firm of Joannes Petreius in Nuremberg, the publisher of Copernicus, brought his work as an astrologer to the attention of a broad public in northern Europe. Early in the 1550s he followed his books north, traveling first to Paris and then to Edinburgh to treat the Scottish cleric John Hamilton, from whom he received an enormous fee. In the course of his time in France and the British Isles, Cardano carried out astrological consultations for a number of very elevated clients: not only his most notorious subject, the young King Edward, for whom he predicted, with reservations, a long life, but also the great French lawyer Aimar de Ranconnet; the French ambassador to England, Claude Baduel; and the English humanist and statesman John Cheke. Where Cardano's geniture collections had offered relatively brief explications of figures, many of which were for celebrities past or present whom he had never met, for these clients he not only cast figures but interpreted them at substantial length, analyzing in detail their physical appearance, their temperaments, and the likely courses of their careers. But in these cases, too, health played only a limited role.[106]

In analyzing Baduel's geniture, for example, Cardano found in it reason to think that his client's mother would have a short life and his father would be troubled by disease and other sorrows; that he would have one brother and three sisters; that he would have difficulties with nourishment in his early life but eventually grow to be very large; that he might, but would not necessarily, die by violence that involved bloodshed; that he would be brilliant, magnanimous, closely linked by friendship to his king and able to serve him effectively as an adviser. He also found it possible to predict, from the presence of Jupiter, the lord of the ascendant, in the humid sign of Gemini, that Baduel would be immensely large and fat. To prove the causal relationship he cited a parallel. Frederick of Saxony, he pointed out, was "so obese and heavy that there is hardly a horse that can carry him." He too had Jupiter in Gemini, in opposition to his ascendant in Sagittarius.[107] The horoscope, in other words, had substantial medical content. But it did not concentrate exclusively on such points.

Cardano's analysis reached its climax, in fact, with a character analysis, not a medical prognosis. He portrayed his client, in a series of Mannerist paradoxes, as a man of contradictions—and also, perhaps, as the ideal embodiment of the late Renaissance ideal ambassador, the man of deep mind sent to lie abroad for his country:

> *On his profession.* He will be desirous of secret things; he will take pleasure in beautiful things, in gems, in clothing, in paintings, in images, and in the liberal disciplines; but, being noble, he will not work at any of these.
>
> *On his journeys.* He will not be more successful at anything, or better fitted, than for journeys, embassies, and expeditions, in the course of which he will endure risks and slanders because of others' envy. But he will attain the highest offices. This horoscope contains some contrary indications: for example, a good temperament and a life full of disease; loss of a spouse even though he is married; a fat body and a sharp intellect; good deeds and suffering slander.[108]

Even when analyzing the horoscope of a client who clearly suffered from a serious weight problem, in other words, Cardano devoted only part of his attention to medical questions. Similarly, his analysis of the horoscope he drew up for the most important medical patient of his career, John Hamilton, dwelt as lovingly on the astral causes for Hamilton's love for Cardano as on those of his respiratory problems.[109] Like the horoscope he also drew up for Gulielmus Casanatus, Hamilton's physician, who had invited him to undertake his great journey to the north in the first place, Cardano's horoscope for Hamilton seems more designed to establish its author's mastery of technique—and to establish a sound celestial basis for his relationship with the other—than to yield results in concrete terms for either man's regimen.[110] At least once, moreover, Cardano explicitly confessed that he had treated a patient suffering from the rare disease of "Diabetes" even though he knew he had almost no chance of success, simply because he wanted so much to have the patient's horoscope—clear evidence of a separation, at least for pragmatic purposes, between medical treatment of a case and the use of astrology to understand it.[111]

Solid precedents supported Cardano's apparent effort to separate astrology from medicine for analytical purposes, and contemporary parallels show that his practice was not unique. Even Ficino, usually hailed as the one who definitively formulated the theory of astrological medicine for the Renaissance, warned his readers to distinguish between what they should seek to learn from the "medicus" and what they should seek to learn from the "astrologus," and at least one reader made a note of the point in his copy.[112] Nostradamus, who ran an astrological boutique for wealthy German merchants and French bourgeois, received numerous letters beseeching him to draw up horoscopes and interpret them. Like Cardano, he was a medical man, trained at Montpellier, a fact that his correspondents often referred to in the respectful headings of their letters. Like Cardano, too, however, Nostradamus analyzed far more than his clients' prospects for health and long life. He was as ready to direct the mining operations of a client in far-off Styria as to analyze the life chances of his sons.[113] True, Nostradamus was no normal astrologer, as he continually insisted and his clients confirmed, when they begged him to express himself more clearly. But a wide range of elaborate horoscopes like Cardano's, drawn up by his rival Luca Gaurico, his well-known central European contemporary Cyprian Leowitz, and others, survive, most of them still in the elegant presentation manuscripts that were evidently the usual form in which the astrologer passed them to his client. These closely resemble Cardano's in their attention to the whole range of questions traditional in horoscopic astrology.[114] Through the mid-1550s, in other words, the evidence of Cardano's astrological practice shows only a professional, not a theoretical, connection between "the art" and the other art of medicine.

The Commentary on Ptolemy's *Tetrabiblos*

Like many of his other works, Cardano's commentary on Ptolemy owed its origins to supernatural inspiration. In this case, an accidental encounter in Lyons resulted in his being given a copy of the work unexpectedly. Then his chance decision to go up the Rhone on a boat, rather than to travel on horseback, provided the necessary time to begin serious work. Cardano knew that these coincidences spelled out the vital point: Higher authority wanted him to write the first proper commentary on Ptolemy's astrology. As so often, Cardano found himself untrammeled by previous writers in the same field, for whom he felt chiefly contempt, as he claimed to be offering his readers something radically new. The finished commentary was as thick with technical detail of many kinds as any of Cardano's earlier works. In it, as never before, he mingled the medical with the astrological in a systematic way.

From the outset, Cardano drew connections between Ptolemy and Galen. He was, of course, hardly the first to do so. The anonymous preface to the Greek commentary on Ptolemy translated into Latin by Giorgio Valla, one of the few earlier sources at Cardano's disposal, made the basic point that the great astronomer and the great physician were contemporaries: "Ptolemy, easily the greatest of all mathematicians, lived, so some have written, in the time of Hadrian and survived to that of Antoninus. They say that Galen, the famous medical author, flourished at this time, as well as Herodian the grammarian and Hermogenes the rhetorician, who left some worthy books on the art of rhetoric."[115] But the editor made nothing further of the point, moving instead into a doxographical account of the early history of Greek astronomy.[116]

Cardano, by contrast, set Ptolemy's text into a richly detailed historical context. When explaining why Ptolemy had found it necessary to argue at such length for the status of astrology as an art, he drew on another learned source of the same period, the *Noctes Atticae* of Aulus Gellius, to identify the nature of the opposition that astrologers faced in the second century A.D. "In Gellius, Phavorinus, who flourished just before Ptolemy, a philosopher of great reputation, made the art of astrology infamous in the way that the ambitious will."[117] If Cardano's sharply polemical account did little justice to the views of the Stoic Favorinus, he nonetheless made clear that Ptolemy conceived his text in a world in which his art had come in for strong criticism from authoritative thinkers. In a more sustained and absorbing discussion of the astrological rhythms of cultural history, Cardano showed that he saw Ptolemy and Galen, as well as other writers, as belonging to a single coherent cultural moment. He quoted the long digression in which the Roman historian Velleius Paterculus described how all the arts and sciences had flourished in the Athens of the fifth century B.C. and in the Rome of the late Republic. Then Cardano pointed out how unusual moments like this were in the human history: "For from the time of Augustus down to the beginning of our own splendid period, which was around A.D. 1440, some 1,400 years passed in which nature produced nothing of miraculously outstanding quality—except in the time of Antoninus. For then there flourished, simultaneously, Alexander of Aphrodisias, Ptolemy of Pelusium, and Galen of Pergamum, and a little before that were Caius Pliny and Plutarch of Chaeronea, the teacher of Trajan. Hence it is clear that such phenomena are caused by the general configurations of the heavens."[118] Like modern historians, in other words, Cardano saw the imperial period as a high point in the development of ancient science and one in which Ptolemy and Galen claimed equally high status as well as simultaneity, even if he invoked astrological rather than sociological considerations to account for their joint success.

Cardano also identified substantial similarities between the arts that Ptolemy and Galen practiced, though his evaluation of them fluctuated in a manner characteristic of this man of supremely protean opinions. At the start of the *Tetrabiblos* commentary, Cardano compared astrology to prognostics, pointing out that neither could properly be described as a science: "This art is the prognostic part of philosophy, which teaches us to know in advance. Accordingly, it is not a science in the true sense, but as the predictive work of Hippocrates or Galen is to the whole body of medicine, so is astrology to the whole body of philosophy. For every art that treats natural phenomena by adopting the method of explicating the causes of present things also teaches the causes of future ones. For future things differ from present ones not in species or in genus, but in time, which is an accidental adjunct of theirs."[119] Astrology, in other words, might reasonably be compared to the part of classical medicine that dealt with prognostics—if not to anatomy or regimen.

But this modest claim did not suffice to carry out the immemorial task of the preface writer: boasting about the unique qualities of the art that he was about to profess. For all Cardano's dedication to medicine—his principal profession throughout his life, and the source of the invitation to travel to the north that led to his writing this commentary—he felt himself compelled in this context to assert that Ptolemy's art was superior to

Galen's. Medicine, after all, was merely a hermeneutical art, one whose practitioner read the signs written on a patient's body. Astrology, by contrast, afforded causal insights into the processes it analyzed, since the stars whose movements its votaries interpreted were not just signs, but causes, of events on earth: "The arts that teach knowledge of the future are agriculture, navigation, medicine, physiognomics and its parts, the interpretation of dreams, and natural magic, and astrology. The noblest of these is astrology. For it deals with everything, while each of the rest has its own specific area of competence. It also always works through causes, and the noblest of them, while none of the others always does so; but it also teaches how to read the future through signs. But those who think that systematic knowledge of astrology is systematic knowledge of fate are wrong. But the configuration of the stars is a major part of fate."[120] Far from being the equivalent to prognostics alone, astrology claimed a universality that even medicine as a whole could not hope to attain (although on another occasion, as already noted, Cardano described medicine as the most universal of studies).

Yet this burst of optimism—or megalomania—did not last. Later in the same paragraph, in fact, Cardano acknowledged that "[t]he science of fate is as obscure as it is certain and noble."[121] In a still later passage, he drew the implications of this concession for astrology. In nobility and certitude of subject matter, he still insisted, astrology reigned supreme among the arts. But mortal men, unfortunately, lacked direct and reliable access to the realm of perfect truths. Accordingly, medicine easily outdid astrology in the human realm of competition for respect and rewards, where mere human standards obtained:

> But when we take account of the weakness of our intellect, matters turn out in the opposite way, not by their nature, but because it is so weak. For, leaving mathematics aside, the most certain of the arts is medicine, then natural philosophy, and then comes astrology, and the last of them all is theology. That explains why in our time medical men of high spirit, like Galen, still have the authority of prophets, and their predictions are taken as so many oracles. That is why the pronouncements of medical men about the future have nothing vain about them, as if they were founded on rock-solid reasoning, whereas those of the natural philosophers, and even more the astrologers and theologians, are empty, and such that not even one of them agrees with another.[122]

Even in the heart of his commentary on Ptolemy, in other words, Cardano admitted that Galen's art of medicine enjoyed higher prestige than astrology.

It is not surprising, then, that he borrowed more than one tool from what he himself acknowledged to be the better-developed discipline.

Cardano imitated Galen, in the first place, in setting out to use the medium of a commentary on a classical text as a central genre of natural science or philosophy. Though numerous commentaries on and summaries of Ptolemy's *Tetrabiblos* had been written in antiquity, the Middle Ages, and the early Renaissance, Cardano had access to few of these documents in the early 1550s. Ptolemy, moreover, gave very little sense of how he himself had read the predecessors from whose work he took both data and models. It was not at all clear what a Ptolemaic commentary would look like. Galen, by contrast, Cardano knew intimately. And Galen, as Vivian Nutton and others have pointed out, was as systematic and prolific a commentator on texts as he was an anatomist and physiologist. He dedicated a large part of his activity as a writer to explicating texts by Hippocrates, often in mind-numbing detail.[123]

Cardano's normal relation to Galen, like that of many Renaissance writers to their ancient models, involved more emulation than adulation. He regularly criticized Galen, in fact, for lacking the solid expertise that would have enabled him to know which of Hippocrates' teachings deserved special support and which did not, for failing to carry out the sort of *experimenta* that would have enabled him to attain to knowledge of "difficult things," and—most remarkably of all—for limiting the task of the commentator to nothing more than explicating the ideas of the author on whom he wrote.[124] In explicating Ptolemy, however, Cardano showed himself more impressed by Galen's utility as a model than by his inadequacy as a scholar. Here, he approvingly quoted Galen's dictum that "the task of the commentator is not to give arguments for what the author says, or to criticize him if what he says is wrong, but to give a clear explanation of his words and sentences," though Cardano then went on to promise his readers precisely the sorts of correction and supplement that Galen had denied an expositor should provide.[125]

He also modeled a number of his own procedures on Galen's. For example, after carefully considering the works ascribed to Ptolemy in his time, he rightly argued that the most popular of them all, the Arabo-Latin *Centiloquium,* could not be authentic. The methods and opinions put forward in it often did not match those found in the *Tetrabiblos* and the *Almagest,* and the prologue to the text, which mentioned Ptolemy's other works, was clearly not by the same author as the genuine ones. Galen had regularly deployed philological arguments like these in his own effort to purify the Hippocratic canon. And Cardano made clear his intellectual debt when he drew on Galen to explain the origin of the forgery: "But Galen explains this when he says: 'In the old days, when kings bought the writings of famous men at vast prices to fit out their libraries, they were responsible for men's attributing their own works to the ancients.'"[126]

Most strikingly of all, Cardano borrowed from Galen what could be described without too much exaggeration as a rule of charity in interpretation. After posing the question why Ptolemy had denied that astrology yielded wealth or glory, when in fact it had often had this effect in antiquity, Cardano answered that he had omitted the point as obvious: "For just as Galen says of Hippocrates, in his commentary on the books on difficulty in breathing, so we too must think of Ptolemy: that is, that he wrote nothing at all which was common, or widely known, but everything that was unique and profound."[127] In another passage, Cardano went further. No one, he said there, who had not worked through Ptolemy's own massive technical work on astronomy, the *Almagest,* could hope to grasp the profundity and subtlety of his thought or the miraculous artfulness of his astrological textbook.[128] Galen had set himself up as the master of a particular Hippocratic tradition in medicine. Cardano portrayed himself as the master of a particular Ptolemaic tradition in astrology. Anyone who wanted to erect a single proper Ptolemaic horoscope, he claimed, must read, mark, and completely master both the *Tetrabiblos* and Cardano's commentary, which formed "a single coherent corpus, without any redundancy."[129]

Cardano's Galenic emphasis, finally, had an impact on the substance as well as the method of his Ptolemaic commentary. For Cardano used Galen's model of a fully developed art again and again to demonstrate in detail the perfection of Ptolemy's astrology. Like Galen, Ptolemy proceeded as a philosopher. He did not simply list the effects of stars and planets randomly, as Firmicus Maternus and the Arabs had, but explained the ways in which different combinations of primary qualities—hot and cold, dry and wet—produced similar results, just as Galen had when examining the properties of simples.[130] Philosophical astrology, like philosophical medicine, was honest about its limitations. Like the author of book 7 of the Hippocratic *Epidemics,* Cardano pointed out, Ptolemy acknowledged that "predictions sometimes fail" but still rightly insisted that individual errors did not detract from the truth and beauty of the art as a whole (1.2).[131] Ptolemy also acknowledged that no single treatise could include all the details of his art, with its spiderweb profusion of intersecting influences and qualities: "It is of course a hopeless and impossible task to mention the proper outcome of every combination [of planetary influences] and to enumerate absolutely all the aspects of whatever kind, since we can conceive of such a variety of them" (2.8). Cardano warmly agreed, citing the parallel acknowledgments of Galen and Avicenna to support his author's decision to remain on the clear level of generality.[132]

Above all, philosophical astrology, like philosophical medicine, was systematic in its procedures. The art of medicine, Galen taught, was established "in two ways, by reason and *experimentum*; just so, Ptolemy teaches by *experimentum* and by reason that the stars have effects on this inferior world." The physician, as Hippocrates and Galen both showed, took his patient's temperament as a baseline, which they used to assess health or illnesses throughout life. Just so, the astrologer took the patient's horoscope (which, of course, accounted for his temperament in the first place) as a baseline for further prediction and analysis. The medical man postponed certain highly refined problems to a relatively late position in his treatment of his subject; so did the astrologer. Ptolemy's astrology, in other words, emerged from Cardano's analysis as an art comparable to classical medicine: comparable in the lucidity of its structure, the solidity of its foundations, and the evident intelligence and good faith of its chief ancient practitioner.

It is certainly possible, moreover, that Cardano had undertaken well before the Ptolemy commentary appeared to reshape astrology on the last offered him by the classical medical tradition. In his commentary on the *Tetrabiblos*, he noted that Ptolemy must have drawn upon the work of early Roman astrologers, like Thrasyllus, who saved his own life by predicting the future for Tiberius. "Unfortunately," Cardano noted, "he did so with such brevity that the bulk of the art is missing. It would have been better if he had followed the example of the great Hippocrates. After he had transformed the data into a systematic art, he should also have written a book of individual examples on the model of the *Epidemics*."[133] Cardano's own collections of genitures may well have represented his attempt to produce the astrological counterpart to the *Epidemics* and thus to reconstitute a vital, lost part of classical astrology. In that case, the medical emphasis of the commentary on the *Tetrabiblos* merely made this side of his enterprise more explicit.

In insisting that Ptolemaic astrology compared well with Galenic medicine (and, like Galenic medicine, needed a dash of Hippocratic reform), Cardano emphatically did not mean that the astrology normally practiced in his day could claim such high status. For a decade and more before the Ptolemy commentary appeared, he had been embroiled in polemic with the most prominent astrologer in Italy, Luca Gaurico, whose work Cardano dismissed as technically unsound. Worse still, he had been surrounded by quacks. Physicians like Cardano—university graduates who belonged to the official Colleges of Physicians in Milan and elsewhere—constantly had to compete with unlicensed empirics (quacks in the eyes of the elite *medici*) who offered their patients not the officially sanctioned remedies of high medicine but the promise of a cheaper, and perhaps a surer, cure.[134] Similarly, astrologers like Cardano—astrologers who believed in applying the full range of techniques and questions sanctioned by Ptolemy, establishing the geniture of a client in all its detail before even beginning to offer advice—constantly had to compete with men they saw as quacks—medical men and astrologers who would quickly compute an interrogation for a given client, using the minimal and fragmentary information it provided to prescribe regimens and remedies.[135]

Cardano declared that he hoped to bring about nothing less than the restoration of an art that had become contemptible, a restoration comparable in drama and pathos to the revival of medicine that had taken place in Galen's time:

It is not only the errors of the practitioners, and their negligence, or also their dishonesty, that has so harmed astronomical divination—so much that it has become an object of contempt in our time. The determined malfeasance of the practitioners has ruined the art. Ordinary men generally agree that what an astronomer says is foolish, stupid, insane. Yet this is no reason to give up hope that the glory of the art can be restored. And no one should think this to be the fault of the art of astrology rather than the times. Galen attests that in his time medicine would have been lost, if some god had not taken pity on the human race and restored it, and yet in the recent past it has revived and is now flourishing. Similarly, it is hardly out of place for this art, which flourished so greatly in the times of the two Gordians, senior and junior, to be restored someday to its original worth. And my efforts, by themselves, have attained at least part of that goal: if the art could not yet be glorious again, given its parlous state, at least it will no longer be a source of shame.[136]

Cardano did not leave matters at that: He also identified at least some of the practices he condemned. The astrologers of his day wrote "long, complicated, problematic texts, full of contradictions and obscurity."[137] They also made snap judgments, like the "ridiculous doctor and incompetent astrologer" who almost killed Cardano in suggesting, on too little evidence, a cure for his enormous flux of urine.[138] Haly, the author of the medieval commentary on the *Tetrabiblos* that had circulated widely among astronomers and astrologers, had taken the *Centiloquium* as a genuine work of Ptolemy's and had wondered why it recommended the practices of interrogations and elections but the *Tetrabiblos* did not mention them. Cardano explained why:

Haly raises a superfluous question: that is, why Ptolemy did not deal with interrogations and elections. For interrogations are entirely magical and unworthy, not only of a Christian, but also of a good man. Elections, similarly, were devised because of the greed of the astrologers rather than because they do any good to the one making the election. And if they have some true content, it is so minor and abstract that it seems unworthy to be part of the art. For the art deals with obvious things and things that can be obtained with profit. Since these have neither, they have no art—just as there is no art of making images in the clouds, as if in a mirror. True, they can be made, but they belong to no art, since they are both very hard and almost useless.[139]

Cardano's whole commentary on Ptolemy, in other words, like the earlier horoscope collections to which he regularly referred, amounted to an attack on what he saw as the standard, and scandalous, practices of other astrologers.

Accordingly, it occasions little surprise that Cardano devoted to interrogations only a short and grudging treatise at the end of his big book. The three figures that Magini pulled from the body of Cardano's work and presented as typical, in some ways, of medical astrology could hardly be less typical of Cardano's own approach to the subject. They illustrate a digression in the Ptolemy commentary. In 2.12 Ptolemy offers three ways of investigating the weather of a given lunar month: by erecting a figure like a horoscope for the new or full moon; by examining the new moons that take place in particular signs of the zodiac, as well as the planetary positions that accompany them; and by observing "even more minutely" the moon's quarters and movements, as well as the new and full moons. Cardano took the third method as offering not only the weather predictions Ptolemy promised, but also "a way of determining the critical days of health or death, the length or brevity of an illness, its wickedness or ease." This, he thought, could reveal "marvelous things" to the astrologer who compared the figures in question over time, following and evaluating the moon's movements.[140] The determination of the moon's deleterious effects on three patients, in other words, formed only part of an excursus, a few pages in Cardano's big book. It offered only a new theoretical basis for the study of critical days, nothing more; and it was hardly large or impressive enough to offset the general bias of Cardano's works toward assigning far more weight to horoscopes than to any other form of astrological inquiry. The fact that the first patient's second disease turned out to have begun before he recovered from the first one, as acute readers will have noticed, shows how theoretical and retrospective Cardano's considerations were. Theoretical in character, small in scale, peripheral to Cardano's main concerns, the figures Magini emphasized do not confirm his view that Cardano regularly practiced astrological medicine.

The Uses of Medical Astrology

In all his concerns, Cardano was profoundly introspective and self referential. Astrology, dream interpretation, physiognomy, and medicine served him not only as predictive and diagnostic arts but also as tools of self-analysis, as both his use of his own experiences as examples in works on these subjects and his well-known *Liber de propria vita* abundantly testify. It comes as no surprise, therefore, to learn that some of his most extended examples of astrological analysis of health and disease occur in narratives that explicate versions of his own horoscope. Beginning in middle age, Cardano seems to have adopted the practice of reviewing his natal horoscope at approximately ten-year intervals, using each occasion to analyze afresh the unfolding history of his own life. He wrote interpretations of his own horoscope in 1545 and again, at much greater length, in 1554 (when he also erected a new figure on different technical principles), and revised the latter interpretation once more about 1564 or 1565.[141] The details of the health history that these narratives relate have recently been discussed elsewhere and need not be repeated here.[142] With their extended portrayal of the changing nature of Cardano's health over time, as he moved from being a somewhat sickly child to a fairly healthy adult to an elderly man "infirm of body," these accounts may provide the best example of Cardano's use of astrology as a tool of medical analysis. They show that he interpreted every episode of acute illness and every minor chronic affliction in terms both of his nativity and of the planetary positions at the time of onset; the latter practice also implies that he habitually noted the times when he fell ill for the purpose of subsequent astrological analysis. But at the same time, they reveal the limitations of the role that Cardano assigned to astrology in medicine. In them, as in other horoscopic narratives, health and disease form only one of several standard categories of analysis. Moreover, they present the influence of the stars on Cardano's individual body as only one factor among several affecting his health. Especially in the two later interpretations, Cardano also assigned an important role to family predisposition or inheritance and the presence of epidemic disease. And in his view epidemic disease was, as we noted earlier, only sometimes caused by the stars.

When faced with a patient who demanded astrological help and reassurance, Cardano could be actively discouraging, even dismissive. One such request came from a patient who suffered from deafness of a type that Cardano diagnosed as incurable. Accordingly, and perhaps conscientiously, he so informed the patient and prescribed only the most general regimen. But the patient wanted to know whether any help could be expected from the stars. The hopeful question reveals the very different expectations of patient and physician. As Gianna Pomata has pointed out with regard to other types of early modern medical practice, patients looked for cure from anyone by any means available, whereas elite physicians generally tried to provide them with rational analysis, scientific explanation, and professional authority.[143] For the physician-astrologer the stars were either causes or signs, which of the two being a subject of academic debate.[144] But in either case, he read a cosmic map solely in order to analyze—and hence predict—its influence on effects or events on earth. Wise use of knowledge thus gained might help him prescribe medication for patients at appropriate times or enable them to avoid the worst effects of malign astral conditions. Since the stars did not control human free will, moreover, but affected only bodies, a wise patient might hope to act in such a way as to avoid astrologically predicted ill health, for as the saying went, "the wise man dominates the stars."[145]

But Cardano's patient's request for help from the stars implies that the heavenly bodies are ensouled beings who can be expected to be responsive to prayer. Versions of this belief, which reflects the association of the planets with the deities of classical mythology and perhaps also some aspects of Christianized neoplatonism, apparently permeated many levels of Renaissance society. Marsilio Ficino's famous discussions of how the ancients called down the spirits, or daimons, of the stars into statues suggest one way—a classicizing, learned, refined, and highly intellectual way—of manipulating ensouled astral powers. Agrippa's *De occulta philosophia* offered less-refined means to achieve the same end, and Benvenuto Cellini's sorcerer friend from Norcia no doubt offered less-refined ones still.[146] Cardano's patient, probably not learned or refined, seems to have hoped that the mere erection of an astrological figure would serve as a magical talisman that might help to persuade the stars to restore his hearing. His or her request for "help from the stars" thus exemplifes a particular kind of hazard involved in the actual practice of astrological medicine. What patients wanted was not learned analysis but help. Probably, most medical practitioners who drew up interrogations did not offer this. Bodier's collection of horoscopes that explained the course and outcomes of his patients' diseases did nothing for his patients beyond analyzing their sufferings. But the kind of help patients expected to get from the stars could only too easily lure the medical astrologer into practices that were both frankly magical (and thus potentially likely to incur religious censure) and plainly unsuitable to the dignity of a learned physician.

For these or other reasons, the patient's request gave Cardano pause. Without addressing the subject of whether or not the stars, or the figure, could be expected to "help," he responded with a little lecture on the inadequacies of current astrological understanding as a predictive system for medicine. Yet in it he could not resist including both his own prescriptions for the reform of astrology and a declaration that the planets and zodiacal signs, even as currently understood, did after all "adumbrate" something about health and disease:

> The astrology of our age is not true because it is not known. For those principles of Ptolemy are different from what is seen and happens. There are besides errors in the movements and places of the stars, as is apparent from the *Prutenic Tables,* which are the only ones that agree with experience. Besides it would be necessary to know the manner of their movements. For I declared in *Paralipemenon* 15 that what appears in the stars cannot be saved either by eccentrics or by epicycles or both, and unless we know why, for example, the moon is diminished by half, we cannot know what it indicates. What is clear, however, is that when the sun and

the rest of the planets are in Pisces they indicate the occurrence of deafness and dumbness, for fish do not hear much and have no voice. . . . [A]nd therefore since the art is imperfect, although it will adumbrate something, yet it is more a cause of errors than help to us. And also this contemplation is outside medicine. And yet I might also consider more diligently whether there were any hope in this.[147]

At this high point in Cardano's career, long after Copernicus and others had undertaken the reform of astronomy that he had once hoped to carry out, his attitude toward medical astrology remained as complex and puzzling as at any other time. No doubt, however, he would agree with us on one point: Generalizations about medical astrology in the Renaissance should be discouraged.

Notes

1. Magini was a professor of mathematics at Padua from 1588 to 1617; see Lynn Thorndike, *A History of Magic and Experimental Science,* 8 vols. (New York: Columbia University Press, 1923–58), 5:250–51.

2. G. A. Magini, *De astrologica ratione ac usu dierum criticorum seu decretoriorum* (Venice, 1607), 81r–v:

Tertia observatio ex Cardano. Decubitus Ioan. Antonii de Campionibus . . . Io. Antonius de Campionibus (ait Cardanus) annorum circiter triginta ex itinere aegrotavit, primum quidem ut visus est leviter usque ad quartam diem, quod Luna esset in sextili Veneris et Venus illam reciperet; nam Venus erat in suis dignitatibus. Et quia Luna erat cursu valde tarda, morbus non est visus accipere incrementum, remorante ut dixi Venere, quae Luna attigit partem vigesimam quintam Geminorum in diebus tribus horis 18. ferme, ob motus tarditatem, unde protracta est quarta dies; tunc vero occurrit Iovis et Martis coniunctio in Leone, et cum stellis humidis, ideo factum est incendium magnum, et turbulentia in urina, quae ob protensam quartam diem visa sunt habere initium in quinta. In septima vero cum nondum pervenisset ad nonagesimum gradum distantiae ob motus tarditatem, sed ad Leonis initium tantum deterrime se habuit, quoniam nulli beneficae occurrit (immo incidit in antiscio Solis existentis in sexta domo, et in caput draconis) similiter in octava et nona morbus augebatur: quia tunc Iovi et Marti iunctis iungebatur, et quia erant inter humidas stellas, sudavit. Calor enim cum humido sudorem creat, et urinam multam, quam reddidit. In undecima sudavit, fuit tamen cum magno labore: nam Luna oppositum Saturni superavit, sed ei successit coniunctio Veneris.

Duodecima aegre se habere visus est, quia multum deliravit. Sed tamen ob Veneris coniunctionem urinae concoctae apparuere. Decimatertia ob Mercurii quadratum (quia inimicus est horoscopi) nihilo deterius se habuit, quia erat motus iam ad salutem; sed nec melius ob Mercurium: decimaquarta sudavit iterum, et melius se habuit. In ea non potuit solvi morbus, quia Luna non peragraverat nisi partes 174. minuti 22. et oportuit extendi morbum ad 17. usque. In decimaquarta tamen occurrit luna sextili Iovis et Martis, et quadrato Veneris, ideo sudavit, decimaseptima liber evasit, cum iam Luna oppositum loci superasset, et trino Veneris applicuisset. Hactenus Cardanus

82r–v:

Quarta observatio ex eodem Cardano. Decubitus eiusdem, qui mortuus est die decima quarta. . . . Hic Lunam ab initio habuit in quadrato Veneris, ex intemperie cibi et potus aegrotavit, celeriter gravatus est ob motum Lunae velocem, in septima deterius valde se habuit, nam Luna erat cum cauda, et vacua a Iovis aspectu ibat ad oppositionem veneris et tunc Sol quadrato Saturni affligebatur. Octava levari visus est fluente e naribus sanguine, sed tamen virtus cecidit ob Veneris oppositum. Nona aliqualiter respirare visus est ob trinum solis. Decima fuit loco undecimae, cum ad angulum, idest Piscium initium pervenisset ad oppositionem Iovis et Martis.

In undecima par erat illum mori, cum ad Saturni coniunctionem hora decima, ad Solis quadratum hora decima octava pervenisset. Mortuus est quinta Iunii h. 3 ante meridiem et fuit initium decimae quartae, et Luna pervenerat ad oppositum loci sui ad unguem. Hactenus Cardanus

82v–83r:

Quinta observatio ex eodem Cardano. Initium aegritudinis ex transfossione brachii Baptistae Cardani, ex qua obiit. Hic alter (ait Cardanus) affinis meus erat et vir sexagenarius cum vulneratus est, Luna a Marte separabatur, et Mars cum capite erat, et Luna cum cauda, et applicabat Saturno et Iovis infelicis opposito, non tamen statim gravatus est, quia Luna ibat ad sextilem Mercurii, et vulnus erat bracchii transfossio. Quarta gravatus est propter Solis quadratum, caruit tamen febre quia nulla malefica oppugnabat, ob id usque ad decimam adeo levatus est, ut surgeret. In undecima gravatus est hora noctis tertia, cum Luna ad oppositum Solis tenderet. Hic autem erat anaereta, dominus siquidem oppositi loci Lunae. Et post Luna ad Mercurii oppositum ibat. Febre igitur, et sanguinis profluvio correptus est 14. die, quae fuit secunda Ianuarii hora diei tertia cum Luna corpore Marti iungebatur, expiravit. Haec Cardanus.

3. The best introduction to the techniques of early modern astrology is J. C. Eade, *The Forgotten Sky* (Oxford: Clarendon Press, 1984); more detailed information is provided by J. D. North in two works of great erudition, *Horoscopes and History* (London: The Warburg Institute, 1984) and *Chaucer's Universe* (Oxford: Clarendon Press, 1986). The fullest analysis of any Renaissance horoscope is W. Hartner, "The Mercury Horoscope of Marcantonio Michiel," *Vistas in Astronomy*, 1 (1955) = Hartner, *Oriens-Occidens* (Hildesheim: G. Olms, 1968–84), 440–95. On medical astrology see Karl Sudhoff, *Iatromathematiker, vornehmlich im 15. und 16. Jahrhundert* (Breslau: Kern, 1902), and Wolf-Dieter Müller-Jahncke, *Astrologisch-magische Theorie und Praxis in der Heilkunde der frühen Neuzeit, Sudhoffs Archiv*, Beiheft 25 (Stuttgart: Steiner Verlag, 1985).

4. Magini, *De astrologia ratione*, 58r–v: "Communi omnium tum Astrologorum, tum Medicorum, qui hanc artem, Astrologiam dico, praeclare exercent, opinione receptum est, construendam esse figuram coelestem ad morbi cuiuscunque initium, ut eiusdem tum essentia, tum dies critici, tum accidentium varietates, et denique exitus praevideri queant; si quidem ex tali coelesti schemate,

sitne aegritudo lethalis vel ad salutem terminans, diuturna vel brevis facillime ratiocinari possumus."

5. Ibid., b3v (an epistle to the candid reader incorporating a lengthy bibliography of earlier works and authors on medical astrology). Of Cardano, he says: "Hieronymum Cardanum Mediolanensem praetereo perinde, ac omnibus notum, quando ipse, cum alibi tum in secundum Ptolemaei librum de Astrorum iudiciis, de schemate octo laterum ad dies criticos morbique progressum cognoscendum, pluribus tractavit."

6. For a list of the areas in which Magini thought the use of medical astrology useful, see ibid., 40r–v.

7. There is no full study of Cardano's work as a natural philosopher. See in general *Dictionary of Scientific Biography*, s. v. Cardano, by M. Gliozzi (New York: Scribner, 1970–80, 1990); *Dizionario biografico degli italiani*, s. v. Cardano, by G. Gliozzi (Rome: Istituto della Encyclopedia Italiana, 1960–); O. Ore, *Cardano, the Gambling Scholar* (Princeton: Princeton University Press, 1954); A. Ingegno, *Saggio sulla filosofia di Cardano* (Florence, La Nuova Italia: 1980); and E. Kessler, ed., *Girolamo Cardano: Philosoph, Naturforscher, Arzt* (Wiesbaden: Harrossowitz Verlag, 1994). On his medical writing and practice, see N. Siraisi, *The Clock and the Mirror* (Princeton: Princeton University Press, 1997); on his astrology see G. Ernst, "'Veritatis amor duleissimus': Aspects of Cardano's Astrology," chap. 2 in this volume, and Anthony Grafton, *Cardano's Cosmos* (Cambridge, MA: Harvard University Press, 1999).

8. Ptolemy, *Tetrabiblos*, ed. G. Cardano (Basel, 1578), 715 = Cardano, *Opera Omnia* (Lyons, 1663; reprinted, New York: Johnson Reprint Corp., 1967), 5:560:

Primum igitur ad praedicendum ne accedas, nisi perfecte instructus in his quae hic traduntur, et illis necessariis ut supra ostensum est. Vt quod cognoscas statim cum Planetae sunt aucti cursu, quod sunt in superiore parvi circuli parte, cum diminuti in inferiore, praeter Lunam: Et plura praedicendi tecum experimenta feceris. Secundum, ut in praedicendo amoveas timorem, odium, et amorem. Illa enim etiam nolentem errare faciunt. Tertium ut ne artem profitearis in triviis, nec coram populo, nec aedas quicquam publicum. Tales enim artem et se ipsos infamia aspergunt, etiamsi vera praedicant. Quanto magis ridiculi fiunt cum falsi deprehenduntur? Quartum ne praedices tentanti nec dubiam habenti genesim, nec sine pretio, nec cum exiguo pretio, nec derideti artem: Nam vilipenditur in omnibus his ars, daturque errandi occasio cum levibus laboribus magna praedicere conamur, et inventu difficilia. Ego bis centum coronatos pro una genesi perficienda respui. Vide modo an tu me sis exercitatior. Quintum ne praedixeris, nisi omnibus diligenter consideratis et bene discussis, ad unguem usque, et ratione habita conditionis hominis, familiae, regionis, legis, aetatis, ac talium. Sextum, homini improbo et malo ne ulla ratione praedixeris quicquam. Ex his sequitur ut ne ignoto: multo minus Principi saevo. Septimum, in praedicendo soli illi qui te rogat dixeris, non in populum praedictiones diffunde. Nec de minutis respondebis, sed maximis tantum et evidentibus. et breviter, non ut impleas folia, nec per ambages, nec contradicentia scribas. Sed pure, nitide, caste, munde, breviter, clare. Exemplum de hoc habes in decima genesi.

9. Ibid., passim.

10. Ibid., 708 = *Opera* 5:556: "*Furtum quis fecerit? an invenietur?*

Adeo desipiunt homines, ut res, non a veritate, sed pollicitis, aestiment. Nulla prorsus est via, ut hoc haberi possit, cum praeter id multa mala subsequantur. Quidam enim ex interrogatione hoc venantur. Sed quid habet interrogatio cum re ipsa, quae iam praecessit? Saepius diximus astra esse causas non signa. Sunt corpora, sunt nobilia, sunt potentia et valida."

11. Ibid., 713 = *Opera* 5:559–60.

12. Ibid., 714 = *Opera* 5:560: "*De medicatione per ferrum vel ignem*

Experimento deprehensum est, quod tales operationes si fiant Luna signum illud, quod membrum respicit corporis, possidente, non sine evidenti noxa opus id fieri. Praesertim si in sectione Saturno aut ustione Marti societur. Et si male dispositi sint, non absque periculo."

13. Ibid.: "nec minore cum gloria ac utilitate quam medici nostri aevi suam artem medicam."

14. Ibid., 712 = *Opera* 5:559: "Tutissima via ex ventris inspectione habetur, inquit enim Hippocrates masculi in dextris, foeminae in sinistris. Quod etiam ex mamillarum differentia dignoscitur. Sed hoc et praecedens signum ut certissimum est, et vix unquam fallit, ita plerisque anceps, cum discrimen hoc tam exiguum sit, ut non nisi longa consuetudine dignosci queat: Estque velut gemmarum notitia. Collocare oportet mulierem supinam ad amussim, et ut in luxatis pedibus exquisitissime laterum comparationem habere. Quae, ut dixi, cum medicos expertos in luxatis fallat, nihil mirum, si ubi obscurius est discrimen, et experientia longe rarior, homines haesitent. Contingit tamen, licet vix in viginti praegnantibus unam, dextrum latus turgidius, et tamen femellam in utero habere, aut converso modo. Idem de mammis. Sed haec extra artem. Ex arte sunt directio, processus et ingressus in loca masculina pro masculinis, in foeminina pro foemininis, pro patre aut matre, tutius pro utroque. Aliud si hoc praesidio careas horam conceptionis habeto, erige ascendens, et vide qui Planetae loco ascendentis et Medii coeli dominentur, et qualiter affecti iuxta duplicem modum sexus. Quidam existimant succedentes fratribus aut sororibus natis, Luna tendente ad novilunium masculos esse, ad plenilunium foeminas. Melior ratio habetur ex nativitate fratrum praecedentium, ad quos Luna applicat, si masculini sint natura, signo, quo ad mundum, et quo ad Solem, aut foeminini. Et similiter dominos loci Lunae et coeli Medii coniectura accipere oportet, non solum secundum quatuor modos sexum decernentes, sed et magnitudinem illorum aut intensionem qualitatis."

15. Ibid.: "Exerceri autem prius in his oportet quam iudices, ne damno tibi sit ars non lucro. Solent enim Antverpiae et Lugduni mercatores magnas circa hoc deponere pecunias, certandi causa." Cf. ibid., 712–13 = *Opera* 5:559, on determining legitimacy, which makes the social context of the drawing of elections even clearer: "Quod ars haec coniecturalis sit, et hominum comparatione ambigua, saepe diximus, quo fit ut apud multos talia damno fuerint. Iniicere autem suspitionem filii spurii, adulterii uxoris, neque tutum est, nec sapientis officium. Propterea etiam si ars ad huiuscemodi extendatur, haud tutum est ei aut prudentis insistere. Inde enim caedes, suspitiones, veneficia, abiectionesque filiorum exoriuntur. Invisa fuit mathematica olim propter hoc, et nunc quoque si talibus

studeamus, astrologia. Quamobrem omnis scientia bona, non omnis suspitio aut coniectura. Hoc igitur unum est ex his quae nescire praestat . . ."

16. Ibid., 708 = *Opera* 5:556–57: *"Electiones an prosint?*
Simili ratione quaesitum est an electiones prosint? Videtur enim quod prosint quoniam figurae quaedam aliquid facere videntur ad renum dolores et calculum. Praeterea in monomachia diximus si dies congruat primae deliberationi, bonum esse: At constat diem eligi posse. Sed figura ut figura est artificiosa, vim agendi non habet: quomodo igitur electio proderit? Dicemus igitur actionem in substantiam cum electione aliquid posse, non quia figura est, sed quia actio. Seu enim canem seu leonem seu montem ea hora sculpseris, idem erit. Si figurae confidas, non solum vanus es, sed superstitiosus: Quod et B. Thomas sensisse videtur. In monomachia aliquid facit electio, sed non secundum diem, verum quia ante vel post adventum directionis aut processus aut ingressus. Sed et utriusque ratio habenda est. Est igitur hoc infinitum et incomprehensibile. Vt amissa occasione saepius damna quam utilitatem afferat. Verum in annis et actionibus naturae etiam momentaneis magnam afferre potest utilitatem. Veluti si hoc anno, non praecedenti, volo quiescere in patria. Est et hoc difficile propter concursus. Ob id in naturalibus electio manifeste prodest. Vt si quis aestate serat aut putet, rideatur. In voluntariis si felix sit sequatur naturae et animi impetum: Si infelix frangat et fugiat: non solum occasionem sed etiam persaepe rem ipsam.

17. See the discussion of the horoscope drawn up by Conrad Heingarter (fl. 1440–after 1483), physician and astrologer to Louis XI of France, to the Duc de Bourbon, and to other notables, for his friend Jean de la Goutte in 1469 in Maxime Préaud, *Les astrologues à la fin du Moyen Age* (Paris: Lattes, 1984), 71–94 and 177–86. On Conrad Heingarter, see also Thorndike, *A History of Magic*, 4:374–85.

18. "Erecta figura (hora qua nuntius ad medicum venit) planetisque in ea cum fortunae, vitae, mortis et planetae interficientis partibus debite collocatis: Ascendens, dominus ascendentis: Luna et dominus domus Lunae aegroto pro significatoribus dandi sunt. Significator etiam qui iudiciariae astrologiae authoribus figurae dominus sive significator caeli praecipuus dicitur, aegroto tribuitur. . . . Sexta vero domus eius dominus et planeta a quo separatur dominus ascendentis, vel is qui ab eo separatur, morbo et morbi causis dantur. Septima cum suo rectore medicum respiciunt." Claude Dariot, *Ad astrorum iudicia facilis introductio . . . Eiusdem tractatus de electionibus principiorum idoneorum rebus inchoandis. Quibus accessit fragmentum de morbis et diebus criticis ex astrorum motu cognoscendis* (Lyon, 1557), 78.

19. See, for example, Magini, title of book 2, chap. 14, of *De astrologia ratione*, "De momento temporis, quo quis primum decubuit, diligenter observando," 58r, and "Constructa itaque ad morbi initium coelestis figura diligenter conferenda est cum figura nativitatis aegri, et cum revolutione anni illius, in quo natus decubuerit. . . ." 58v, which, though late, is representative.

20. According to Auger Ferrier, "Ubi hora primi insultus haberi non potest, confugiant astrologi ad quaestionum, seu interrogationum artem. Allata enim aegrotantis urina, aut accedente aliquo, qui pro aegrotante medicum accersat, eo momento figuram coelestem erigunt." Auger Ferrier, *De diebus decretoriis* (Lyon,

1549), 120. Ferrier was perhaps somewhat uneasy about this recommendation, since he added on 121–22 that erudite doctors of the Hebrews, Ptolemy, the Christian Albertus Magnus, the Platonist Marsilio Ficino, the physician Arnald of Villanova, and the distinguished philosopher Pietro d'Abano had all followed this practice and that it was legitimate provided questions involving the soul or free will were avoided; he reinforced his orthodoxy by continuing on 123–30 with a denunciation of medical magic using charms said over herbs, geomancy, and seals. Magini would have no truck with Ferrier's relaxed view of interrogations of this kind:

Sed aliorum quorundam curiosam superstitionem maxime redarguendam censeo, qui aegritudinis principium ignorantes, ad damnatam quaestionum seu interrogationum artem confugiunt, et figuram coelestem interrogationis erigunt ad illud temporis momentum, quo urina aegri ipsis oblata est, aut quo de morbi auxilio ab aegroto consuluntur; et de hac figura non secus iudicium ferunt, quam si exactum primi decubitus punctum obtinuissent. Atque haec Arabum quondam et Iudaeorum Astrologorum fuit amentia, quos nescio quo consilio neotherici complures sectantur, ut Albertus Magnus, Marsilius Ficinus, Villanovanus, Apponensis, Boderius, aliique. Quinimo et Augerius Ferrerius perversa aliorum opinione induci se est passus, ut crederet tolerandam esse hanc doctrinam, dummodo Medicus horam interrogationis non deliberato animo, sed quasi interrupto sortiatur. Scribit enim in haec verba in libello de diebus decretoriis. Putant Astrologi repentinos illos incautosque animi motus e Coelo promanare, Coelumque et ad accersendum Medicum, et aegrotantis amicos et ministros eo modo impellere. Verum enimvero nos superstitiones hasce, apparenti aliqua veritatis umbra, seu superinducto illinimento contectas, omnino repudiandas censemus, et rationabilia artis principia tantum retinenda, non minus enim hanc interrogationum Astrologicorum vanitatem, quam Geomantiae frivolam superstitionem a bonis omnibus merito damnatam, explodendam iudicamus. (Magini, *De astrologica ratione*, 58v.)

21. See Vivian Nutton, "Idle Old Trots, Coblers and Costardmongers: Pieter Van Foreest on Quackery," in *Petrus Forestus Medicus*, ed. Henriette A. Bosman-Jelgersma (Amsterdam: Stichting AD&L, 1996), 243–56.

22. "Opera quidem ad salutem mira, quae a medicis in astrologia peritis per res ex multa compositas, id est pulveres, liquores, unguenta, electuaria fieri possunt, probabiliorem in se rationem et notiorem quam imagines habere videntur, tum quia pulveres, liquores, unguenta, electuaria opportune confecta coelestes influxus facilius citiusque suscipiunt quam materiae duriores ex quibus imagines fieri consueverunt . . ." Marsilio Ficino, *De vita libri tres* 3.13, in idem, *Three Books on Life*, ed. Carol V. Kaske and John R. Clarke (Binghamton, N.Y.: Medieval and Renaissance Texts and Studies in Conjunction with the Renaissance Society of America, 1989), 306.

23. See Müller-Jahncke, *Astrologisch-magische Theorie*, 153–84.

24. "Marsilius Fiscinus, Florentin, grant philosophe, medecin et astrologien et le plus que l'on sache de sont temps sçavant divers langaiges, comme grec, caldée, arabic, ebreu et latin. Cestui a bien monstré en ses euvres qu'il estoit souverain astrologien, part expecial en ung traicté qu'il a composé et intitulé *De vita sana, de*

vita longa et de vita celesti." Simon de Phares, *Le recueil des plus celebres astrologues et quelques hommes doctes,* ed. Ernst Wickersheimer (Paris: H. Champion, 1929), 266 (new ed., ed. Jean-Patrice Boudet, [Paris: H. Champion, 1997], 1).

25. Andrew Wear, "Galen in the Renaissance," in *Galen: Problems and Prospects,* ed. Vivian Nutton (London, Wellcome Institute for the History of Medicine 1981), 229–62 at 245–50, is one of the very few scholars to have drawn attention to the restricted role of astrology in *general* works on medicine, even during the period in which European culture was most thoroughly astrological and the usefulness of astrology for medicine was most frequently and emphatically asserted.

26. See Pietro d'Abano, *Conciliator: Ristampa fotomeccanica dell'edizione Venetiis apud Iuntas 1565,* ed. Ezio Riondato and Luigi Olivieri (Padua: Editrice Antenore, 1985), differentia 10, "Utrum quis medicus existens per scientiam astronomiae, possit conferre in salutem aegroti necne," 15v–18r, and differentiae 103–6, 153r–57v, all of which are on critical days. See also Graziella Federici Vescovini, "La place privilégiée de l'astronomie-astrologie dans l'encyclopédie des sciences théoriques de Pierre d'Abano," in *Historia philosophiae Medii Aevi: Studien zur Geschichte der Philosophie des Mittelalters,* ed. Burkhard Mojsisch and Olaf Pluta, 2 vols. (Amsterdam: Gruner, 1991), 1:259–69.

27. *De vita* was first published in 1489. On Ficino's ideas regarding astrology, see G. Zanier, *La medicina astrologica e la sua teoria: Marsilio Ficino e i suoi critici contemporanei* (Rome: Edizioni dell'atteneo & Bizzarri, 1977), and Melissa M. Bullard, "The Inward Zodiac: A Development in Ficino's Thought on Astrology," *Renaissance Quarterly,* 43 (1990): 687–708, which offers a full bibliography of earlier studies.

28. The Hippocratic corpus as a whole contains only a very slight amount of astronomical material, but of what there is, much is concentrated in the *Epidemics* and *Airs Waters Places;* see Otta Wenskus, *Astronomische Zeitangaben von Homer bis Theophrast. Hermes: Zeitschrift für klassische Philologie,* Einzelschriften 55 (Stuttgart: Steiner, 1990), 90–123.

29. On these debates see Vivian Nutton, "The Seeds of Disease: An Explanation of Contagion and Infection from the Greeks to the Renaissance," *Medical History,* 27 (1983): 1–34, and idem, "The Reception of Fracastoro's Theory of Contagion: The Seed That Fell among Thorns," *Osiris,* 6 (1990): 196–234; also Jon Arrizabalaga, John Henderson, and Roger French, *The Great Pox: The French Disease in Renaissance Europe* (New Haven: Yale University Press, 1997), 56–126.

30. See Giovanni Pico della Mirandola, *Disputationes adversus astrologiam divinatricem* 3.16–19, ed. Eugenio Garin (Florence: Vallecchi, 1946), 2:322–63. The work was first published in 1496.

31. A considerable number of Hippocratic treatises were available in Latin translation in the Middle Ages, but by no means the whole corpus; see Pearl Kibre, *Hippocrates latinus* (New York: Fordham University Press, 1985). Some humanist translations of individual treatises began to appear in the late fifteenth century, but the first Latin translation of the Hippocratic corpus as a whole was Hippocrates, *Octaginta volumina . . . per M. Fabium Calvum . . . Latinitate donata* (Rome, 1525). The Aldine edition of the Greek text followed in 1526. Subsequently, an-

other translation was made by Janus Cornarius: *Opera quae apud nos extant omnia*. (Basel, 1546). The complete *Epidemics* was one of the most important works first made widely available in print by Calvi's translation.

32. Girolamo Cardano, *In Hippocratis Coi prognostica, opus divinum . . . Item in libro Hippocratis de septimestri et octomestri partu . . .* (Basel, 1568) (no commentary on *De octomestri partu* is included; *De septimestri partu* = *Opera* 9:1–35); idem, *Commentarii in Hippocratis de aere, aquis et locis opus* (Basel, 1570) = *Opera* 8:1–212. Cardano's commentary on *Epidemics* 1 and 2 was not printed in his lifetime; it survives in manuscript (Vatican City: Biblioteca Apostolica Vaticana vat.lat. 5848) as well as in his *Opera* 10:168–387. In their final form, almost all of Cardano's Hippocratic commentaries reflect his lectures at Bologna in the 1560s, but they were also part of a larger project on which he had begun working many years earlier. See Siraisi, *The Clock and the Mirror,* chap. 6.

33. Cardano, *Encomium medicinae,* in his *Quaedam opuscula, artem medicam exercentibus utilissima* (Basel, 1559), 135–36 (*Opera* 6:6–7, quoted with translation, in Siraisi, *The Clock and the Mirror,* 228–29).

34. "At meo iudicio, Hippocrates nunquam diceret aerem divinum esse, sed coelum: Id enim aperte dicit in libro de Carnibus (Pag. 2) dicitque sempiternum, et quod est calidum, et omnia novit. In libro vero de Aere, Aquis ac locis (Pag. 2) tum in primo de Diaeta (Pag. 2) dicit, Astrologiam esse Medico necessariam, oportereque illum noscere ortus, et occasus syderum, plurimumque hoc conducere ad artis gloriam." Cardano, comm. *Prognostica* 1.5, 6, 7 [division of Cardano's commentary; he is commenting on *Prognostic* 1, according to the reckoning of modern editors], 15 = *Opera* 8:594. Cardano often cited *De carnibus* and *De diaeta* (= *Regimen*) as important authorities for some of his philosophical ideas; see Ingegno, *Saggio sulla filosofia di Cardano,* 226–27, and Siraisi, *The Clock and the Mirror,* 67 and note.

35. Cardano, *Contradicentium medicorum liber continens contradictiones centum octo* (Venice, 1545), 1.6.10, 165r = *Opera* 6:415. On the publication history of this work, see Siraisi, *The Clock and the Mirror,* 43–44.

36. "Sola enim medicina constantem facit praedictionem, modumque docet, tempus ac eorum omnium certas et evidentes causas affert, atque ostendit." Cardano, comm. *Prognostica,* dedicatory epistle, *3v *Opera* = 8:583. Cf.: "Et ob hoc intelligimus, medicinam esse certiorem naturali philosophia, cum naturalis philosophia semper procedat ab effectibus ad causas, medicina vero persaepe a causis supra effectus. Et ob id dicebat Galenus primo artis curandi (cap. 4) quod demonstrationes medicae sunt similes mathematicis, et a causis." Cardano, comm. *Prognostica* proem. 2 (*Opera* 8:585).

37. "Reliquum est, ut videamus de stellis, physiognomia, chyromantia, somniis, et fato, seu ordine causarum: sub his enim quinque continentur omnia genera praedictionum communia. Et dico, quod de stellis eatenus, quatenus agunt calore et lumine, dubium non est: sed haec, si rite extendantur, etiam influxus continent. Fatum quoque pro ordine causarum admittitur ab Augustino, in opere de Civitate Dei. Physiognomia quoque haud dubie magnam continet veritatem. Chyromantia est alterius rationis, nec potest adeo reduci ad causam seu causas naturales. Somnia

probat Hippocrates, et lex utraque, antiqua et nova. Ideo videntur, posse facere praecognitionem adeo certam, ut medicina. Itaque in universum quo scientia est unius generis, eo certior est in praedicendo." Ibid., 8 (*Opera* 8:589).

38. *Opera* 10:194.

39. Cardano, comm. *De aere, aquis et locis,* 103 [mispagination for 119]–20 = *Opera* 8:102–3.

40. "Secundo devenit ad placita medicorum, utpote quod in purgando debeamus observare quod luna sit in signis aqueis, et hoc credo non intellexisse Hippocratem, sed potius de his tribus ut apparet ab illo, et est Astrologia naturalis rationi conveniens et vera: primum, ratione temporum. . . . Secundum quod recipitur ab Hippocrate est ex aeris qualitate quae manifeste est calida, frigida, humida et sicca, quia in temporibus humidis securiores sunt omnes operationes evacuandi. . . . Tertium est aliquo modo extra medicinam, non tamen prorsus, velut quod prima quadra lunae a coniunctione est calida et humida, secunda calida et sicca, tertia frigida et sicca, quarta frigida et humida. Et ideo in prima et quarta maxime conveniunt evacuationes scilicet in prima per sectionem venae, in quarta per purgationem (2. Apothelesmatum cap)." Cardano, comm. *De aeris, aquis et locis* 1.12, p. 25 = *Opera* 8:21–22.

41. "Primum igitur quantum nostris temporibus astrifera illa sphaera senserit mutationis, diligenter explanandum est; Quod tametsi viri aliquot ante nos percelebres, ut Regiomontanus Mathematicorum omnium decus et ornamentum eximium, Iacobus Zieglerus Laudanus, Gemma etiam Phrysius, ac Ruellius Suessionensis, accurata diligentia sint persecuti, quoniam ipsorum monumenta non ad manum cuivis sunt, maximum sane operae precium me facturum sum arbitratus, si quod longa animadversione observatum a me sedulo est, scriptis mandaretur." Pedro Jaime Esteve, *Hippocratis Coi medicorum omnium principis Epidemion liber secundus . . . Latinitate donatus, et fusissimis commentariis illustratus . . .* (Valencia, 1551), 5v. The entire discussion occupies 4r–13v. According to Esteve, citing Galen, "erubescere quivis [medicus] huius scientiae [astronomiae] ignarus deberet." Ibid., 4v.

42. Adrien L'Alemant, *Hippocratis medicorum omnium principis, de aere, aquis et locis liber . . .* (Paris, 1557), 30v–42v. On Ptolemy's *Phaseis aplanon asteron,* a work transmitting one of the oldest forms of Greek interest in the stars, which long predated planetary astronomy, see *Dictionary of Scientific Biography,* s.v. Ptolemy, by G. J. Toomer.

43. Cardano, comm. *De aeris, aquis et locis* 4.1–2 (*lectiones* 55–56), 103 [mispagination for 119]–124 (*Opera* 8:102–4); regarding the occult influence of the heavens, ibid., 1.11–12 (*lectio* 12), 25–26 (*Opera* 8.21); in *Septem Aphorismorum Hippocratis particulas commentaria . . .* (Basel, 1564), 3.14, 253–62 (*Opera,* 8:322–27). "Quae vero nugatur Brasavola circa astrorum ortum et occasum, quanto plura, tanto magis risum moveat," ibid., 4.5 (*Opera* 8:353). On Brasavola, see *Dizionario biografico degli Italiani,* s.v. Brasavola (Brasavoli), Antonio, by G. Gliozzi. On Brasavola's astronomical or astrological interests centered on the work of Manilius, see Anna Maranini, "La tradizione degli 'Astronomica' di Manilio nell'ambiente Ferrarese," in *Alla corte degli Estensi,* ed. Marco Bertozzi (Ferrara: Università degli Studi, 1994), 425–45.

44. "Febres pestilentes fiunt, aqua vel aere corruptis. Aer corrumpitur ob siderum dispositionem, aut ventos, aut aquas, aut casum, ut in strage hominum, in abundantia locustarum mortuarum, in copia reptilium. et quod ad sidera attinet, Ptolemaeus hoc refert maxime in eclipses luminarium: et hoc, cum loco dominantur maleficae. Et cum haec scriberem, facta est eclipsis Solis vigesima Iunii, et luna fuit in nodo australi, et fuit in octava parte Cancri: et Iupiter ac Saturnus fuerunt occidentales in ipso Cancro, et procul dubio non est bona. Sed alii referunt in magnas coniunctiones. Et utrique dicunt verum. Nam vidimus ex coitu Saturni et Iovis in Cancro, anno 1504, inchoatam febrem pestilentem cum maculis, pulicum maculis similibus: quae, ut ait Fracastoreus, satis bonus observator talium, erat morbus antea solum endemius Cypro et vicinis insulis, ut anno 1505 saevierit in Italia. Et ex alio congresso eorundem cum aliis omnibus planetis, anno 1524 per totum orbem, adeo anno 1528 invaluere, ut memoria illius anni multis saeculis sit celebranda. Cum vero congressus fuerit in signo piscium patet trigonum aqueum totum esse huiusmodi morbis corruptionis obnoxium. Ab anno autem 1544 in Scorpione conversa est ad dissidia potius legis, ex quibus eversa est tota Gallia. Modo praesenti anno transit haec magna coniunctio ad trigonum igneum signumque leonis. Est igitur eclipsis haec ex his quae pertinent ad morbos pestilentes in proprios locis. Et anno 1484 ostendit adventum pestis Indicae, quae subsecuta est generalis, et invisa nostro toti orbi." Cardano, *De venenorum differentiis, viribus, et adversus ea remediorum praesidiis* . . . (Basel, 1564), 1.9, 876 (*Opera* 7:285–86). The work is printed in 1564 in the same volume as Cardano's commentary on the *Aphorisms*, with continuous pagination.

45. For example, Symon Pistoris, a medical professor at the University of Leipzig, who was the particular target of one of Giovanni Mainardi's sharpest attacks on medical astrology. On this dispute see Paola Zambelli, "Giovanni Mainardi e la polemica sull'astrologia," in *L'opera e il pensiero di Giovanni Pico della Mirandola nella storia dell'umanesimo* (Florence: Nella sede dell'Instituto, 1965), 2:205–79, with an edition of the text of Mainardi's letter to Martin Mellerstadt at 260–79. On astrological explanations of plague in the fourteenth century, see Anna M. Campbell, *The Black Death and Men of Learning* (New York: AMS, 1966, reprint of 1931 edition), 14–17 and 37–44; Campbell attributes the spread of astrological explanations of plague largely to the influence of the *Compendium de epidemia* produced in October 1348, in response to the request of Philip VI of France, by the faculty of medicine of Paris, which emphasized the role of a conjunction of the three superior planets. On the debate over the astrological causes of syphilis, see Müller-Jahncke, *Astrologisch-magische Theorie*, 193–207; see also Arrizabalaga Henderson, and French, *The Great Pox*, 107–12.

46. "Ultimo vero et magis serio reditum coniunctionis duorum superiorum qui fit per trigona singulis viginti annis, ita ut in uno trigono maneat plusquam per ducentos annos, velut nostra aetate, et hoc etiam anno iuxta veram supputationem tabularum Prutenicarum facta est ultima coniunctio trigoni aquei, Saturni et Iovis in fine cancri, et per hos ducentos et amplius generati sunt morbi contagiosi ut lues Indica. . . . At in trigono igneo fiunt morbi proprii magis non serpentes graviores, tamen et non venenosi. . . . Erunt ergo sterilitates et siccitates maximae et bella, quia Leo et Sagittarius et Aries sunt signa pugnacia, et Leo significat cor. Et ita

erunt morbi pernitiosi absque veneno tamen. Et incipient ab anno MDLXXXIV quo anno transibit haec coniunctio a trigono aqueo ad igneum, et erit in Ariete; nec statim sentietur, sed ut reliqua naturalia sensim invalescet, incipiet incrementum post annum domini MDCXXIV." Cardano, comm. *Epidemics* 1, 2.40 (*Opera* 10:272).

47. "Historia morbi validissimi," Cardano, *Consilia* (*Opera* 9:245–46).

48. "sed si intelligit quod influant, hoc est solum dogma astrologorum, quos pro nunc mittamus: quandoquidem sufficiunt nobis praecepta Hippocratis ad veritatem eruendam." Cardano, *De providentia ex anni constitutione liber* (Basel, 1564), 1042 (*Opera* 5:16). The work is printed in 1564 in the same volume with Cardano's commentary on the *Aphorisms* and *De venenis*, with continuous pagination.

49. Statutes of the University of Arts and Medicine of Bologna dated 1405 require the professor of *astrologia* to prepare a *iudicium anni* for each year; see Carlo Malagola, ed., *Statuti delle Università e dei collegi dello studio bolognese* (Bologna, 1888), 264. The practice was still being observed at Bologna in the sixteenth century; see Thorndike, *A History of Magic,* 5:235–45, which describes a collection of such prognostications made during the years 1500–40, as well as scattered similar items from later decades. See also Robert Westman, "Copernicus and the Prognosticators: The Bologna Period, 1496–1500," *Universitas: Newsletter of the International Centre for the History of Universities and Science* [Bologna], no. 5 (December 1993): 1–5.

50. "Propter hoc satis erit, si dicam de Roma; excusatum tamen me habebunt, quod Romam non viderim. sed si vera est imago antiquae et novae urbis. . . ." *De providentia ex anni constitutione liber,* 1048 (*Opera* 5:18).

51. See Siraisi, *The Clock and the Mirror,* chap. 6.

52. Cardano, *De providentia ex anni constitutione liber,* 1051–53 (*Opera* 5:19–20). On Cardano and anatomy, see Siraisi, *The Clock and the Mirror,* chap. 5. These cases, not discussed in that chapter, supplement the account there. Gabriel Cuneo was the first holder of a chair in anatomy at the University of Pavia, being appointed in 1554. See *Memorie e documenti per la storia dell'Università di Pavia e degli uomini più illustri che v'insegnano* (Pavia, 1878), 127. On the readiness with which elite families in Renaissance Italy had recourse to postmortem dissection for family members, see Katharine Park, "The Criminal and the Saintly Body: Autopsy and Dissection in Renaissance Italy," *Renaissance Quarterly,* 47 (1994): 1–33, at 8–9.

53. "Hieronymus Fracastoreus (vir nostra aetate insignis, et praeter medicinam, etiam in mathematicis rebus subtiliter exercitatus, et huic negocio totus intentus, ut qui tres libros in hac materia luculenter scripserit, inscriptos de Morbis contagiosis). . . ." Cardano, *De providentia ex anni constitutione,* 1055 (*Opera* 5:21).

54. "Quare concludo, quod generalis pestilentia aeris non potest generaliter facere pestem buboniam, sed potest ex effectibus producere causam pestis buboniae: id est, quae ex conversatione contrahitur. Et hoc fecit Deus Gloriosus: quia si pestis communis posset per se transire in buboniam, cum quae ab aere fit aliquando sit communis toti orbi (ut de illa quae fuit tempore Antonini et Galieni, et

ea quae fuit anno mcccxxxix et perseveravit quindecim annis) posset et secunda esse communis toti orbi: et ita interiret totum humanum genus, quia non possent [sic] se tueri ab uno et altero simul." Ibid., 1057 (*Opera* 5:22).

55. "Ex quo patet, quod sunt quatuor genera pestis: Commune, quod fit ex corruptione aeris, vel aquarum, quod (ut dixi) est contagiosum, id est ab aere: et etiam a consuetudine, sed leviter, et est impressio ab astris, et non potest gigni (ut putant stulti) neque ob inopiam rei frumentariae, neque alia causa, sed est a Deo, et coelo. Secundum est ex cadaverum multitudine, exhalatione terrae, et locustis: et hoc fit ex vaporibus et est contagiosum ab aere, non a consuetudine, sed persaepe generat pestem buboniam. Differt a primo, quia est in vaporibus, non in aere, vel aqua. Tertium est ex consuetudine, et contactu, et est bubonia pestis: et est cum ultima putredine, et potest generare secundum genus, et inficere aerem vaporibus lethalibus, et est pessimum celeritate mortis, et multitudine morientium: et quia pauciores servantur. Quartum est, quod fit ex fame: et si comedant cibaria agrestia, non permutatur, nisi quandoque in secundum genus. Si autem comedant res corruptas, et maxime ex animalibus, permutatur in buboniam . . . Et sicut primum genus non potest fieri ab aliis unquam, sed potest esse causa omnium aliorum, ut dictum est: ita bubonia pestis necessario fit ab una aliarum semper, et per accidens. Et hoc declaratur, quia cum non fiat a coelo, nec aere, nec sit perpetua igitur oportet ut oriatur casu ex aliis, vel etiam per putredinem linteorum, sepultorum, vel exhalatione alicuius corruptione." Ibid., 1066–67 (*Opera* 5:25).

56. See Ann Carmichael, "Diseases of the Renaissance and Early Modern Europe," in *The Cambridge World History of Human Disease,* ed. Kenneth F. Kiple (Cambridge: Cambridge University Press, 1993), 279–87. Of course, Cardano's use of the term *bubonio pestis* does not imply a modern understanding of the disease as an entity.

57. "Oportet autem solum tamen eas observare, quae evidentes habent qualitates, et conspicuas constitutiones validas et firmas, ac diuturnas: ne pulcherrimum inventum astrologorum more foedetur." Cardano, *De providentia ex anni constitutione liber,* 1072 (*Opera* 5:27).

58. "Septenorum quartus est index. Alterius septimane octavus principium. Est autem et undecimus consideratione dignus: ipse enim est quartus secundae septimanae. Rursus vero et decimus septimus considerabitur: ipse siquidem quartus est a quartodecimo, septimus vero ab undecimo." *Aphorisms* 2.24, according to the lemma from the text of the Latin version used by Cardano in his commentary (Cardano, comm. *Aphorisms,* 157 [*Opera* 8:282–83]). A modern English translation from the Greek runs as follows: "The fourth day is indicative of the seven; the eighth is the beginning of another week; the eleventh is to be watched, as being the fourth day of the second week; again the seventeenth is to be watched, being the fourth from the fourteenth and the seventh from the eleventh." (Trans. W. H. S. Jones, *Hippocrates,* Loeb Classical Library [Cambridge, Mass.: Harvard University Press, 1979, first printed 1931] 4:115). For a collection of some of the remarks in the *Epidemics* that were held to provide empirical examples, see Galen, *De diebus decretoriis* 2.3, in his *Opera omnia,* ed. C. G. Kühn, 20 vols. (Leipzig, 1821–33), 9:848–52.

59. Galen, *De diebus decretoriis* 3, in *Opera omnia,* 9:900–41. The first two books of this treatise (ibid., 769–899) establish the existence of critical days on the

basis of cases in the Hippocratic *Epidemics* and attack other views on the subject. The astrological material is mostly to be found in chapter 9 of book 3 (ibid., 928–33). In addition to being included in Renaissance editions of Galen's *Opera*, the work was several times published as an independent treatise during the sixteenth century in a translation by Guenther of Andernach. A thorough historical study of the development of the doctrine of critical days is Karl Sudhoff, "Zur Geschichte der Lehre von den kritischen Tagen im Krankheitsverlaufe," *Wiener medizinische Wochenschrift*, 52 (1902): 210–13, 272–75, 321–25, 371–74.

60. "Et super medico quidem non inest, nisi ut cognoscat quod egreditur cum crisibus pluribus: et non pertineat ei ut sciat quae sit causa eius, quum declaratio illius causae extrahat tum ad artem aliam: imo oportet, ut sit sermo de diebus crisi, sermo quum loquimur secundum semitam experientiae aut secundum viam positionis et axiomatum." Avicenna, [*Canon*] 4.2.2.2, ed. Giovanni Costeo and Giovanni Paolo Mongio (Venice, 1595), 2:103.

61. See the index entries under "critical days" in Lynn Thorndike and Pearl Kibre, *A Catalogue of Incipits of Mediaeval Scientific Writings in Latin*, 2nd ed. (Cambridge, Mass.: The Medieval Academy of America, 1963). For an example of this literature, see Cornelius O'Boyle, *Medieval Prognosis and Astrology: A Working Edition of the* Aggregationes de crisi et creticis diebus: *With Introduction and English Summary* (Cambridge: Wellcome Unit for the History of Medicine, 1991). According to the editor, this treatise survives in eleven manuscripts and was probably composed shortly after Galen's *De diebus decretoriis* was translated into Latin for the first time in the second half of the thirteenth century, possibly by William of Moerbeke. But long before the transmission of Galen's treatise, short texts associating the days of the lunar month with health and disease had circulated in Europe; see Christoph Weisser, *Studien zum mittelalterlichen Krankheitslunar: Ein Beitrag zur Geschichte laienastrologischer Fachprosa, Würzburger medizinhistorische Forschungen* 21 (Pattensen: H. Wellm, 1981), which lists more than 130 manuscripts dating from between the ninth and the sixteenth century in various European languages. On astronomical knowledge and astrological beliefs in the early Middle Ages, see Bruce S. Eastwood, *Astronomy and Optics from Pliny to Descartes: Texts, Diagrams, and Conceptual Structures* (London: Variorum, 1989), and Valerie J. Flint, *The Rise of Magic in Early Medieval Europe* (Oxford: Clarendon Press, 1991).

62. "Quapropter Galenus et alios multos medicorum motum hunc lunae non taliter distinguentes, sed simpliciter confundentes tanquam uniformem in eorum crisibus plerumque errare contingit. . . . Tertius [mensis] autem compositus est, qui secundum phantasiam Galeni ex mense peragrationis, sive propriae impressionis et mense communis ill[umin]ationis conficitur: coniungunt namque isti duo menses invicem: et quod resultat ex eis sunt 53 dies, et 22 horae: huius quidem aggregationis medietas sunt 26 dies, et 22 horae. Et hic mensis medicinalis a Galeno appellatus existit." Pietro d'Abano, *Conciliator*, differentia 104, fols. 154v–155r. See also *De diebus decretoriis* 3.9, in *Opera omnia*, 9:932.

63. Avicenna's *Canon* continued to be a standard medical authority and was frequently reprinted during the sixteenth century; see Nancy G. Siraisi, *Avicenna in Renaissance Italy* (Princeton: Princeton University Press, 1987). The astrological works

of Avenezra (Abraham ben Meir Ibn Ezra, ca. 1090–ca. 1164–67) were translated into Latin during the thirteenth century (via an intermediary French translation made by a Jewish scholar) and appeared in early printed editions; see *Dictionary of Scientific Biography,* s. v. Ibn Ezra, Abraham ben Meir, by Martin Levy. Pietro d'Abano's *Conciliator* retained its status as an authoritative work on many topics, and, unlike most other works of scholastic medicine, continued to be reprinted into the second half of the sixteenth century (as in the example in note 26).

64. For example, the *Amicus medicorum* of Jean Ganivet (fl. ca. 1431–34) and the *Iatromatematicae* of Girolamo Manfredi (1455–92), each issued in a number of early printed editions. See Müller-Jahncke, *Astrologisch-magische Theorie,* 137.

65. "Sed si quis aegrotet pridie quam Luna coeat habeatque diem septimam creticam, quomodo id erit ex influentia potestatis, tam a Sole receptae, quam a signis? Cum maximam partem eorum dierum, sicuti non viderit, ita solari virtute Luna nos non affecerit, auctore ipso Galeno." Pico, *Disputationes adversus astrologiam divinatricem,* 3.16, 1:330–32. The whole chapter, which is entirely devoted to demolishing Galen's astrological theory relating critical days of illness to the moon, occupies 322–48.

66. At different times in his career Nifo (ca. 1469/70–1538) taught philosophy, medicine, or both at the universities of Padua, Naples, Salerno, and Rome. He wrote a number of commentaries on Aristotle and was involved in philosophical controversies, as well as being a member of the humanist circle of Giovanni Pontano at Naples. See "Nifo," *Dictionary of Scientific Biography,* s. v. Nifo, Agostino, by Edward P. Mahoney. At the time of Nifo's attendance on the Grand Captain in 1504–5 (ibid., 10:122), the task must have been a demanding one, as Gonzalo Hernandez, then just taking up his position as the new Spanish viceroy of Naples, was a sick man; see Gerald de Gaury, *The Grand Captain: Gonzalo de Cordoba* (London: Longmans, Green, 1955), 110–11.

67. Agostino Nifo, *De diebus criticis seu decretoriis aureus liber* (Venice, 1519), 2r. An earlier edition of the work appeared in 1504.

68. "Multa Apponensis et Picus contra Galenum scribunt quae frivola sunt et nostro libro deiecta." Ibid., 8v. The repudiation of Pico's demolition of the planets/humors theory is on 6v.

69. Agostino Nifo, *Ad Apotelesmata Ptolemaei eruditiones* (Naples, 1513). This work, too, repudiates Pico.

70. "Et hi [dies critici] sunt in primo mense lunari omnes septenarii. 7. 14. 20. 27. Ut experientia gravissimorum medicorum Hippocratis in primis.... Similiter et aliorum experientia qui praxim exercuerunt et exercent," Cesare Ottato, *Opus tripartitum de crisi, de diebus criticis: et de causis criticorum* (Venice, 1517), 7v.

71. "Et vere mihi videtur etiam quod opinio Conciliatoris et astrologorum esset vera. quod nullum possit fieri iudicium de crisi bona vel mala nisi habita ratione nativitatis et geniture." Ibid., 10v.

72. "Sed nolumus admittere aspectus illos astrorum inimicorum. non angulos non masculinitatem vel feminitatem. non frigiditatem, etc., que omnia vana sunt cum illis suis mathematicis precisionibus." Ibid., 11v.

73. "sequitur dicta medicorum esse arbitraria et in pronosticis vana valde de crisi," ibid., 12r; and "medici exercitati in magnis civitatibus et hospitiis publicis dicunt nullam penitus invenisse differentiam sensibilem in pharmacando: vel flobotomando. sive in plenilunio sive in coitu lune sive alio tempore oppositionis: vel consummationis lune etc. vel aliorum astrorum," ibid., 12r.

74. "Quero tamen ab ipso: si ille sit unus de illis sic exercitatis et bonis medicis: qui non adinvenit diversitatem in flobotomando et pharmacando suos infirmos in plenilunio vel novilunio. Caveat dicere quod sic: quia viles muliercule levantes [lavantes] pannos: essent magis prudentes ipso: quae novilunium vel coitum lune cum sole observant: ne panni in illa hora loti corrodantur." Federico Grisogono, *De modo collegiandi: prognosticandi: et curandi febres* (Venice, 1528), 7v. The work is dedicated to Doge Andrea Gritti.

75. In addition to the items in note 29, see Wear, "Galen in the Renaissance," 246–49, and Alessandra Preda, "La peste astrologica, ovvero il dibattito circa la 'scienza dei cieli' tra Symphorien Champier e Giovanni Mainardi," in Bertozzi, *Alla corte degli Estensi*, 323–43. On the printing history of Mainardi's epistles, see Zambelli, "Giovanni Mainardi," 256–58. For Leoniceno's views, see Daniela Mugnai Carrara, *La biblioteca di Nicolo Leoniceno* (Florence: Leo S. Olschlei, 1991), 41–42, 73, 82–84. But as the same author demonstrates, although Leoniceno repudiated astrological prediction in medicine, he was deeply interested in the stars.

76. "Semel evenit ut, cum praestantis cuiusdam viri curationi, una cum Hieronymo Manfredo, astrologo sui temporis famatissimo, Bononiae interesset, immineretque simul et solvendi occasio et luminarium coitus, astrologo reclamante et aegrotanti mortem minitante, potio tamen ad Bencii imperium exhibita est infirmusque per eam e gravi morbo convaluit." Giovanni Mainardi, letter to Martin Mellerstadt, in Zambelli, "Giovanni Mainardi," 278. Francesco Benzi was the third son of one of the most celebrated of all scholastic physicians, Ugo Benzi (d. 1439).

77. "nedum si Aegyptius aliquis astrologus, sed si Aegyptius aliquis deus, aut Anubis, aut Osiris haec mihi dixerit, non facile credam." Girolamo Fracastoro, *De causis criticorum dierum libellus,* in his *Opera omnia* (Venice, 1584), 48v–56r, at 48v.

78. Fracastoro's theory is analyzed in detail in Sudhoff, "Zur Geschichte der Lehre von kritischen Tagen," 322–23.

79. For example, Federico Grisogono, as in note 74; Johannes de Indagine, *Canones astrologici, de iudiciis aegritudinum,* part 3 of his *Chiromantia* ([NP], 1547); Auger Ferrier, as in note 20; Thomas Bodier, *De ratione et usu dierum criticorum opus* (Paris, 1555); Claude Dariot, as in note 18; and Giovanni Antonio Magini, as in note 2. On the work of Indagine, Bodier, Dariot, and Magini (and others), see Müller-Jahncke, *Astrologisch-magische Theorie*, 146–53; see also P. Ulvioni, "Astrologia, astronomia e medicina nella Repubblica veneta tra Cinque e Seicento," *Studi Trentini di Scienze Storiche*, 61 (1982): 1–69 at 15–23, on the defense of astrology, and especially medical astrology, against Pico, by the empiric medical practitioner Tommaso Zefriele Bovio (1521–1609).

80. Passages on critical days in Cardano's medical works include *Contradicentium medicorum liber* 1.3.2–7, 55v–66v (*Opera* 6:337–44) (except in the first of this set, much of the discussion is, however, purely medical and includes citations of such medieval medical authorities as Rasis, Taddeo Alderotti, and Gentile da Foligno on 65v–66r [344]); comm. *Aphorisms,* 2.24, 157–58, and 4.36, 387–92 (*Opera* 8:283 and 381–83); comm. *Prognostic,* 3.1–5 (fourth section of commentary), 462–76 (*Opera* 8:741–46); comm. *Epidemics,* 1, 2.33, 2.52, and 3.53 (*Opera* 10:267, 282–84, and 324); comm. *Epidemics,* 2, 2.23 (*Opera* 10:375). The foregoing is not intended to be an exhaustive list.

81. "At vero miscere rem imaginariam cum re naturali, adeo puerile est ac indignum Galeni autoritate, ut malim totum quod superest a capite octavo supra tertii libri de Diebus iudicatoriis superadditum ab aliquo existimare, quam autoritatem tam gravis viri violatam videri." Cardano, *Contradicentium . . . liber* 1.3.2, 56r–v (*Opera* 6:337). Cardano was probably wrong, since Galen mentioned writing the work in three books in *De libris propriis* chap. 5, in *Opera omnia,* 19:32. We are grateful to Vivian Nutton for drawing our attention to this reference.

82. "Habes igitur totam rationem deductam ex loco solis, nam ut centesimusvigesimus dies est tertia pars totius anni, ita quadragesimus centesimavigesimae, et vigesimus dimidium quadraginta, et quatuordecim tertia pars, et septem dimidium quatuordecim. Quae omnia cum ignoraverit Galenus, longe petitis auxiliis a luna hanc doctrinam dierum criticorum totam confudit, nec sibi nec Hippocrati concors. Constat ergo, Hippocratem a circuitu solis hanc rationem deducere." Cardano, comm. *Aphorisms,* 4.36, 390–91 (*Opera* 8:383).

83. "Haec igitur est summa dierum criticorum, numerus, ordo, explicatio, causa, iuxta veritatem, et Hippocratis sententiam, quam Galenus tot nugis involuit false, confuse, inconstanter, adeo ut qui eam secuti sint, finem nullum invenire potuerunt." Cardano, comm. *Prognostic,* 3.1 (fourth commentary), 465 (*Opera* 8: 741). And "Dico ergo, quod nemo potest dissolvere hanc difficultatem, nisi qui calleat artem Hippocraticam Ptolomei." Ibid., 3.5, 476 (*Opera* 8:741, 746).

84. "Stultum haec medicorum vulgus quod nostro seculo sic videmus insolescere, atque imponere nobis purpura sua, cum sibi ab autoribus suis praeceptum sciant, qui absque astrorum consilio nulli medicetur: et tam longe ab medicina est, qui astrologiae ignarus est, ut non medicus dici debeat sed et impostor: tamen eo nunc ventum est, ut e centum vix unum aut alterum reperias, qui vere sciat diiudicare, quo tempore quaelibet medicina adhibenda sit." Johannes de Indagine, *Canones astrologici, de iudiciis aegritudinum,* 63v–71r, in his *Chiromantia* ([Paris], 1546), at 63v.

85. Cardano, *De malo recentiorum medicorum medendi usu libellus, centum errores illorum continens* (Venice, 1536), chap. 24 and 55 (32–33 and 59). The work was subsequently incorporated as part 1 of Cardano's *De methodo medendi,* and as such these chapters appear in *Opera* 7:211 and 220. On the publication history of this work, see Siraisi, *The Clock and the Mirror,* xiv and 28–9.

86. "Utrum vero verum illud sit quod luna in ariete capite minetur, et in Tauro collo, et ita in cancro pectore, in leone cordi, et in virgine visceribus, vere fateor nondum diligenter adeo animadvertere potui ut affirmare possim vel negare.

Quamobrem tutius est non contemnere, maxime cum ea opinio etiam apud nos percrebuerit." Cardano, comm. *De aeris aquis et locis,* 1.11–12 *(lectio* 12), 25 *(Opera* 8:22).

87. "Indignabuntur imperiti, quod tam aperte lacessam Galenum suum. . . . Unde illi tanta gloria? Quid hic affert nisi suos illos dies Iudicatorios, quorum causam ne novit (ut alias saepe docui) quidem: Et si verae millies essent, quid medico utilitatis afferunt?" Cardano, comm. *Epidemics,* 1, 3.53 *(Opera* 10:324).

88. "Circa causam huius Alemanus nihil dicit, sed multa tamen (15) e quibus unum ridiculum, quod data cassia nigra cum manna dum luna esset in XV parte Capricorni mulier una in operatione periit. Sed melius fuisset quaerere causam in medicamento aut in morbo, et non in astris. Fieri potest ut materia veneni non expers mota sit, aut quod inclitum sit venenum, aut quod laboraret abscessu aliquo prope cor aut in cerebro ei incognito, e tribus enim unum fuit non ex astris, neque enim novit vim astrorum quomodo sit sumenda, ut video ex omnibus primum aut saltem potentioribus et convenientibus inter se, et postmodum comparatis ad hominem hunc, et in hoc tempore: dico iuxta genethliace, ut quatuor necesse sit concurerre ad tales effectus repentinos. Et licet hoc acciderit anno MDLI, ut colligitur ab eo, quia dicit quod fuit die XI Augusti, tum Mars esset infelix in Libra respiciens quadrato Lunam, est tamen causa absurda et indigna tanto viro, cum mille homines acceperint medicamenta ea die ex scammonio, etiam qui ne ungue quidem laesi sunt. Ideo doleo illius causa quod occasionem dederit imperitis irridendi, cum alioquin sit peritus. . . ." Cardano, comm. *De aere aquis et locis,* 4.13 *(lectio* 65), 139 *(Opera* 8:120).

89. Bodier, *De ratione et usu,* 17v–51v. On Bodier, see Thorndike, *A History of Magic,* 5:301–3. Bodier's work is dedicated to Oronce Fine.

90. Cardano, *De methodo medendi sectiones quatuor* (Paris, 1565), section 3, *curatio* 18, p. 230 *(Opera* 7:253–64, at 256 and 258). Cardano collected accounts of his cures throughout his life, publishing them in several different versions. See Siraisi, *The Clock and the Mirror* 29 and 235–36.

91. Cardano, *Consilia,* nos. 19 and 22 *(Opera* 9:102 and 124). The largest collection of Cardano's *consilia* is the fifty-three included in his *Opera,* of which only a selection had previously appeared in works published during Cardano's lifetime. For example, three lengthy *consilia* formed one section of his collection of short medical treatises, *Quaedam opuscula* (Basel, 1559); in this copiously indexed volume there are no index entries for *astra, crisis, dies critici, sidera* [*sydera*]*, luna,* or *stellae.*

92. See Siraisi, *The Clock and the Mirror,* 4–5, with references to the earlier literature on Cardano's career.

93. See J. Dupèbe's introduction to his edition of Nostradamus, *Lettres inédites* (Geneva: Librarie Droz, 1983), 18 and n. 43, and P. Brind'Amour, *Nostradamus Astrophile* (Ottawa: Presses de l'Université d'Ottawa, 1993), 318–19, 372–73.

94. For Nostradamus's incompetence, see Brind'Amour, *Nostradamus,* esp. 70–78.

95. G. Cardano, *Pronostico o vero judicio generale* (Venice, 1535), ed. G. Ernst, in Marialuisa Baldi and Guido Canziani, *Girolamo Cardano: le opere, le fonti, la vita* (Milan: F. Angeli, 1999); cf. Ernst's discussion of this important document.

96. I. Maclean, "Cardano and His Publishers 1534–1663," in Kessler, *Girolamo Cardano*, 313.

97. Cardano published ten genitures with his *Libelli duo* (Milan, 1538); sixty-seven in the reprint of this work (Nuremberg, 1543), and 100 in the expanded *Libelli quinque* (Nuremberg, 1547). The cases cited here are numbers 8, 56, and 59 in the collection of 100 genitures, to be found in *Opera* 5:458–502.

98. Ibid., genitures 5 and 21. On Corbetta and Capella, see S. Albonico, *Il ruginoso stile* (Milan: F. Angeli, 1990).

99. Geniture 55 (*Opera* 5:486): "Ioan. Antonius Castiloneus, civis noster, regiusque medicus et vir insignis, cum hanc vidisset genituram, dixit puerum hunc nutriri non posse . . ."

100. *Opera* 5:436.

101. See Thorndike, *A History of Magic*, 5:244–47, though he understandably failed to see that Niccolò and Costanzo were one and the same astrologer.

102. Cardano, *Aphorismi astronomici* (*Opera* 5:56): "Hic debilis ex reiectatione sanguinis ex pulmone, et quasi tabificus, cum consuluisset Constantium Bononiensem de Symis, an adepturus esset a Caesare regnum loco fratris, dixi ego illo ostendente figuram, moriturum eo anno, nam Luna inter Pleiades erat in sexta, in quadrato Martis et Iovis in fixa Saturni domo peregrinantium, et sic repente in itinere suffocatus est."

103. ". . . multi invidi dicerent me literas medicinae nescire quod totus mathematicis viderer intentus," Cardano, *Ephemerus, de libris propriis*, 1544 (*Opera* 1:57).

104. The following genitures have medical content (often minimal or tangential): 3, 4, 5, 7, 8, 12, 19, 20, 21, 22, 43, 45, 50, 51, 52, 55, 56, 57, 59, 67, 68, 73, 75, 77, 81, 86, 92, and 94.

105. Geniture 67 (*Opera*, 5:491).

106. Cardano's *Liber xii geniturarum* contains his surviving large-scale horoscopes and analyses. It first appeared with his edition of Ptolemy's *Tetrabiblos* in 1554 and is to be found in *Opera* 5:503–52.

107. Ibid., 510–11.

108. Ibid., 511: "*De opificio*. Erit cupidus secretarum rerum, delectabitur pulchris rebus, gemmis, vestibus, picturis, imaginibus: et liberalibus disciplinis, nec tamen quicquam horum ob nobilitatem exercebit.

De itineribus. Nulla in re aut felicior, aut aptior quam in itineribus, legationibus et expeditionibus, in quibus et pericula et calumnias patietur ob invidiam. Ad maximas tamen administrationes perveniet. Complectitur autem haec genesis contraria quaedam simul, velut bonam temperiem et vitam morbosam, orbitatem cum sit uxoratus, obesum habitum et ingenium acre, benefacere et calumniam pati."

109. Ibid., 508–10.

110. Ibid., 510: "Causa concordiae cum octava genesi et amoris erga nos quintuplex est . . ."; 541–44 at 544. Cardano's advice for Casanatus sounds exactly like the sort of vague astrologer's chat that he elsewhere condemned. He predicted that

he had chances "Ab ascendente autem iuxta 28. annum rixae peregrinationis acutae febris et iuxta 50. submersionis, Mortem quidem evadere poterit, periculum non. Vel erit suffocatio ex morbo, ut asthma, vel attonitus. Generaliter autem si non violenta morte moriatur erit satis longaevus" (545)—a fair sample of Cardano's predictions of life chances, mingling the medical freely with the accidental and hedging both.

111. Cardano, *Aphorismi astronomici,* 1547 (*Opera* 5:85): "Hunc ego curandum suscepi, levi spe, praesertim cum immensis distinear negotiis. Desiderium tamen habendae geniturae ad hoc me impulit."

112. "Vitam vero longam non solum ab initio semel fata promittunt, sed nostra etiam diligentia praestat. Quod et astrologi confitentur, ut de electionibus et imaginibus agunt, et medicorum cura diligens experientiaque confirmat." Ficino, *Three Books on Life,* 2.1, p. 166. A reader of a copy of the first edition of *De vita* (Florence, 1489); Houghton, Inc 6151 (A), emphasized the words "Quotiens septimo cuilibet propinques anno, consule diligenter astrologum" (2.19, [e v] recto [Ficino, *Three Books on Life,* 2.20, p. 232]) by writing a marginal note: "Consulere astrologum."

113. Nostradamus, *Lettres,* 94–101.

114. For comparison we have used, inter alia, Luca Gaurico's revolution for 1532–33 for Ferdinand, King of the Romans (Vienna, Österreichische Nationalbibliothek, MS 7433); his horoscope for one Stephen of Nuremberg (Paris, Bibliothèque Nationale, MS lat. 7385, 332r–370v); Bartholomäus Reischer's horoscopes for the members of the house of Austria (Österreichische Nationalbibliothek, MS 10754); Cyprian Leowitz's horoscope for Adam von Dietrichstein (Bibliothèque Nationale, MS lat. 7443A). Thorndike, *A History of Magic,* vols. 5–6, still offers the fullest guidance through the jungles of this unstudied literature. For published samples of period horoscopes and revolutions, see the fascinating work of R. Castagnola, *I Guicciardini e le scienze occulte* (Florence: Leo S. Olschlei, 1990), which prints, among other documents, the horoscope drawn up for Francesco Guicciardini by Ramberto Malatesta, and W. Pirckheimer, *Briefwechsel,* ed. E. Reicke et al. (Munich: Beck, 1940–89), 2:362–73.

115. *Praeclarissimi viri Georgii Valle Commentationes in Ptolemei Quadripartitum inque Ciceronis Partitiones et Tusculanas questiones ac Plinii Naturalis historie librum secundum* (Venice, 1502), ep. ded., [A vo]: "Ptolemaeus mathematicorum omnium facile princeps ut quidam scripsere Adriani vixit temporibus ad Antoniumque usque pervenit: quo tempore Galenum inclitum medicinae auctorem perhibent floruisse: necnon Herodianum grammaticum et Hermogonem rhetorem: qui de arte rhetorica libros reliquit non contemnendos."

116. Ibid.: "Primus autem apud Graecos traditur Chius Oenopides de astrologia scripsisse nonnulla . . ."

117. Ptolemy, *Tetrabiblos,* ed. G. Cardano (Lyons, 1554), 3 = *Opera* 5:94: "Namque ut apud Gellium Phavorinus qui parum ante Ptolemaeum floruit famosus philosophus, more ambitiosorum artem astrologiae infamem reddiderat." For Favorinus's critique of astrology, see A. Bouché-Leclercq, *L'astrologie grecque* (Paris, 1899), 571ff.

118. Ptolemy, *Tetrabiblos*, 28 = *Opera* 5:113: "Cum igitur eodem tempore non unum genus artis floreat, utpote militaris disciplina, poesis, eloquentia, pictura, plastica, musica, medicina, philosophia, simul quoque desinant, ut Alexandri et Augusti et nostra etiam aetate, cum tamen ab Augusto ad initium nostrae florentis aetatis, quod fuit circa annum salutis 1440, fluxerunt anni intermedii circiter mille quadringenti, nihil egregium pro miraculo natura produxerit in lucem, praeterquam Antonini tempore cum tunc etiam floruissent simul Alexander Aphrodisaeus, Ptolemaeus Pelusiensis et Galenus Pergamenus, et parum ante id etiam C. Plinius et Plutarchus Cheroneus Traiani magister, manifestum est ex generalibus constitutionibus coeli principaliter ista pendere." Here Cardano introduced astrological questions into a discussion usually conducted on a different level, as Michael Baxandall has shown. See his *Giotto and the Orators* (Oxford: Clarendon Press, 1971) and A. Grafton, *Defenders of the Text* (Cambridge, Mass.: Harvard University Press, 1991), chap. 7.

119. Ptolemy, *Tetrabiblos*, "Prooemium expositoris," 1 = *Opera* 5:93: "Est autem ars haec philosophiae pars prognostica et praecognoscere docens, unde non vere scientia, sed ut ad medicinam se habet liber praedictionum Hippocratis aut Galeni, ita hic ad totam Philosophiam. Vnaquaeque enim ars quae de naturalibus tractat ea ratione, quae causas praesentium rerum explicat, docet et futurorum, nam futura a praesentibus non specie nec genere differunt, sed tempore, quod illis ut accidens adiungitur."

120. Ibid. 5:94: "Artes autem quae futura hoc modo cognoscere docent, sunt Agricultura, Nautica, Medicina, Physiognomia et illius partes, Somniorum interpretatio, et Magia naturalis, ac Astrologia. Harum nobilissima astrologia est, quia de omnibus est, alia autem sunt certi generis. Est etiam per causas semper atque eas nobilissimas, reliquarum nulla semper, sed etiam per signa docet futura praedicere. Non est autem ut quidam existimant scientia Astrologiae, Fati scientia, sed Fati pars constitutio astrorum est."

121. Ibid., 2 = *Opera* 5:94: "Porro quanto Fati scientia certior atque nobilior, tanto etiam obscurior."

122. Ibid., 77 = *Opera* 5:206.

123. Vivian Nutton, "'Prisci dissectionum professores': Renaissance Humanists and Anatomy," in *The Uses of Greek and Latin*, ed. Carlotta Dionisotti, Anthony Grafton, and Jill Kraye (London: The Warburg Institute, 1988), 11–26. W. Smith, *The Hippocratic Tradition* (Ithaca and London: Cornell University Press, 1979), 122–72.

124. Siraisi, *The Clock and the Mirror* 138–41.

125. Ptolemy, *Tetrabiblos*, "Prooemium expositoris," 2 = *Opera* 5:93–94: "Non est officium expositoris, ut Galen. inquit, vel reddere rationem dictorum ab authore, vel si qua non placeant reprehendere, sed verba illius ac sententias dilucide explicare; nos tamen non propriam faciemus expositionem, sed conabimur declarare causas dictorum a Ptolemaeo, et quae ab ipso obscure dicta sunt, clariora reddere, quae breviter ac circumcise, perficere, quae dubie, determinare." But with his usual indifference to consistency, on another occasion Cardano sharply criticized Galen for precisely the same approach to commentary: "Interrogo te, o

Galene, in quo opus est medico ad interpretationem librorum medicinae si solum verba auctoris sunt explicanda? Nam si de dictionibus agendum est ac sensibus grammatici hoc opus est." Cardano, comm. *Aph.* 6.42 [6.44] (*Opera* 8:513).

126. Ptolemy, *Tetrabiblos,* 198 = *Opera* 5:242: "Sed huius rationem reddit Galenus, dicens: Olim reges ut instruerent bibliothecas magno pretio emendo illustrium virorum scripta, causam dedisse ut sua veteribus attribuerent."

127. Ibid., 24 = *Opera* 5:110: "Nam quemadmodum Galenus de Hippocrate inquit, in libros quos de difficultate spirandi conscripserat, ita nos de Ptolemaeo existimare oportet: nihil, scilicet, illum scripsisse commune aut vulgare, sed omnia singularia atque recondita."

128. Ibid., 64 = *Opera* 5:141–42; 218 = *Opera* 5:257: "quod Ptolemaeus mira arte usus sit in hoc libro."

129. Ibid., 257: ". . . unum corpus absque repetitione. Vt siquis velit per eum facere genituram, necesse est, ut totum bene perlegerit, intellexerit, et mente teneat, aliter millies errabit."

130. Ibid., 4 = *Opera* 5:95–96.

131. Ibid., 17 = *Opera* 5:105: "Nam et in septimo Epidemiarum quicumque illius libri author fuit (et certe vir egregius fuit) ingenue fassus est deceptum se sutura capitis."

132. Ibid., 146 = *Opera* 5:204: "Dicit ergo quod hanc mistionem vel remissionem qualitatum aut argumentum impossibile est penitus describere: quod etiam Galenus in libris methodi et Avicenna quarta primi testati sunt."

133. Ibid., 33–35 = *Opera* 5:117.

134. G. Pomata, *La promessa di guarigione* (Rome and Bari: Laterza, 1994); translated into English as *Contracting a Cure* (Baltimore: Johns Hopkins University Press, 1998).

135. There must have been many practitioners in Italy whose work consisted largely of drawing up interrogations for a mixed clientele, like Richard Napier, whose work is studied by Michael Macdonald in *Mystical Bedlam* (Cambridge: 1981), though without discussion of the astrological details of his technique. We know of no published studies, however, indicating the survival from Italy of any notebooks of rapid interrogations like Napier's, preserved in the Bodleian library.

136. Ptolemy, *Tetrabiblos,* 17 = *Opera* 5:105.

137. Ibid., 230 = *Opera* 5:267: "post longa scripta perplexa dubia qualia solent facere astrologi nostri temporis, plena contradictionibus et obscuritate"—a criticism that certainly held for the eminently fashionable Nostradamus; see Dupèbe's edition of Nostradamus's *Lettres* and Brind'Amour, *Nostradamus Astrophile.*

138. Ptolemy, *Tetrabiblos,* 252 = *Opera* 5:282.

139. Ibid., 198 = *Opera* 5:242.

140. Ibid., 173–75 = *Opera* 5:234–36.

141. Cardano's first horoscope is included in *Libelli quinque. . . . V. De exemplis centum geniturarum* (Nuremberg, 1547), no. 19 (*Opera* 5:468–72); the date of

composition is given by the remark "pervenisset caput ad Lunam in 43 annis [of Cardano's life] iam, id est anno praeterito." The second is no. 8 in . . . *Geniturarum XII . . . utilia exempla*, printed with his commentary on Ptolemy's *Tetrabiblos* in the Basel, 1554, edition. Cardano revised both figure and narrative once more, stating, "Patior etiam in sinistro oculo lachrymam, quae eo anno coepit quo et podagra. Anno 64" (*Opera* 5:523, the entire horoscope occupying 517–41).

142. See Siraisi, *The Clock and the Mirror* chap. 10.

143. Pomata, *La promessa di guarigione*, chaps. 3 and 5.

144. See Zanier, *La medecina astrologica*.

145. For a helpful account of theological positions on this issue, see Laura Smoller, *History, Prophecy and the Stars: The Christian Astrology of Pierre d'Ailly, 1350–1420* (Princeton: Princeton University Press, 1994), chap. 2.

146. See the classic work of D. P. Walker, *Spiritual and Demonic Magic from Ficino to Campanella* (London: The Warburg Institute, 1958). The precise nature of Ficino's views on talismans and their use is hotly disputed by the leading authorities, B. P. Copenhaver and P. Zambelli, whose *L'ambigua natura della magia* (Milan: Il Saggiatore, 1991) should be read against Copenhaver's "Lorenzo de'Medici, Marsilio Ficino, and the Domesticated Hermes," in *Lorenzo il magnifico e il suo mondo*, ed. G. C. Garfagnini (Florence: Leo S. Olschlei, 1994), 14–34.

147. "Et quia requiris ultra hoc, an ex astris hoc decernatur, et an ab illis passio petat auxilium? Respondeo quod astrologia nostrae aetatis non est vera, quia non est cognita. Principia enim illa Ptolemei evariant ab eo quod videtur et contingit. Sunt praeterea errores in motibus ac locis astrorum, ut apparet ex tabulis Prutenicis, quae solae consentiunt cum experimento. Praeterea oporteret scire modum motuum illorum. Declaratum est enim a nobis in Quinrodecimo [sic] Paralipomenon quod neque per excentrices [sic] neque parvos circulos neque per utrosque possunt servari quae apparent in astris, et nisi sciamus cur, gratia exempli luna sit dimidia, non possumus scire quid decernat. Quod tamen est perspicuum, est quod sol et caeteri Planetae cum sint in piscibus decernunt surditates et mutitates, Pisces enim parum audiunt et voce carent. Et sol in oppposito Saturni infelicis tale incommodum significabat. Et effectus apparuerunt annis MDXXII, XXIII, et XXX, cum Saturnus premebat Iovem ac Mercurium et ideo cum ars sit imperfecta licet adumbret aliquid, potest tamen magis esse causa erroris quam iuvare nos. Et etiam hanc contemplatio est extra medicinam. Nec video auxilium ex directione Iovis ad Venerem, quae enim apud nos sunt tametsi illis subiiciuntur tamen sunt certiora. Et si spes etiam aliqua in hoc diligentius etiam considerassem." Cardano, *Consilia*, no. 12 (*Opera* 9:78). In Cardano's *Opera*, book 15 of his *Paralipomena* (not printed in his lifetime) is about the lives of illustrious men; book 8 is about the stars but does not appear to contain anything relating to this *consilium*.

4

Celestial Offerings: Astrological Motifs in the Dedicatory Letters of Kepler's *Astronomia Nova* and Galileo's *Sidereus Nuncius*

H. Darrel Rutkin

The years 1609 and 1610 saw the publication of two epoch-making works in the history of astronomy and in the history of science overall: Johannes Kepler's *Astronomia Nova* (1609) and Galileo Galilei's *Sidereus Nuncius* (1610). The revolutionary contributions of these works are too well known to require retelling here.[1] In this chapter, I shall focus, rather, on their dedicatory letters.

Mario Biagioli has recently called attention to Galileo's scientific production and its presentation within the context of the overall design of his scientific career—his socioprofessional self-fashioning—in the courtly milieu of an absolute prince.[2] Biagioli pays particular attention to reconstructing Galileo's patronage situations within the broader culture of early modern Europe, where patronage concerns to a great extent conditioned the social system within which the practice of science took place.[3] Specifically, Biagioli reconstructs the patronage situation most relevant to our purposes, in which Galileo tried to woo Cosimo II de Medici, who had just succeeded to the grand duchy of Tuscany. Galileo had been his tutor in mathematics for the prior several years, during the summers, in the time off from his teaching duties as a professor of mathematics at the University of Padua in the Venetian Republic.[4]

In chapter 2 of his book, Biagioli focuses his account on the patronage strategies that surrounded the publication—and presentation—of the *Sidereus Nuncius* to Cosimo II. In the dedicatory letter, Galileo fashioned his discovery of the satellites of Jupiter into Medicean stars in a spectacularly successful attempt to raise his status from that of professor of mathematics at a university to that of court mathematician and philosopher of an absolute prince.[5] Biagioli argues that Galileo's receipt of patronage from an absolute prince significantly augmented the epistemological legitimization he so greatly desired.[6] Even if one does not accept every detail of his attempt to tie the contents of the preface into a purported Medicean dynastic mythology,[7] Biagioli has been quite successful in showing how Galileo skillfully related his "gift"[8] of Jupiter's stars to Cosimo II personally by the device of relating Jupiter, and thereby Galileo's discoveries, to Cosimo's natal horoscope.

Galileo thus associated his epoch-making telescopic discovery of the moons of Jupiter with the astrological nativity of his intended patron. Not only was Galileo, then, ingenious in his artful prefatory invention, he was also brilliantly successful in his intended grab for patronage. His gambit worked spectacularly well, but was it original? Did he invent this brilliant literary conceit, or did he, rather, skillfully adapt an already existing model in an act of literary *imitatio?*[9] I will argue in this chapter that Galileo probably did borrow at least one of the central structures of his preface: the association of his planetary astronomical discovery with his patron's astrological nativity. Indeed, he seems to have borrowed this device from the almost exactly contemporaneous—and also epoch-making—astronomical contribution of the imperial mathematician, Johannes Kepler: the *Astronomia Nova* of 1609. The chronology, as we will see in more detail below, admits the possibility; the content and structure argue the probability.

Furthermore, in looking for other examples of the particular astrological device in question,[10] it seems obvious to examine Tycho Brahe's dedicatory letters, since it is well known that he was a serious astrologer[11] (as well as an alchemist).[12] In particular, one thinks of his dedicatory letter to the *Astronomiae Instauratae Mechanica* (1598), also dedicated to Rudolf II. Although there is nothing obviously astrological there, let alone an earlier example of the device in question, the letter does contain what may be the model Galileo used for the first part of his dedicatory letter to Cosimo, for which Kepler provided the model of the second part, especially of its astrological centerpiece.

The chapter will begin with a detailed treatment of the first part of Galileo's dedicatory letter; then I turn to Tycho's preface and argue that it provides Galileo's model therefor. Next I discuss the second part of Galileo's preface, where we meet our crucial astrological passage. I then present Kepler's dedicatory letter, also in some detail, and argue on several counts that it provides the model for the central device around which Galileo structured his dedicatory rhetorical tour de force. The concluding section will shore up these arguments with some chronological considerations. The fact that these prefaces happen to have been written by three of the most important astronomers in the history of science makes an investigation into their literary qualities a worthwhile endeavor;[13] the additional fact that those by Galileo and Kepler also have astrological motifs at their centers makes their interest even more compelling.[14]

Excursus on Dedicatory Prefaces

Dedicatory letters played many roles in Renaissance books.[15] They could justify the work as a whole, suggest a way to read it, or create a relationship between the author and the dedicatee—or carry out all three tasks at once. They thus offered authors a kind of performance space in which they had more room for innovation and creativity than was often available in the text proper.[16] What follows is meant to be preliminary, to indicate some basic patterns that may then allow a fuller picture to emerge in considering the possibilities of this literary genre.[17]

I will first characterize certain features of what may transpire in this performative space—this public stage,[18] as it were—which is placed at the very beginning, before the hard work of the treatise (or dialogue)[19] proper. The dedicatory letter, first of all, is just that, a letter.[20] There appear to be two quite different kinds or styles—"modes" in Brian Vickers's sense—of dedicatory letters: (1) a more straightforward style in which, for example, the author either tries to defend what has been done in the body of the work or introduces it by describing some of the circumstances of its investigation and composition,[21] and (2) a more artificial, highly elaborated composition in which the mode appears to be much less that of forensic oratory than of a type of epideictic "show" oratory.[22] Several of Kepler's other dedicatory letters are of the first, more straightforward type, where he describes *in propria persona* the nature of his research directly to the patron in question;[23] he often treats the issue of financial support openly in his dedicatory letters.[24] Neither of our featured prefaces by Galileo or Kepler,[25] however, falls into this category; rather, they are both highly ornate literary productions of a rather tall order, which, though they function quite effectively as dedicatory letters, have much more of a novelistic or dramatic flavor.

One of the most important social functions that took place within the dedicatory letter's communicative space concerned patronage dynamics, many fine discussions of which have appeared in recent years.[26] We shall not explore directly in this chapter the intricacies of patronage dynamics beyond attempting to gain a deeper appreciation of one of the central literary spaces within which patronage relations were expressed, and thereby constituted, in the public sphere.[27] These dedicatory letters seem, furthermore, not only to provide a public demonstration of their authors' abilities, but also to publicize the magnificence of the dedicatee. This feature, especially when a prince is the dedicatee, seems to drive the dedicatory letter almost into the domain of courtly—and more public—entertainments, much like those reconstructed by Roy Strong.[28]

We shall also be well served by paying close attention to the rhetorical practice whereby the author persuades the reader by (selectively) informing him and thereby shaping how he reads the work overall.[29] This awareness is essential for understanding how the author purposefully crafts the narrative to inform the particular way that the intended primary reader—the dedicatee—will read and react to what the narrator presents. We shall find some striking examples in what follows. These, then, are some of the basic patterns that can help us see more clearly what the dedicatory letter was and how it functioned, and, most importantly, how our historical authors used them.

Galileo (I)

Let us now examine the dedicatory letter to Galileo's *Sidereus Nuncius*. The only historical context I wish to provide at the outset is simply that Galileo, the author/dedicator, presented this work to the historical dedicatee, the nineteen-year-old newly crowned grand duke of Tuscany, Cosimo II de Medici, whom Galileo had recently tutored in mathematics.[30] That Galileo in his mid-forties was writing to his teenage ex-pupil is not without significance for the narrative strategy he develops in the dedicatory letter.

First, let us begin to characterize the narrative situation of the dedicatory letter.[31] Some of its distinctive features will come out more clearly by contrasting them with the narrative situation of the dedicatory letter to Kepler's *Astronomia Nova*.[32] The historical author, Galileo, begins by adressing the dedicatory letter to Cosimo II de Medici, the fourth grand duke of Tuscany; he does this in the dative case as is normal in epistolary addresses.[33] This address to the historical Cosimo has no explicit connection, qua narrative, with the narrative itself; it simply announces the historical dedicatee, who then does not figure in the narrative for quite a while. The dedicatee leaves no explicit linguistic trace in the text until the thirty-third line of the dedication. We shall explore the rhetorical effect of such a strategy as we go through the text. Kepler by contrast develops a rather different structure in the dedicatory letter of the *Astronomia Nova*, where he also at first addresses the historical dedicatee, Rudolf II, in the dative case.[34] But then the narrator immediately addresses Rudolf in the vocative case (in the first line of the text of the dedicatory letter proper [7,5]).[35] Kepler then refers to him three more times in the next four lines (7,6–9) in the genitive case, using a standard imperial formula.[36]

After this disconnected address, Galileo begins the dedicatory letter proper with a highly impersonal construction. Van Helden captures this well:[37] "A most excellent and kind service has been performed by those who defend from envy the great deeds of excellent men and have taken it upon themselves to preserve from oblivion and ruin names deserving of immortality" (29).[38] Indeed, the entire first paragraph and, in fact, the entire first part of the dedicatory letter (2,5–3,5) is presented in this completely impersonal narrative style, which is wholly uncommitted, within the text itself, with respect to the identities of both the narrator and the intended audience. The narrator speaks only in impersonal, third-person utterances,[39] in the manner of one speaking general truths. The tone, furthermore, is quite didactic, much like that of an older, more experienced teacher instructing a pupil. We happen to know, of course, that Galileo, more than twice Cosimo's age, had recently been his tutor, so this didactic tone appears to be perfectly in keeping with the tenor of their historical relationship. Furthermore, the subject at hand—how to memorialize effectively a great ruler's actions and his name—would be of great concern to a young prince. This strikes us, then, as a perfectly sound strategy on the historical Galileo's part to capture the attention and goodwill of the historical dedicatee.

The actual content of the narrative, on the other hand, the picture that Galileo builds up of how great rulers' names and deeds are to be memorialized, is also significant in shaping how Cosimo is to read and understand what follows in the remainder of the dedicatory letter. This is, after all, a rhetorical structure designed to influence and persuade.[40] Indeed, Galileo presents a graduated model of how this feature of the patronage game has been played in the past and is played in the present, that is, the different ways in which patrons' virtues can be memorialized and the relative value, closely related to permanence, of these different ways. Kepler also touches on these themes, but in a much less overtly didactic manner.[41]

Let us now look at the graduated model: "Because of this (*hinc*), images sculpted in marble or cast in bronze are passed down for the memory of posterity; because of this (*hinc*), statues, pedestrian as well as equestrian, are erected; because of this (*hinc*), too, the cost of columns and pyramids,[42] as the poet says,[43] rises to the stars; and because of this, finally (*hinc denique*), cities are built and adorned by the names of those who grateful posterity thought should be commended to eternity" (29).[44] Thus we have a fourfold series, with each stage marked by its own reiterated *hinc*. First, the material nature of the monument is emphasized, then the types of material monuments to memory, in ascending order of magnitude: statues, columns or pyramids, and cities.

The narrator then drives home the point with a reflection on human nature, much like our own saying "out of sight, out of mind": "For such is the condition of the human mind that unless continuously struck by images of things rushing into it from outside, all memories easily flee from it" (29).[45]

Let us now turn to the second series in Galileo's model (2,16–3,5). Here, the narrator presents a second type of memorialization, which he contrasts with the first—literally monumental—series. He contrasts literal monuments with literary monuments, beginning this second series with a strong contrast—"But others (*Verum alii*) looking to more permanent and longer lasting things, have entrusted the eternal celebration of the greatest men not to marbles and metals but rather to the care of the Muses and to incorruptible monuments of letters" (29).[46]

Having evoked this contrast, which serves as an introduction to the second series, the narrator peeks his head out from behind the wings, as it were (or lifts his eyes up to the audience), and speaks in his own voice for the very first time: *At quid ego ista commemoro?* (But why do I recall these things?) He then immediately disappears again as unobtrusively as before and returns to his objective, authoritative, didactic voice, deepening the theme just mentioned: "But why do I recall these things as though human ingenuity, content with these [earthly] realms, has not dared to go beyond them? Indeed, looking far ahead, and knowing full well that all human monuments (*omnia humana monumenta*) perish in the end through violence (*vi*), weather (*tempestate*) or old age (*vetustate*), it [human ingenuity] contrived more incorruptible symbols (*incorruptiora signa*[47] *excogitavit*) against which voracious time and envious old age (*Tempus edax atque invidiosa Vetustas*)[48] can lay no claim" (29-30).[49]

Now that Galileo has begun to reorient our thinking about commemorations from the material and perishable to the literary and—paradoxically—less perishable, we should look more closely at the brief poetic passage that Galileo inserted, with no explicit attribution,[50] in the third stage of the series of material commemorations.[51] I will not tell a definitive story here, which would take us too far afield for our present purposes.[52] Even penetrating a little below the surface, however, can provide a deeper appreciation of Galileo's artistry.

Let us look at some of the resonances of the poem that Galileo has quoted: Propertius's *Elegies* III,2.[53] One of the ironies resounding in Galileo's quotation of Propertius in this context is that Propertius described not a literary monument to the name or deeds of a man renowned for *virtus*, but one to his girlfriend, and in a programmatic poem for a book of erotic elegies. Nevertheless, the ideas presented in both are indeed parallel: Propertius considers his poems to be a monument to her beauty (*carmina erunt formae tot monumenta tuae*, 18), where, for Galileo, the eternal celebration of the best men (*aeternum summorum virorum praeconium*) are immortalized in the incorruptible monuments of letters (*incorruptis litterarum monumentis*).

Galileo quotes line 19 from Propertius: *nam neque pyramidum sumptus ad sidera ducti*.[54] Indeed, the full line of Galileo's: *hinc Columnarum atque Pyramidum, ut inquit ille, sumptus ad sidera ducti* (2,10-11) also has a resonance with line 11 of Propertius: *quod non Taenariis domus est mihi fulta columnis*.[55] Likewise Galileo's *sed Musarum custodiae, et incorruptis litterarum monumentis consecrarunt* (2,17-18) has much similarity to Propertius's lines 15-18 (which occur directly before the line Galileo quoted):

at Musae comites et carmina cara legenti,
   et defessa choris Calliopea meis.

fortunata, meo si qua es celebrata libello!
    carmina erunt formae tot monumenta tuae.[56]

Finally, some of the ideas in the last section of Galileo we examined, *omnia humana monumenta vi, tempestate, ac vetustate tamen interire . . . Tempus edax atque invidiosa Vetustas nullum sibi ius vindicaret* (2,21–24), are closely paralleled in Propertius's lines 23–24:

aut illis flamma aut imber subducet honores
    annorum aut ictu, pondere victa, ruent.[57]

The ideas are similar, to be sure, but the language reads more like a paraphrase than an exact reminiscense. But when we look at the poem by Horace on which Propertius's elegy was modeled,[58] *Ode* III,30,[59] we see that the language in Horace's poem oddly seems to be more exactly the language that Galileo used—and to reflect the ideas more precisely—than that from Propertius's poem, which he actually quoted. The lines in question from Horace are the first five:[60]

Exegi monumentum aere[61] perennius
regalique situ pyramidum altius,
quod non imber edax, non Aquilo impotens
possit diruere aut innumerabilis
annorum series et fuga temporum.[62]

Why would Galileo have done this? Was it to dazzle us with his virtuosity: to quote a line from a poem, one of whose invoked resonances in the mind of a well-informed reader would be not only to the poem actually quoted, but also to the poem on which the quoted poem was modeled? Yes, but not only. A further motivation, I think, is that the line from Propertius's poem contains a term utterly central to Galileo's concerns that is not found in Horace's poem and that the narrator had not yet mentioned up to the time of the quotation. The same term then also serves to foreshadow the second, and most significant, series in Galileo's continuing informative description of his model: the term *sydera*. That this term, furthermore, was introduced in the only text explicitly quoted in the narration further adds to its emphasis.

Let us return to Galileo's preface. His next phrase, *In coelum itaque migrans*, directs our attention to the ultimate realm where Galileo, via the narrator, has been heading for the entire time. Now that he has directed us there—toward the heavens—and has prepared the ground of our understanding, he goes into much more detail than before. He informs us explicitly what the *incorruptiora signa* are that have just been contrasted with the corruptible *omnia humana monumenta*: "And thus, moving to the heavens, it assigned to the familiar and eternal orbs of the most brilliant stars the names of those who, because of their illustrious and almost divine exploits, were judged worthy to enjoy with the stars an eternal life" (30).[63]

Human ingenuity (*humana solertia*) continues to be the subject of this new sentence, carrying over from the beginning of the prior rather long sentence (2,15–20), but without being explicitly reiterated. This human ingenuity consigns the names (*nomina consignavit*) of those who are deemed worthy (*digni habiti sunt*)—based on their extraordinary, almost divine deeds (*ob egregia ac prope divina facinora*)—to the stars. The narrator thus picks up again and reiterates the central theme of this entire first part of the dedicatory letter: how to commemorate the deeds (*res gestae*) of men who excel in *virtus* (*excellentium virtute virorum*) and whose names are worthy of immortality (*immortalitate digna nomina*).[64] Galileo's narrator has now linked the names and deeds of great men with the stars, which are not man-made—in contrast to *omnia humana monumenta*—and which have perpetual (i.e., incorruptible) orbits.

The narrator then explicitly spells this out. Galileo in his didactic narrative voice does not want his privileged audience to miss the point: "For this reason, the fame of Jupiter, Mars, Mercury, Hercules, and other heroes after whom the stars are named will not be obscured before the splendor of the stars themselves is extinguished" (30).[65] The fame of those heroes, for whom the stars are named (*quorum nominibus stellae appellantur*), will thus not fade out before the radiance of their stars (*ipsorum syderum*) fades out. Galileo seems to be following here a euhemerist tradition that would have been well known to him.[66] We should note once again that, besides the brief aside at 2,18–19, our narrator is still speaking in a completely impersonal, objective, third-person narrative voice.

Our narrator then presents a very conspicuous example of this process of naming, conspicuous both in its protagonists—Julius and Augustus Caesar—and in its failure. The relationship that he establishes here is central to Galileo's rhetorico-didactic purposes, central, that is, to the narrative strategy that informs how Cosimo is supposed to understand his own relationship with Galileo. Indeed, this relationship is one virtually guaranteed to appeal to the sense of historical destiny of a nineteen-year-old grand duke, especially of such a distinguished city-state: "This especially noble and admirable invention of human sagacity, however, has been out of use for many generations, with the pristine heroes occupying those bright places and keeping them as though by right. In vain, Augustus's affection tried to place Julius Caesar in their number, for when he wished to name a star (one of those the Greeks call *Cometa* and we call hairy) that had appeared in his time the Julian star, it mocked the hope of so much desire by disappearing shortly" (30).[67]

Having established this relationship, then, and the failed attempt at providing such an incorruptible commemoratory gift, and with this failed gift presented by one of the greatest rulers the world had ever known, Galileo makes his next crucial rhetorical move, which thus inaugurates the second (and final) part of the preface. The narrator at this point, after speaking only in the impersonal mode, finally turns toward and explicitly addresses the dedicatee, Cosimo II: *Atqui longe veriora ac feliciora, Princeps Serenissime, Celsitudini tuae possumus augurari.* Nor is this change in tone transitory. But before we turn in detail to this second part of Galileo's dedicatory letter—and the passage of most concern for our central argument—let us consider a text that Galileo may well have used as the model for the first part of his dedicatory letter to Cosimo.

Tycho

With Tycho Brahe's dedicatory letter to Rudolf II in his *Astronomiae Instauratae Mechanica*[68] of 1598, I will not go into nearly as much detail. I would only like to suggest that Galileo may have taken over certain patterns from Tycho's preface for his own purposes. I begin with Tycho's narrative strategy, which is even more extreme than Galileo's. After the address to Rudolf II, for which he used an *ad* + accusative construction instead of the usual dative case construction,[69] Tycho begins with a similar objective, didactic narrative tone in which neither the narrator nor the dedicatee appears in any significant way until 6,25 for the narrator, after 60 lines of folio text, but for two faint premonitions,[70] and not until 9,8, after 160 lines of text, for the dedicatee, Rudolf II.

In addition, Tycho, in the early part of his objective, didactic narrative, brings up columns (5,19) and expensive pyramids (*sumptuosissimae Pyramides*; 5,22), which some *memoriae causa ad posteros inscripsisse* (5,21).

The concern with memory is not in itself particularly revealing, but that it appears together with columns and pyramids suggests that it may relate more closely to Galileo's preface. And to top it all off, the passage even begins with *hinc*.[71]

Further, apropos of memory, there is another striking similarity to Galileo's preface where Tycho as narrator, in discussing the *memoria* and *fama* of Rudolf II, says that his will endure as long as the sun and stars will: "In addition, may your Imperial Majesty's memory and fame, since they are so excellent and altogether the most important among worldly concerns, not be weighed down by the tasks of preserving, protecting, and promoting, and, for this reason, may they shine brightly and endure for all posterity, as long as the sun and stars remain."[72] To be sure, this could easily have been a common trope that any astronomer would have placed close to the front of his rhetorical arsenal, but I am not familiar with any other examples.

Finally, only when Tycho made the analogy between Rudolf's memory and the celestial bodies, both of which would endure forever, did he first explicitly mention Rudolf. Similarly, Galileo compared the longevity of Cosimo's fame to that of Jupiter and its satellites soon after his master first appeared in the text. But Tycho's comparison was cast in general terms; he did not discuss any individual planet or Rudolf's geniture.

To summarize, none of these themes, taken in isolation, would provide a strong argument for Galileo's use of Tycho's dedicatory letter as a model for the first part of his own; collectively, however, the similarities are suggestive enough to warrant serious consideration of the possibility. In addition, the subject of Tycho's work overall was instrument making, a central concern of Galileo's during his Padua years. Indeed, *Sidereus Nuncius* would not have been possible without Galileo's own significant improvement on a recent, epoch-making astronomical instrument, the telescope. This adds external support to my argument, in that it makes it likely that Galileo would have read Tycho's book, published twelve years before, with close attention.[73]

Galileo (2)

Let us now return to Galileo's preface to the *Sidereus Nuncius*: "But now, Most Serene Prince, we are able to augur truer and more felicitous things for Your Highness, for scarcely have the immortal graces of your soul begun to shine forth on earth than bright stars offer themselves in the heavens which, like tongues, will speak of and celebrate your most excellent virtues for all times" (30–31).[74] In this opening section of part 2, Galileo's narrator directly addresses Cosimo II for the first time. In this first passage, he describes the bright stars that have offered themselves like tongues, which will sing out his outstanding virtues (*praestantissimas virtutes*[75] *tuas*) for all time (*in omne tempus*). The stars, which he has not yet described, have thus appeared and offered themselves just at the time when Cosimo ascended to the grand duchy. Biagioli calls this a fateful conjuncture.[76] Galileo thus begins to associate the stars with Cosimo, and in a manner that relates closely to the patterns set up at the end of the didactic first part of the dedicatory letter, where the stars and planets were associated also with praising gods and heroes in euhemerist fashion and where we met a significant historical example of a failed attempt at such an association.

In the next passage, Galileo gives us our first information about the stars themselves: "Behold, therefore, four stars reserved for your illustrious name, and not of the common sort and multitude of the less notable fixed stars, but of the illustrious order of wandering stars, which, indeed, make their journeys and orbits with a marvellous speed around the star of Jupiter, the most noble of them all, with mutually different motions, like children of the same family, while meanwhile all together, in mutual harmony, complete their great revolutions every twelve years about the center of the world, that is, about the sun itself" (31).[77] There are four stars and they circle around Jupiter, the most noble of the planets (*stella . . . nobilissima*). Furthermore, these stars have been reserved for his glorious name (*tuo inclyto nomine reservata*). By whom they have been reserved is nowhere mentioned; we can only assume that it is by the *Syderum Opifex* of the following passage.

We should note also the long delay before Galileo has the narrator actually mention what the gift is. Nevertheless, Galileo has not yet fully conditioned and personalized his gift for his intended patron; but he is almost there. First he set up the situation by educating Cosimo as to the ranks of gifts that may be given. Now he alludes to the gift, which he has almost completely finished preparing. Kepler, too, in his preface, delays for quite a long while the revelation of his planetary gift to his princely patron.

Galileo as narrator then emphasizes his own not insignificant role as he provides the last preparation for our crucial passage: "Indeed, it appears that the Maker of the Stars himself, by clear arguments, admonished me to call these new planets by the illustrious name of Your Highness (*inclito Celsitudinis tuae nomini*) before all others" (31).[78] Thus, none other than the maker of the stars himself has persuaded our narrator with crystal clear arguments (*perspicuis argumentis*) that he should affix to these new planets Cosimo's illustrious name.[79] Biagioli seems to have gotten Galileo's role here as intermediary between God and Cosimo just right.[80] Furthermore, Galileo has placed himself in the role of Augustus Caesar in so far as he attempted (albeit in vain) to honor his divine forebear with a stellar commemoration. Galileo even lets Cosimo know, however subtly, that he could well have given this supremely noble gift to some other patron.

What then are these crystal clear arguments? Galileo presents the first as follows, and with a somewhat complex, highly rhetorical structure (3,19–33): "For as these stars, like the offspring worthy of Jupiter, never depart from his side except for the smallest distance, so who does not know the clemency, the gentleness of spirit, the agreeableness of manners, the splendor of the royal blood, the majesty in actions, and the breadth of authority and rule over others, all of which qualities have found a domicile and exaltation for themselves in Your Highness. Who, I say, does not know that all these emanate from the most benign star of Jupiter, after God the source of all good?" (31).[81]

Galileo here, at the beginning of the first perspicuous argument, associates the spatial proximity of the new planets to Jupiter with the noble virtues that likewise surround Cosimo. He then further associates Cosimo with Jupiter by identifying those very virtues, which everyone knows Cosimo possesses, with the virtues that everyone also knows emanate precisely from Jupiter himself. Galileo uses a *tamquam-ita* construction for the basic simile, which he embellishes with a long two-part rhetorical question for the *ita* clause on Cosimo and his virtues. The rhetorical question is of the *quis ignorat* type. This is a very powerful form of rhetorical argument, for very few people will nod internal assent to a question such as: Who is so ignorant as to *x*? He repeats this question again within the same sentence for further emphasis as he identifies Cosimo's virtues as those that indeed come from Jupiter. This passage, further, provides the sort of fulsome rhetoric that one would expect from a courtier. Indeed, this sort of praise is what characterizes epideictic rhetoric, concerned as it is precisely with praise and blame, as we find in its classic definition.[82]

With this twofold association of Cosimo with Jupiter and their associated virtues and satellites, respectively, Galileo now presents the crux of his first argument and our crucial passage: "It was Jupiter, Jupiter I say,[83] who at Your Highness's birth, having already passed through the murky vapors of the horizon, and occupying the midheaven and illuminating the eastern angle from his royal house, looked down upon Your most fortunate birth from that sublime throne and poured out all his splendor and grandeur into the most pure air, so that with its first breath Your tender little body and Your soul, already decorated by God with noble ornaments, could drink in this universal power and authority" (31–32).[84]

Galileo further emphasizes here the relationship between Cosimo and Jupiter developed in the first movement of this argument by emphasizing Jupiter's role in Cosimo II's natal horoscope. Evidently Galileo drew up two horoscopes for Cosimo II: He drew the extant one, the one that is described in the text here and was published by Righini, on the back of one of his drawings of the mountains on our moon.[85] Furthermore, Galileo here associates Jupiter only with Cosimo II personally vis-à-vis his horoscope and not with any sort of Medicean dynastic imagery.[86] Indeed, the only mention whatsoever of Cosimo's illustrious forebears occurs toward the very end of the preface, where the narrator explicitly states that he will remain silent about them (4, 12–15).

Galileo has thus, in a rhetorically emphatic way, associated Jupiter, and his own discoveries, with Cosimo's natal horoscope, and thereby personalized the gift for his patron. But then Galileo makes a rather odd rhetorical move, at least to my lights. He almost makes it a throwaway: "But why do I use probable arguments when I can deduce and demonstrate it from all but necessary reason?" (32).[87] The narrator makes this transition with his most prominent intrusion into the text as well—with three instances: a first-person personal pronoun (*ego*) and two first-person singular verbal forms (*utor, queam*); he also expresses it, once again, as a rhetorical question. But what is even more odd than the throwaway nature of the astrological crux is what he throws it away for. We might expect, indeed as we find in Kepler, that Galileo would turn from a conjectural astrological argument to a certain astronomical argument of some sort. This relationship between astrology and astronomy concerning their relative certainty had been central to their disciplinary configuration already from the time of Ptolemy.[88] Further, the certitude of mathematics in general, including mathematical astronomy, was a central concern of sixteenth- and seventeenth-century mathematicians and figured prominently in their justifications for the epistemological superiority of their discipline in relation to the less certain but more powerful discipline of natural philosophy.[89]

Contrary to this expectation, however, Galileo presents his necessary argument in providential, existential terms: for example, that it was evidently by divine inspiration that Galileo became Cosimo's tutor and that it was under Cosimo's auspices that Galileo made his astronomical discoveries.[90] Even though the particulars of this existential argument are not of central concern to us, the rhetorical structure of Galileo's presentation of this astrological motif and then his abrupt turning away from it for a more certain form of demonstration certainly is of paramount interest in our attempt to ascertain Galileo's use of his dedicatory predecessors.

Kepler

Let us now turn to Kepler's preface. Kepler, unlike Galileo, immediately provides a great deal of information toward setting the dramatic stage of his dedicatory letter: "Most August Emperor. In order that there be happiness and prosperity for the most serene Name of Your Holy Imperial Majesty and for the entire House of Austria, I am now finally, at long last, exhibiting for public view a most Noble Captive, who has long since been captured by me in a difficult and laborious war waged under Your Majesty's auspices" (30).[91] In this first sentence, Kepler establishes his most important structures and themes: (1) his privileged reader, the dedicatee, the Most August Emperor, Rudolf II, whom he had just named above in the formal address. Then, in rapid succession,[92] he presents (3[a]) the narrator, Kepler's dramatic personality for the purposes of this dedicatory letter, (3[b]) exhibiting (4) a noble captive, not yet defined,[93] (2) who is to be publicly viewed, that is, by a much broader audience in addition to the emperor.

As we saw, Galileo, in contrast, did not focus on the narrator nor on the dedicatee until well into his preface. Likewise, he did not mention the gift itself, except in the most allusive way, until rather far along in the preface. Kepler, on the other hand, although neither specific nor full of information, at least points to his gift (albeit under cover) as a noble captive; he, too, holds off fully revealing the nature of the captive for quite a while as he artfully builds up quite a good measure of rhetorical anticipation in the reader, spinning metaphor after playful metaphor on a martial theme. In retrospect we will be able to see that Kepler also dropped many hints as to where he was going.

I would also like to note the explicitly public nature of Kepler's exhibition of the gift. He develops this theme further just below, and in a way that relates directly to Galileo's model of public commemorations of the names and deeds of great men:[94] "The renown (*celebritas*) of this spectacle could not be greater than if I were to write a panegyric upon this most distinguished captive, and proclaim it publicly" (30).[95] Here our narrator discusses both composing and publicly proclaiming[96] an epideictic oration in praise of the captive to publicize the spectacle[97] in the most effective way possible. To be sure, although Kepler is not here making a contrast between different types of commemorations, that is, between material and literary, he certainly is valuing a type of literary commemoration very highly indeed: panegyric oratory.[98]

Kepler then turns to the second section of his preface, where he treats another type of literary composition, historiography, in quite a clever and rhetorically powerful way. Historiography, like panegyric, is also deeply concerned with commemorating both the names and deeds of great men. It is worthwhile to look at these passages in some depth: They develop the military theme further, and, more importantly, in a way that would appeal directly both to the emperor's current grave concerns and to the deeper patterns of his mentality; their rhetorical presentation also is quite striking.

Kepler as narrator now makes a segue to the historians' treatment: "I therefore leave it to the writers of history books to describe the greatness of our Stranger, which he acquired in the art of war" (30).[99] He follows this introduction with two more-detailed passages, each of which begins with *dicant* and proceeds, in *oratio obliqua*, with *hunc esse . . .* , where *hunc* refers to the noble captive. *Dicant* is, of course, in the subjunctive mood and indicates what the historians would say in Kepler's hypothetical situation. At the end of the two small historiographical paragraphs, the narrator then returns the spotlight to himself and thus to the primary level in the narrative.

Let us now hear what these historians would say about the greatness of Kepler's stranger:

> They would certainly say that it is he through whom all armies conquer, all military leaders triumph, and all kings rule, without whose aid no one ever honorably took a single captive. Let them now feast their eyes with looking at him, captured through my martial effort (*meo Marte captum*).
>
> Those who admire Roman greatness would say that he is the begetter of the Kings Romulus and Remus, the preserver of the City, protector of the Citizens, Supporter of the Empire, by whose favor the Romans discovered military discipline, improved and perfected it, and subjugated the orb of the world. Let them therefore give thanks at his being confined and at his being acquired as a happy omen for the House of Austria.[100] (30–31)

This captive is, then, he through whom (*hunc per quem*) all armies are victorious (*omnes exercitus vincant*), all leaders in war triumph (*omnes belli duces triumphent*) and all kings rule (*omnes Reges imperent*). In addition to the fact that these three parallel structures reflect good rhetorical technique, they also directly address issues of central concern to the historical dedicatee at this particular, increasingly beleaguered moment in his long reign, which was soon to end.[101] Furthermore, by casting the captive as the preserver of the City (*conservatorem Urbis*), protector of the Citizens (*protectorem Quiritium*), and Supporter of the Empire (*Statorem Imperii*), and as he by whose favor the Romans discovered, improved, and perfected military discipline and conquered the world, Kepler would be magnifying Rudolf's interest in this gift to quite an extent. That Rudolf as Holy Roman Emperor considered himself heir to the Roman emperors would have further focused his desire in that he would have considered this gift, fashioned thus, as his own proper inheritance, and one that would be coming at a particularly propitious time (1608–9) in relation to both external affairs (that ever present menace, the Turk) and internal (the deeply troubling situation with his brother Matthias).[102] Kepler, as one of Rudolf's trusted advisers, would have known this very well.[103] As it turned out, Rudolf was forcibly relieved of his empire in May 1611,[104] which presented its own set of difficult ramifications for Kepler's career.[105]

Furthermore, these military and political concerns would have been directly connected to another central feature of Rudolf's psyche: his intense interest in magic and the other so-called occult sciences, including astrology and alchemy.[106] Indeed, Kepler seems to be shaping his gift to Rudolf as a sort of magical talisman, which he describes as having been a source of great power for the Romans and which would also, therefore, be so for Rudolf himself. That this power is also deeply astrological might further evoke a resonance in Kepler's treatment with one of the most important genres of medieval prose literature, the mirror of princes, especially its most popular representative, the pseudo-Aristotelian *Secret of Secrets,* which had as its most central secret precisely the use of astrologically charged talismans to achieve political ends.[107] Rudolf's curiosity, piqued to such an intensity, would surely be wondering at this point what precisely Kepler had in mind; the imperial mathematician knew his patron very well indeed.

After this hypothetical historical brief, the second form of commemorative rhetoric touched upon (along with panegyric), Kepler now turns to his more familiar domain of professional expertise. Here we meet our crucial passage. The rhetorical legerdemain here especially, but also throughout the preface more generally, seems to me worthy in many respects of Cicero's finest performances.[108] Kepler now makes the transition from the historical section—and thus from the first, introductory part of the dedicatory letter—into the main body of the preface. The narrator intrudes himself rather strongly into the text at this moment to pull the reader out of the historians' rhetorical grasp and, now, back into his own. *Ego me* are the first two words: "I, for my part, retreat hence to other ground better suited to my powers. Nor will I make a stand in that part of my profession in which strife arises between me and my fellow soldiers" (31).[109] We can also see that Kepler continues to develop his military metaphor. We should also note that even though Kepler explicitly referred to Rudolf II immediately at the start of the preface, he has not actually resurfaced overtly since the first four lines of the first paragraph. Nevertheless, Kepler has certainly done his best to keep Rudolf's interest deeply engaged, as we have seen. He will reappear directly in the text proper, and in a rhetorically powerful manner guaranteed to continue keeping Rudolf's attention fully captivated.

So Kepler has moved from a discussion of historians to that of members of his own profession, which, as he says, is more suited to his own powers. He then immediately sets up two camps within his own discipline and proceeds forthwith to present the views of his fellow metaphorical comrades in arms. We will note that he presents their views in much the same style as he presented the historians' views:[110] "They, for their part, would surely rejoice with a different joy: he has been restrained by the bonds of

Calculation, who, so often escaping their hands and eyes, was accustomed to deliver vain prophecies of the greatest moment, concerning War, Victory, Empire, Military Greatness, Civil Authority, Sport, and even the cutting off or calling forth of Life itself" (31).[111] Here Kepler begins the narrator's presentation of the astrologers. He treats here again certain of the themes and the language of the passage on the historians: the prognostications are made *de Bello, de Victoria, de Imperio, de Dignitate militari*. Moreover, in the opening moment of our crucial passage, Kepler uses exactly the same verb in exactly the same form—*gratulentur* (8,3)—to introduce the horoscope passage, just as he did to end the historical (7,30), thus further tying the two sections together.

The second movement presents us with our crucial passage: "Let them congratulate Your Majesty that the lord of Your geniture has been brought under control and even made to be friendly, for by their account Mars rules Scorpio, which has the Heart of Heaven [i.e., the midheaven]; in Capricorn, which is rising, he is exalted; in Cancer, into which the moon has entered, he customarily plays the triangular game with knucklebones;[112] in Leo, where the Sun plays host, he is recognized as being one of the family; and finally, he is the ruler of Aries, beneath whose power Germany is supposed to be, over which he rules in complete harmony with Your Holy Imperial Majesty" (31).[113] In this discussion of Rudolf's nativity Kepler finally identifies the noble captive for the first time.[114] But what is even more important for our purposes is the way that Kepler, even if not in the narrator's own voice,[115] associates Mars with Rudolf's nativity. Indeed, he has personalized his gift to Rudolf by the device of discussing certain features of Rudolf's horoscope[116] in a way that is strikingly similar to, but by no means identical with, the way that Galileo will also do so a scant six months later. This is not the only similarity, however, as we shall see just below.

But before we examine these other important similarities, we should first look more closely at the details of the different ways that Kepler and Galileo related their respective planetary gifts to their respective patrons. How, particularly, does Kepler relate his gift of Mars to Rudolf's nativity? He says that the (hypothetical) astrologers will congratulate him (Rudolf) that the lord of his geniture has been returned into his power (brought under control) and even made friendly.[117] Determining the lord of the geniture, that is, the planet that "rules" the horoscope overall, and thus the native himself, is one of the most important procedures used in interpreting a horoscope. The procedure itself is based on a series of simple calculations meant to determine the overall strengths and weaknesses (dignities and debilities) of each planet. These strengths and weaknesses are measured, based essentially on where each planet falls in a chart—what sign and house it is in and also what angular relationship it bears with its fellow planets—and also, as in Kepler's treatment of Rudolf's chart, on which planet rules the sign of the ascendent and the midheaven. The planet that acquires the largest relative score is thus determined the lord of the geniture, the ruling planet of the nativity overall.[118]

Indeed, Kepler's treatment of Rudolf's horoscope centers completely on his establishing the fact that Mars—his gift, after all—is indeed also the lord of Rudolf's geniture. He works to establish this important fact by enumerating five positive features that Mars has in Rudolf's chart: (1) Mars rules Scorpio, which is in the midheaven;[119] (2) Mars is exalted in Capricorn, which is Rudolf's rising sign;[120] (3) Mars, which rules Scorpio, is thus in a trine (120°) relationship to Rudolf's Moon, which is in Cancer;[121] (4) Mars, which also rules Aries,[122] a fire sign, is in a familial relationship with the Sun, which rules Leo, another fire sign, that is, they are both members of the fiery triplicity, along with Sagittarius.[123] Finally (5) Mars rules Aries, and Aries is the sign that rules Germany; thus both Rudolf and Aries co-rule Germany together in harmony.[124] I must note at this point one very conspicuous absence from Kepler's presentation of some of the details of Rudolf's horoscope: He has somehow neglected to mention one minor detail, that is, where the planet Mars actually is in Rudolf's nativity! True enough, his first two points are indeed strong indicators for the influence of Mars as lord of Rudolf's geniture, but his failure to mention where Mars actually was points rather strongly toward a rhetorical cover-up. But Kepler is not interested in presenting a complete and accurate picture of Rudolf's nativity, warts and all. He has a well-conceived rhetorical strategy in which Mars plays a leading role in both his own researches and in their presentation to Rudolf; both then interconnect in this rhetorically conditioned representation of Rudolf's nativity.

Galileo uses a rather different approach, and here, perhaps, we might catch him red-handed in the act of astrologico-literary *variatio*. Galileo had better luck than Kepler, since the astrological details of his patron's geniture corresponded precisely with his rhetorical intentions. Jupiter was perfectly placed in Cosimo's horoscope for Galileo's purposes: in the midheaven.[125] Galileo points to only this one significant feature of Cosimo's nativity—and with a bright rhetorical spotlight—which he then describes with colorful rhetorical adornment as pouring forth his benevolent influences on Cosimo's tender newborn body (*tenerum corpusculum*) at the same time as God provided Cosimo's soul.[126]

Nevertheless, even though Kepler and Galileo constructed their astrological devices rather differently, the similarities—that they both personalized their epoch-making planetary gifts to their respective patrons by means of a rhetorically conditioned astrological device that related their respective gifts to important features of their patrons' respective nativities—are far more significant than the comparatively minor rhetorico-astrological *variatio*.

With respect to Rudolf's nativity, once again, we should also note that it is precisely in this passage that Rudolf also reappears explicitly in the text for the first time since the opening paragraph. What then is the narrative structure into which Kepler has so effectively led us? The narrator does not narrate this astrological passage in his own voice, the emphatic *ego me* notwithstanding. It is *illi*—the astrologers—who would rejoice (in the subjunctive, *gaudeant*) and who would congratulate (*gratulentur*) Rudolf for realizing these specific features of his birth chart.

The narrator then leads us back from where he had taken us, back, that is, to his own primary narrative voice: "Let them be occupied in this part of the triumph; I do not mind. I shall give them no cause for quarreling on such a festive day: let this impertinence pass as a soldiers' joke. I myself shall occupy myself with Astronomy, and, riding in the triumphal chariot, will display the remaining glory of our captive, which is known particularly to me, and every aspect of the war, as it was waged and completed" (31).[127] Having thus presented their views, he then distances himself from his hypothetical copractitioners: "Let the astrologers have that part of the triumph." In this way, Kepler skillfully moves the rhetorical structure of the narrative forward from the different ways that the captive could be praised (i.e., by historians and astrologers) to the particular way that Kepler will in fact do so. At the same time, Kepler also thickens the play on the military theme[128] by describing himself and his professional fellow soldiers as all participating in a triumph. Triumphs, which were central to Roman military pageantry in both the Republic and the Empire, were hardly only of antiquarian interest by Rudolf's time. Indeed, triumphs *à l'antique* had become central to early modern ceremonial displays of power, especially since the time of the Emperor Charles V, as Roy Strong has brought out so vividly.[129]

After this brief transitional aside, however, Kepler turns to his main point: *Ipse ad Astronomiam vertar, curruque triumphali invectus* (8,13). Kepler now portrays himself playfully as the Triumphator himself, but with the important courtier's caveat, as he emphasized before—but not now—that he waged the war under Rudolf's patronal auspices (7,8–9). He will now reveal the rest of his captive's glory, which is known especially to him. Kepler here sets up the same type of special relationship between Rudolf and his noble captive as did Galileo with Cosimo II and Jupiter's stars, that of a privileged mediator.

Indeed, Kepler then immediately invokes the *aeternus mundi huius architectus, communisque Siderum Hominumque Pater Jova* (8,16–17), much like Galileo's invocation of the *Syderum Opifex* (3,18), but we shall not follow him any further in his splendid rhetorical performance. Let us rather look in detail at what he has just done and compare it with Galileo's strikingly similar use of the very same literary device and examine also his use of a very similar structural presentation.

Kepler turns from astrology to astronomy, but not without first associating his noble captive, his scholarly gift, with the details of Rudolf's nativity, exactly as Galileo does with Cosimo II's. At the same time, however, he distances himself from this very same astrological approach by the narrative technique of having the narrator not describe the astrological situation *in propria persona*, but rather by having him present it "as the astrologers would say," using the subjunctive mood. No one should be deceived by what Kepler does here. He employs a very effective rhetorical device, *praeteritio*, to create an effect much like the one whereby the orator says that he is not going to do something, and in the process of saying that he is not going to do it actually does what he said he was not going to do: in this case, provide an astrological interpretation that would link the noble captive, Mars, to the emperor's nativity.[130] Cicero often made this sort of move in his courtroom oratory.

In addition to this rhetorical and narratological distancing, Kepler also moves away from the astrological association of his gift with Rudolf II in a structure that is exactly parallel to the way in which Galileo makes the same kind of transition. They both personalize their gifts for their patrons by means of their patrons' horoscopes; then they both turn away from this device to pursue rhetorical strategies that they both consider more powerful. Kepler turns from astrology to astronomy, which is where he will play his role in the triumph; but he also graciously permits the astrologers their place, however less exalted. Galileo, on the other hand—and here, I think, we can catch him once again in the act of literary *variatio*—makes the same structural move as Kepler, also distancing himself from the astrological argument, which Galileo characterized as a conjectural type of argument, by moving to a necessary argument, albeit of a very different stripe than Kepler's. Galileo's move, then, is similar in structure to Kepler's, not only in the movement away from astrology per se, but also in the movement away from a conjectural art (as astrology is) to a necessary or certain art (as mathematical astronomy is). But Galileo's necessary argument in the preface decidedly does not proceed from mathematico-astronomical bases, even though the content of the work overall is solidly based on observational astronomy. His argument proceeds from more circumstantial and existential considerations, specifically, that God had established him as Cosimo's tutor, on the one hand, and, on the other, that he had discovered the satellites of Jupiter very soon after Cosimo became grand duke, thus establishing the certainty—rhetorical, at any rate—of his argument. This would be Galileo's literary transformation (*variatio*) of Kepler's content and structure, if I am correct in the overall thrust of my argument, that is, that Galileo did indeed borrow and adapt this material from Kepler's *Astronomia Nova*. In fine, the similarity in content and structure of Galileo's use of an astrological device to link his astronomical gift to Cosimo II is so strikingly similar to Kepler's that it is hard to believe that this was a historical accident.

Chronological Considerations

If we now consider the chronological circumstances of the composition and publication of these two works, we will find that they definitely allow for the historical possibility of my argument. Certain circumstances push this possibility further into the realm of plausibility. The internal evidence from the dedicatory letters themselves provides further weight. Whatever the historical situation actually was, however, it is striking that both

Galileo and Kepler used extraordinarily similar astrological motifs in presenting their works to their respective absolute princes.

It is difficult to find precise information on the publication and immediate reception of *Astronomia Nova*[131] beyond the well-known fact that Kepler had still not received it back from the printer by 1 September 1609, even though it had apparently reached the Frankfurt book fair by that point.[132] Evidently there were difficulties both in the publication and the distribution, with Kepler ultimately selling the stock of the small print run to the printer himself.[133] There is, however, some interesting evidence in Kepler's letters (both to and from him) that provides very helpful, precise information concerning the earliest reception of *Astronomia Nova*. We shall find this very useful indeed for our purposes.

Our first bit of evidence overall, and our *terminus post quem*, is Kepler's letter to Thomas Harriot (September 1, 1609), where we find that the book is for sale at the Frankfurt book fair, but that Kepler does not yet have a copy. It appeared in the catalogue for the Fall book fair.[134]

In a letter of September 25, 1609, from the rector, chancellor and doctors of the University of Tübingen, Kepler's beloved alma mater, we get our first evidence for the actual reception of *Astronomia Nova*. They wrote a very friendly letter about having received the book that day in which they confirm that they immediately dispatched a remittance of the five-ducat price.[135]

Our second piece of evidence for the reception and our first evidence for an actual reader comes from a letter of December 3, 1609, from Nicholas Vicke to Kepler.[136] He found part of the *Astronomia Nova* difficult to understand, so he made a somewhat detailed query to Kepler, the details of which do not concern us.

Martin Horky's letter of January 12, 1610, provides our third piece of unambiguous[137] evidence for the reception of *Astronomia Nova*, although it appears unlikely that he actually read it. Horky wrote from Bologna, where, after much travel, he was currently staying with Giovanni Antonio Magini. While there, he was able to see a copy,[138] most likely the same copy about which Magini wrote three days later.

Our last and most important piece of evidence, for both reception and reading, comes from Magini himself, also from Bologna, dated January 15, 1610.[139] The volume he saw was brought to Bologna *from Venice* for a nobleman by a Bolognese bookseller. Magini examined it quickly, during the one day that he had access to it. He also discusses a problem he found in the *Astronomia Nova,* the details of which, once again, do not concern us. But having solid evidence that Magini, professor of mathematics at Bologna, had access to the *Astronomia Nova,* in Bologna, by January 15, 1610, certainly does concern us, as we shall see in some detail below.

Galileo's side of the historical equation, on the other hand, is much better known.[140] During the composition and publication of the *Sidereus Nuncius,* Galileo was living and working in Padua, where he had been a professor of mathematics at the university for almost twenty years (from 1592). We have good evidence that Galileo turned his new, improved telescope[141] from a military[142] to an astronomical purpose sometime in the autumn of 1609.[143] He first paid most attention to the irregular surface of our moon;[144] soon after, on January 7, 1610, with a telescope that magnified thirty times, he discovered the satellites of Jupiter: "By 15 January at the latest he had the solution . . . : Jupiter had four moons!"[145] By January 30, he had composed *Sidereus Nuncius* and gone to Venice to have it published.[146] It was only "after February 13," however, that he fashioned the satellites of Jupiter into Medicean stars. *Sidereus Nuncius* was published in early March 1610. The last observation was dated March 2.[147] The dedicatory letter was dated March 12, 1610. On the next day he sent an unbound copy to the Tuscan court with a letter. Finally, on March 19 he sent off a properly bound copy in company with the very telescope he had used to make his epoch-making discoveries.[148]

The only external evidence I know of for Galileo's having access to *Astronomia Nova* is circumstantial; it is not, however, insignificant.[149] Padua, where Galileo had lived and worked for almost twenty years, was, of course, the main university for the Venetian Republic, with the city of Venice itself one of the major centers of book production and trade.[150] Further, considering that Kepler was probably the most famous mathematician/astronomer in all of Europe at that time,[151] it would be highly probable that Galileo would have had at least some access to Kepler's work. We know that Galileo had known of Kepler's work since at least 1597, when Kepler sent him a copy of *Mysterium Cosmographicum,* to the substance of which Galileo apparently never replied.[152] Galileo undoubtedly knew about *Astronomia Nova* by late April 1610, upon receiving Kepler's reply to the *Sidereus Nuncius.*[153] One piece of possibly useful positive evidence (although by itself not very strong) comes from Galileo's letter to Giuliano de Medici, Florentine resident ambassador at Prague,[154] dated October 1, 1610. Galileo requests that Giuliano procure for him two of Kepler's works mentioned in Kepler's personal letter to Galileo, dated April 19, 1610, just before the *Dissertatio cum nuncio sidereo* was published: the *De stella nova* (1606) and the *Optica* (1604),[155] to which he stated he did not then have access.[156] We can reasonably infer that Galileo did not also ask in the same letter for the *Astronomia Nova* because he already had access to it by then (October 1610); on the basis of this evidence we cannot speculate soundly any further. Perhaps he had access to it much earlier—that is, at some time after September 1, 1609—and perhaps not. Favaro notes that Galileo did indeed possess a copy of *Astronomia Nova* at some point, but he provides no indication as to when Galileo had it, or if he annotated it.[157]

But if Magini—professor of mathematics at the University of Bologna—had access to *Astronomia Nova* by January 15, 1610, in Bologna, it seems to me even more likely that Galileo—professor of mathematics at the University of Padua—would also have had access to it at that time in Venice, especially since the Bologna copy had itself been brought there from Venice. Likewise, Galileo would almost certainly have known of *Astronomia Nova* earlier from its advertisement in the general catalogue of the fall 1609 Frankfurt book fair,[158] where it had been for sale since at least September 1, 1609. We know, further, that Galileo was keenly interested in Kepler's response to his own *Sidereus Nuncius*. In this light, it seems rather likely that Galileo, during the prior six months, would have had some rather intense interest in Kepler's most recent work, which promised nothing less than a new astronomy.

The combined weight of this admittedly circumstantial evidence makes it highly likely that Galileo both knew about and had access to *Astronomia Nova* at the time he wrote his dedicatory letter to the *Sidereus Nuncius*. Indeed, the internal evidence from the dedicatory letters themselves seems to be the strongest evidence that Galileo did have access to, and actually read, *Astronomia Nova* between September 1, 1609, and March 12, 1610. But perhaps this motif of associating an astronomical discovery, or something else for that matter, with a dedicatee's horoscope was common in early modern dedicatory letters.[159] Kepler mentions astrology in several of his pre-1609 prefaces,[160] but there is no mention whatsoever of a patron's horoscope, besides the *De stella nova* of 1606, which I discussed above. At any rate, the *De stella nova*, apparently, could not have been a model for Galileo's *Sidereus Nuncius* because as late as October 1610, as we saw, he claims that he did not yet have access to it in Padua or Venice. Tycho also discusses astrology in some of his prefaces, but again with no reference to a patron's horoscope.

Indeed, it might even be more striking to find out, especially in lieu of any examples to the contrary, that Galileo had in fact *not* had access to the *Astronomia Nova* before the composition and publication of *Sidereus Nuncius*; that somehow he and Kepler both came up with an extraordinarily similar astrological device employing distinctive structural similarities and yet in complete and utter ignorance of each other's work. This would be very striking indeed, however unlikely, and points to the need to investigate further the dedicatory letters of early modern writings, scientific and otherwise. Be the precise details of the actual historical situation as they may, we are left with the singular fact that these two epoch-making works in the history of astronomy used prominent astrological devices of essentially similar content and structure in the dedicatory letters to their princely patrons.

Acknowledgments

My thanks to Mario Biagioli and Robert Westman for their helpful answers to inquiries, to Daniel Stolzenberg for the title, and especially to Domenico Bertoloni-Meli and the editors of this volume for their careful reading and insightful criticism.

Notes

1. Inter (multa) alia, for Kepler: Alexandre Koyré, *The Astronomical Revolution: Copernicus–Kepler–Borelli*, trans. R. E. W. Maddison (New York: Dover, 1992), 119ff.; for Galileo: Galileo Galilei, *Sidereus Nuncius or the Sidereal Messenger*, trans. with introduction, conclusion, and notes by Albert van Helden (Chicago: University of Chicago Press, 1989), intro., vii–viii.

2. Mario Biagioli, *Galileo Courtier: The Practice of Science in an Age of Absolutism* (Chicago: University of Chicago Press, 1993). See also Nicholas Jardine's review essay, "A Trial of Galileo's," *Isis*, 85 (1994): 279–83.

3. Especially in chap. 1, "Galileo's Self-Fashioning," 11–101.

4. See also Richard S. Westfall, "Science and Patronage: Galileo and the Telescope," *Isis*, 76 (1985): 11–30 at 14–16.

5. Biagioli, *Galileo Courtier,* 127–33.

6. Ibid., 1–10; see also Robert S. Westman, "The Astronomer's Role in the Sixteenth Century: A Preliminary Study," *History of Science,* 18 (1980): 105–47.

7. Biagioli *Galileo Courtier,* esp. 106–12. Cf. Michael H. Shank, "Galileo's Day in Court," *Journal for the History of Astronomy,* 25 (1994): 236–43; Mario Biagioli, "Playing with the Evidence," *Early Science and Medicine,* 1 (1996): 70–105; Michael H. Shank, "How Shall We Practice History? The Case of Mario Biagioli's *Galileo Courtier,*" *Early Science and Medicine,* 1 (1996): 106–50, esp. 110–40. For another perspective, see Horst Bredekamp, *Florentiner Fußball. Die Renaissance der Spiele: Calcio als Feste der Medici* (Frankfurt: Campus, 1985).

8. Specifically, 130; at 36–54 Biagioli reconstructs the cultural context of gift exchange in early modern Europe. See also Karl Schottenloher, *Die Widmungsvorrede im Buch des 16. Jahrhunderts* (Münster: Aschendorf, 1953), 194. Westfall, "Science and Patronage," 22 and 26 speaks in these terms as well.

9. I am unaware of any such discussion in the secondary literature; cf. Galileo, *Siderius Nuncius;* Galileo Galilei, *Le Messager Celeste,* ed. and trans. Isabelle Pantin (Paris: Belles lettres, 1992); Biagioli, *Galileo Courtier.*

10. I have yet to find any prior examples; the authors I have examined so far are discussed below.

11. See Victor E. Thoren, *The Lord of Uraniborg: A Biography of Tycho Brahe* (Cambridge: Cambridge University Press, 1990), index, s. v. "astrology."

12. Indeed, a question on the astrological impact of a new star took him away from his alchemical work, as he describes in a different preface—to *De nova stella,* 1573 (*Tychonis Brahe Dani opera omnia,* ed. J. L. E. Dreyer, 15 vols. [Copenhagen: Gyldendal, 1913–29], 1:9, 27–31). For Tycho as an alchemist, see, in addition to Thoren, *The Lord of Uraniborg,* Alain P. Segonds, "Tycho Brahe et l'Alchimie," in J.-C. Margolin and Sylvain Matton, eds., *Alchimie et Philosophie à la Renaissance* (Paris: J. Vrin, 1993), 365–78.

13. Galileo Literator is well known; see, e.g., Leonardo Olschki, "Galileo's Literary Formation," in E. McMullin, ed., *Galileo: Man of Science* (New York: Basic Books, 1967), 140–59. For Kepler, see Anthony Grafton, "Humanism and Science in Rudolphine Prague: Kepler in Context," in his *Defenders of the Text: The Traditions of Scholarship in an Age of Science, 1450–1800* (Cambridge: Harvard University Press, 1991), 178–203.

14. I should note that I will not discuss in this chapter the important topic of Kepler's and Galileo's actual views about, and practice of, astrology. For Kepler see J. V. Field, "A Lutheran Astrologer: Johannes Kepler," *Archive for History of Exact Sciences,* 31 (1984): 189–272; and G. Simon, *Kepler Astronome Astrologue* (Paris: Gallimard, 1979). The situation with Galileo is less well known; see now my "Galileo, Astrologer: Astrology and Mathematical Practice in the Late-Sixteenth and Early-Seventeenth Centuries," Galilaeana 2 (2005): 107–43.

15. Considering the importance of this literary genre for both literary and historical concerns, it seems surprisingly understudied. I found Schottenloher, *Die Widmungsvorrede,* and Robert S. Westman, "Proof, Poetics, and Patronage: Copernicus's Preface to *De revolutionibus,*" in D. C. Lindberg and R. S. Westman,

eds., *Reappraisals of the Scientific Revolution* (Cambridge: Cambridge University Press, 1990), 167–205, the most useful. See also Kevin Dunn, *Pretexts of Authority: The Rhetoric of Authorship in the Renaissance Preface* (Stanford: Stanford University Press, 1994); Jean Dietz Moss, *Novelties in the Heavens: Rhetoric and Science in the Copernican Controversy* (Chicago: University of Chicago Press, 1993); and Biagioli, *Galileo Courtier,* 128–32. Furthermore, with respect to the study of prefatory matter in scientific works by historians of science, even Asger Aaboe admits that much may be learned from a nuanced appreciation thereof; see his comments in R. Palter, ed., *The* Annus Mirabilis *of Sir Isaac Newton, 1666–1966* (Cambridge: MIT Press, 1970), 86–87.

16. See Schottenloher, *Die Widmungsvorrede,* 2 and 4.

17. Schottenloher's first chapter in *Die Widmungsvorrede,* "Die Widmungsvorrede als Humanistisch-Literarisch Erscheinung," 1–4, provides basic information on the structure of dedicatory letters, based on his extensive acquaintance with the printed book of the sixteenth century. His emphatic introductory statement is well worth heeding:

Wer sich viel mit dem Schrifttum des 16. Jahrhunderts beschäftigt, begegnet auf Schritt und Tritt der Widmungsvorrede, die in den meisten Schriften der eigentlichen Veröffentlichung wie ein Herold vorausgeht und diese in Gestalt eines erläuternden Briefes ankündigt, darüber aber hinaus mit allen möglichen Mitteilungen ihre eigene Wege geht und damit eine gewisse Selbstständigkeit erlangt. Sie kehrt so häufig wieder und hat so ausgeprägte Formen, daß sie als ständige Zugabe des damaligen Buches, ja mit ihrem häufig recht bedeutsamen Inhalt als selbstständige literarische Erscheinung des 16. Jahrhunderts mit eigenen Lebensgesetzen bezeichnet werden kann und damit unsere volle Aufmerksamkeit verdient. (1)

18. See Schottenloher, *Die Widmungsvorrede,* 194.

19. For a stimulating, sophisticated discussion of the dialogue in the Renaissance, see Virginia Cox, *The Renaissance Dialogue: Literary Dialogue in Its Social and Political Contexts, Castiglione to Galileo* (Cambridge: Cambridge University Press, 1992).

20. Schottenloher, *Die Widmungsvorrede,* 2; see also Paul Grendler, *Schooling in Renaissance Italy: Literacy and Learning, 1300–1600* (Baltimore: Johns Hopkins University Press, 1989), esp. chap. 8, "Rhetoric," 203–34. I subscribe to Brian Vickers's further refinement, which applies perfectly as well to dedicatory letters (Brian Vickers, "Epideictic Rhetoric in Galileo's *Dialogo*," *Annali dell'Istituto e Museo di Storia della Scienza di Firenze,* 8 [1983]: 69–102 at 70): "'Form' is not quite enough, however, as an analytic concept. We must also distinguish 'mode.' The form or genre is obviously the dialogue [epistle] . . . but within this form the mode is that of epideictic rhetoric, the techniques of praise and blame."

21. Schottenloher, *Die Widmungsvorrede,* 2: "Sie (sc. die Widmungsvorrede) kann die eigentliche Veröffentlichung erläutern und begründen"; and it can show "seine Stellungnahme zu wissenschaftlichen Fragen." This applies closely to Copernicus's dedicatory letter to *De rev.*

22. Schottenloher does not discuss this beyond referring in general terms to the possibilities of the genre. Epideictic as a rhetorical/literary genre is much more

characteristic of the Roman Empire, especially the Second Sophistic (2nd century C.E.), than of the Roman Republic, where forensic and deliberative oratory held the field; for our period, see Vickers, "Epideictic Rhetoric," 69–102, esp. 71–77, where he provides a useful history and a rich bibliography.

23. Of the dedicatory letters Kepler wrote before that for *Astronomia Nova* (1609), those for the *Ad Vitellionem paralipomena* (1604) and the *De stella nova* (1606) fit this description most closely. The dedicatory letter to *De fundamentis astrologiae certioribus* (1602) fits somewhat less well, in that it is a defense, after a fashion, of what Kepler intends to do in the treatise, but the narrative structure is not as straightforward. That for *Mysterium Cosmographicum* (1596), on the other hand, is more a small, elaborate essay than a defense speech—yet another "mode" perhaps. There is nothing straightforward, however, in Kepler's highly elaborate prose style in the dedicatory letters.

24. E.g., *Myst. Cosm.* 8,16; *Ad Vit.* par. 7,35; *De stella nova*, 151,11.

25. Nor that by Tycho.

26. See e.g., Westfall, "Science and Patronage"; Westman, "The Astronomer's Role" and "Proof, Poetics, and Patronage"; and Biagioli, *Galileo Courtier*. Schottenloher, *Die Widmungsvorrede,* also has much of interest; see 175, 195 and chap. 5, "Widmungsempfänger," and 6, "Widmungen und Mäzenatentum" (177–194, passim).

27. See Schottenloher, *Die Widmungsvorrede,* 3, 194–96.

28. Roy Strong, *Splendor at Court: Renaissance Spectacle and the Theater of Power* (Boston: Houghton Mifflin, 1973). Schottenloher develops this public side of the dedicatory letters at some length in *Die Widmungsvorrede*, 1–3.

29. For stimulating comment, see Westman, "Proof, Poetics, and Patronage."

30. See Westfall, "Science and Patronage," 15–16. Cosimo became grand duke in February 1609.

31. I use Pantin's Latin text (Galileo, *Le Messager Celeste*) and her very useful notes. Also helpful is Van Helden's English translation (Galileo, *Siderius Nuncius*) with introductory and supplementary essays.

32. I have found it useful to analyze these dedicatory letters as a species of narrative. For a brief exposition of narrative theory with further bibliography, see John J. Winkler, *Auctor and Actor: A Narratological Reading of Apuleius's* Golden Ass (Berkeley and Los Angeles: University of California Press, 1985).

33. "Serenissimo/Cosmo Medices II./Magno Haetruriae/Duci IIII." (2,1–4).

34. "D. Rudolpho II./Romanorum Imperatori Semper Augusto./Germaniae, Hungariae, Bohemiae &c. Regi./Archiduci Austriae &c." (7,1–4). References to Kepler's Latin texts are to *Johannes Kepler Gesammelte Werke* (Munich: C. H. Beck, 1937–). *Astronomia Nova* is in vol. 3.

35. "Augustissime Imperator."

36. "S^{ae}. C^{ae}. M^{tis}. V^{ae}. (Sacrae Caesareae Maiestatis Vestrae)" (7,6); "M^{tis}. V^{ae}." (7,7); "M^{tis}. V^{ae}." (7,9).

37. I follow Van Helden's translation for the most part, sometimes with noted, sometimes with silent changes.

38. "Praeclarum sane, atque humanitatis plenum eorum fuit institutum, qui excellentium virtute virorum res praeclare gestas ab invidia tutari, eorumque immortalitate digna nomina ab oblivione atque interitu vindicare conati sunt" (2,5–8).

39. With one, very minor exception, as we will see below.

40. See Westman, "Proof, Poetics, and Patronage."

41. This less didactic tone seems appropriate, too, given the historical actors' relative ages and statures. The situations are very different.

42. This is probably meant to refer to obelisks, as often in the sixteenth century.

43. This is a slight overtranslation of the Latin *ut inquit ille;* I discuss this poetic allusion below.

44. "Hinc ad memoriam posteritatis proditae Imagines, vel marmore insculptae, vel ex aere fictae; hinc positae Statuae tam pedestres, quam equestres; hinc Columnarum, atque Pyramidum, ut inquit ille, sumptus ad Sydera ducti; hinc denique urbes aedificatae, eorumque insignitae nominibus, quos grata posteritas aeternitati commendandos existimavit" (2,8–13).

45. "Eiusmodi est enim humanae mentis conditio, ut nisi assiduis rerum simulacris in eam extrinsecus irrumpentibus pulsetur, omnis ex illa recordatio facile effluat" (2,13–15).

46. "Verum alii firmiora, ac diuturniora spectantes, aeternum summorum virorum praeconium non saxis, ac metallis, sed Musarum custodiae, et incorruptis litterarum monumentis consecrarunt" (2,16–18).

47. It should be noted that *signa* can also refer to physical memorials.

48. In a fine chiastic structure.

49. "Quasi vero humana solertia his contenta regionibus, ulterius progredi non sit ausa; attamen longius illa prospiciens, cum optime intelligeret omnia humana monumenta vi, tempestate, ac vetustate tandem interire, incorruptiora signa excogitavit, in qua Tempus edax atque invidiosa Vetustas nullum sibi ius vindicaret" (2,19–24; the first set of brackets in the translation is Van Helden's).

50. *ut inquit ille* is all Galileo wrote.

51. Neither Pantin nor Van Helden discusses the significance of the quotation nor its placement at this point in the narrative. Pantin simply gives a reference in her note *ad loc.* to a passage of Propertius: "Properce, III,2,17" (51, n. 11). Van Helden gives a slightly fuller reference and then quotes a relevant passage (lines 19–26) in translation (29, n. 6).

52. This would require checking contemporary editions for the allusions and references therein, etc., as Westman ("Proof, Poetics, and Patronage," 182ff.) did so effectively in discussing Copernicus's unacknowledged use of Horace's *Ars Poetica* in the dedicatory letter to *De rev.*

53. The textual references are to *Sexti Properti Elegiarum Libri IV,* ed. Paulus Fedeli (Stuttgart: B. G. Teubner, 1984). The translation, with minor modifications, comes from Propertius, *The Poems,* trans. Guy Lee (Oxford: Clarendon Press, 1994), 73–74.

54. "Neither the expense of Pyramids raised to the stars/"

55. "Though my house is not supported on Taenarian columns/"

56. "Still the Muses befriend me, my songs are dear to readers / And Calliope unwearied by my dances. / Lucky you, the girl who is celebrated in my book; / Each song will be a monument to your beauty."

57. "Or flame or rain will dispossess their honour, or / They'll fall by thrust of years and their own weight."

58. "II, On the power of poetry to immortalize its subjects and its author. . . . Lines 19ff. echo Hor. *Od.* III, XXX, 1ff." From the commentary to Propertius, *Elegies,* book 3, ed. W. A. Camps (Cambridge: Cambridge University Press, 1966), 59.

59. *Q. Horati Flacci Opera,* ed. Fridericus Klingner (Leipzig: B. G. Teubner, 1959). The translation is from *The Odes of Horace,* trans. Lord Dunsany (London: Heinemann, 1947), 132.

60. In a first asclepiadic meter.

61. Compare this with the line from the first movement of the material model: *"Hinc ad memoriam posteritatis proditae Imagines, vel marmore insculptae, vel ex aere fictae"* (2,8–9).

62. "A monument more durable than bronze, / Rising above the regal pyramids, / have I erected, which no rain nor wind, / Nor centuries unnumbered, could destroy, / Nor all the flights of seasons."

63. "In Coelum itaque migrans, clarissimorum Syderum notis, sempiternis illis orbibus eorum nomina consignavit, qui ob egregia, ac prope divina facinora digni habiti sunt, qui una cum Astris aevo sempiterno fruerentur" (2,24–27).

64. 2,6–7.

65. "Quam ob rem non prius Iovis, Martis, Mercurii, Herculis, caeterorumque heroum, quorum nominibus Stellae appellantur, fama obscurabitur, quam ipsorum Syderum splendor extinguatur" (2,27–29).

66. For a rich historical discussion of euhemerism, see Jean Seznec, *The Survival of the Pagan Gods: The Mythological Tradition and Its Place in Renaissance Humanism and Art,* trans. B. F. Sessions (New York: Harper and Brothers, 1961).

67. "Hoc autem humanae sagacitatis inventum cum primis nobile, ac mirandum multorum iam saeculorum intervallo exolevit, priscis heroibus lucidas illas sedes occupantibus, ac suo quasi iure tenentibus: in quorum coetum frustra pietas Augusti Iulium Caesarem coaptare conata est: nam cum Stellam suo tempore exortam, ex iis, quas Graeci Cometas, nostri Crinitas vocant, Iulium Sydus nuncupari voluisset, brevi illa evanescens, tantae cupiditatis spem delusit" (2,30–3,5; Van Helden's parentheses in translation).

68. Tycho Brahe, *Opera omnia,* v. V, 5–10.

69. "Ad Augustissimum Imperato-/rem Rudolphum Secundum/Tychonis Brahe/Praefatio" (5,1–4). *Ad* + accusative is equivalent in construction to using the dative case for an address; this is Tycho's substantively insignificant *variatio.* Stylistically, however, it might have raised a few eyebrows.

70. *inquam* at 5,8 and *invenio* at 6,1.

71. 5,19–22: "Hinc sunt illae columnae, quas Iosephus Iudaicarum rerum scriptor refert, Adae Nepotes in Syria extruxisse, iisque sua inventa memoriae causa ad Posteros inscripsisse. Huc pertinent Aegyptiorum et aliarum gentium altissimae et sumptuosissimae Pyramides[.]"

72. "Tuaeque insuper Caes. Majestatis memoria et fama, quod haec tam excellentia, et in rebus Mundanis ferme praecipua, conservare, tueri, et promovere non degravetur, ad omnem Posteritatem, quoad Sol et Sidera durabunt, eo illustrior fulgeat et perduret" (9,30–34; my translation).

73. Antonio Favaro, "La Libreria di Galileo Galilei," *Bullettino di Bibliografia e Storia delle Scienze Matematiche e Fisiche,* 19 (1886): 219–93, notes that Galileo had a copy at some point, but with no further details (no. 168).

74. "Atqui longe veriora ac feliciora, Princeps Serenissime, Celsitudini tuae possumus augurari; nam vix dum in terris immortalia animi tui decora fulgere coeperunt, cum in Coelis lucida Sydera sese offerunt, quae tanquam linguae praestantissimas virtutes tuas in omne tempus loquantur ac celebrent" (3,5–10).

75. We will recall that the preface overall is motivated in the opening sentence by a discussion of the kinds of memorials for the deeds of men excellent in *virtus*: "*excellentium virtute virorum*" (2,6).

76. Biagioli, *Galileo Courtier,* 128.

77. "En igitur quattuor Sydera tuo inclyto nomine reservata, neque illa de gregario, ac minus insigni inerrantium numero, sed ex illustri vagantium ordine, quae quidem disparibus inter se motibus circum Iovis Stellam caeterarum nobilissimam, tanquam germana eius progenies, cursus suos, orbesque conficiunt celeritate mirabili interea dum unanimi concordia circa mundi centrum, circa Solem nempe ipsum, omnia simul duodecimo quoque anno magnas convolutiones absolvunt" (3,10–17).

78. "Ut autem inclito Celsitudinis tuae nomini prae caeteris novos hosce Planetas destinarem, ipsemet Syderum Opifex perspicuis argumentis me admonere visus est" (3,17–19).

79. Galileo uses a similar manner of speaking in his letters of January 30, 1610, and February 13, 1610, (quoted in Galileo, *Sidereus Nuncius,* 17–18). He also refers to his power as discoverer to name them. There are quite a few thematic similarities with the dedicatory letter, especially in the letter of February 13, but decidedly not with respect to the astrological device.

80. Biagioli, *Galileo Courtier,* 129.

81. "Etenim quemadmodum hae stellae tamquam Iove digna proles nunquam ab illius latere, nisi exiguo intervallo discedunt; ita quis ignorat clementiam, animi mansuetudinem, morum suavitatem, regii sanguinis splendorem in actionibus maiestatem, authoritatis, et Imperii in alios amplitudinem, quae quidem omnia in tua Celsitudine sibi domicilium ac sedem collocarunt, quis inquam ignorat haec omnia ex benignissimo Iovis Astro, secundum Deum omnium bonorum fontem, emanare?" (3,19–26).

82. "Nunc ad demonstrativum genus causae transeamus. Quoniam haec causa dividitur in laudem et vituperationem. . . ." *Rhetorica ad Herennium,* trans. (with

copious annotations) Harry Caplan (Cambridge: Harvard University Press [The Loeb Classical Library], 1954), III,vi,10 (172). The author discusses epideictic at 172–85.

83. *Jupiter inquam* here picks up the *quis inquam ignorat* of two lines before, tying this passage to the previous rhetorical question (and especially its second movement) and at the same time moving the stream of thought into the next emphatic, highly charged passage.

84. "Iuppiter, Iuppiter inquam, a primo Celsitudinis tuae ortu turbidos Horizontis vapores iam transgressus mediumque coeli cardinem occupans, Orientalemque angulum sua Regia illustrans, foelicissimum partum ex sublimi illo trono prospexit, omnemque splendorem, atque amplitudinem suam in purissimum aerem profudit, ut universam illam vim, ac potestatem tenerum corpusculum una cum animo nobilioribus ornamentis iam a Deo decorato, primo spiritu hauriret" (3,26–33).

85. G. Righini, "L'oroscopo Galileiano di Cosimo II de'Medici," *Annali* (1976): 29–36. Pantin corrects his account on several points of detail (Galileo, *Le Messager Celeste*, n. 22, 53–54). I discuss this further below.

86. See note 7.

87. "Verum quid ego probabilibus utor argumentationibus, cum id necessaria propemodum ratione ac demonstrare queam?" (3,33–35).

88. *Tetrabiblos* I,1. This is an important—and understudied—issue that requires further treatment.

89. See Paolo Mancosu, *Philosophy of Mathematics and Mathematical Practice in the Seventeenth Century* (New York: Oxford University Press, 1996), esp. 10–33.

90. 3,35ff.

91. I use Donahue's translation, sometimes with minor, sometimes with major modifications. The Latin text is from Kepler, *Gesammelte Werke*, vol. 3: "AUGUSTISSIME IMPERATOR, Quod S$^{ae.}$ C$^{ae.}$ M$^{tis.}$ V$^{ae.}$ [Sacrae Caesareae Maiestatis Vestrae], totiusque adeo Domus Austriacae serenissimo Nomini foelix faustumque sit, imperiis M$^{tis.}$ V$^{ae.}$ tandem aliquando publice spectandum exhibeo Captivum Nobilissimum, jam pridem auspiciis M$^{tis.}$ V$^{ae.}$ bello difficili et laborioso a me acquisitum" (7,5–9).

92. I present them here in what seems a more natural order of exposition. The numbering refers to the actual order in the sentence: (2) *publice spectandum* (3) *exhibeo* (4) *Captivum Nobilissimum*.

93. It will turn out to be Kepler's "gift" to Rudolf once it has been fully rhetorically conditioned and personalized.

94. I should note that Kepler also emphasized Rudolf and his house's name in the first sentence.

95. "Huius vero spectaculi non major poterit esse celebritas, quam si panegyricum captivo praestantissimo scribam publicaque voce pronunciem" (7,19–20).

96. Donahue rather overtranslates *publica voce pronunciem* as "shout it out loudly and publicly."

97. *spectaculum* here picks up *publice spectandum* in the first sentence.

98. On panegyrics, see F. J. Stopp, *The Emblems of the Altdorf Academy: Medals and Medal Orations, 1577–1626* (London: Modern Humanities Research Association, 1974). For an earlier period, see also Sabine MacCormack, *Art and Ceremony in Late Antiquity* (Berkeley and Los Angeles: University of California Press, 1981).

99. "Itaque relinquo scriptoribus historiarum explicandam Hospitis nostri magnitudinem, re bellica comparatam" (7,19–20).

100. "Dicant illi sane, hunc esse, per quem omnes exercitus vincant, omnes belli duces triumphent, omnes Reges imperent; sine cuius ope nemo unquam quenquam captivum cum laude abduxerit. Hunc jam meo Marte captum spectando, suos illi oculos exsatient.

"Dicant Romanae magnitudinis admiratores, hunc esse satorem Regum Romuli et Remi, conservatorem Urbis, protectorem Quiritium, Statorem Imperii: quo propitio Romani militarem disciplinam invenerint, auxerint, perfecerint, Orbemque Terrarum subjugaverint. Hunc igitur circumscriptum, Domuique Austriacae foelici omine nunc acquisitum gratulentur" (7,21–30; Donahue's capitalizations, which follow the typography of Caspar's text).

101. For Rudolf II overall, see R. J. W. Evans, *Rudolf II and His World: A Study in Intellectual History, 1576–1612,* corrected ed. (Oxford: Clarendon Press, 1984).

102. See Barbara Bauer, "Die Rolle des Hofastrologen und Hofmathematicus als fürstlicher Berater," in A. Buck, ed., *Höfischer Humanismus* (Wienheim: VCH, 1989), 93–117 at 105ff.; Evans, *Rudolf II and His World,* index s. v. "Turks" and "Matthias, H.R.E. 1612–1619" and Max Caspar, *Kepler,* trans. and ed. C. Doris Hellman, with new introduction and references by Owen Gingerich (New York: Dover, 1993), 186–89, 203–4.

103. See esp. Bauer, "Die Rolle des Hofastrologen," particularly 102ff.

104. Caspar, *Kepler,* 203.

105. Caspar, *Kepler,* 186–88, 204ff. This could be another reason why Kepler was so full of praise for Galileo in the *Dissertatio cum nuncio sidereo* (1610), where Kepler seems to have been seeking Giuliano de Medici's patronage. Notwithstanding the difficult political situation at the time, it should also be noted that there is an important difference between Kepler's and Galileo's overall patronage strategies: Galileo was trying to move up in the world by his particular patron-grabbing strategy; Kepler, on the other hand, already had his top position as imperial mathematician, even though this position was becoming increasingly precarious.

106. Evans, *Rudolf II and His World,* esp. chaps. 6 and 7: "Rudolf and the Occult Arts" (196–242), and "Prague Mannerism and the Magic Universe" (243–74); see also Bauer, "Die Rolle des Hofastrologen," 102ff.

107. See W. F. Ryan and Charles B. Schmitt, eds., *Pseudo-Aristotle, The* Secret of Secrets: *Sources and Influences* (London: Warburg Institute, 1982), with its extensive bibliography.

108. For a penetrating evaluation of Kepler as a humanist, see Grafton, "Humanism and Science," 178–203.

109. "Ego me hinc ad alia recipio, quae sunt viribus meis accommodatiora. Neque tamen in ea professionis meae parte pedem figam, in qua mihi simultas intercedit cum commilitonibus" (7,31–33).

110. The narrator began his presentation of the historians thus: *Dicant illi sane,* etc. He now begins to present the astrologers thus: *Illi sane gaudium aliud licet gaudeant.*

111. "Illi sane gaudium aliud licet gaudeant: constrictum vinculis Calculi, qui toties ipsorum manus et oculos effugiens, irrita solitus est reddere vaticinia maximi momenti: quippe de Bello, de Victoria, de Imperio, de Dignitate militari, de Magisterio, de Lusu, de ipsa denique Vita abscindenda vel proroganda" (7, 34–8,3).

112. The translator acknowledges *ad loc.* (n. 1) that he could not find information on the *"astragalis lusum trigonicum."*

113. "Illi M$^{ti.}$ V$^{ae.}$ gratulentur de Domino Geniturae in potestatem redacto, imo vero conciliato; quippe illis testibus Mars Scorpioni dominatur, qui cor Coeli habet; in Capricorno exaltatur, qui oritur; in Cancro, in quem Luna ingressus est, ludere solet astragalis lusum trigonicum; in Leone, quo Sol utitur hospitio, familiariter notus est; Ille denique et Arietis est dominus, cui subesse creditur Germania, planeque concurrens cum S$^{a.}$ C$^{a.}$ M$^{te.}$ V$^{a.}$ habet imperium" (8,3–10).

114. Actually the first time directly: Kepler mentions Mars furtively at 7,24 when he notes that the captive was *meo Marte captum.*

115. As I discuss more fully below.

116. Kepler also discusses Rudolf's nativity in the dedicatory letter, also to Rudolf, of *De stella nova* (1606). Although the basic idea is similar—associating something with Rudolf's nativity—it is executed rather differently. The treatment in *De stella nova* is in the most general terms, whereas, in *Astronomia Nova,* Kepler treats Rudolf's horoscope in some detail: "Nullum umquam coeleste Thema Genethliacum tam pulchre adumbrare creditum est cujusquam hominis fortunam, quam ad S. C. Majest. V$^{ae.}$ gravissimarum occupationum successus, studiorum Astronomicorum cursus et fortuna sese accommodavit hactenus" (I,152,3–6). Kepler then proceeds to correlate the phenomena in the heavens with the events of Rudolf's public life, but without discussing the details of his nativity. This preface is quite interesting in itself and worthy of further study.

117. "Illi M$^{ti.}$ V$^{ae.}$ gratulentur de Domino Geniturae in potestatem redacto, imo vero conciliato" (8,3–4).

118. See J. C. Eade, *The Forgotten Sky: A Guide to Astrology in English Literature* (Oxford: Clarendon Press, 1984). For a simplified method of arriving at the "lord of the geniture," see 88–89; for the details of how to calculate dignities and debilities, see 59–88.

119. "Mars Scorpioni dominatur, qui cor Coeli habet" (8,5).

120. "in Capricorno exaltatur, qui oritur" (8,5–6).

121. "in Cancro, in quem Luna ingressa est, ludere solet astragalis lusum trigonicum" (8,6–7). The *lusum trigonicum* refers to the trine relationship. Kepler seems to be stretching here: He is not saying that Mars in Rudolf's chart is actually

in Scorpio and thus trines Rudolf's Moon in Cancer; rather, that Rudolf's Moon, by virtue of being in Cancer, trines Rudolf's midheaven in Scorpio, and since Mars rules Scorpio, Mars gets some extra points thereby. This appears to be a rhetorical bending of the rules.

122. Each of the planets, except the Sun and Moon, rule two signs each: Mars rules Aries and Scorpio.

123. "in Leone, quo Sol utitur hospitio, familiariter notus est" (8,7–8). Kepler is stretching here again, trying to associate Mars with the other of the two luminaries, but here he stretches even further, because he is not even talking (apparently) about the placement of the Sun in Rudolf's horoscope, but rather about the rulership of the Sun in general; nor is he talking about either the placement of Mars or the location of Aries in Rudolf's horoscope, but just about the fact that Mars rules Aries in general. Mars apparently gets these extra points in Rudolf's chart as a fringe benefit of his having Scorpio in the midheaven, which Mars co-rules. We should probably refer this rather creative "dignity" accounting to Kepler's rhetorical license. No wonder he distances himself from the practice of such astrologers.

124. "Ille denique et Arietis est dominus, cui subesse creditur Germania, planeque concurrens cum S$^{a.}$ C$^{a.}$ M$^{te.}$ V$^{a.}$ habet imperium" (8,8–10). The study of astrological rulership of geographical regions goes back to antiquity. Ptolemy treats this topic in book 2 of the *Tetrabiblos*. Franz Boll discusses this in some depth; "Studien über Claudius Ptolemäus. Ein Beitrag zur Geschichte der Griechischen Philosophie und Astrologie," *Jahrbuch für Klassische Philologie,* Supplementband 21: 51–243.

125. At least this was the case in the second horoscope that Galileo cast. One wonders what he found the first time.

126. "Juppiter, Juppiter inquam, a primo Celsitudinis tuae ortu turbidos Horizontis vapores iam transgressus mediumque coeli cardinem occupans, Orientalemque angulum sua Regia illustrans, foelicissimum partum ex sublimi illo trono prospexit, omnemque splendorem, atque amplitudinem suam in purissimum aerem profudit, ut universam illam vim, ac potestatem tenerum corpusculum una cum animo nobilioribus ornamentis iam a Deo decorato, primo spiritu hauriret" (3,26–33). One of the compromises with astrological doctrine in the Middle Ages was that astrological influences could affect the body but not the soul; for a recent orientation, see Laura A. Smoller, *History, Prophecy, and the Stars: The Christian Astrology of Pierre d'Ailly, 1350–1420* (Princeton: Princeton University Press, 1994), 29–32.

127. "Hanc igitur triumphi partem illi licet occupent; nullam ipsis tam festo die rixandi causam exhibebo: transeat haec licentia inter jocos militares. Ipse ad Astronomiam vertar, curruque triumphali invectus, reliquam captivi nostri gloriam, mihi peculiariter notam, omnesque adeo belli gesti confectique rationes explicabo" (8,11–15).

128. The entire preface can be considered a *iocus militaris,* which is indeed how Kepler himself describes it in the preface to his *Dissertatio cum nuncio sidereo* (IV, 286, 19–23), esp. 21–22: "Lusus enim seu iocus militaris, quo sum usus in opere illo publico (sc. *Astronomia Nova*)[.]"

129. Strong, *Splendor at Court*, 25–37, with many illustrations. Furthermore, it is worth noting the dynamic nature of Kepler's preface, with all its movement and celebration, which is in stark contrast to Galileo's much more static preface. Indeed, Kepler's preface seems to be almost a literary rendering of the magnificent illustrations of recent imperial triumphal processions, but with Kepler as Triumphator instead of Rudolf.

130. Perhaps Kepler distanced himself also because astrologically speaking, as he would have well known, the case for Mars really being the lord of Rudolf's geniture is questionable at best, as our analysis indicated.

131. For the publication history, see Friedrich Seck, "Johannes Kepler und der Buchdruck: Zur äußeren Entstehungsgeschichte seiner Werke," *Archiv für Geschichte des Buchwesens*, 11 (1970): 610–728, esp. 643–48; and Caspar, *Kepler*, 139–42, 177, 187 and 194, which is essentially a minor expansion of his treatment in *Bibliographia Kepleriana*. Wilbur Applebaum, "Keplerian Astronomy after Kepler: Researches and Problems," *History of Science*, 34 (1996): 451–504 at 456ff., is quite helpful on the reception in general, although not for our particular questions; see also Massimo Bucciantini, "Dopo il *Sidereus Nuncius*: Il Copernicanesimo in Italia tra Galileo e Keplero," *Nuncius*, 9 (1994): 15–35, also for a slightly later period of the reception, i.e., after the publication of *Sidereus Nuncius*.

132. In Kepler's letter (no. 536) to Thomas Harriot, dated September 1, 1609 (in *Gesammelte Werke*, 16:251, ll. 49–51): "Quaeris de studiis meis. Commentaria de Marte titulo Astronomiae novae *aitiologetou*, seu Physicae coelestis, prostant jam Francofurti. Exempla nondum habeo."

133. This last information comes from Kepler's letter to Magini of February 1, 1610 (letter 551, in *Gesammelte werke*, 16:279, ll. 33–35): "At quia strenue me esurire patitur, coactus sum vendere typographo, sine exceptione. Pro tribus tamen florenis hic Pragae habere possum unum." Caspar tells of the publication problems, especially Tengnagel's obstructions and Rudolf's inconsistent financing, in *Kepler*, 139–42.

134. Seck, "Johannes Kepler und der Buchdruck," 645: "Am 1. September hat Kepler seine Exemplare noch nicht bekommen; das Buch ist aber zu dieser Zeit schon in Frankfurt käuflich und erscheint im Kataloge der Herbstmesse. Es war demnach im Juli oder August 1609 ausgedruckt."

135. Letter 540 (in *Gesammelte werke*, 16:254–55): "Es hat unns diser tagen der Ehrwürdig unnd hochgelert Herr Matthias Hafenreffer, der heiligen Schrifft Doctor unnd Professor, unnßer freündtlicher lieber Collega, deß Herrn in offnen Truckh außgeverttigten Commentarium de motibus Stellae Martis praesentirt unnd angezeigt, daß wir selbigen von deß Herrn wegen in gemeiner Universitet Bibliothec zu guttem angedenckhen verwahren und offhallten sollen.

Wann wir dann hierauß deß Herrn gegen unns habende gutthertzige affection im Werckh verspüren: Allß thun wir unns solches verehrten Commentarii ganz freündtlich bedanckhen, unnd demselben hingegen eingeschloßne fünff Ducaten verwahrlich ubersenden" (ll. 3–12).

136. Letter 542 (in *Gesammelte werke*, 16:256–59): "Incidi in doctissimum tuum librum de motu Martis, ex cuius lectione tanta affectus sum voluptate, ut vix

supra; sed cum quaedam intellectu mihi difficilia occurrerint, te Virum doctissimum amicum meum honorandum consulendum duxi" (ll. 8–11).

137. I say unambiguous because letter 545, written by Kepler to an anonymous recipient at an indeterminate time (but we may assume after September 1, 1609), refers to *Astronomia mea* (1) and immediately begins discussing a problem that the current recipient of Kepler's letter had apparently raised originally.

138. Letter 547 (in *Gesammelte werke*, 16:267–70): "Vestrae Excellentiae opus insigne de motu Martis oculis meis maxime placet, sed loculis displicet: Nimis enim care venit in Italiis, et pro uno exemplari librarii nostri 6 aureos demandant. Reversus *sun theoi* ad patres lares credo me precio viliori adsecuturum." Apparently he thinks he can get a better price back home.

139. Letter 548 (in *Gesammelte werke*, 16:270–74): "Vidi nuper insigne tuum opus de motu Martis a librario quodam Bononiensi huc pro nobili viro Venetia allatum, et mutuo quidem mihi ad unicam diem concessum percurri breviter, quantum scilicet per angustiam temporis mihi concessum fuit" (ll. 2–5).

140. This paragraph depends heavily on the treatments in the introduction to Galileo, *Siderius Nuncius,* and on Westfall, "Science and Patronage."

141. Galileo, *Siderius Nuncius,* 6.

142. Ibid., 7 and esp. 9.

143. Westfall, "Science and Patronage," 18, n. 23: "probably near the end of November."

144. Galileo, *Siderius Nuncius,* 9: "Between 30 November and 18 December he observed and drew our satellite as it went through its phases, leaving no fewer than eight drawings."

145. Van Helden tells this story on ibid., 15ff. This quotation is from 16.

146. Westfall, "Science and Patronage," 19.

147. Galileo, *Siderius nuncius,* 19.

148. Westfall, "Science and Patronage," 21.

149. Galileo's letters during this period cast no light whatsoever on our central questions; there are no references at all to Kepler, to *Astronomia Nova,* to Cosimo's horoscope, or to the Frankfurt Book Fair in the correspondence (to or from Galileo) between September 1, 1609 (letter 235 is the earliest, September 4), and March 12, 1610 (letter 268); A. Favaro, ed., *Edizione Nazionale delle Opere di Galileo Galilei,* 20 vols (Florence: G. Barbara, 1890–1909), 10:256–87.

150. See, e.g., Martin Lowry, *Power, Print and Profit: Nicholas Jenson and the Rise of Venetian Publishing in Renaissance Europe* (Oxford: B. Blackwell, 1991) and *The World of Aldus Manutius: Business and Scholarship in Renaissance Venice* (Ithaca: Cornell University Press, 1979); Brian Richardson, *Print Culture in Renaissance Italy: The Editor and the Vernacular Text, 1470–1600* (Cambridge: Cambridge University Press, 1994); and Paul F. Grendler, *The Roman Inquisition and the Venetian Press, 1540–1605* (Princeton: Princeton University Press, 1977).

151. See Applebaum, "Keplerian Astronomy," 455: "By virtue of his title of court astronomer to the Holy Roman Emperor and successor to Tycho Brahe, he was regarded as the leading astronomer in Europe." See also 455 for some general notes

on the relation of Galileo and Kepler, none of which are directly relevant to our concerns.

152. Caspar, *Kepler*, 69–70; and for much more detail on the reception of the *Mysterium Cosmographicum*, see James R. Voelkel, "The Development and Reception of Kepler's Physical Astronomy" (Ph.D. diss., Indiana University, 1994).

153. Kepler presented Giuliano de Medici with his epistolary response to Galileo on April 19, 1610; it was then delivered to Galileo. On May 3, Kepler published it, with additional prefatory matter, including a dedication to Giuliano himself, for the benefit of a broader readership; see *Kepler's Conversation with Galileo's Sidereal Messenger*, trans. E. Rosen (New York: Johnson Reprint Corp., 1965), xiv.

154. This same Giuliano had mediated Galileo's earlier letter to Kepler requesting his response to *Sidereus Nuncius;* see ibid.

155. Galileo's reference is to the *Ad Vitellionem paralipomena, quibus astronomiae pars optica traditur* (1604) and not to the *Dioptrice* of 1601; pace Favaro, *Edizione* 10:322, n. 1.

156. "Io prego V.S. Illma a favorirmi di mandarmi l'Optica del S. Keplero e il trattato sopra la Stella Nuova, perche ne in Venezia ne qua gli ho potuti trovare. Desidererei insieme un libro che lessi due anni sono sul catalogo di Francofort, il quale, per diligenza fatta con librari di Venezia, che mi promessero farlo venire, non ho mai potuto havere: io non mi ricordo del nome dell'autore, ma la materia e *de motu terrae;* et il S. Keplero ne havere notizia. Mi fara insieme favore avvisarmi della spesa, la quale rimborsero qua in casa sua, o dove mi ordinera" (ibid., 10: 402, ll. 45–52).

157. Favaro, "La Libreria di Galileo Galilei," no. 115.

158. From the letter to Giuliano quoted above, we have good evidence based on Galileo's actual scholarly practice that he read the Frankfurt catalogues and then tried to locate in Venice the books mentioned therein that he found of interest. The dating of the letter (October 1610) makes this evidence directly pertinent to our question. The advertisement for the fall 1609 book fair was in no sense conspicuous among the listing of *Libri Philosophici*. The entire entry ran as follows: *Astronomia nova AEtiologitos seu Physica coelestis tradita comentariis de motibus Stella* [sic] *Martis ex observationibus Tychonis Brahe, jussu & sumptibus Rudolphi II. Rom. Imp. elaborata a Ioanne Keplero Mathematico Caesareo. Prostat Francof. apud Godf. Tambach, & Pragae in taberna Marneana in fol.*

159. Schottenloher, *Die Widmungsvorrede*, does not mention any such motif. So far I have examined the pre-1609 prefaces of Kepler, Tycho, Regiomontanus, Schöner, Cardano, Magini, and Maurolico that I have had access to. The conclusions presented here are admittedly tentative; only further research will reveal additional examples, if there be any.

160. Most obviously in *De fundamentis astrologiae certioribus* (1602), passim, but also in *Mysterium Cosmographicum* (1596), 5,15. And since Kepler's astrology is based on a geometrical optical model of astrological action, the *De stella nova*'s dedicatory letter (1606) also has material relevant to his astrology, at least to its natural philosophical foundations, at 8,19; 9,13; etc.

5

Astronomia inferior: Legacies of Johannes Trithemius and John Dee

N. H. Clulee

John Dee's *Monas hieroglyphica* of 1564 confronts us with a particularly sharp paradox. The *Monas,* with its symbol of the monas, was clearly Dee's best known and most influential work, yet in his day as well as ours it has been noted for its difficulty, opacity, and obscurity. Besides the fascination it has elicited among modern occultists, alchemists, and faddists, Dee's symbol of the monas was borrowed, the text was cited and quoted, and the ideas and concepts were employed surprisingly often in the century after its publication.[1] On the other hand, not only have modern authors sympathized with Josten's experience that the *Monas* "resisted the onslaught of historical research" and concurred in his conclusion that "the specific message which Dee tried to convey by his symbol of the monad, and by the treatise thereon, is lost," contemporaries of Dee roundly criticized the work for its unintelligibility.[2] Clearly, for the *Monas* to have had the impact it did, some readers must have found it significant despite its difficulties. A text as opaque as the *Monas* may attract attention for a variety of reasons because it lends itself to a number of significations. This chapter will examine one of these reasons: Dee's elaboration of alchemy as an *astronomia inferior* and some peculiarities of its transmission and reception in the early seventeenth century.

The idea that alchemy was a type of "inferior" or earthly astronomy, or even a terrestrial astrology, was a standard theme in alchemy, but it took on a more developed and potent connotation in the sixteenth century as a result of Johannes Trithemius's interpretation of the *Emerald Tablet,* one of the touchstone texts of Western alchemy.[3] Trithemius's concept of alchemy as an *astronomia inferior* was based on his interpretation of the *Emerald Tablet* of Hermes Trismegistus as representing a cosmological process and not just an alchemical recipe. John Dee was one of Trithemius's most creative heirs. This chapter will look first at Dee's *Monas* and how it embodied a peculiarly sharp, rich, and graphic development of Trithemius's concept of alchemy as a terrestrial astrology. It will then examine one of Dee's legacies as a transmitter of Trithemius's ideas: the *Secretioris philosophie consideratio brevis* of Philipp à Gabella. Associated with the Rosicrucian tracts of 1614–15, Gabella reflects one of the curious paths of the influence of Dee's *Monas hieroglyphica* as well as the limitations Dee's readers had in fully understanding that text.

John Dee, the Monas, and *astronomia inferior*

Throughout his life Dee assumed that there was a relationship between the heavens and the earth, between astronomy and *astronomia inferior;* that is, between astronomy/astrology and alchemy. In his first work, the *Propaedeumata aphoristica* of 1558, he developed a theory of the operation of astrological influences on the terrestrial sphere based on the emanation of celestial virtues as rays that propagate in the same way as visible light. Because of this, celestial virtues may be studied and manipulated through the science of optics. This feature of nature makes possible the "greatest part of the natural magic of the ancient wise men"—the imprinting of heavenly rays upon terrestrial matter—which Dee links with "the very august astronomy of the philosophers, called inferior."[4] Alchemy, therefore, as *astronomia inferior,* or lower/terrestrial astronomy, is a branch of natural magic. Dee also indicates that the symbols of this terrestrial astronomy are included in his monad represented on the title page of the *Propaedeumata* (figure 5.1).[5] Subsequently, in the 1568 edition of the *Propaedeumata,* Dee indicated that this symbol was previously explained in another work, the *Monas hieroglyphica* of 1564 (figure 5.2).[6] In the twenty-four quasi-Euclidian "theorems" of Dee's *Monas* this symbol of the monad, or monas (figure 5.3), is geometrically constructed, and then its disassembled parts, both singly and as variously recombined, are shown to have cosmological, astronomical, numerological, alchemical, magical, and mystically spiritual meanings.

What made and still makes the *Monas* such a challenging text to understand is that it combines in novel ways and gives idiosyncratic twists to a number of very common themes in intellectual fashion during the

Astronomia inferior 175

Figure 5.1
Title page, John Dee, *Propaedeumata aphoristica*, 1558 (Beinecke Rare Book and Manuscript Library, Yale University)

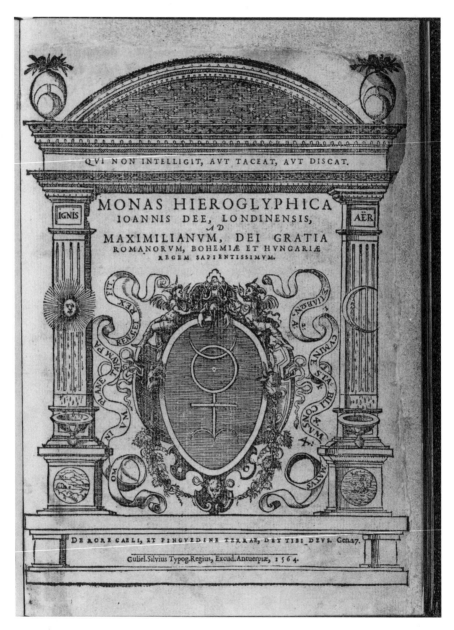

Figure 5.2
Title page, John Dee, *Monas hieroglyphica*, 1564 (Beinecke Rare Book and Manuscript Library, Yale University)

Figure 5.3
The monas hieroglyphica, John Dee, *Monas hieroglyphica*, fol. 12 (Rare Book and Special Collections Division, Library of Congress)

Renaissance. The explication of the symbol of the monas in the *Monas* involved several levels. Dee considered the most novel contribution of the *Monas* to be its presentation of a new and sacred art of writing, a writing that embodied the pristine divine language through which the essential reality of creation was communicated.[7] To emphasize the unique character of this writing embodied in the monas, which is a "holy language," Dee calls it a "real" cabala. Whereas other languages and Hebrew cabala are merely grammars of "that which is said," Dee's is a cabala of "that which is" because it corresponds to the "written memorial . . . which from the Creation has been inscribed by God's own fingers on all Creatures" and which therefore speaks of "all things visible and invisible, manifest and most occult, emanating by nature or art from God himself."[8]

At the second level, Dee claims that his new art of writing transcends and reforms all that is currently known in all the intellectual and scientific disciplines. Not only does this new language supersede and replace the "vulgar" linguistic disciplines of grammar and Hebrew cabala; it transcends and almost makes obsolete the traditionally legitimate disciplines of arithmetic, geometry, music, astronomy, optics, and so on while at the same time legitimating and elevating in status esoteric disciplines, including alchemy, divination, and magic, that were traditionally considered illegitimate and marginalized.[9] At the root of this reform is the claim that the new writing of the monas reforms the basic cosmological framework in which all specific intellectual disciplines operate: Dee's uniquely modified symbol of Mercury is the "rebuilder and restorer of all astronomy."[10]

By implication all astronomy encompasses both the celestial and the terrestrial, both astronomy and alchemy. Fundamental in accomplishing this reform, the basic precepts of the *Tabula smaragdina* can be heard speaking through this new language of the monas.

Monas, Language, and Cosmos

The ultimate basis for Dee's claim that the monas is the sacred art of writing the pristine divine language was his conviction that the shapes of the "first and mystical letters of the Hebrews, the Greeks, and the Latins," given directly to mankind by God, "were produced from points, straight lines, and the circumferences of circles."[11] Likewise, Dee's monas is generated from a point, lines, a circle, and semicircles (figure 5.4), but it can claim priority over all other languages because it bears a more direct correspondence to creation (figure 5.5). This is because the construction of the monas has a clearly cosmological character. The point represents the earth and the circle represents the sun and also the entire frame of the heavens surrounding the earth. The semicircle represents the moon, and the double semicircle at the base represents Aries, the first sign of the zodiac and the sign under which creation took place, and can be taken as an analog of the entire zodiac and the fixed stars. Whereas the circular components of the monas relate to the heavens, Dee relates the cross, composed of straight lines, to the sublunary realm of the elements.[12]

The structure of the monas not only epitomizes the structure of the cosmos; it also embodies a cosmogony insofar as the genesis of the symbol mirrors the mathematical genesis of the universe. This correspondence of both the construction of the monas to divine creation and of the derived components and meanings to constituents and processes of the natural

Figure 5.4
Point, line, and circle, John Dee, *Monas hieroglyphica*, fol. 12 (Rare Book and Special Collections Division, Library of Congress)

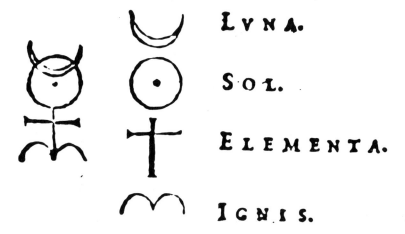

Figure 5.5
Monas and cosmos, John Dee, *Monas hieroglyphica*, fol. 13v (Rare Book and Special Collections Division, Library of Congress)

world is the key to Dee's central claim to have discovered a new and sacred art of writing or language that is an alphabet of nature and a "writing of things." The straight line and the circle represent nature because "the first and most simple manifestation" of things happened by means of the straight line and the circle, but, since the line is generated by the point and the circle by a line rotated around a point, "things first began to be by way of a point and a monad."[13]

As a single symbol, the monas represented to Dee a powerful hieroglyph revealing the unity of created nature and embodying the unity of knowledge about the unity of creation that had clear parallels with the *Tabula smaragdina* (the *Emerald Tablet*) of Hermes. Just as the *Tabula* indicates that "the world was created" and "all things were made from the one by the contemplation of the one, so all things are born of this one thing by adaptation," the *Monas* represents the cosmos as a monadic unit both in the genesis of all things from the point or monad and in its presentation of all things in a single symbol. Yet, although all parts of the monas derive from the point and the line just as all things in the cosmos have a common origin, the symbol also reflects the qualitative division of the universe into the celestial and the elemental realms. Whereas the geometric progression from point to line to circle culminates in the heavenly spheres, genesis in the terrestrial realm progressed only so far as the line, which, as the analog to the dyad, pertains to imperfection and change, but the celestial realm progressed beyond the line to the circle, which is monadic and therefore perfect.

The elemental realm, however, is the domain of a pattern of numerical relations familiar as the "Pythagorean" tetractys and frequently associated with a numerological portrayal of cosmogony in the Renaissance.[14] Just as rectilinear motion is proper to sublunary bodies, so the lines of the elements (figure 5.5) are produced by the "flowing of a point" and arranged as a cross corresponding to the pattern of the four elements (earth, air, fire, and water) and the four qualities (cold, hot, dry, and moist). With four arms and four right angles, the cross embodies the number four, or the quaternary, which in Pythagorean cosmology was the source of hosts of patterns of fours.[15] Besides the number four, the cross also embodies the preceding numbers. The point is a monad and corresponds to one; the flowing point produces a line that, as bounded by two points and capable of division, is a dyad and corresponds to two; and the two segments of the line, when crossed, have one point in common and so correspond to the ternary, or three. The cross of the elements thus corresponds to the sequence from one to four, the sum of which is the denary, or ten.[16]

As an embodiment of the Pythagorean tetractys $(1 + 2 + 3 + 4 = 10)$, Dee's cross of the elements resonates with the host of meanings that were associated with it. The monad, as the primeval unit that is the source of all numbers but not a number itself, and the analogous point, which is a dimensionless unit with position, were associated with God. The number two, which is even and divisible, corresponds to the unlimited, formless original matter and to the single dimension of the line. Three points define a triangle and the first surface, and the number four, besides corresponding to numerous quaternaries, defines the first solid body through the four points that mark out a solid angle. The progression from one through four thus corresponds to the generation of the physical world and also sets the limit to that creation, because through addition this series produces ten, by means of which the progression returns to unity, perfection, and ultimate stability $(10 = 1 + 0 = 1)$. These Pythagorean associations were also associated with a conception of the created world as the product and manifestation of the ideas in the mind of the divine creator. The universe was actually created by numbers, which are the intelligible principles underlying the flux and imperfections of the sensible world, and through the progression of the tetractys the actual unfolding of creation took place. The tetractys, therefore, is not merely a symbol; in a very real way it is the universe, and understanding it provides access to the actual thoughts of God.[17]

The cabalistic exegetical techniques of *notarikon, tsiruf,* and *gematria* can be applied to the monas and its components, as a "cabala of the real" or a "cabala of that which exists," to reveal an esoteric knowledge of creation.[18] With written texts, *notarikon* finds in letters and punctuation marks abbreviations that point to other meanings. In the case of the monad, *notarikon* involves taking geometrical components of the total symbol as abbreviations representing words or concepts. So when Dee has the point represent the earth, this is a kind of *notarikon*. Likewise, in the construction of the monas, the circle with the point at the center represents both the sun and the geocentric universe, the upper semicircle represents the moon, the cross represents the elements, and the double semicircle represents Aries (figure 5.5). In addition to being signs of astronomical things, these same components also had alchemical significance. Thus, in alchemical discourse the sun was commonly the symbol for gold and the moon represented silver. More fundamentally, the sun and moon also referred to the philosophical or sophic sulfur and mercury that were the fundamental principles of all metals according to the dominant alchemical theory of the West.[19] Because Aries is also the first sign in the triad of constellations corresponding to the element fire, the double semicircle also represents fire.[20]

Whereas *notarikon* considers parts of words, or in Dee's case parts of the monas, as abbreviations, through *tsiruf* the individual letters of words are rearranged to discover other words, or with the monas, parts can be recombined to yield other symbols and meanings.[21] The upper part of the monas, combining the sun and the moon, represents the evening and morning of the day of genesis on which the "light of the philosophers was made."[22] Dee considers the same part of the monas to represent the sign of Taurus and the exaltation of the moon.[23] Through various other recombinations of components of the monas Dee is able to construct the signs for all the planets in addition to the sun and the moon (figure 5.6).[24]

Dee employs the final cabalistic technique, *gematria*, the use of the numerical equivalents of letters to reveal hidden meanings, extensively.

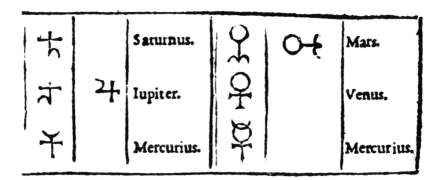

Figure 5.6
Signs of the planets generated from the monas, John Dee, *Monas hieroglyphica*, fol. 14 (Rare Book and Special Collections Division, Library of Congress)

Numbers associated with individual letters not only had mystical significance, but if the sum of the numerical values of the letters of two words was the same, the two words could be considered identical in meaning. In the monas, numerical interpretation is applied largely to the cross of the elements. We have already noticed how the cross represents various numbers, but through these numbers, according to Dee, the cross also reveals the rationale behind both the shape and the placement of various letters in the Latin alphabet, establishing the roots of the alphabet in the fundamentals of creation.[25] Through the numerical equivalents of the cross, Dee also discovers which "useful offices in Nature were assigned by God to the numbers" derived from the monas "when elements are to be weighed, when measures of time are to be determined, and finally when the power and virtue of things have to be expressed in certain degrees."[26] Thus, by *gematria*, the numerical equivalents of the monas and their permutations yield numbers corresponding to natural processes and reveal hidden explanations of nature's mysteries.

Monas and *astronomia inferior*

Since the construction of the monas mirrors divine creation and the components and the meanings derived from the monas correspond to the constituents and the processes of the natural world, this new form of writing is a cabala of that which exists. As it turns out, this cabala serves to explicate the alchemical cosmology of the *Emerald Tablet*. As a mirror of the cosmos, the very arrangement of the components of the monas not only develops more explicitly alchemy's dependence on the heavens, which is only a cliché in most alchemical texts, but also reveals the true natures and interrelations of the planets in the light of their correspondence to alchemical processes. Dee's claim that "celestial astronomy is like a parent and teacher to inferior astronomy" evokes the *Tabula*'s precept that "what is below is like that which is above, and what is above is like that which is below, working the miracle of the one thing."[27] The earth, at the center, is encircled by the sun and the orbits of all the planets on whose influences it is dependent. The moon is represented as a semicircle, because it emulates and is dependent on the sun.[28] Of all the planets these two are explicitly represented because the "Sun and Moon infuse their corporeal virtues into all inferior bodies that consist of elements in far stronger manner than do all the other planets."[29] This strongly echoes a similar aphorism in Dee's earlier *Propaedeumata,* but here the corporeal virtues of the sun and the moon are not merely the vivifying heat and moisture of the previous work.[30] In the monas the "aqueous moisture of the Moon" and the "fiery liquid of the Sun" are revealed through the "magic" of the elements: the analysis by fire of corporeal things, symbolized by the sign of Aries.[31] This separation of the sun and the moon by the "magic" of the four elements of the cross adds an alchemical dimension to the monas. The sun and moon not only are symbols of those planets but are also alchemical symbols for both gold and silver and, as a "fiery liquid" and an "aqueous moisture," for the sulfur and mercury that are the principles of gold and silver and of the philosophers' stone. Dee's title page evokes the four elements and depicts drops of liquid descending from the sun and the moon, and the quotation from Genesis 27 at the foot of the title page referring to the "dew of heaven" and "the fat of the earth," which are alchemical references to mercury and sulfur, respectively, completes the alchemical motif (figure 5.2).[32] Incorporating a direct quotation from the *Tabula*, Dee summarizes these celestial-alchemical relations when he says "this whole magisterial work depends upon the Sun and the Moon, which a long time ago that thrice-great Hermes admonished us when he asserted that the Sun is its father, and the Moon its mother; and we know that it is nourished in Lemnian earth by lunar and solar rays which exert a singular influence around it."[33]

Although quite abstract, these associations reflect the concrete alchemical background to the *Monas,* which is the sulfur-mercury theory of the generation of metals. This mercury and sulfur are not the ordinary substances of those names, but hypothetical intermediary substances, often called philosophical or sophic mercury and sulfur, whose purity and nature are only approximated by those of ordinary mercury and sulfur. Under the influence of the planets, different metals result from the combination of mercury and sulfur depending on their relative purity and differences in the proportion of the two principles in the combination. If perfectly pure and combined in perfect equilibrium, they produce gold, otherwise one of the inferior metals results. Yet, since all metals have the same constituents as gold, purification and readjustment of the proportion of the constituents by means of suitable elixirs should transform the inferior metals into gold.[34]

The problem for the alchemist was how to imitate the natural process and speed up nature in its production of gold. In most views this involved the creation of an elixir or the "philosophers' stone," which had the power to rapidly transform large quantities of imperfect metals by rectifying their imperfect composition.[35] The philosophers' stone is either a blend of philosophical mercury and philosophical sulfur, or, as the "mercury-alone" theory of the pseudo-Geber suggests, it is philosophical mercury containing an inner and nonvolatile or nonflammable sulfur.[36] In either case, philosophical mercury is a mercury from which the fluidity and humidity have been removed. Philosophical or inner sulfur likewise has had its flammability and earthiness removed.[37]

The actual process involved taking some substance, which could be any common substance, breaking it down into its constituent qualities, and subjecting these qualities to a series of operations through which accidental imperfections are purged and the remaining purified substances are combined and unified, first into philosophical sulfur and mercury and then into the stone. As pseudo-Geber expressed it: "imperfect bodies have superfluous humidities and combustible sulfurity, with blackness corrupting them, an unclean, feculent, combustible and very gross earthiness," but the "spoliation" of these accidental parts of bodies through the use of fire will yield a substance in which only mercury and sulfur remain.[38]

The exact process for the production of the stone varied from author to author, usually involving some sequence of standard chemical operations, including such things as calcination, solution, sublimation, distillation, and fermentation.[39] This sequence was frequently presented as following some kind of cycle that corresponded to some natural pattern. The text that comes closest to the *Monas* is Thomas Norton's presentation of the process in the *The Ordinall of Alchemy*. Presaging Dee's discussion of symbolic numbers as revealing numbers, weights, and measures, Norton's process begins with breaking down the beginning matter into the four elements, whose qualities are then recombined

... by ponders right,
With Number and Measure wisely sought,
In which there resteth all that *God* wrought:
For *God* made all things, and set it sure,
In Number Ponder and in Measure,
Which numbers if you do chaunge and breake,
Upon *Nature* you must doe wreake.[40]

The process by which this recomposition occurs involves seven circulations of the elements presided over by the astrological influences of the planets. The seven circulations are divided into two sequences. The first begins with (1) fire acting on (2) earth producing (3) pure water, leading to (4) air. The second sequence begins with (5) air and leads through (6) clean earth to return to (7) fire.[41]

In carrying out this process the alchemist is thus an imitator of the creator and alchemy a replication of creation on a local scale. Dee's quotation/paraphrase from the *Tabula,* indicating that the "una res," the philosophers' stone, "is nourished in Lemnian earth by lunar and solar rays" (the stone's mother and father), suggests that in the *Monas,* elemental earth is the basic matter of the alchemical process, through which philosophical mercury (the moon) and philosophical sulfur (the sun) are drawn forth, refined, and generate the "stone." More than just a concatenation of astronomical/alchemical symbols, Dee derives from the writing of his monas an account of the *process* of the alchemical work resembling Norton's model. In his first ten theorems, for instance, the hieroglyphic writing of the monas yields the message that "the sun and the moon of this monad desire their elements, in which the denarian proportion will be strong, to be separated, and that this be done with the aid of fire."[42]

This message results from applying the technique of *tsiruf* to the monas, by which the symbols for all the planets are constructed from components of the monas. This technique yields alchemical significance in revealing by analysis both the astral and the elemental components of the philosophers' stone and astronomical significance in revealing the character and interrelationships of the planets. Dee divides the planets into a lunar group and

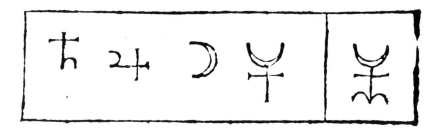

Figure 5.7
The genesis of lunar mercury, John Dee, *Monas hieroglyphica*, fol. 14 (Rare Book and Special Collections Division, Library of Congress)

a solar group based on the presence of the symbol of the moon or of the sun in their symbols. The lunar group, displaying the cross of the elements and a semicircle, comprise the sequence of Saturn, Jupiter, Moon, and Mercury, represented by the unconventional symbol of a semicircle on top of a cross (figure 5.7). As the diagram indicates, each successive symbol in the series is related to the one before it in being a simple rotation of the symbol or the addition or subtraction of parts, implying, according to Dee, that these four lunar planets constitute a hierarchy in which the shared lunar quality is progressively enhanced. Reference to this sequence as the result of four revolutions of the lunar nature around the earth, in which the work of "albification" (whitening) is carried out by applying the moon to the elements, invokes the alchemical dimension. The moon has already been identified with mercury, and what seems to be at work here is the separation and purification (albification) of the mercurial or lunar principle from the elements to yield lunar mercury, represented by the cross topped by the lunar crescent.[43]

Mars and Venus, along with the Sun, are solar planets interrelated by a shared characteristic and their sequence, from Mars through Venus, involves a similar progressive enhancement of the solar principle, or sulfur, inherent in the elements. These three solar revolutions of the elements, when joined to the previous lunar revolutions, unite lunar mercury with solar sulfur to yield the conventional symbol for mercury, containing both the lunar semicircle and the full solar circle. Driving home the theme of *astronomia inferior,* Dee labels the diagram illustrating this process the "principal monadic anatomy of the totality of astronomia inferior" (fig-

Astronomia inferior 187

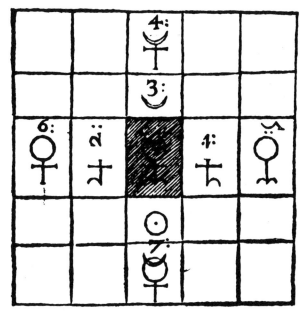

Figure 5.8
The seven stages: the "principal monadic anatomy of the totality of astronomia inferior," John Dee, *Monas hieroglyphica*, fol. 14v (Rare Book and Special Collections Division, Library of Congress)

ure 5.8).[44] Through an egg-shaped figure, Dee evokes both a common alchemical image and the notion found in some sixteenth-century astronomical texts that Mercury's deferent was oval shaped (figure 5.9).[45] Within the egg the seven planets are in their Ptolemaic order and follow geocentric paths, but in addition the Sun and the solar planets Mars and Venus are shown within the yolk, whereas the Moon, lunar Mercury, Jupiter, and Saturn are shown within the white. In Dee's suggested interpretation, the shell, which commonly represented earth, is dissolved by heat and compounded with the lunar mercury of the white, then that mixture is saturated with the solar sulfur of the yolk through repeated rotation.[46] These rotations, echoing the seven revolutions previously discussed, are represented in another figure (figure 5.10) as a spiral through which the "terrestrial center" of the monas ascends through seven stages corresponding to the planets. Thus, in representing the planets and the metals and embodying the essence of the alchemical work, the egg is an analog of the

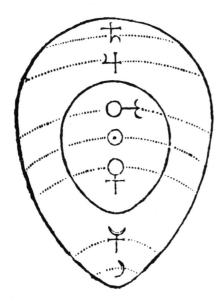

Figure 5.9
The celestial egg, John Dee, *Monas hieroglyphica*, fol. 17 (Rare Book and Special Collections Division, Library of Congress)

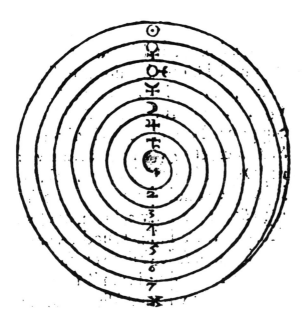

Figure 5.10
The seven celestial revolutions, John Dee, *Monas hieroglyphica*, fol. 18 (Rare Book and Special Collections Division, Library of Congress)

monas, which itself is a hieroglyph of the cosmos and the alchemical work rather than a component of either.

The murkiest astronomical/alchemical interpretation of the monas comes when Dee inverts the symbol. Again he is able to derive the signs of the planets from parts of the symbol, but in this case they emerge in the order Saturn, Jupiter, Mars, Venus, Mercury, Sun, and Moon, which he says is the order Plato ascribes to them.[47] In this form Dee is able to derive the order of the planets accepted by Plato, Aristotle, and others earlier than the second century B.C.E. Although placing Venus and Mercury above the Sun runs counter to the traditional Ptolemaic order, there was in fact fair leeway in the order of the planets in a geocentric universe, because there was no reliable way of determining the relative sizes and distances of the planets.[48] It is difficult to know what Dee intends by this rearrangement, since he defers treating the astronomical issue to another place, but he does say that this inverted arrangement is meaningful.[49] He places the Sun and Moon, which as sulfur and mercury are the most powerful influences both astrologically and alchemically, closest to the earth, and the other planets are farther away in decreasing order of importance. He also shows how, in addition to those of Jupiter and Saturn, the sign for Venus can be made from the inverted cross by closing the semicircles of Aries at its top. Since this circle is smaller than that of the Sun, however, this appears to imply that Venus by itself cannot yield true gold (Sun).[50] The enigmas of the inverted monas are not yet exhausted, for they reappear later in the context of the mystical dimension of the *Monas*.

The monas also contains the sign for Aries, which is the house of Mars (strength) and the exaltation of the Sun, and the sign of Taurus, which is the house of Venus (love) and the exaltation of the Moon (figure 5.11). Thus, after telling us to separate the elements of the Sun and the Moon by means of fire, the monas summarizes the remainder of the alchemical process as the "exaltations of the Moon and the Sun by means of the science of the elements."[51] This idea of the monas as a hieroglyphic writing containing a discourse on alchemy and its celestial correspondences also emerges in the "magic parable" of the letter of dedication to King Maximilian II Habsburg. Here Dee says that the monas "teaches without words" how the terrestrial body at its center is to be actuated by a divine force and united with the generative lunar and solar influences that have been separated both in the heavens and on earth.[52]

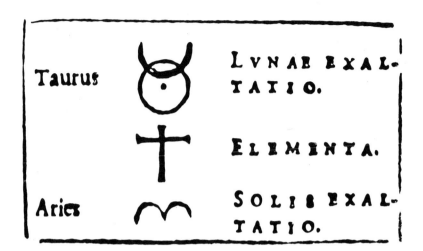

Figure 5.11
The "exaltation of the moon and the sun by the science of the elements," Dee, *Monas hieroglyphica*, fol. 15 (Rare Book and Special Collections Division, Library of Congress)

In just this fashion Dee's *Monas* is rooted in the *Tabula smaragdina*. Beyond the idea of alchemy as an *astronomia inferior* reflecting the precept of the *Tabula* that "what is below is like that which is above, and what is above is like that which is below," Dee's monas, as the monad, the *una res*, "is the father of all works of wonder in the world" from which "proceed the miracles." It "has the supreme power because it conquers all subtle things and penetrates all solid things." In particular, Dee's monas prescribes that it turn "toward the earth," "separate the earth from the fire, the subtle from the gross," "ascend from the earth to heaven, and then again descend to earth and unite together the power of the superiors and the inferiors."[53]

Serving as an "astronomical messenger" of a new art of writing, the monas therefore rebuilds and restores *all* astronomy, both inferior and superior, by making explicit the intimate similitude and correspondence between celestial astronomy and elemental alchemy.[54] With these correspondences revealed, the "common astronomical symbols of the planets (instead of being dead, dumb, or, up to the present hour at least, quasi-barbaric signs)" assume the status of "characters imbued with immortal life and should now be able to express their properties most eloquently in any tongue and to any nation."[55]

Trithemius and the *Tabula smaragdina*

In its alchemical dimension, the *Monas hieroglyphica* is clearly a presentation of Geberian alchemy in the framework of a cosmology based on the *Tabula smaragdina*. Dee did not come to this entirely on his own. His interest in the *Tabula* was undoubtedly related to the fact that the text of the *Tabula* occurs within Roger Bacon's text and commentary on the *Secretum secretorum*, from which Dee derived some of his idea of the adept and certainly his aspiration to the status of the "British Aristotle."[56] The fundamental inspiration for the formulation of alchemy in the *Monas* was, nonetheless, Johannes Trithemius's concept of an alchemical magic developed in the context of a commentary on the *Tabula*, presented in a well-published and frequently cited letter of August 24, 1505, to Germanus de Ganay.[57] Trithemius's treatment of the *Tabula* has rightly been seen as significant in a number of contexts. Noel Brann found it to be a fundamental expression of Trithemius's progression from a mystical theology to a magical theology incorporating an esoteric occultism and the association of the alchemical process with the purification of the soul as it ascends the ladder of being.[58] This is a magical theology, because the miraculous ascent of the soul is presented as a series of "inward alchemical transformations" embodying the magical power and "miracle of the one thing" at the center of the elevation of the earth to the heavens in the *Tabula*.[59] In this fashion, Trithemius develops what Brann calls an "introverted Hermeticism" focused more on internal spiritual transformation, where any external operation on the world of nature depends on a previous internal transformation of the mind through illumination by divine light.[60] Thus, Trithemius says, "our philosophy is celestial, not worldly, in order that we may faithfully behold that principle which we call God, by an intuition through faith and knowledge, as Father, Son, and Holy Spirit, one principle, one God, and one highest good in a trinity of eternal persons."[61] Focusing on Trithemius's import for alchemy itself, William Newman has emphasized Trithemius's transformation of the conception of the *Tabula* from a cryptic recipe into an embodiment of the alchemical work as a cosmic process.[62] The ascent of earth to the heavens and its return, in this interpretation of the *Tabula*, also underpinned the central position assumed by elemental earth as the basic ingredient in alchemy throughout the Renaissance.[63]

Dee's *Monas* reflects both of these features of Trithemius's alchemical magic/magical theology. We have seen how the alchemical discourse generated by the writing of the monas conveys the process by which the "terrestrial center" (elemental earth) of the monas is actuated. Although Dee claimed that this writing was a "cabala of the real" that applied to physical phenomena, the *Monas,* in the style of Trithemius's "introverted Hermeticism," is almost entirely confined to the internal and abstract operations of the monas symbol within the intellect, which must be divinely inspired.[64] Further, Dee associates with this alchemical process a spiritual ascent of the adept from the terrestrial to the supercelestial spiritual realm above the "horizon of eternity."[65] Beyond these two features, there is a further key element in Trithemius's magical theology that is perhaps most central to the attraction Trithemius had for Dee's and to Dee's expression of the cosmological interpretation of the *Tabula* in the *Monas:* Trithemius's numerology.

The occasion for Trithemius's letter to Ganay was a request from Ganay for Trithemius to explain the "very rare and admirable philosophy, shrouded in numbers, elements, and enigmas, and abstruse with arcane words" that Ganay had read of in a letter he had come upon from Trithemius to a disciple, Johann Steinmoel.[66] In reply, Trithemius proclaims that understanding the return of all things to unity by the ternary encompasses the mysteries of the "fundamentum arcani."[67] It is to explicate this core secret of his philosophy that Trithemius introduces the *Tabula smaragdina,* giving a verbatim version of the text interspersed with additions and emendations relating the precepts of the *Tabula* to the process of recovering the simplicity and purity of unity. He begins by saying that "the unit is not a number, but from it all numbers arise. By discarding the binary, the ternary will be convertible to unity." Continuing, with italics indicating Trithemius' additions and brackets indicating modifications to the original precepts of the *Tabula,* Trithemius says that this

truth, Germanus, as Hermes said, [is] without falsehood, [the] certain and most true *relative of unity.* . . . [1] What is below is like that which is above, and what is above is like that which is below, *because all numbers consist only of unities,* working the miracle of the one thing.[2] *Is it not true that all things flow from one thing, from the goodness of the One, and that whatever is joined to Unity cannot be diverse, but rather fructifies by means of the simplicity and adaptability of the One?*[3] *What is born from Unity? Is it not the ternary? Take note: Unity is unmixed, the binary is compounded, and the ternary is reduced to the simplicity of Unity. I, Trithemius, am not of three minds, but persist in a single integrated mind taking pleasure in the ternary,* which gives birth to a marvelous offspring. Its father is the Sun, its mother the Moon; the wind carried it in her breast; the earth nourished it.[4] It is the father of all works of wonder in the world.[5] Its power is complete *and immense*. If it is turned toward the earth[6] it will separate the earth from the fire, the subtle from the gross,[7] *and when the ternary has at last returned to itself it may, by an inner disposition and great delight,* ascend from the earth to heaven, and again, *after it has been adorned with virtue and beauty,* descend to the earth, and receive the powers of things superior and things inferior; *thus it will be made powerful and glorious in the clarity of Unity, demonstrate its ability to bring forth every number,* and put to flight all obscurity.[8] *The One is the pure origin of all things, the binary, by departing from unity, is compounded, so the binary cannot be a principle. Only, therefore, when the sacred, excellent, and potent ternary surmounts the binary and returns to unity, not in its original nature, but by participation in similitude, does the mind understand without contradiction all the mysteries of the excellently arranged arcanum.* This one thing, which has the supreme power and *the most noble virtue,* conquers all worldly things and penetrates every solid body, *touching all with its desirable excellence.*[9][68]

Trithemius thus enhances the centrality of the *una res* of the *Tabula* and suggests that the ascent and descent of the one thing, and its acquisition thereby of magical power by uniting the celestial and the terrestrial, carries out the fundamental arcanum of the restoration of diversity to unity through the ternary. This establishes the importance of considering the heavens (*astronomia superior*) for true alchemy and distinguishes it from false alchemy.[69] This is also the context in which Trithemius states that "our philosophy is celestial, not terrestrial," and invokes "Father, Son, and Holy Spirit, one principle, one God, and one highest good in a trinity of eternal persons," thus linking alchemy and magic through his numerology to his mystical theology.[70]

Later in the same letter to Ganay, Trithemius develops the importance of number for alchemy and natural magic further, since number is the root of the order and measure that govern the celestial harmonies and establish the concord between terrestrial and heavenly things.[71] Magic—the philosophy that is celestial—is nothing other than a wisdom founded upon the celestial harmonies of number, measure, and order.[72] In another letter to Johannes von Westenburg, Trithemius makes the connection of these ideas to alchemy even more concrete. Alchemy may imitate nature only through an understanding of numbers, weights, and measures, because the descent from unity by the binary to the ternary and the return to simplicity through the ternary and the quaternary govern all natural operations.[73] The first principle of magic, through which all natural wonders are produced, is the restoration of purity in unity, which in essence is the recapturing of the monadic origin of things through the Pythagorean progression of the tetractys.[74] Applied to the magician, "whoever has become elevated to the uncompounded pure state of utter simplicity may be perfect in every natural science, may bring to pass marvelous works, and may discover amazing effects."[75] Applied to alchemy, this natural magic entails the reduction of composites of the elements by fire to purity, simplicity, and unity from the diversity and compoundedness arising from the binary.[76] In this context Trithemius specifically discusses earth, which in being composed of pure, simple, and unitary elements, is itself compound, diverse, and impure but can be reduced to simplicity by fire.[77]

John Dee and Trithemius

In his first indication of an interest in Trithemius in 1563, Dee coupled him with his own search for insight into "the Science *De numeris formalibus,* the Science *De ponderibus mysticis,* and the Science *De mensuris divinis:* by which three the huge frame of the world is fashioned," an interest that reached fruition in the numerology of the *Monas.*[78] This numerology yielded the numbers useful "when elements are to be weighed, when measures of time are to be determined, and finally when the power and virtue of things have to be expressed in certain degrees," which echo both Norton's "by ponders right, With Number and Measure wisely sought," and Trithemius's celestial harmonies of number, measure, and order as the basis of his natural magic.[79] Dee was clearly familiar with Trithemius's letters, and in an annotation to Trithemius's phrase, "a ternario in vnitatem per binarium divisum," he associates the symbols for mercury, salt, and sulfur with the ternary, compounds of mercury-salt and sulfur-salt with the two units of the binary, and his monas symbol with the unity that results form the union of these.[80] In the *Monas* Dee evokes not only themes from the *Tabula* but also the numerological magic of Trithemius by showing how the elements of the earth, through the separation of the sun and the moon and the elimination of impurities, are restored through the binary, ternary, and quaternary to a purified unity.

Going far beyond mere numbers, the most involved integration of Trithemius's numerological alchemy with Dee's concept of the monas emerges in Dee's geometrical discussion of the cross of the elements, where the Trithemian idea of the restoration of unity from the binary through the ternary that resides in the binary is related to the elements and alchemy. The two lines that represent the binary of material creation are formed into a cross to signify the four elements. This cross contains the ternary hidden within it, because what makes it a cross is not just the two lines, but also the point where they cross. This point, making the ternary, is intrinsic to the binary nature of the cross, because removing it would destroy the binary and produce the quaternary.[81]

From another perspective, considering the cross as a quaternary, the central point is superfluous. When this central point is removed, the quaternary "in the realm of the four elements" is more distinct. The point, therefore, which is not superfluous in the "divine" ternary, becomes, in the terrestrial sphere of the four elements, "feculent, corruptible, and full of darkness" and must be removed.[82] By proceeding in these ways, the "monas is restored through the binary and ternary to its oneness in a purified quaternary. . . . During this process our monad does not admit any external units or numbers, because it suffices to itself most precisely, complete in all its numbers. . . . It is restored to its first and own matter, while in the meantime the impurities that have nothing to do with its genuine and inherited proportion have in every way and carefully been cut off and removed forever."[83]

Besides explicating the core alchemical process of the cosmos and thereby revealing the unity of astronomy and alchemy, the sacred art of writing of the cabala of the real also encompasses the dimension of spiritual ascent in Trithemius's magical/mystical theology. By mastering this language the natural philosopher gains access to the innermost secrets of the cosmos, elevating the philosopher to the level of "adeptship." Adeptship grants the philosopher command of a cabalistic magic that includes mastery of the "magic of the elements" (alchemy) as well as a spiritual magic directed toward the healing of the soul that opens the way to the celestial rather than the terrestrial.[84] Dee presents a diagram (figure 5.12) presenting the mystical ascent to the supercelestial realm as an integral part of this magic. Various parts of this scheme are disposed according to the ternary, the septenary, the octonary, and the denary derived from the quaternary of the cross of the elements, which also figures prominently here. This diagram graphically represents a metamorphosis leaping from the temporal to the eternal and supercelestial realm beyond the "horizon aeternitatis" when the monas, after being "correctly, wholly, and physically restored to itself" in perfect unity, undergoes four "supercelestial revolutions."[85] Dee's philosophy in the *Monas* thus bears a resemblance to

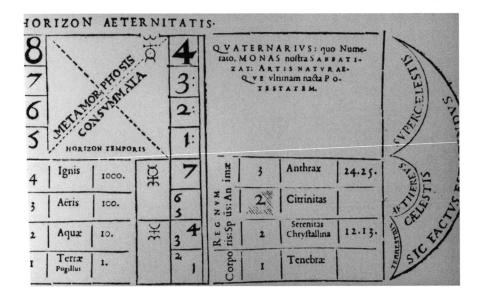

Figure 5.12
"Horizon aeternitatis," John Dee, *Monas hieroglyphica*, fol. 27 (Rare Book and Special Collections Division, Library of Congress)

Trithemius's "introverted Hermeticism," in which internal illumination takes precedence over the external world and mystical ascent to and knowledge of God is an integral part of and the culmination of the attainment of an integral knowledge of the cosmos.

Trithemius's philosophy of natural magic, which includes alchemy, and his magical theology are together, therefore, one of the foundations of the *Monas*. To this Dee has added the other major foundation: the idea of a geometrically based hieroglyphic writing. Inspired by his study of cabala, Dee claims this writing to be a divine language of creation because of its derivation from the geometrical and numerological processes through which creation took place. As a "real character" and a "cabala of that which exists," this writing reforms all the disciplines because it reveals the unity of all knowledge, particularly the unity among astronomy, alchemy, and magic in an occult philosophy. The monas's ability to explicate the materials and sequential process of traditional alchemy by reference to astronomy as a celestial alchemy, as well as providing insight into cosmology as analogous to alchemy, in a way that reflects the *Emerald Tablet* serves to validate for Dee his claim about the power of his new language and links these two foundations.

John Dee *redivivus:* Philipp à Gabella's *Consideratio Brevis*

Philipp à Gabella's *Secretioris Philosophiae Consideratio Brevis* is perhaps the most extensive but least studied instance of the influence of Dee's Monas.[86] This text is quite rare—I have found no copy of it in North America—and notable because it was published in 1615 with the original publication of the *Confessio Fraternitatis R.C.*, the second of the two Rosicrucian "manifestos."[87] Nothing is known of Gabella himself, and most writers on the Rosicrucian phenomenon, considering Gabella as not integral to the original tracts, quite rightly as we will see, do not discuss the text at all.[88] Frances Yates has made rather more of this text. Pointing out that the *Consideratio* quotes from the first thirteen theorems of Dee's *Monas,* Yates adduces these quotations to support the inference that Gabella's *Consideratio* was the product of a school of disciples that remained from Dee's activities in eastern Europe in the 1580s. She has also considered this text to be integral to the Rosicrucian "movement" as the articulation of the "more secret philosophy" behind the manifestos.[89] In the combination of these two premises Yates finds the key evidence that Dee was the original inspiration for the Rosicrucian movement and its "mentality."[90] Yates did not study the *Consideratio* or its relation to Dee in any depth, so the *Consideratio* may have been just another text that the editors or publishers of the *Confessio,* for reasons now known only to themselves, included with the manifestos despite only marginal connections to the original Rosicrucian documents.[91]

Yates's suggestions, however, have subsequently provoked more serious consideration of Gabella's place within the Rosicrucian context. The name Philipp à Gabella, with its apparent allusion to Cabala, has the markings of a pseudonym. The publication of the *Consideratio* by Wilhelm Wessel, who was court printer to Landgraf Moritz of Hesse-Kassel, and its dedication to Bruno Carolus von Uffel, a Hessen nobleman who once proposed an alchemical recipe to Moritz and later held an administrative post at court, suggest more than a hint of a connection with Moritz's court.[92] Bruce Moran has argued that Raphael Eglinus, an alchemist Moritz patronized, was the *Consideratio*'s author, whereas Carlos Gilly has suggested Johannes Rhenanus, another member of Moritz's scientific entourage.[93] Besides writing to Moritz in 1595, Dee had direct connections with the court at Kassel under Wilhelm IV, Moritz's predecessor, so there is a basis for a Dee legacy at Moritz's court, including some of Dee's writings, that would have been available to Eglinus or Rhenanus, although neither likely ever met Dee personally.[94] Moran has found these associations supportive of Yates's contentions, whereas Gilly has been more dismissive, verdicts deriving from differing readings of the *Consideratio*.[95] For our purposes here, the character of the *Consideratio*'s Rosicrucian connections are of only slight importance; I shall be primarily interested in how the *Consideratio* uses Dee and reflects the theme of *astronomia inferior* of the *Tabula* and Trithemius as mediated by Dee. We will return to the Rosicrucian connection at the end of the chapter, because the Dee connection does bear on Yates's claim that Dee was the progenitor of Rosicrucianism. It turns out that any lingering suspicions of a formative connection of either Gabella or Dee with Rosicrucianism will ultimately dissipate.

Gabella's text is most interesting as an attempt to explicate alchemical ideas through an elaboration of the *Emerald Tablet* by means of very close borrowings from John Dee, borrowings so extensive that the *Consideratio* may appear almost a reincarnation of John Dee's early natural philosophy. Because of Gabella's use of Dee, the *Consideratio* evokes something of Trithemius's interpretation of the *Tabula* but with significant limitations because Gabella did not know Trithemius directly and had only a narrow understanding of Dee's ideas limited by his own concrete focus on alchemy.

Although the *Consideratio* closes with a prayer signed "Philemon Philadelphiae R.C.," the dedication to Bruno Carolus von Uffel expresses the hope that this treatise "may be decorated by the deeds, the enthusiasm and the diligence of the Rosicrucian Brotherhood." This suggests that, rather than being a member of the proclaimed brotherhood or associated with the authorship of the *Fama* and the *Confessio*, Gabella was publishing one of the earliest expressions of interest in the supposed brotherhood that appeared in response to the *Fama Fraternitatis* of 1614.[96] This would explain the signs of great haste in the composition of the *Consideratio*. The first indication of this haste is that the text seems to have little coherent structure. Although the entire text deals with alchemy or topics of alchemical import and each of the nine chapters is related to alchemy in a general way, the flow from one to the next of the nine chapters is not always evident, and there is little sense of a progressive development within the text as a whole. It is almost as if it were a series of separate considerations, reminiscent of texts such as the *Rosarium philosophorum* and the *Turba philosophorum*, each of which is a compilation of excerpts, or apparent excerpts, from ancient and classical alchemical and "philosophical" sources, all dealing with alchemy, but often without clear interconnections or progression. These analogies point to another feature of the *Consideratio* that contributes significantly to its disjointed character: It is highly derivative. Major sections, as we shall explore later, are direct quotations from other authors. Like those in the *Rosarium* and the *Turba*, these extracts often seem to shift topics abruptly, but unlike them, Gabella rarely attributes his extracts to their actual authors. Other indications of haste are the crude character of the figures and careless typography and orthography.

The *Consideratio* is a primer by Gabella, as a teacher of the "pyronomic art," on alchemical wisdom. In it, he exhorts "whoever wishes to know the daughter of alchemical wisdom, resplendent in her brilliant white dress," to prepare for an arduous struggle[97] in which alchemical books contain the foundations of wisdom and can drive out ignorance but are, Gabella admits, frequently obscure and require the guidance of a teacher. Still, he covers his bases by saying that if he leaves anything obscure, it is the result of the sources, of whom he mentions Hermes, Plato, and Seneca as well as "many other philosophers," upon which his contemplation is founded.[98]

Gabella's demurral is both well taken and disingenuous: well taken because the *Consideratio* is in large part a compilation of massive quotations from others' writings rather than Gabella's expert explication of those books and disingenuous because the sources of Gabella's quotations are mostly modern authors; any lingering obscurity should not be laid at the feet of the authors Gabella mentions, such as Hermes and Plato, but attributed to the sources of Gabella's extensive quotations, who are rarely cited. Though bowing to contemporary cultural icons and citing "ancient philosophers," Gabella prefers modern authors on principle. Despite his professed homage to the ancients, Gabella says that "if Hermes, the father of philosophy, were to be brought back to life today, there is no doubt that he would be laughed at by the alchemists,"[99] because skill and knowledge are enhanced over time; so "the wise of today far excel their predecessors."[100] This produces the very curious product of a text patterned on a very ancient but unacknowledged structure, the *Tabula smaragdina*, pretending to be an original explication of the ideas of "ancient philosophers," such as Hermes and Plato, but constituted almost entirely of a patchwork of quotations from unacknowledged modern authors.

Although Gabella's direct expropriation of other texts raises red flags to modern sensibilities, it may well be only an example of a work constructed by drawing upon the compilation of commonplaces, which was a standard part of humanist education. Commonplace books, recording quotations, facts, and observations of various sorts, became important tools for storing and organizing information in the Renaissance.[101] Ann Blair and others have pointed to the "reprocessing" of others' texts from synthesis to quotation that this method of compilation facilitated and to the attendant juxtaposition of positions, inconsistencies, and credulity coexisting with observation and cautious assessment that was apparent in the resultant works.[102] One problem with commonplace collections is that their quotations were often so removed from their original contexts that they often bear meanings alien to the original when they are used in the production of new texts.[103] Gabella's *Consideratio* exhibits many of these characteristics.

Although the *Consideratio* may be a patchwork of extracts and echoes of other texts, Gabella was not a slave to their meaning. For instance, his comment on Hermes being laughed at by the alchemists has a possible foundation in Michael Sendivogius's *Novum lumen chemicum*:[104]

Consideratio sig. D1ᵛ	*Novum lumen chemicum* 465
Quod si hodie Hermes, philosophorum pater, reviviscat, proculdubio ab Alchymistis irrideretur. Quemadmodum ipsum Daedalum, sculptores ajunt, si reviviscens, talia fabricaret, qualia quondam, ex quibus sibi gloriam comparavit, ridiculum fore.	Si hodie revivisceret ipse Philosophorum pater Hermes, & subtilis ingenii Geber, cum profundissimo RAIMUNDO LULLIO, non pro Philosophis, sed potius pro discipulis a nostris Chemistis, haberentur: Nescirent tot hodie usitatas distillationes, tot circulationes, tot calcinationes, tot alia innumerabilia Artistarum opera, quae ex illorum scriptis hujus saeculi homines invenerunt & excogitarunt.

Not only has Gabella abbreviated Sendivogius's passage, he has eliminated the irony, and ultimately the intent, of the original as well. This should caution us that, just because Gabella may have extracted passages from other authors, he has not always done so with the intention of communicating their meaning faithfully.

Consideratio and *Tabula*

These observations aside, there is a larger structure behind the scattered appearance of Gabella's considerations. Gabella presents himself to the reader as a teacher who will guide the reader in the quest for the "pyronomic art."[105] In carrying this out, Gabella shaped the general structure of the *Consideratio* to reflect the basic themes of the *Tabula smaragdina*. Because of its extremely concise and pithy character, the *Tabula* is often cryptic and subject to many interpretations, but Gabella develops several themes among the scattered statements of the *Tabula* reflecting the cosmological foundations of the alchemical process developed by Trithemius and embodied in Dee's *Monas*. The precepts, indicated in brackets, that Gabella develops as his themes are

1. Truth: The precepts of the *Tabula* are true and certain [1].
2. Cosmology: All things are related to a primal monad (the *una res*) from which all has been created by the action of the one [supreme] being [3]; the terrestrial world and the celestial world are mutually interrelated in carrying out the miracles of this one thing [2], which was born of the Sun (its father), the Moon (its mother), the air (its womb), and the earth (its nurse) [4]; the world originated from this process [10].
3. Alchemy: This primal monad with its magical powers is the implied aim of the alchemical work; it requires uniting the celestial and the terrestrial to accomplish the miracle of the monad [2], turning it toward the earth [6] and separating the earth from fire and the subtle from the gross with prudence and ingenuity [7].

Although the *Consideratio* does not precisely follow the *Tabula* and at times jumps around among these themes, thinking of it as developing considerations on them helps draw out some of what might have been Gabella's objective.

In actually developing his instruction in the "pyronomic art," Gabella begins his first chapter echoing the first precept of the *Tabula* with an assertion that he is presenting a Truth that will end uncertainty:

Consideratio	*Tabula*
sig. B1ʳ	no. 1
Veritatem, (Mortales) cujus splendor ambiguitates omnes propellit, non mendacium quod eam in profundo penitus abstrusit, vobis trado. . . .	Verum, sine mendacio, certum et verissimum.[106]

With confidence that time reveals all things and that all secret things will be brought to light, Gabella states that his aim is to investigate the causes and reasons of secret matters and above all to attain knowledge of "M."[107] What this "M" might be is not the last of the obscurities of Gabella's text. It may well be the philosophers' stone or the *una res* of the *Tabula*, since later in the text when discussing philosophical mercury as the prime material of all things Gabella says that "without it M. cannot exist."[108] What "M." may be an abbreviation for is less certain. Possibilities include Dee's *Monas,* which figures so prominently later in the text and has associations with the philosophers' stone and the *una res,* or the *magisterium,* referring to the culmination of the alchemical art. Since the *Consideratio* may be a response to the *Fama Fraternitatis,* this may also refer to the "Liber M" referred to in the *Fama*, where "M" most likely denotes "mundi."[109] Whatever the case may be, Gabella says that this "M." has "its origins in the heavens," so in conjunction with his conviction that understanding of the hidden and secret origins of natural things requires tracing their origin, derivation, and development, this leads to a consideration of cosmology.[110]

Cosmology, the *Tabula*, and Dee

Gabella's exposition develops a number of cosmological themes reflecting the *Tabula*'s precepts that the world and all things are derived from the primal monad through the action of the supreme being, that the terrestrial world and the celestial world are mutually interrelated and their connections are essential to the miracles of the *una res,* and that the Sun, the Moon, the air, and the earth are key elements in the process.[111] Gabella starts by dealing with the heavens, based on John Dee's *Propaedeumata aphoristica,* and immediately presents us with how massively Gabella has excerpted others' writings. Dee was particularly sensitive to others' unacknowledged use of his work, having criticized Offusius for plagiarizing his *Propaedeumata* and taken offense at Gerhard Dorn's use of his monas symbol.[112] Imagine, then, his reaction upon reading the opening passage of chapter 2 of the *Consideratio:*

Consideratio	*Propaedeumata*
sig. B2[r/v]	CII (183)[113]
A capite autem mortales, arcessere, sic accipite. Lux & motus, sunt coelestium corporum maxime propria: Inter planetas, Sol, luce propria, omnes alios superat: Et Luna, proprii motus pernicitate reliquos omnes vincit. Merito igitur hi duo, omnium planetarum, excellentissimi censentur.	Ut Lux & Motus sunt caelestium corporum maxime propria, ita inter planetas, SOL, LUCE propria omnes alios superat: & LUNA, proprii MOTUS pernicitate, reliquos omnes vincit. Hi ergo duo, omnium planetarum excellentissimi, merito censentur.

All of chapter 2, with the exception of one passage and a few transitional words, is in fact entirely made up of quotations from Dee's *Propaedeumata aphoristica,* originally published in 1558. Dee reissued this in 1568 with changes reflecting the ideas of the *Monas,* and Gabella uses this edition. There is very little evidence that the *Propaedeumata* had much of an audience, so it is of some interest to find it so extensively used almost fifty years after its last printing. Although Gabella's extensive quotation—some might be tempted to say plagiarism—makes the *Consideratio* highly derivative, there is nonetheless a formative process at work, for Gabella does not slavishly follow just one work or author, but in the first half of the Consideratio he at least pieces passages together to develop his own presentation of alchemy and the *Tabula.*

Gabella's use of the *Propaedeumata* is exemplary of this process. He does not attempt to present all of Dee's ideas and material; there is no treatment of the technical aspects of astronomy, or of the need to calculate the sizes and distances of the planets and stars, or of the principles and mechanics of calculating the relative power of celestial influences, or of the physics of light that is the basis of Dee's astrological theory. Gabella begins not at the start of the *Propaedeumata,* where Dee lays down the foundations of his astrology, but toward the end, with a group of Dee's aphorisms that echo the "Pater ejus est Sol, mater ejus Luna" of the *Tabula.* Following the opening quotation indicating that the Sun and Moon are the most excellent among the planets because they surpass all others respectively in their light and motion, which "are the most distinctive properties of heavenly bodies," Gabella continues by developing their influence. Whereas the Sun, through its light, governs vital heat, the Moon's dominion over humidity is linked to its swift motion:

Consideratio
sig. B2ᵛ

Luna potentissima est, humidarum rerum moderatrix, humiditatisque excitatrix, & effectrix.

Propaedeumata
CIII (184)

LUNA, potentissima est humidarum rerum moderatrix: humiditatisque excitatrix & effectrix.

Consideratio
sig. B2ᵛ

Ut igitur Solis, excellentem lucem, praecipuum vitalis caloris comitatur: ita, cum Lunae motu, mira quadam analogia, conjuncta est ejus vis humiditatis effectiva, & moderatrix.[114]

Propaedeumata
CIIII (184)

UT Solis excellentem LUCEM, praecipuum vitalis caloris moderamen comitatur: ita cum LUNAE MOTU, mira quadam analogia, coniuncta est eius vis, humiditatis effectiva & moderatrix.

Gabella concludes this by quoting Dee's extended aphorism in which the generation of all primary qualities is attributed to the sun's heat, which regulates the cycle of the year as well as the analogous course of the day from morning (spring), through afternoon (summer), evening (autumn), and night (winter).[115] Presaging what increasingly appears to be his mechanical quotation of other sources, Gabella even quotes, without any effort at elucidation, Dee's obscure injunction at the end of this aphorism for those "who investigate the physical mysteries in the unity of the Trinity" and those who seek to hide their work to "apply this aphorism to higher matters":

Consideratio
sig. B3ʳ

Has philosophorum considerationes, ad altiora traducas, & maximum secretum habes, tu, qui trinitatis in unitate Mysteria tractas Physica, & ad noctis multi coloris, nigredinem opus tuum involvendum, ratione, anhelas.

Propaedeumata
CVII (186)

Aphorismum istum ad altiora traducas, & maximum Secretum habes, Tu, qui Trinitatis in unitate, mysteria tractas physica: & ad Noctis multicoloris Nigredine, Opus involvendum tuum, anhelas.

Gabella jumps from this to a very early group of Dee's aphorisms dealing with the behavior of created substance that is the foundation of the mutual interrelation of all creation, including the interdependence of terrestrial and celestial. Creation includes not only what is apparent but also occult and seminal things. All things spherically emit rays of their species so that the entire universe is filled with the rays of all things. The efficacy of these species varies, with substantial species exceeding those of accidents, and the species of spiritual substance exceeding those of corporeal substance:

Consideratio
sig. B3ᵛ

Non solum enim ea tantum esse, asserendum est, quae actu in rerum natura sunt conspicua notaque: Sed illa quoque, quae quasi seminaliter, in naturae latebris extare, sapientes, non ignorantes, docere possunt.

Propaedeumata
III (122)

NOn solum ea Esse asserendum est, quae Actu in rerum natura sunt conspicua, notaque: Sed & illa quoque quae quasi Seminaliter, in naturae latebris, Extare, Sapientes docere possunt.

Consideratio
sig. B3ᵛ

Nam quicquid actu existit, radios orbiculariter ejaculatur, in singulas mundi partes, qui universum mundum suo modo replent. Unde omnis locus mundi radios continet, omnium rerum in eo actu existentium.

Propaedeumata
IIII (122)

QUicquid Actu existit, Radios orbiculariter eiaculatur in singulas mundi partes, qui universum mundum suo modo replent. Unde omnis locus mundi radios continet omnium rerum in eo Actu existentium.

Consideratio
sig. B3ᵛ–B4ʳ

Sed, tam substantia, quam accidens, suam a se speciem exerunt? substantia vero omnis, multo excellentius est quam accidens: substantiarum, enim illa quae corporea & spiritualis est, (vel quae spiritualis facta est) in hoc munere, longe superat illam, quae est corporea, ac ex fluxis coagmentata impuris elementis.

Propaedeumata
V (122–24)

TAm Substantia quam Accidens, suam a se Speciem exerunt: Sed Substantia omnis, excellentius multo quam accidens. Et Substantiarum quidem, illa quae incorporea & spiritalis est, (vel quae Spiritalis facta est) in hoc munere longe superat illam quae est corporea, ac ex fluxis coagmentata elementis.

The foundation for these interrelationships via rays derives from the fundamentally geometrical nature of creation, since the first manifestation of things was produced from the straight line and the circle. This observation marks the transition between the first part of Gabella's chapter dealing with the sun and moon and this second part. This is also the only passage not quoted directly from Dee's *Propaedeumata*. Gabella has not, however, been struck with a fit of creativity; he is merely quoting from his other major Dee source, the *Monas hieroglyphica*:

Consideratio
sig. B3ʳ

Sed porro ex hac consideratione, per lineam circulumque, Prima, simplicissimaque fuit rerum, tum non existentium, tum in natura latentium, involucris, in lucem productio ac repraesentatio.

Monas hieroglyphica
Theorema I, fol. 12r
(154/55)

Per Lineam rectam, Circulumque, Prima, Simplicissimaque fuit Rerum, tum, non existentium, tum in Naturae latentium Inuolucris, in Lucem Productio, representatioque.

Not only does this process of creation link the action of the sun's heat in generating the primary qualities with the mechanism of occult interaction among the species of created things, Gabella also sees it as the basis for the ability to produce wonders by artificially manipulating nature through the principles of pyronomia:

Consideratio
sig. B3ʳ/ᵛ

Nam ex linea & circulo, mirabiles rerum naturalium metamorphoses, fieri a nobis in rei veritate possent, si artificiose naturam ex Pyronomiae institutis recte urgeremus.

Propaedeumata
II (122)

Mirabiles ergo rerum naturalium Metamorphoses fieri a nobis, in rei veritate possent, si artificiose Naturam ex pyronomiae Institutis urgeremus.

This capability resonates with the suggestions of magic in the *Tabula*, where the miracle of the one thing is "the father of all works of wonder in the world" and "has the supreme power because it conquers all subtle things and penetrates all solid things."[116] These powers, however, are not supernatural; they only apply to nature, which is only what has been created by the will of God. To emphasize that such power works within and only changes nature, but can never overturn creation or rival divine power, Gabella quotes Dee's very first aphorism:

Consideratio
sig. B4ʳ

Sed ut Deus (quod nostrum non est considerare) ex nihilo, contra rationes & naturae leges, cuncta creavit, ita in nihilum rerum, aliqua nunquam potest, nisi contra rationis naturaeque leges, per super naturalem ejus potentiam fiat.

Propaedeumata
I (122)

UT DEUS, EX NIHILO, CONTRA rationes & naturae leges, cuncta creavit: ita in Nihilum abire, rerum creaturum aliqua nunquam potest, nisi contra rationis Naturaeque leges, per Supernaturalem Dei potentiam fiat.

Nature has not existed from all eternity, nor is it the result of accident or some natural mechanism; it is the miraculous work of the will of God.

Equally important, however, is the implication that unless God intervenes, nature and creation have a stability and are entirely regular and predictable in the normal course of their operation because they are governed by rational natural laws.[117]

To incorporate the earth along with the sun, the moon, and the heavens into this geometrical cosmology derived from and governed by the straight line and the circle, Gabella turns to Dee's *Monas*. Again there is no acknowledgment of Dee, and Gabella gives a more concrete cast to the ideas by replacing the term "monas" with "star" and "hieroglyphic star," represented by his own diagram instead of Dee's famous symbol (figure 5.13; cf. figure 5.3). This is clearly intentional, because later in the *Consideratio* Gabella employs figures that are quite similar to Dee's (figure 5.20, combining elements of figure 5.5, and figure 5.18, which can be found in figure 5.8). It would appear, therefore, that Gabella was not without models of Dee's original figures. Carlos Gilly has noted the similarity of Gabella's "stella" with the symbol for vitriol, an association that seems to have been intentional, because later in the text Gabella gives a formula for the discovery of the true medicine, the initials of which spell VITRIOLVM.[118] Replacing Dee's symbol of the monas with that of vitriol, however, has some curious effects on Gabella's use of the *Monas*. First of all, eliminating the very term "monas" obscures the core idea of unity at the heart of creation, even though this is what Gabella seeks to establish through his use of Dee. Second, the various numerological analogies Dee develops in the *Monas* have some rationale in relation to the fundamental unit of the monas, but they seem arbitrary when Gabella uses them. Gabella's

Figure 5.13
Stella hieroglyphica, Gabella, *Secretioris Philosophiae Consideratio Brevis,* sig. B4r (by permission of the British Library, from shelfmark 1033.H.6.(4.))

different symbol also seems impoverished; without the separate lunar semicircle and the Aries element, it does not as clearly support some of the manipulations and interpretations as did Dee's original.

Continuing from the previous reference to "the first and most simple manifestation" of things happening by means of the straight line and the circle, Gabella takes up his "star," saying

Consideratio sig. B4$^{r/v}$	*Monas* II (154/55)
... nec sine recta circulus nec sine puncti [*sic*], recta artificiose fieri potest linea. Puncti proinde stellae, ratione res & esse cæperunt primo & quae periphera sunt affectae (quantocunque fuerint) centralis puncti, nullo modo carere possunt ministerio.	At nec sine Recta, Circulus; nec sine Puncto, Recta artificiose fieri potest. Puncti proinde, Monadisque ratione, Res, & esse coeperunt primo: et quae peripheria sunt affectae, (quantaecunque fuerint) Centralis Puncti nullo modo carere possunt Ministerio.
Consideratio sig. B4^v	*Monas* III (154/55)
Stellae itaque hyeroglyphicae, conspicuum centrale punctum, terram refert, circa quam, tum Sol, tum Luna, reliquique planetae, suos conficiunt cursus, & impressiones.	Monadis, Igitur, Hieroglyphicae Conspicuum Centrale Punctum, Terram refert, circa quam, tum Sol tum Luna, reliquique Planetae suos conficiunt Cursus.

Gabella here echoes the clearly cosmological character of the construction of Dee's monas: The point represents the earth, and the circle represents both the sun and also the entire frame of the heavens surrounding the earth. Dee then followed this with a discussion of the lunar semicircle, but since Gabella has eliminated this part of the symbol, he truncates and modifies Dee's next aphorism in a way that makes it quite ambiguous:

Consideratio sig. B4^v	*Monas* IIII (156/57)
Solaribus ita tandem imbui radiis appetat, ut in eundem quasi transformata, toto dispareat Caelo: donec aliquot post diebus, omnino hac qua depinximus figura appareat.	Solaribusque ita tandem inbui Radijs appetat, vt in eundem quasi Transformata, toto dispareat Caelo: donec aliquot post Diebus, omnino hac qua depinximus, appareat corniculata figura.

In the context of Gabella's text the central point, or earth, is the most proximate object desiring to be imbued by solar rays, but the disappearance and reappearance of the earth and its depiction by Gabella's lens-shaped

Figure 5.14
Lens-shaped luna, Gabella, *Secretioris Philosophiae Consideratio Brevis*, sig. B4v (by permission of the British Library, from shelfmark 1033.H.6.(4.))

figure makes little sense (figure 5.14). Only by quoting in modified form Dee's next aphorism does Gabella make it clear that he has the moon in mind.

Consideratio sig. B4ᵛ	*Monas* V (156/57)
Ex Lunari certa simulacro, & solare complementum perducto: factum sit vespere & mane dies unus, qui est primus, quo lux, philosophorum apparuit.	Et Lunari certe Semicirculo ad Solare complementum perducto: Factum est Vespere & Mane Dies vnus. Sit ergo Primus, quo Lvx est facta Philosophorum.

This observation links the genesis of the "star" with the first day of Genesis and marks the conclusion of the creation of the universe's circular and celestial components as well as of light, which is so preeminently connected with the role of the sun and moon as Gabella has already developed them. In fact, at this point he quite cleverly explicates the "light of the philosophers" with a reference to Dee's *Propaedeumata*:

Consideratio sig. B4ᵛ	*Propaedeumata* XXII (130)
Nam sicut primi motus privilegium est, ut sine eo, torpeant omnes reliqui, sic primae & praecipuae formae sensibilis (nimirum lucis) ea est facultas, ut sine ea ceterae formae omnes agere nihil possint.	Sicut primi motus privilegium est, ut sine eo torpeant omnes reliqui, sic primae & praecipuae Formae sensibilis, (nimirum LUCIS) ea est facultas, ut sine ea caeterae formae omnes agere nihil possint.

Gabella quotes here the crucial passage in the *Propaedeumata* evoking the physics of light, which Dee derived from Roger Bacon, al-Kindi, and Robert Grosseteste. This physics was the basis for Dee's theory of astrology, because it establishes that all influences, including the occult, propagate and behave just as light does and so can be studied using the principles of geometric optics.[119] The diffusion and interaction of rays of influence from all created objects, both celestial and terrestrial, is for Dee the key mechanism interconnecting and unifying all creation. Gabella's usage of this in conjunction with Dee's *Monas* serves as a quite appropriate embodiment of the precepts from the *Tabula* that "what is below is like that which is above, and what is above is like that which is below," "and as all things were made from the one by the contemplation of the one, so all things are born of this one thing by adaptation."[120]

Although the circular components of the Gabella's star and Dee's monas relate to the heavens, Gabella continues with his direct quotations from the *Monas* and follows Dee in relating the sublunar elements to straight lines, for just as the flowing of a point defines a line, the elements follow straight lines as they return to their natural seats when displaced. Gabella's hieroglyphic star thus includes a cross, representing the four elements (earth, air, fire, water) and the four qualities (cold, hot, dry, moist) (figures 5.15 and 5.13). Because of the relation of the lines composing the cross of the elements to the diameter of the solar circle, the proportions of Gabella's star embody the relation of the elements to the sun and moon. Here we get the first hints of alchemical implications, since these lines, by dividing the circle, suggest how the "magic of these four elements" serves to separate the sun and the moon into their own lines.[121]

In quoting Dee's aphorisms on the cross of the elements in their entirety, Gabella repeats Dee's numerological interpretation of the cross, in which the four arms and the point joining them can variously refer to the ternary, the quaternary, the septenary, the octonary, and, through the "Pythagorean"

Figure 5.15
Star with cross, Gabella, *Secretioris Philosophiae Consideratio Brevis*, sig. B4v (by permission of the British Library, from shelfmark 1033.H.6.(4.))

tetractys, the denary.¹²² Here Gabella's rather mechanical compilation of extracts begins to appear arbitrary. Although the "Pythagorean" tetractys and the numerological portrayal of cosmogony popular in the Renaissance are not entirely foreign in the framework of Gabella's cosmology, they are much less intrinsically motivated than in Dee's case, in which a numerological dimension is implicit in the concept of the monas from the start. Gabella's repetition of Dee's association of the numerological values of various Roman numerals with the rationale for the placement of these letters within the alphabet makes even less sense. Gabella completely misses or ignores the dimension of the *Monas* presenting a new kind of writing embodied in the symbol of the monas and the rules for its manipulation. The numerology also does not carry through and inform the rest of Gabella's discussion as it does in the *Monas,* and Gabella misses the "a ternario in vnitatem per binarium divisum" that links Dee's discussion of the tetractys to the Trithemian *Tabula,* which suggests that Gabella was not directly aware of Trithemius's interpretation. Rather, the majority of his embodiment of the Trithemian sense of the *Tabula* derives from Dee's *Monas:* By adopting parts of the *Monas* he has also picked up some of the resonance of Trithemius, but because he shows no direct knowledge of Trithemius and he has used Dee so selectively, he misses the full import of both Dee's *Monas* and the Trithemian *Tabula.*

The final piece of Dee's monas is the symbol for Aries, evoking the beginning of creation as the first sign of the zodiac, the vernal equinox, and the fiery trigon, signifying the use of fire. Gabella's "hieroglyphic star" lacks this piece, but, continuing to quote from the *Monas,* he follows through paralleling the construction of the monas to its conclusion by generating this element from the two halves of his solar circle joined by a point (figure 5.16).¹²³ The culmination of Gabella's "hieroglyphic star"

Figure 5.16
Aries, Gabella, *Secretioris Philosophiae Consideratio Brevis,* sig. C4v (by permission of the British Library, from shelfmark 1033.H.6.(4.))

as embodying all of cosmology in its combination of sun, moon, elements, and Aries is that the hieroglyphic signs of all of the other planets may be constructed from combinations of components of the "star," as taught by the wise magicians (actually John Dee):

Consideratio sig. D1^r	*Monas* XII (160)
Unde sapientes magi, recte nobis quinque planetarum tradidere notas hieroglyphicas, compositas autem omnes, ex Lunae vel Solis characteribus cum elementorum aut arietis hieroglyphico signo.	Antiquissimi Sapientes Magi, quinque Planetarum, nobis tradidere Notas Hieroglyphicas: Compositas quidem omnes, ex Lunae vel Solis Characteribus: cum Elementorum aut Arietis Hieroglyphico Signo.

How this actually works is not clear from Gabella's cryptic symbols (figure 5.17). Dee follows this with a detailed diagram (figure 5.6) and discussion, but, in contrast to the aphorisms on the numerology of the cross where Gabella includes material he never develops, in one of the rare times Gabella does not quote the entirety of Dee's aphorism, he seems to cut things short prematurely and provides less development than available in Dee.

Gabella does not exclusively quote from Dee; in certain passages he seems to develop his own discussion, although in some of these instances it turns out he is also excerpting other sources. If the *Consideratio* is the hastily composed text that it seems to be and draws liberally on Gabella's compilations of commonplaces, much of the *Consideratio* may be entirely a pastiche of extracts, some of which I have been able to recognize or to find and some of which have, until now, eluded me but may be identified in the future. Dee's *Monas* treats what are supposed to be concrete natural processes in an extremely abstract fashion, so in these other passages, Gabella

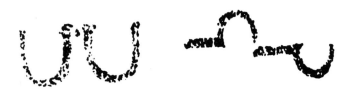

Figure 5.17
Signs of planets? Gabella, *Secretioris Philosophiae Consideratio Brevis*, sig. D1r (by permission of the British Library, from shelfmark 1033.H.6.(4.))

elaborates on the implications of this cosmology, "fleshing out" the *Tabula*'s very sparse precepts and the abstract concepts that Trithemius and Dee built on those precepts. The first and foremost of these implications is the unity and interconnectedness of creation: The abundance and multiplicity found in nature are encompassed within a larger unity because, echoing the *Tabula*'s "all things were made from the one," "nothing is born without unity or without the point."[124] Despite the separation and the apparently radical difference between the elements of the terrestrial sphere and the heavens, both are equally part of the universe, without which it could not exist.[125] Serving as the medium joining these is the element air. Although this element separates the heavens and the earth, it also joins them and is the basis for their concourse. Air "receives a virtue from the earth below, and at the same time hermetically transfuses the strength of the stars to the earth."[126] Gabella here echoes Cornelius Agrippa's comments on air, which played a prominent role in some pseudo-Paracelsian tracts as well, within Agrippa's discussion of the features applicable to the practice of magic within the elemental world. Agrippa notes that air has been considered the medium of the other elements and the "spiritus mundi" in that it accepts and transmits the influx of celestial and other species and thus serves as the instrument by which the "anima mundi" exercises its influence throughout the world.[127]

In all, Gabella's cosmology seems to embody the *Tabula*'s assertion that "what is below is like that which is above, and what is above is like that which is below" and its prescription to "ascend from the earth to heaven, and then again descend to earth, and unite together the power of the superiors and the inferiors." This interconnectedness on the cosmic level is also reflected in the microcosm, that is, humankind, in which the various parts are united by blood, spirit, and other humors flowing through veins and arteries. Similarly, the earth itself is also something of a microcosm writ large, since it has its own channels like veins and arteries through which flow various humors.[128]

Alchemy
This last observation provides a transition to the alchemical parts of the *Consideratio*, for the flow of these humors through the interior of the earth evokes an idea of the formation of the metals that was at the heart of several related conceptions of the foundation of alchemy in the West.

Gabella says that in containing the four elements, these humors contain the seeds of all things, some of which are turned to stone by a petrifying liquid and some of which are hardened to become the earth of the metals.[129] What Gabella echoes here evokes several possible strands of alchemical theory available in the early seventeenth century. Because of the shifting and derivative character of Gabella's text, it is difficult to identify a definitive commitment on his part to a single theory of alchemy. Like his quotations, Gabella's theoretical evocations span a range of ideas available in the early seventeenth century.

At his most basic, Gabella's ideas reflect the hint in Aristotle that minerals and metals are the product of the exhalation of an "earthy smoke" and a "watery vapor" from inside the earth. When these become imprisoned within the earth and combined, they formed various metals and minerals according to the different proportions of smoke and vapor.[130] Arabic alchemists of the eighth and ninth centuries elaborated these ideas into what became known as the mercury-sulfur theory of the metals. In the thirteenth and fourteenth centuries these ideas were taken up in the West and became the core of Latin alchemy, most importantly in the *Summa perfectionis* of pseudo-Geber, which is perhaps one of Gabella's major alchemical sources.

According to the mercury-sulfur theory, Aristotle's earthy smoke and watery vapor were first transformed into sulfur and mercury, respectively, which then combined in the earth under the influence of the planets to form metals.[131] Reflecting this more developed theory, Gabella says that the search for the One Medicine should begin with the sources of metals and minerals: sulfur and mercury, which are essential principles, not ordinary elements.[132] From these the alchemist attempts to produce an elixir or the "philosophers' stone," or the "Universal Medicine," the "One Medicine," or "M" as Gabella variously calls it. Gabella's "philosophers' mercury," as the first matter of all things suggests a view similar to that of the "mercury-alone" theory of pseudo-Geber that the philosophers' stone is philosophical mercury containing an inner and nonvolatile or nonflammable sulfur, rather than a mixture of philosophical mercury and philosophical sulfur, which was the more common belief. In either case, to transform them into their philosophical forms, the earthly forms of mercury and sulfur had to be rid, through spoliation, of their impure characteristics (fluidity and humidity for mercury, flammability and earthiness for sulfur). As pseudo-Geber expressed it: "imperfect bodies have superfluous humidities and combustible sulfurity, with blackness corrupting them, an unclean, feculent, combustible and very gross earthiness." Spoliation yields a substance in which only mercury and sulfur remain.[133]

The alchemist's job was, thus, to use a series of standard chemical operations through which accidental imperfections were purged and the remaining purified substances were combined and unified, first into philosophical sulfur and mercury and then into the stone. This is the essence of Gabella's definition of alchemy, or the "spagyric art," as that which "teaches how to distinguish the pure from the unpure, . . . and how to separate and recompose substances according to the course of nature." It "distinguishes the clear from the confused, the subtle from the gross, the light from the heavy, fire from air, air from water, and water from earth" just as the creator did in the first creation. The alchemist is thus an imitator of the creator, and alchemy a replication of creation on a local scale.[134] The association of alchemy and creation brings us back to the cosmological dimension of the *Tabula*.

This cosmological dimension is also prominent in Michael Sendivogius's *Novum lumen chemicum*, another of Gabella's alchemical sources. Mercury and sulfur are key elements in Sendivogius's work as well, with a philosophical mercury taking the leading role as the product of the seed (*sperma*) injected into the central cavity of the earth by the elements that yield a "sal nitrum," or univarsal salt, as the principle of all things. Everything originates from this universal seed, because nature is "one, true, simple, and self-contained," reflecting the "una res" of the *Tabula*.[135] This seed, after digestion in the earth's womb, is driven up through pores and veins as a vapor—philosophical mercury—leaving behind rocks, minerals, and metals depending upon its contact with philosophical sulfur in the soil, differences in heat, and the presence of impurities. On the earth's surface this philosophical mercury nourishes plants and, under the influence of celestial rays, is drawn up into the atmosphere where it receives "the power of life from the air." This process produces a "water" in which the *sal nitrum* of the earth is imbued with the "power of life" and returns to earth to combine with "the fatness of the earth"; that is, philosophical sulfur. Sendivogius thus introduces an important vitalist element to alchemy and through this becomes the root of a school of alchemical thought that sought the principle of the metals in a "universal salt" usually identified with saltpeter or *sal nitrum*.[136]

I have noted above a possible reflection of Sendivogius in Gabella's comment on Hermes' being eclipsed by modern alchemy. Gabella's definition of nature as the will of God and the role he attributes to celestial rays, which he elaborates through the material from Dee's *Propaedeumata*, are also found in Sendivogius.[137] Gabella reflects Sendivogius's alchemy when he advises the alchemist not to seek the essential principles of things in existing substances because they are already "dead." The alchemist, in contradiction to what was implied above, must not extract this principle through the decomposition of existing things. He must, instead, find it in the "dew of heaven": the vapors that have been generated from the "semina" of the elements in the center of the earth, driven to the surface purified of their excrements, elevated through the atmosphere and exposed to celestial rays, from which they return to earth actuated by the heavens to work the miracle of the "one thing."[138]

In a comment summarizing the genesis of the "one thing"—the philosophers' stone—that integrates the alchemical process with his evocation of the *Tabula* through Dee's *Propaedeumata* and *Monas*, Gabella indicates that fully refined philosophical mercury, "the mercury of Hermes and of all the Philosophers . . . is water, the water that falls from the sky as rain and which the Sun, as its father, extracts from the earth each day in a very fine vapor and takes up into the region of the air where impressions are made, and here by the same force by which the moon, its mother, controls things below, it condenses into rainwater and falls in drops by its own weight. It is moved around willy-nilly by the air or the wind (which is, after all nothing but the movement of the air) until it lands upon its central point, that is the earth, its nursemaid, who then carries it in her lap."[139] This process is similar in a general way to that depicted in Sendivogius's discussion of the ascent and celestial actuation of philosophical mercury, which he presents in a context where he also quotes the *Tabula*'s "its father is the sun, its mother is the moon, and the wind has borne it in her womb."[140] Gabella concludes his observations on water as philosophical mercury with a passage drawn directly from Sendivogius to the effect that "all water that is without spirit may be congealed by heat, and that which has spirit may be congealed by cold. He who understands how water can be congealed by heat and how the spirit can be joined with it, will certainly discover something a thousand times more precious than gold or anything else. Therefore, the alchemist should separate the spirit from the water and allow it to decay so its seed appears. After discarding the waste, he should reintroduce the spirit into the water from above, and effect a conjunction between the two, which will generate an offspring different from its parents."[141]

Consideratio sig. E1[r]	Novum lumen chemicum 467
Sed ex praxi certum est, quod omnis aqua, calore congeletur, si est sine spiritu, & congeletur frigore, si habet spiritus. Qui vero scit aquam congelare calido, & spiritum cum ea jungere, certe, rem inveniet millefies pretiosiorem, auro & omni re. Igitur efficiat Spagyrus, ut separetur spiritus ab aqua, & putrescat, ut granum appareat, postea rejectis faecibus, reducat spiritum ex alto in aquam, & faciat conjungere simul, ista enim conjunctio, sive digestio, generabit ramum dissimilem forma parentibus.	Congelatur enim omnis aqua calore si est sine spiritu, congelatur frigore si habet spiritum; sed qui scit congelare aquam calido, & spiritum cum ea jungere, certe rem inveniet millefies pretiosiorem auro & omni re. Efficiat igitur ut separetur spiritus ab aqua, ut putrescat, & appareat granum; postea rejectis faecibus reducat spiritum ex altro in aquam, & faciat conjungere simul: illa enim conjunctio generabit ramum dissimilem forma parentibus.

This presentation of Gabella's discussion of alchemy is somewhat misleading, because it may give the impression of coherence. It is, however, difficult to determine what he is presenting. Identification of his sources, as in the case of Sendivogius, is no reliable guide to what he believed. As with the parallel references to Hermes that have very different meanings, Gabella ignores or strips Sendivogius's alchemy of key elements. In particular, he makes no reference to a "central salt" or *sal nitrum* or aerial nitre as the first matter of metals.[142] We cannot assign him with any assurance to a Sendivogian school that sought the ultimate principle in a "universal salt."[143] The last several chapters of the *Consideratio* actually cover what seem to be the same ground several times with slight variations interspersed with passages that sound like recipes, analogies to natural phenomena that serve as examples of the "Great Work," comments on concealing alchemical secrets, and criticism of Galenists and academic doctors. This further suggests that the later parts of *Consideratio* are as much a compilation of extracts or paraphrases of other sources as are the sections derived from Dee. A particularly striking instance of this follows Gabella's criticism of the Galenists' rejection of distillation and his rejection of what he considered Paracelsus's cures of disease by magical "characters, words, and spoken formulae."[144] After these comments comes an extended quotation from the commentary of Jacques Gohory, writing as Leo Suavius, on the fifth chapter of the fifth book of Paracelsus's *De vita longa,* a section of an obscure work that Gohory admits is particularly difficult and obscure.[145]

This quotation is odd in a number of ways. It seems to be about the extent of Gabella's knowledge of Paracelsus; that is, a second-hand knowledge through fragmentary extracts from a commentary on Paracelsus. Despite several mentions of Paracelsus, these extracts are not always complimentary, and when they are, they seem to pander to the references to Paracelsus in the Rosicrucian tracts. This selective and isolated use of Gohory also reinforces Gabella's lack of a direct knowledge of Trithemius's treatment of the *Tabula,* because Gohory defends Trithemius extensively against detractors who charged him with illicit magic, in the course of which he also mentions Trithemius's concept of the numeric progression from and return to unity: "per similitudinem binarij a ternario redierit in vnitatem, deinceps iam facile in denarium consurget."[146] Gohory admittedly does not discuss this aspect of Trithemius extensively; Gohory was always more interested in the magic of the *Steganographia* and the *Polygraphia* than in Trithemius's "alchemical magic."[147] Nonetheless, Gohory's work contains indications of Trithemius's numerological alchemical magic, but Gabella makes no use of them, just as he does not follow through on the numerology of Dee's monas to Dee's evocation of the Trithemian association of alchemy with the progression of the tetractys to the restoration of unity.

In the passage from Gohory quoted by Gabella, commenting on Paracelsus's treatment of the magical preservation of health of the human soul—man's "praeternatural" body—Gohory relates the highly figurative and metaphorical language Paracelsus uses in dealing with this magic to the alchemical process and Geber's idea of three orders of the Great Work:

in the fourth chapter it says that the Necrolii are forbidden a long life, that is, they are barred from the Great Work, which Geber calls the Third Order. The elemental substances in their crude state of blackness (according to Raymond Lull they are of a blackness blacker even than black) can produce a solution for the dead. The Scaiolae are the four elements in the vitriol of Venus after they have been purified. In the Necrolii, that is the First Order of the Work, are contained ridiculous travesties, sophistical preparations indeed, that do not withstand the test of fire. Yet they do shed light on the Cyphant, in other words, on the formation of the embryo or infant (as Arnold and Lull refer to it), and which Geber refers to when he says that the instruction is not complete until the preparations of the first order have been made. [Gabella inserts the aside that these preparations were adequately shown in the previous chapter.] Those who get to this stage who do not advance to the other orders and therefore do not produce the pyraustae are referred to as Alloeani by Paracelsus, since they are superficial changers of the form and sophistical white-washers of the tinctures of Venus and the Moon.[148]

Consideratio sig. E4^v–F1^r	Gohory, *Scholia* 313
Necrolii sive Necrolici, capite iiii excluduntur cancellis longae vitae, id est operis magni qui tertii ordinis a Gebro nuncupatur. Nam materiae elementares, rudes & nigrae, a Reymundo, Nigrum nigrius nigro, solutionem mortis ferunt. Scaiolae, sunt elementa quatuor in vitriolo Veneris, depurata. In Necroliis, id est, primo ordine in sunt hypermenica figmenta, seu praeparationes Sophisticae, quae judicium cineritii non sustinent: aperiunt tamen fenestram in Cyphanto, id est, ad formationem nostri Embrionis seu infantis (ut Arnoldus & Lullius loquuntur) quod sic Gebor exponit: Sine praeparationibus primi ordinis, non perficitur Magisterium. . . . Qui in his figunt pedem, nec ad alios ordines progrediuntur, ut pyraustas habeant, hi appellantur a Paracelso Alloeani id est, immutatores formarum superficiales & dealbatores Veneris & Tinctores Lunae sophistici.	Necrolij siue Necrolici cap. iiij. excluduntur cancellis longae vitae, id est, operis magni qui tertij ordinis a Gebro nuncupatur. Nam materiae elementares rudes & nigrae, a Raymundo, Nigrum nigrius nigro, solutionem mortis ferunt. Scaiolae videntur elementa quatuor depurata. In Necrolijs, id est, primo ordine in sunt ypermenica figmenta, seu praeparationes sophisticae, quae iudicium cineritij non sustinent: aperiunt tamen fenestram in Cyphanto, id est, ad formationem nostri Embryonis seu infantis (ut Arnaldus & Lullius loquuntur) quod sic Geber exponit: Sine praeparationibus primi ordinis, non perficitur magisterium. Qui in his figunt pedem, nec ad alios ordines progrediuntur, appellantur a Paracelso Allaeani, id est, immutatores formarum superficiales, vt dealbatores veneris & tinctores lunae sophistici.

I quote at length to give some sense of the character of the language that Gabella seems to adopt without much prior preparation, suggesting little prior thought. Although this material is opaque and seems to veer in a very different direction from earlier parts of the *Consideratio,* and Gabella suggests that the preparations of the first order are those he has previously discussed (even though they seemed complete in his earlier discussions), he appears to have found in Gohory's comments an expression of some progression of stages in the great work corresponding to Geber's three stages. So we find, among others, a further veiled indication relating to the advancement of mercury to the second and third orders: "but if the Sun or the Moon is to be added to this crude preparation something must first be removed, in other words, the receiving material must be prepared by transmutation: this is the extent of the medicine of the second order. But the greatest Adech exceeds even this with the medicine of the third order, for Mercury is first prepared philosophically and then accurately and fully gathered together."[149]

Consideratio sig. F1ᵛ–F2ʳ	Gohory, *Scholia* 316
Quibus si rudi praeparatione inseratur Sol, vel Luna, esset aliquid excipiendum, id est, necessaria esset praeparatio materiae recipientis projectionem, nempe in medicina secundi ordinis tantum. Sed maximus Adech antevertit, id est, in tertii ordinis medicina, Mercurius philosophice praeparatus, & circulatus ad amussim omnibus suis numeris, deducit propositum nostrum, quippe eandem materiam propositam (secundum Gebrum) ducit in operis progressum. . . .	Quibus si rudi praeparatione inseratur Sol vel Luna, esset aliquid excipiendum, id est, necessaria esset praeparatio materiae recipientis proiectionem, nempe in medicina secundi ordinis tantum. Sed maximus Adech anteuertit. Id est, in tertii ordinis medicina Mercurius philosophice praeparatus & circulatus ad amussim omnibus suis numeris deducit propositam (secundum Gebrum) ducit in operis progressum. . . .

Gabella's extract continues with what seem to be several additional stages and transformations of the "Nymphidic spring water," the actions of the "water of the Scaolii" and the "lightening of the Scaolii," the cooperation of the "White Sun" with the Moon, and the "King" turning red.[150] We are then brought up short when Gabella seems to undercut the entire Gohory extract by including an indication of Gohory's despair over making sense of Paracelsus's statements, since Gohory says that "all that is written at the close of the book concerning travesties and the Nymphidic spring water lead to obscurity, since they pervert the traditional order."[151]

Or is he just mechanically quoting? Gabella does not seem to acknowledge Gohory's skepticism. He links Gohory's hints of some progression to a celestial cycle as the model for terrestrial astronomy by returning to Dee's *Monas,* picking up where he had left off some ways back with the idea that the signs of all five planets can be derived from elements of the monas symbol suitably arranged (figure 5.18).[152] It is almost as if Gabella returned to Dee for an intelligible explanation of the mystifications of the Gohory passages, because his quotations from Dee include Dee's association of the sequence of these planetary transformations of the monas with

Figure 5.18
Lunar mercury, Gabella, *Secretioris Philosophiae Consideratio Brevis*, sig. F3r (by permission of the British Library, from shelfmark 1033.H.6.(4.))

Figure 5.19
"Albification"—Lunar venus, Gabella, *Secretioris Philosophiae Consideratio Brevis*, sig. F3r (by permission of the British Library, from shelfmark 1033.H.6.(4.))

the series of revolutions of the alchemical process. He thus repeats the connection of the sequence of Saturn, Jupiter, Moon, and Mercury with the four revolutions of the lunar nature around the earth, by which application of the moon to the elements produces "albification" (figure 5.19):

Consideratio sig. F2ᵛ–F3ʳ	*Monas* XII (162)
Lunaris enim aquosa natura, dum per Scaiolorum scientiam, circa nostram sit semel revoluta terram, Saturnus mystice dicitur. Et eadem de causa Jovis quoque habet nomen. Sed Lunam, tertia elementatam vice, obscurius sic notamus, quem Mercurium Vocare so-	Lunaris nostra Natura, dum per Elementorum scientiam, circa nostram sit semel reuoluta Terram, Saturnus mystice dicebatur. Et eadem de causa, Iovis quoque habebat nomen: istamque retinebat figuram secretiorem. Et Lunam, tertia elementatam vice,

lent: qui quam sit Lunaris videtis. Istum quarta revolutione produci, licet quidem Velint Sophi, nostro Secreto proposito tamen non erit id contrarium. Modo purissimus spiritus Magicus, loco Lunae, τῆς λευκάνσεος administrabit opus. Et sua virtute spirituali, nobiscum Solus, per medium quasi naturalem diem, sine verbis, hyeroglyphice loquatur: in purissimam simplicissimamque & albissimam a nobis praeparatam terram, geogamicas istas quatuor introducens, imprimensque figuras. Vel illarum loco illam alteram proximam.

obscurius sic notabant. Quem, Mercurium vocare solent. Qui, quam sit Lunaris, videtis. Istum, Quarta Reuolutione produci, licet Quidem velint Sophi: nostro Secreto proposito tamen, non erit id Contrarium: Modo Spiritus Purissimus Magicus, loco Lunae, τῆς λευκάνσεος administrabit Opus: & sua virtute Spirituali, nobiscum Solus, per Mediam quasi Naturalem diem sine verbis, Hieroglyphice loquatur: in Purissimam Simplicissimamque, a nobis praeparatam Terram, Geogamicas, istas 4 introducens, Imprimensque figuras: vel illaram loco, illam alteram.

Gabella continues by excerpting the subsequent aphorism (XIII) of Dee's *Monas* that continues the work into three solar revolutions, from Mars to Venus and the Sun, which unite the lunar mercury with solar sulfur, completing the lunar and solar magic of the elements (figure 5.20).[153] He thus returns, through Dee, to the inspiration of a cosmological interpretation of the *Tabula smaragdina* for an understanding of the alchemical process.

He also links this with Sendivogius, by incorporating at the end of this same chapter a quotation from the *Novum lumen,* in which he inserts an explicit reference to Vitriol:

care, therefore, must be taken when such an operation takes place in the Vitriol to ensure that the central heat can change water into air, so that it can spread out over the flat earth and scatter residue, with the aid of the rain, throughout the channels of the earth. Finally, the opposite will also come about: the air will turn to water of a particularly fine type. This occures if you bring about the overwhelming of the gold and silver by the Old Man, that is, our aqueous Mercury, so that the water consumes them: eventually he will die and be consumed as well. The ashes of the

Figure 5.20
Rehabilitation of the metals by the Solar revolution, Gabella, *Secretioris Philosophiae Consideratio Brevis,* sig. F3v (by permission of the British Library, from shelfmark 1033.H.6.(4.))

gold are then to be sprinkled on the water, and the water boiled until it is ready. You will then have a medicine for curing leprosy. But take care that you do not use cold instead of hot, or hot instead of cold. Mix like natures togeather, but if you must use a substance that does not occur in nature, then separate it until it resembles a natural substance. In the end—by the Will of God—the Great Work is achieved not by hand but by fire.[154]

Consideratio sig. F4[r/v]	*Novum lumen chemicum* 471
Igitur Cura, ut operatio in Vitriolo talis sit, nimirum ut Calor Centralis, aquam possit mutare in aerem, ut in planitiem mundi egrediatur, & residuum beneficio Diluvii, per poros Terrae spargat, Et tunc in opposito, aer Vertetur in aquam, multo subtiliorem. Hoc nimirum sit, si Seni sive Mercurio nostro aquoso, aurum, & argentum deglutire dabis, ut aqua consumat illa & tandem ille etiam moriturus comburatur: Cineres vero auri spargantur in aquam, postea coquito ea, donec satis est. Et habebis medicinam Curandi lepram. Cave tamen ne frigidum pro calido aut calidum pro frigido sumas. Naturas naturis misce, si aliquid est, quod contarium naturae, (si quidem una tibi est necessaria) separa illud, ut natura naturae similis sit, Quod tandem Volente Deo Op. M. fit igne non manu.	Fac igitur ut operatio in talis in terra nostra sit, ut calor centralis aquam possit mutare in aerem, ut in planitiem mundi egrediatur, & residuum, ut dixi, per poros terrae spargat, & tunc in opposito aer vertetur in aquam multo subtiliorem quam fuit prima: & hoc sic fiet, si seni nostro aurum, & argentum deglutire dabis, ut ipse consumat illa, & tandem ille etiam moriturus comburatur, cineres ejus spargantur in aquam, coquito eam donec satis est, & habes medicinam curandi lepram. Hoc saltem animadverte, ne frigidum pro calido, vel calidum pro frigido accipias, naturas naturis misce, si aliquid est quod contarium naturae, (siquidem una tibi est necessaria) separa illud, ut natura naturae similis sit, hoc fac igne non manu, & scito si non sequeris naturam vana esse omnia: & hic tibi dixi per sanctum Deum quod pater filio debet: qui habet aures audiat, & qui habet sensus animum advertat.

Once again, Gabella has borrowed directly from Sendivogius but blended the extract into his own structure, ignoring or expunging ideas unique to Sendivogius and incorporating phrasing concordant with his own apparent commitment to a theory centered on vitriol.

This is driven home in Gabella's last chapter, devoted entirely to vitriol or "chalcantum" as the "lunarium" of the philosophers, reflecting the importance of vitriol in the alchemical thought at the court of Hesse.[155] "All metals can be reduced to a vitriol, resembling their aqueous source," thus emphasizing the importance of water as the fundamental material. Within the earth, metals are formed from the vapors released as this vitriol boils. Through this stone—vitriol—therefore, the alchemist knows the bowels of the earth, and by purifying it the "Hidden Stone," the "True Medicine," is discovered.[156]

Conclusion

Dee's *Monas hieroglyphica* and Gabella's *Consideratio brevis*, both embodying the idea of alchemy as an *astronomia inferior*, provide interesting interrelated cases of the transmission and development of ideas in the Renaissance. The idea of an interrelation of the celestial and the terrestrial provided a powerful inspiration for Dee from the beginning of his career. His discovery of Trithemius's idea of alchemy as a magic grounded in this interrelation was a galvanizing experience. Trithemius's commentary on the *Tabula smaragdina* confirmed Dee's idea that the arrangement and course of the celestial bodies directed terrestrial events and modeled the alchemical processes in the elemental realm, but that the alchemical processes of the terrestrial sphere revealed fundamentals of cosmic genesis, structure, and interrelationships. What Dee perhaps focused on most in Trithemius was his numerology: Trithemius's numerology provided a springboard for Dee's pursuit, as a mathematician convinced of the importance of numbers as revealing the essential features of the cosmos, of the numbers behind creation. In the *Monas* Dee built on the *Tabula* and Trithemius's interpretation but added elements of his own interest in mathematics, particularly geometry, of which Trithemius had no idea, as well as Dee's recent studies of Hebraic cabala. With these elements, Dee constructed a natural philosophy integrating a language of creation through his own "cabala of the real," which provided both a total reform of all intellectual disciplines based on a knowledge of nature and, from this, both a formula for carrying out alchemical work and the key to the structure of the cosmos.

Gabella, despite his haphazard and derivative construction of the *Consideratio* by reprocessing others' texts for his own program, pursued a project similar to Dee's in that his overall objective seems to have been to ground traditional ideas of the alchemical process in a larger cosmic framework derived from the nature of creation. That Gabella based his project on works of Dee, which have never been considered to have had the wide readership and influence of the better-known works of Renaissance occultism, is of some interest. Most interesting is Gabella's use of Dee to evoke a cosmological interpretation of the *Tabula smaragdina* deriving ultimately from Trithemius but without direct knowledge of Trithemius himself. Gabella clearly knew the *Tabula* itself and was quite perceptive in picking up its development in the *Monas* and in relating this to his awareness of Sendivogius. Even more surprising, he knew Dee's *Propaedeumata* and saw how it could be related to the *Monas* and even to the *Novum lumen*. But it is also clear that Gabella's major interest was in alchemy alone, because he drew very selectively from Dee, showing no interest in or even understanding of Dee's ideas of the cabala of the real, the monas as a writing, the reform of the disciplines, the spiritual and religious themes associated with adeptship, or even of astronomy as a celestial alchemy. His adoption of Sendivogius was equally selective.

This observation of Gabella's limited understanding of Dee and the complete reinterpretation of Dee's symbol that Gilly notes call into question Frances Yates's suggestion that Dee and Gabella had connections with the Rosicrucian phenomenon.[157] If the *Consideratio brevis* was the product of a disciple Dee recruited during his travels in eastern Europe, or even of the student of such a disciple, we ought to expect a much more faithful sense of the really unique elements of Dee's natural philosophy. Gabella's use of the *Monas*, however, seems to draw on elements that accord most easily with readily available understandings of alchemy through the *Tabula* while ignoring or not fully understanding Dee's most novel contributions. Although Gabella does include elements of Dee's numerology, this is a result of apparently hasty copying, since he does not follow through on it, indicating that he has neither an interest in how it functions in the *Monas* nor any inkling of how central it was in Trithemius's conception of the *Tabula*.

The other part of the Rosicrucian connection is equally tenuous. Not only is there little evidence of a direct connection of Gabella with Dee, there is likewise little evidence that Gabella was an integral part of the genesis of the Rosicrucian texts. He shows little interest in medicine and healing, which are an important part of the humanitarian agenda of the Rosicrucian texts. Whereas the neo-Paracelsian character of the alchemy and medicine of the Rosicrucian texts is well established, Gabella's relationship to Paracelsus is notably ambiguous. Although he pays homage to the name at strategic points in his text, Gabella criticizes Paracelsus on critical points and in others presents a post-Paracelsian alchemy despite any homage he may pay to the name.

The Rosicrucian manifestos also suggest that the movement that they proclaim was grounded in a spiritual renewal based in some special divine revelation, but Gabella seems more interested in concrete alchemical processes. Ultimately, to use Brann's terminology, Gabella's hermeticism is extroverted rather than introverted. Although the sources he uses hint at the need for divine illumination, Gabella does not emphasize this, nor does he emphasize that the process he presents has any spiritual dimension either in transforming the soul of the alchemist or in facilitating spiritual ascent to divine illumination. He is most interested in translating the ideas of the *Tabula* into concrete operations in the material sphere, thus reverting to the older notion of the *Tabula* as a veiled recipe.

Notes

1. Originally published in 1564, the *Monas* was reprinted at Frankfurt in 1591 and was included in both the 1602 and the 1659 editions of the *Theatrum Chemicum*. Besides this, there are a number of manuscript copies of the Latin text as well as manuscript translations into German and English, suggesting that the *Monas* was sought after and cherished in certain circles. The work, or more especially, Dee's unique symbol of the hieroglyphic monad, seen inside the oval form on the title page (figure 5.2), was widely cited and adopted. C. H. Josten, "Introduction," in "A Translation of John Dee's *Monas Hieroglyphica,* Antwerp, 1564, with an Introduction and Annotations," *Ambix,* 12 (1964): 90–99. The following list of manuscript copies and translations is not necessarily exhaustive: Milan, Ambrosiana MS S97 Sup; Florence, Biblioteca Nazionale MS Magl. XVI 65; Rome, Vatican MSS Reg. Lat. 1266 & 1344; Austria, Schlierbach MS 8; and Glasgow, Ferguson MS 21.

2. Josten, "Introduction," pp. 84–85; Nicholas H. Clulee, *John Dee's Natural Philosophy: Between Science and Religion* (London: Routledge, 1988), 120.

3. William R. Newman, "Thomas Vaughan as an Interpreter of Agrippa von Nettesheim," *Ambix,* 29 (1982): 129–30; Noel Brann, "George Ripley and the Abbot Trithemius: An Inquiry into Contrasting Medical Attitudes," *Ambix,* 26 (1979): 213–15.

4. John Dee, *Propaedeumata aphoristica* (London, 1558), LII; in Wayne Shumaker, ed. and trans., *John Dee on Astronomy: "Propaedeumata aphoristica" (1558 and 1568) Latin and English,* (Berkeley and Los Angeles: University of California Press, 1978), 148–49.

5. Ibid.

6. Dee, *Propaedeumata,* note to LII: Shumaker, *John Dee on Astronomy,* 224.

7. John Dee, *Monas Hieroglypica* (Antwerp, 1564), in Josten, "A Translation of John Dee's *Monas Hieroglyphica,* Antwerp, 1564, with an Introduction and An-

notations," *Ambix,* 12 (1964): 114–21. Clulee, *John Dee's Natural Philosophy,* 82–86.

8. Dee, *Monas,* 124–25.

9. Ibid., 136–41. Clulee, *John Dee's Natural Philosophy,* 83–86.

10. Dee, *Monas,* 122–23.

11. Ibid., 126.

12. Ibid., 154–61.

13. Ibid., 154–55.

14. S. K. Heninger, Jr., *Touches of Sweet Harmony: Pythagorean Cosmology and Renaissance Poetics* (San Marino: Huntington Library, 1974), 78–84.

15. Heninger, *Sweet Harmony,* 158–77.

16. Dee, *Monas,* 156–59, 180–87.

17. F. M. Cornford, "Mysticism and Science in the Pythagorean Tradition," *Classical Quarterly,* 17 (1923): 1–3; Heninger, *Sweet Harmony,* 78–84, 146–94.

18. Dee, *Monas,* 132–35.

19. Clulee, *John Dee's Natural Philosophy,* 97–99.

20. Dee, *Monas,* 160–61.

21. Ibid., 160–65, 194–97.

22. Ibid., 156–57.

23. Ibid., 166–67.

24. Ibid., 160–63.

25. Ibid., 158–59, 168–73.

26. Ibid., 172–75, 208–13.

27. Ibid., 174–75; Julius Ruska, *Tabula smaragdina: Ein Beitrag zur Geschichte der Hermetischen Literatur* (Heidelberg: Carl Winter, 1926), 2; John Read, *Prelude to Chemistry: An Outline of Alchemy, Its Literature, and Relationships* (London: G. Bell and Sons, 1961), 54, gives a translation.

28. Dee, *Monas,* 156–59.

29. Ibid., 180–81.

30. John Dee, *Propaedeumata,* in Shumaker, *John Dee on Astronomy,* 179–85.

31. Dee, *Monas,* 180–81, 160–61.

32. Ibid., 164–67.

33. Ibid.

34. William R. Newman, *Gehennical Fire: The Lives of George Starkey, an American Alchemist in the Scientific Revolution* (Cambridge: Harvard University Press, 1994), 96; Geber, *Summa perfectionis,* in William R. Newman, *The "Summa perfectionis" of Pseudo-Geber: A Critical Edition, Translation and Study* (Leiden: E. J. Brill, 1991), 645–46. See also Read, *Prelude to Chemistry,* 17–18, 119–20; E. J. Holmyard, *Alchemy* (Baltimore: Penguin 1957), 74–75; idem,

"Introduction," in Geber, *The Works of Geber*, trans. Richard Russell (London: Dent 1928), xi–xiii.

35. Read, *Prelude to Chemistry*, 130–33.

36. Newman, *Gehennical Fire*, 96–97.

37. [Pseudo-]Geber, *The Investigation of Perfection*, in *The Works of Geber*, 9–11.

38. Geber, *Summa perfectionis*, 664–68, 749–51; [Pseudo-]Geber, *The Investigation*, 9.

39. Read, *Prelude to Chemistry*, 136–42.

40. Thomas Norton, *The Ordinall of Alchemy*, in Elias Ashmole, ed., *Theatrum Chemicum Britannicum* (London, 1652), 57–58.

41. Norton, *The Ordinall*, 82–85.

42. Dee, *Monas*, 160–61.

43. Ibid., 160–63.

44. Ibid., 164.

45. Ibid., 176–77; J. Peter Zetterberg, "Hermetic Geocentricity: John Dee's Celestial Egg," *Isis*, 70 (1979): 391–92.

46. Dee, *Monas*, 176–79; H. J. Sheppard, "Egg Symbolism in Alchemy," *Ambix*, 6 (1958): 140–48.

47. Dee, *Monas*, 186–87.

48. Albert van Helden, *Measuring the Universe: Cosmic Dimensions from Aristarchus to Haley* (Chicago: University of Chicago Press, 1985), 9, 20–23.

49. This order is found in Plato, *Timaeus*, in *Plato's Cosmology*, trans. F. M. Cornford (Indianapolis: Bobbs-Merrill, 1937), 38C–D; and Plato, *Republic*, trans. Paul Shorey, in Edith Hamilton and Huntington Cairns, eds., *The Collected Dialogues of Plato*, Bollingen Series, vol. 71 (Princeton: Princeton University Press, 1961), 616E–617B. See also J. L. E. Dreyer, *A History of Astronomy from Thales to Kepler*, 2d ed. (New York: Dover Publications, 1953), 44, 168–69.

50. Dee, *Monas*, 188–93.

51. Ibid., 168–69.

52. Ibid., 134–35.

53. Ruska, *Tabula smaragdina*, 2; Read, *Prelude to Chemistry*, 54, for a translation.

54. Dee, *Monas*, 122–23.

55. Ibid., 120–21.

56. Clulee, *John Dee's Natural Philosophy*, 125–28.

57. Johannes Trithemius, letter to Germanus de Ganay, August 24, 1505, in Johannes Trithemius, *De septem secundadeis, id est intelligentiis sive spiritibus orbis post deum moventibus* (Nuremberg, 1522), 65–76. This is also included in Johannes Trithemius, *Epistolarum familiarium libri duo* (Haganau, 1536), 89–94. For Trithemius, see Noel L. Brann's recent book, *Trithemius and Magical Theol-*

ogy: A Chapter in the Controversy over Occult Studies in Early Modern Europe (Albany: State University of New York Press, 1999), which appeared too late for extensive use in this study.

58. Noel L. Brann, "The Shift from Mystical to Magical Theology in the Abbot Trithemius," in *Studies in Medieval Culture* (Western Michigan University Medieval Institute), 11, (1977): 148.

59. Ibid., 154.

60. Noel L. Brann, "Was Paracelsus a Disciple of Trithemius?" *Sixteenth Century Journal,* 10 (1979): 76–77, 80.

61. Trithemius to Ganay, 69–70; translated by Brann, "The Shift," 154.

62. Newman, "Thomas Vaughan," 130; *Gehennical Fire,* 215–16.

63. Newman, "Thomas Vaughan," 128, 129.

64. Dee, *Monas,* 122–23, 198–201, 218–19.

65. Ibid., 214–17.

66. Brann, "George Ripley," 213.

67. Trithemius to Ganay, 66.

68. Trithemius to Ganay, 66–69; Brann, "George Ripley," 213–14, from which I have adapted the translation.

69. Trithemius to Ganay, 69.

70. Ibid., 69–70.

71. Ibid., 72–73.

72. Ibid., 72.

73. Trithemius, letter to Johannes von Westenburg, May 10, 1503, in *De septem secundadeis,* 86–87.

74. Ibid., 92–93.

75. Ibid., 82–83; translation from Brann, "The Shift," 155.

76. Trithemius to von Westenburg, 94–95.

77. Ibid., 83–84.

78. John Dee, letter to Sir William Cecil, February 16, 1562, *Bibliographical and Historical Miscellanies* (Philobiblion Society), 1, no. 12 (1854): 1–16.

79. Dee, *Monas,* 212–13.

80. Jacques Gohory, *De usu & mysteriis notarum* (Paris, 1550), sig. H.iv.

81. Dee, *Monas,* 180–83.

82. Ibid., 184–85.

83. Ibid., 184–87.

84. Clulee, *John Dee's Natural Philosophy,* 110–14.

85. Dee, *Monas,* 214–17; Clulee, *John Dee's Natural Philosophy,* 111–14.

86. Philippus à Gabella, *Secretioris Philosophiae Consideratio Brevis* (Cassel: William Wessel, 1615). I have used a microfilm of the copy in the British Library, shelf mark 1033.H.6(4.). An English translation of this by Christopher Atton is

available at <http://www.levity.com/alchemy/consider.html>. It is a bare translation, not providing the original text, any of the diagrams, or any annotations.

87. The full title is *Secretioris Philosophiae Consideratio Brevis à Philippo à Gabella Philosophiae St. conscripta, & nunc primùm unà cum Confessione Fraternitatis R.C. in lucem edita.*

88. Arthur Edward Waite, *The Brotherhood of the Rosy Cross* (Secaucus, N.J.: University Books, 1961), 17–18, 143–46, says that "the *Consideratio* itself is not of our especial concern"; Will-Erich Peuckert, *Das Rosenkreutz*, 2d ed. (Berlin: Erich Schmidt, 1973), does not mention Gabella's *Consideratio*.

89. Yates, *Rosicrucian Enlightenment*, 45–47.

90. Ibid., 30–40, 220–22.

91. George M. Ross, "Rosicrucianism and the English Connection," *Studia leibnitiana* 5 (1973): 244.

92. Bruce Moran, *The Alchemical World of the German Court: Occult Philosophy and Chemical Medicine in the Circle of Moritz of Hessen (1572–1632)*, Beiheft 29 of *Sudhoff's Archiv*, (Stuttgart: Franz Steiner, 1991), 94, 48–49, 100.

93. Moran, *Alchemical World*, 92–101 and 40–49; Carlos Gilly, ed., *Cimelia Rhodostaurotica: Die Rosenkreuzer im Spiegel der zwischen 1610 und 1660 entstandenen Handschriften und Drucke*, 2d ed. (Amsterdam: In de Pelikaan, 1995), 73–74.

94. Clulee, *John Dee's Natural Philosophy*, 226–27; Moran, *Alchemical World*, 93–101.

95. Moran, *Alchemical World*, 98–100; Gilly, *Cimelia Rhodostaurotica*, 22, 74.

96. Gabella, *Consideratio*, sig. G3v, A2v.

97. Ibid., sig. A4r.

98. Ibid.

99. Ibid., sig. D1v.

100. Ibid., sig. B1r; D1r–D2r.

101. Ann Blair, "Humanist Methods in Natural Philosophy: The Commonplace Book," *Journal of the History of Ideas*, 53 (1992): 541–42; Ann Moss, *Printed Commonplace-Books and the Structuring of Renaissance Thought* (Oxford: Clarendon Press, 1996), vii, 134.

102. Blair, "Humanist Methods," 544–50.

103. Moss, *Printed Common place-Books*, 198.

104. I want to thank William R. Newman for suggesting the possibility of references to Sendivogius. On Sendivogius, see Rafat T. Prinke, "The Twelfth Adept," in Ralph White, ed., *The Rosicrucian Enlightenment Revisited* (Hudson, N.Y.: Lindisfarne Books, 1999), 141–92. Prinke suggests (p. 182–83) that Sendivogius might actually have been the author of the *Consideratio*. The echoes of Sendivogius are actually more extensive than Prinke realises, as we shall see.

105. Gabella, *Consideratio*, sig. A4r.

106. Ruska, *Tabula smaragdina*, 2.

107. Gabella, *Consideratio*, sig. B1v: "Nam, Philosophare, aliud mihi non est, quam Caelum intueri, rerum abditarum & secretarum, investigare causas, & rationes, ante omnia autem notitiam M. habere. . . ."

108. Ibid., sig. G3v.

109. "[T]he year following he translated the book M. into good Latin," *Fama Fraternitatis*, trans. Thomas Vaughan, in Yates, *Rosicrucian Enlightenment*, 239. Roland Edighoffer, "Le 'Liber M.'" *Aries*, 15 (1993): 78–83.

110. Gabella, *Consideratio*, sig. A2r, B1v–B2r.

111. Ruska, *Tabula smaragdina*, 2: "2. Quod est inferius, est sicut quod est superius, et quod est superius, est sicut quod est inferius, ad perpetranda miracula rei unius. 3. Et sicut omnes res fuerunt ab uno, meditatione unius: sic omnes res natae fuerunt ab hac una re, adaptatione. 4. Pater ejus est Sol, mater ejus Luna; portavit illud ventus in ventre suo; nutrix ejus terra est. 10. Sic mundus creatus est."

112. Nicholas H. Clulee, "John Dee and the Paracelsians," in Allen G. Debus and Michael T. Walton, eds., *Reading the Book of Nature: The Other Side of the Scientific Revolution*, vol. 41 of *Sixteenth Century Essays & Studies*, (St. Louis, Mo.: Sixteenth Century Journal Publishers, 1998), 111–32.

113. References to the *Propaedeumata* are to the aphorism number and the page number in Shumaker, *John Dee on Astronomy*.

114. Gabella's text in *Consideratio* lacks "moderamen."

115. Gabella, *Consideratio*, sig. B2v–B3r; quoting Dee, *Propaedeumata*, CVII (184–86).

116. Ruska, *Tabula smaragdina*, 2: "5. Pater omnis thelesmi totius mundi est hic. . . . 9. Hic est totius fortitudinis fortitudo fortis: quia vincet omnem rem subtilem, omnemque solidam penetrabit."

117. Gabella, *Consideratio*, sig. E1r–E2v; Clulee, *John Dee's Natural Philosophy*, 43.

118. Gilly, *Cimelia Rhodostaurotica*, 74; Gabella, *Consideratio*, sig. G1r: "Visitetis Interiora, Terrae, Rectificando, Invenietis Occultum Lapidem Veram Medicinam."

119. Clulee, *John Dee's Natural Philosophy*, 46–47, 52–56.

120. Ruska, *Tabula smaragdina*, 2.

121. Gabella, *Consideratio*, sig. C3r, C4r; cf. Dee, *Monas*, 158–61.

122. Gabella, *Consideratio*, sig. C1r; Dee, *Monas*, 156–59.

123. Gabella, *Consideratio*, sig. C4v–D1r; quoting Dee, *Monas*, X (160) and XII (160).

124. Gabella, *Consideratio*, sig. C1v–C2r.

125. Ibid., sig. C2r.

126. Ibid., sig. C1v.

127. Cornelius Agrippa, *De occulta philosophia, Libri Tres*, ed. V. Perrone Compagni, vol. 48 of *Studies in the History of Christian Thought* (Leiden: E. J. Brill, 1992), 96–98; Newman, *Gehennical Fire*, 214–15; Walter Pagel, *Paracelsus: An*

Introduction to Philosophical Medicine in the Era of the Renaissance, 2d rev. ed. (Basel, Switzerland: Karger, 1982), 140, 298–99.

128. Gabella, *Consideratio,* sig. C2v.

129. Ibid., sig. C3r.

130. Read, *Prelude to Chemistry,* 17–18.

131. Geber, *Summa perfectionis,* 645–46.

132. Gabella, *Consideratio,* sig. C3r, D3r–v, G1r–v.

133. Geber, *Summa perfectionis,* 664–68, 749–51; [Pseudo-]Geber, *The Investigation,* 9.

134. Gabella, *Consideratio,* sig. D2r–v.

135. Michael Sendivogius, *Novum lumen chemicum,* in Jean-Jaques Manget, ed., *Bibliotheca Chemica Curiosa* (Geneva, 1702), 465.

136. Newman, *Gehennical Fire,* 87–89, 212.

137. Sendivogius, *Novum lumen chemicum,* 465, 466.

138. Gabella, *Consideratio,* sig. D3r–D4v.

139. Ibid., sig. D4v.

140. Sendivogius, *Novum lumen chemicum,* 472.

141. Gabella, *Consideratio,* sig. E1r.

142. Newman, *Gehennical Fire,* 87–89, 212, 216–17.

143. Ibid., 212.

144. Gabella, *Consideratio,* sig. E3v–E4r.

145. Jacques Gohory, *In Libros Quatvor Ph. Theoph. Paracelsi de Vita Longa Scholia,* in Theophrastus Paracelsus . . . *Compendium. Ex optimis quibusque eius libris: Cum scholijs in libros IIII eiusdem De vita longa, Plenos mysteriorum, parabolarum, ænigmatum,* (Basel, 1568), 304.

146. Gohory, *Scholia,* 195.

147. D. P. Walker, *Spiritual and Demonic Magic from Ficino to Campanella* (Notre Dame: University of Notre Dame Press, 1975; reprint of original 1958 edition), 96–106.

148. Gabella, *Consideratio,* sig. E4v.

149. Ibid., sig. F1v–F2r.

150. Ibid., sig. F2r.

151. Gohory, *Scholia,* 317: "Quae iam disserit in exitu libri de figmentis, deinde de Nymphidica, ὕστερον πρότερον est, ordine traditionis praepostero ad tenebras offundendas. . . ." Cf. Gabella, *Consideratio,* sig. F2r.

152. Gabella, *Consideratio,* sig. F2v; Dee, *Monas,* 160–62.

153. Gabella, *Consideratio,* sig. F2v–F4r; Dee, *Monas,* 162–64.

154. Gabella, *Consideratio,* sig. F4r–v.

155. Moran, *Alchemical World,* 40–46.

156. Gabella, *Consideratio*, sig. G1r–v.

157. Gilly, *Cimelia Rhodostaurotica*, 22, 74. T. M. Luhrmann, "An Interpretation of the *Fama Fraternitatis* with Respect to Dee's *Monas Hieroglyphica*," *Ambix*, 33 (1986): 1–16, recognizes that Yates did not successfully support her claims with her references to Gabella's *Consideratio brevis* in her *Rosicrucian Enlightenment*. Luhrmann attempts to rectify this by arguing a connection with the actual Rosicrucian texts, specifically the *Fama Fraternitatis*. This approach, however, has two limitations. It requires relying on evidence from analogies rather than any direct influences. It also requires considering Rosicrucianism as deriving from a Hermetic-Cabalistic philosophy, which has been disputed: see Ross, "Rosicrucianism," 242.

6

The Rosicrucian Hoax in France (1623–24)

Didier Kahn

There was a young man . . . who was looking for us everywhere. He had said first that he was one of our own and that, having lingered along the way, he had gotten lost, but then he had begun to cry and admitted that he had been unfaithful to us and had fled, but that now he saw that he could no longer live outside of the Order. . . . This story was told to us here and there, more or less everywhere; wherever we went, the unfortunate man had just passed through there. We asked the speaker what he thought of this and what was going to come of it. "I don't believe that he will find us," was all the speaker said.
—Hermann Hesse, *Die Morgenland Fahrt* (1932)

The incident of the Rosicrucian placards of Paris is well known. During the summer of 1623, posters were put up at the crossroads and upon the church doors of Paris, proclaiming the presence of some "representatives of the Principal College of the Brothers of the Rose-Cross" endowed with marvelous powers and desirous of saving their fellow men from "error and death."

This episode, and the confusion that apparently ensued in the French capital, has been touched upon for decades, if not for centuries, in nearly all the works devoted to the Rose-Cross, as well as in many other works devoted to Descartes. Each author favored his own point of view, however, and this variety of critical sources, combined with the historical vagueness resulting from the small number of established certainties as to the fictitious fraternity, has dissuaded most authors from proposing an interpretation of the event, compromising until now the possibility of offering a satisfactory synthesis of this case.[1] The recent discoveries of Carlos Gilly on the German Rosicrucian movement, as well as François Secret's older discovery concerning this modern "incident of the placards" of Paris, allow us henceforth to venture a consideration of the event as a whole, with the intention of placing it in the various contexts from which it derives its meaning. Who wrote and who put up the 1623 placards, and with what aim? Were there several texts, distinct from one another? What exactly were the reactions that they provoked? The many responses to this last question make us take into account the stakes of the affair on the sociopolitical level, particularly in the context of the libertine hunt, while at the same time considering this phenomenon as an episode and a caricature of the reception of Paracelsus in France. An investigation into the different sources—literary, historical, medical, and scientific—posterior to 1624 allows us, moreover, to follow the destiny of the Rosicrucian episode in France until the eighteenth century and to try to measure its impact. Perhaps, once the affair has been put back into the historical continuum that witnessed its birth, it will then be possible to propose an interpretation of it.

Before all else, it seems prudent to take stock of the certainties acquired up to now concerning the Rosicrucian movement and its twists and turns.

Present State of Research on the Rosicrucian Movement

Even if the many works of Frances Yates have often shown themselves to be beneficial, and even if several of these works are now considered classics, there is no choice but to accept that *The Rosicrucian Enlightenment,* published in 1972, has scarcely done anything but add to the reigning confusion on the topic. In this book, Yates assigned British origins to the German movement, relying especially on the symbolism of the Order of the Garter and on the influence exerted by the *Monas hieroglyphica* of John Dee in Germany at the beginning of the seventeenth century. What is more, the marriage of the future "Winter King," Friedrich V von der Pfalz, with King James I of England's own daughter Elizabeth (1613) raised great hopes, in the Calvinist party. And through these sometimes millenarian hopes in the events that preceded the Thirty Years War, the Rosicrucian movement came to constitute, according to Yates, the mystical background "of a vast reform movement whose nature was magical, hermetic and alchemical, in every respect comparable to that which John Dee had propagated in Bohemia [between 1583 and 1589]."[2] Yates thus discovered an active sympathy at the level of politics between the Calvinists and the Rosicrucian movement. Dee became "a towering figure in the European scene," the German Rosicrucian movement, "in one sense an export of the

Elizabethan period and of the inspirations behind it, scientific, mystical, poetic." The word "Rosicrucian" itself henceforth evoked "both English chivalric influences and a Dee influence behind them."[3]

Equipped with all the authority that her preceding works had justly earned her, the author of *Giordano Bruno* succeeded, with *The Rosicrucian Enlightenment,* in establishing in a large part of the scholarly, non-specialist world a theory that was at least stimulating, if not original,[4] but unfortunately founded on too many hypotheses and too little critical documentary research to resist for long the assaults of scholars who, in the 1970s and 1980s, embarked upon shedding an ever more precise light upon the origins and the true stakes of the Rosicrucian manifestos. Ten years after Yates's work, Roland Edighoffer expressed his criticism as follows: "How to delimit and define with precision that which is elusive? With an accomplished art of persuasion, Frances Yates expounds theories that are capable of seducing the reader, but which remain hypotheses."[5]

More recently, Gilly, incontestably the best present-day specialist on the topic,[6] has taken care to refute Yates's theories each time that the opportunity for doing so has presented itself—an eminently beneficial task, given the very wide audience that these theories have received until now, if not as a result of Dee's magic, then because of the very name of the author, hastily promoted to the rank of indisputable authority on the subject.

Synthesizing the different contributions of the most recent research and adding to it his own contributions, Gilly has now established on a sound basis the reconstruction of a rather large part of the genesis and history of the movement. Entirely rejecting the idea according to which the very term "Rosicrucian" was supposedly exported from Elizabethan England, Gilly takes away from John Dee the central position too often accorded to him in the birth of the movement since *The Rosicrucian Enlightenment* and brings his role back to its proper place, "scarcely more important" in this genesis, Gilly says, "than that of a Paracelsus, a Lautensack, a Suchten, a Brocardo, a Castellion, an Arndt, a Croll or the author of the *Cyclopaedia Paracelsica Christiana,*" all authors who constitute, with many others, the vast hydrographic network of wellsprings that irrigate the native compost of the Rosicrucian manifestos.[7]

We know that with the "general reform of the divine and the human" announced by the Rosicrucian manifestos, "it was, in the last analysis, very much a question of a radical change of values in religion, in science and in politics"; in fact, "between the dates of 1614 and 1625 alone, more than 400 writings appeared which, coming from the most diverse circles, assumed their position in relation to the Rosicrucian phenomenon."[8] We know as well that these manifestos, which are completely imbued with a "harmonious philosophy,"[9] were produced by a circle of friends of whom the best known, from the point of view of the Rosicrucian movement, were named Tobias Hess (1568–1614), Christoph Besold (1577–1638) and Johann Valentin Andreae (1586–1654). If the conception of the first of the Rosicrucian manifestos, the *Fama Fraternitatis (Echoes of the Fraternity of the Admirable Order of the Rose-Cross, Written to All the Scholars and Leaders of Europe)*, must be attributed in a collective manner to the circle of Tobias Hess, its composition, on the other hand, is the work of Andreae alone, who is also the author of the *Confessio Fraternitatis* and the *Chymische Hochzeit (The Chemical Wedding of Christian Rosenkreutz, in the year 1459)*.[10] This last treatise, published in 1616, which exerted only a very small influence, is the first in the order of composition; Andreae probably composed it in 1607.[11] The composition of the *Fama* can be dated to approximately 1608;[12] although the *Confessio* affirmed that the *Fama* had been diffused in five languages, only the original German version was actually circulated.[13] Adam Haslmayr, the very first witness of the propagation of this text, stated that he saw it in Tyrol as early as 1610; in 1611, he referred to it implicitly and on New Year's Day of 1612, he gave a copy of it to his protector August von Anhalt (1575–1653), declaring that he had received it from Tobias Hess, a doctor from Tübingen.[14] For his part, the Danish doctor Ole Worm (1588–1654), at that time traveling in Europe, received a copy of it in 1611, probably from Johann Hartmann (1561–1631), professor of *chymiatria* at the University of Marburg since 1609. We are currently aware of four handwritten copies prior to the *editio princeps* (1614), which permits the reconstitution of the text in its original form, since this edition was made without the authors' knowledge.[15]

In 1612, August von Anhalt, having heard about the Rosicrucians from Haslmayr and having received from him a copy of the *Fama*, was seized with a passion for the Fraternity and had a letter of inquiry sent to Tobias Hess to find out more, particularly about the unattainable *Confessio* announced first in the *Fama Fraternitatis*. This step remained fruitless, "perhaps for the very simple reason that the *Confessio Fraternitatis* had not yet been written."[16] Gilly has shown that Haslmayr—who presented himself as a disciple of Paracelsus and cultivated a rather heterodox doctrine called *"Theophrastia Sancta,"* tying together mystic theology, philosophy of nature, medicine, and alchemy, which he sometimes epitomized by affirming his knowledge of the *elementi verbique mirifici sacramentum*, "the secret of the element of nature and of the word of Creation"[17]—tried with all his might during that time to persuade August von Anhalt to assume the role of political champion of the future universal reformation advocated by the Rosicrucians. Anhalt, however, quite lucid concerning his own lack of power on the political level, did not wish in any way to do so. From 1603, the prince had indeed renounced the direction of the principality of Anhalt to the advantage of his brother Christian, in order that he might dedicate himself in all tranquility to alchemy and the circulation of Paracelsian, Weigelian, and mystical writings; a general reform of the world seemed impossible to him without great bloodshed, thus he preferred to leave that up to God.[18] In fact, "Prince August entertained the project of publishing, while managing to bypass any censorship, not only the works of Valentin Weigel, but also the theological writings of Paracelsus, the commentary on the Apocalypse of Paul Lautensack, the *Offenbahrung Göttlicher Majestät* by Aegidius Gutman or the prophetic texts of Helisaeus Röslin." He then had the idea in 1611 of installing a secret printing house next to his alchemy laboratory in Zerbst, whose supervision he envisioned entrusting to Haslmayr.[19] We know little about the activity of this secret printing house, for August von Anhalt was forced to use great caution; but it is certain that he himself, in March 1612, had printed at his own expense scarcely 100 copies of Haslmayr's *Response* to the *Fama Fraternitatis*, which Haslmayr had sent him two months prior. Although August ultimately gave up supporting this publication by enhancing it with his own princely authority, he printed it in the hope of inciting the brothers of the Rose-Cross finally to emerge from their reticence—at least if the brotherhood did really exist, which August wondered about early on, although his doubts were subsequently erased little by little.[20]

As for Haslmayr, fortified by a blind trust in his local lord, Archduke Maximilian of Austria—who was completely devoted to the Jesuits and the Inquisition—he naively told him about his many Paracelsian, theosophical and Rosicrucian speculations to clear himself of accusations of heresy and charlatanism leveled at him by a former student of the Jesuits,

Hippolytus Guarinoni, doctor in the city of Hall and fiercely anti-Paracelsian. Haslmayr went so far as to ask the archduke, in August 1612, for a "small travel fund" to go in search of the brothers of the Rose-Cross, whom he thought he would find mainly around Montpellier. "Haslmayr," Gilly writes, "received these small provisions, this travel money, from the Archduke, not to seek out the Rosicrucians in the region of Montpellier, but rather to row as a galley slave for four and a half years on the open seas of Genoa."[21] Haslmayr later refused to describe "out of consideration for chaste ears," "what sort of existence one leads on the galleys, savage, without rest, monstrous, hopeless and sodomitical,"[22] "and all that only," he added, "for having admitted myself to be a poor Theophrastic Christian," or more precisely, for having been held in suspicion, notably on the strength of his printed response to the *Fama Fraternitatis*, of having sought "to introduce into Tyrol a new sect or heresy."[23]

While Haslmayr was rowing on the open seas of Genoa, the seizure of his papers allowed the Tyrolian authorities to discover several letters from one of his friends, the German Paracelsian Benedictus Figulus (1567–1624), who was at that time staying at Freiburg-im-Breisgau, a city under the control of the Austrian Habsburgs, which permitted an order for his arrest to be issued. Figulus took flight, abandoning in spite of himself his project to edit the "cabalistic and theological books" of Paracelsus, as well as his project for a continuation of the greatest alchemical printed collection of the time, namely the famous *Theatrum Chemicum*, and leaving some of his manuscripts at Marburg and Strasbourg before finally being imprisoned until 1617 in an undisclosed place.[24] Meanwhile, in March 1614, the *Fama Fraternitatis* had been published in Kassel, against the will of its authors, who found themselves forced to confront the new situation its publication engendered.[25] The *Fama* was accompanied by Haslmayr's *Response*, which was announced right from the title page: "With a short response by Mr. Haselmeyer, who was because of it imprisoned by the Jesuits and chained up on a galley."[26] It is not known who edited the work, but given the place and the printer, it could only have been with the consent of Moritz, the landgrave of Hessen-Kassel.[27] It is also not known how the editor knew Haslmayr's fate. Carlos Gilly has put forward the hypothesis that Figulus could have left in Kassel a copy of the *Fama* made from the manuscript Haslmayr possessed, which would explain the long-lasting wrath of Johann Valentin Andreae toward Figulus.[28] As for the *Confessio*,

August von Anhalt obtained a manuscript of it as early as September 1614, which constitutes chronologically the first account of its actual existence, from an individual whom Gilly is strongly tempted to identify as Johannes Rhenanus, a doctor of the landgrave of Hessen-Kassel. This Rhenanus could well be the "Philippus a Gabella" who published in March 1615 the first edition of the *Confessio,* once again at Kassel, on printing presses controlled by the landgrave.[29]

Studying the implications of these historical data, Gilly has refuted the idea—dear to Frances Yates and, he specifies, to some contemporaries, Lutheran as well as Catholic, of Andreae—that the "Rose-Croix manifestos allegedly prepared ideologically the Calvinist Friedrich V von der Pfalz's political venture in Bohemia." Indeed, if in the *Confessio* the eagle's feathers obviously designate "the house of Austria, that is above all the monarchy of Spain in its role as the only and last supporter of the faltering Papacy," the lion, on the other hand, is "neither that of the Palatinate nor that of Bohemia, nor that of the Low Countries (nor is it any more the lion of alchemy who engulfs the eagle's head), but rather the lion of the Bible, that spoken of by Isaiah and Micah, the Fourth Book of Esdras and the Apocalypse: the one who will come to open the book, break the seven seals and introduce a New Age." This is indeed the lion that appeared to Tobias Hess in 1605, in a vision that nourished his later chiliastic concerns, which subsequently earned him accusations from the Faculty of Theology of Tübingen. The prophetic lines of the *Confessio* concerning the role of this lion—putting an end to the tyranny of the papacy—coincide with those of Hess's vision, even the detail of the eagle's losing its feathers little by little.[30]

According to Gilly, there could not be any sort of original sympathy, and still less a complicity on the political level, between the Calvinists and the Rosicrucian and Weigelian movement.[31] Gilly first points out that the Calvinists who from 1618 intervened in the Rosicrucian quarrels by taking sides in favor of Friedrich V von der Pfalz—in pamphlets, moreover, that Frances Yates does not account for—would have supported any other prince as well, provided that he was opposed to the rise of power of the Habsburgs and the Jesuits in central Europe. Gilly then indicates that this propaganda, which moreover came belatedly, played no role in the birth of the Rosicrucian movement: quite the contrary, since the Lutheran, not Calvinist, princes Friedrich von Württemberg and August von Anhalt, respectively, were the ones whom Tobias Hess in 1605 and Adam Haslmayr in 1611 endeavored to set up as political leaders of the future universal reform.[32] As for the court of Heidelberg, it was scarcely, if at all, interested in the Rosicrucian movement, and when it became interested, it was not to support it, but to condemn it as a whole. It is moreover in the reformed University of Marburg and on the order of the Calvinist landgrave Moritz von Hessen-Kassel that in 1619–20, the first trial for Weigelianism, Rosicrucianism and other "errors" stemming from the "Theophrastic mob" took place. Three years later (1623), a new trial took place at the Lutheran university in Giessen, brought against Heinrich Nollius for Rosicrucianism and Weigelian fanaticism.[33] The first of these trials concluded with the chief person accused, Philipp Homagius—who was moreover the son-in-law of the first printer of the Rosicrucian manifestos—being condemned to life in prison, upon the express order of the landgrave Moritz, "in order to discourage all those who pass themselves off as highly enlightened beings and as prophets or apostles sent to us, or who promise by a false and vain hope the philosophers' stone and other great secrets, the great delight and blessing of God." We see that the position of the landgrave with regard to the Rose-Cross, far from being uniform, deserves to be carefully reconsidered.[34]

Faced with the surprising complexity of these doctrines, these inextricable entanglements of facts and publications, we can only feel a heavy disappointment when we now turn to the tragicomic Parisian episode of the Rosicrucian placards. This episode has nevertheless caused too much ink to flow for us to allow ourselves to neglect it, as much in the framework of the study of Paracelsianism as in that of the history of ideas in France at the time of Louis XIII.

The Incident of the Parisian Placards of the Rose-Cross

One morning in June or July of 1623, Parisians discovered to their surprise posters put up at the street corners and on the church doors, written in these words: "Nous deputez du College principal des Freres de la Roze-Croix, faisons sejour visible & invisible en cette ville, par la grace du Tres-haut, vers lequel se tourne le coeur des Justes. Nous monstrons & enseignons sans livres ny marques à parler toutes sortes de langues des pays où voulons estre, pour tirer les hommes nos semblables d'erreur <&> de mort."[35]

The exact date of these posters is difficult to pinpoint. Gabriel Naudé, who obtained the King's privilege for his *Instruction à la France* on November 13, 1623, situates the event as occurring "approximately three months ago," which would place it in the month of August.[36] If we submit to the evidence, this dating must be pushed back slightly, for on exactly August 3, 1623, the scholar Peiresc writes from Paris to his friend the painter Rubens, who is living in Anvers: "nous avons eu ici certains autres sectaires nouveaux de la Rose-Croix, assez célèbres en Allemagne, qui sont peut-être les mêmes que ceux de Séville. Ils songèrent à afficher sur les murs dans la rue certains papiers ou avis dans lesquels ils promettaient au public en quatre lignes la connaissance de la divinité et l'invisibilité personnelle et cela au nom des frères de la Rose-Croix."[37] There is no doubt: It is definitely on the subject of the placards of Paris, and these lines situate them in July at the latest, on an unspecified date. The month of July was also suggested, although without justification, by the editors of the *Correspondence* of Mersenne, perhaps based on the pamphlet *Effroyables Pactions,* rather eccentric but published shortly after the event, in November or December 1623, which placed the arrival of the supposed "Invisible ones" in Paris around July 14 and the posting of the placards some time later.[38] A manuscript of Peiresc again places the posting of the text that he reproduced in July.[39] However, a Latin document preserved in the manuscripts of Théodore Turquet de Mayerne is dated June 13, 1623,[40] and the *Mercure françois* places the circulation of the posters "in the spring," without any further precision: "Il s'est tousjours trouvé des esprits curieux de sçavoir toutes sortes de nouvelles, ce qui leur fait passer autant de temps; & d'autres qui leur en donnent à garder, & s'en donnent du plaisir. Ceux-cy voyant le change du Palais de Paris sans nouvelles,[41] au Printemps de ceste annee, s'adviserent d'y faire trotter de main en main plusieurs petits billets manuscrits, & en afficherent aux Carrefours, qui contenoient, *Nous Deputez du College Principal* [etc.]."[42]

Visibly inspired by Naudé in many details, the article of the *Mercure,* published in 1624, is the most belated of the accounts of the time; it cannot be neglected, but can it be considered a model of accuracy? It must certainly be admitted that using the month of July alone would simplify everything. Indeed, retaining the months of June and July as the most probable dates would amount to admitting that the placards were posted several times in the capital with an interval of a few weeks between postings; yet, once a first posting had been carried out, would it not have been excessively dangerous to be caught, even by a local inhabitant, in the process of a second posting campaign? Still, the case is such that in the absence of other criteria of assessment, there is no choice but to confine ourselves to the hypothesis of a posting in several stages, carried out between June 13 and the end of July 1623.

In what form were these posters presented? According to Naudé's account, confirmed by the author of *Effroyables Pactions,* they were simply a "note containing six handwritten lines."[43] As for the *Mercure François,* which also refers to "small notes" that it says trotted "from hand to hand," it specifies in a similar fashion that they were "handwritten."[44] It is therefore useless to look for printed copies of them. If one explores, on the other hand, some handwritten collections and some contemporary accounts, one obtains curious results, of which this divergence among sources lets us have a premonition: Several posters seem to have circulated. Before examining them, it is worthwhile to ask questions concerning their source, since this is an area that we can clarify.

The Author of the Posters

The first commentators of the incident immediately thought it was a trick: So said the *Mercure François* in 1624. The preceding year, Naudé had already written:

> Toutesfois si nous voulons passer plus avant, & rechercher precisément la premiere cause de cette bourrasque, laquelle souffle maintenant dans nos campagnes, nous trouverons que le bruit de cette confraternité s'estant espandu depuis peu par l'Allemagne, quelques Professeurs, Medecins & personnes studieuses de cette ville, avoient eu cette curiosité que d'en rechercher la cognoissance, par le moyen des livres nouveaux qui leur estoient communiquez par les Libraires apres leur retour de la foire de Francfort, lesquels neantmoins n'y recognoissans rien que des chimeres & fanfaronneries, aimoient beaucoup mieux en attendant la farce prendre le plaisir de cette Comedie,
>
> *quam protinus urbi / Pandere, res alta sylva & caligine mersas,*[45]
>
> & mettre leurs renommees en compromis pour en estre les premiers denonciateurs, jugeans qu'il y avoit assez de fols dans Paris pour ne laisser croupir cette marote. Et de faict il y a environ trois mois que quelqu'un d'iceux voyant que le Roy estant à Fontainebleau, le Royaume tranquille, & Mansfeld trop esloigné pour en avoir tous les jours des nouvelles,[46] l'on manquoit de discours sur le Change,[47] & par toutes les compagnies, s'advisa pour vous en fournir de placarder par les carrefours ce billet contenant six lignes manuscrites, duquel j'ay jugé estre à propos de vous communiquer la copie, pour soulager une infinité de personnes qui ne l'ont veuë, d'en barboüiller leurs tablettes.[48]

By the same token, at the end of the century, Adrien Baillet did not hesitate to say that the poster was the product of "some buffoon's imagination."[49] No serious mind was therefore prepared to see in this affair anything else but a trick. The poster's author, however, remained unknown.

It is no longer a mystery, nor has it been one for more than a quarter of a century. As soon as 1971, one year before *The Rosicrucian Enlightenment*, François Secret exhumed an account by Nicolas Chorier that established the origin of the placards.[50] Secret himself, it is true, expressed some doubts about the authenticity of this account, dictated more by his historian's prudence than by the quality of the testimony. Furthermore, the account was given in Latin, without a French translation and accompanied by very little commentary. Finally, Secret's article was published in a journal devoted not to the seventeenth century, but rather to the Renaissance, which can be easily justified as long as one understands the rather vague notion of the Renaissance in its broadest sense. For all that, it is still not any less surprising that since 1971, only *two* researchers, Carlos Gilly and Lorenzo Bianchi, have taken notice of this fortunate discovery.[51]

As Secret has shown, then, in 1680 the scholar Nicolas Chorier, a lawyer and historian, perhaps also the clandestine author of a famous Neo-Latin erotic poem, *Aloysia Sigea*, devoted a work to his protector, the duelist and academician Pierre de Boissat (1603–62), as well as to de Boissat's friends.[52] Among these was found a certain Étienne Chaume, a doctor from Montpellier, to whom Chorier devoted an entire article:

ÉTIENNE CHAUME died some years before Boissat, who would still be alive, if Chaume had lived. This man, admirably well-versed in the art of medicine, used to wrest away from death those whom the force of illness condemned to die. Born in Vienne, he had learned medicine in Montpellier and had practiced it in Paris for a few years; it is in this capital of the kingdom of France that he had resolved to settle. The story got in the way.

At the time many things were being said on the subject of the enlightened brothers of the Rose-Cross, (as they were called):[53] that these men had knowledge of whatever they wished to know—learned and popular languages, arts and sciences—at will, without any work and in a few days; that starting with the most base metals, in an admirable manner they fabricated from them more noble ones, and even that they changed them into gold; that finally, since the eternal spirit of the divinity refused them nothing, they immediately obtained what they desired, whether it be for themselves or for others. Pure nonsense. But treatises written on this matter by some authors, among whom were included Paulus Didisus,[54] Theophilus Schweighart,[55] Eucharius Cygnaeus (fictitious names) and Heinrich Neuhus from Dantzig,[56] brought to these imaginary stories the weight of a great authority.

Driven by youthful jocularity and a juvenile spirit, Chaume added to this a great deal of his own invention, particularly the fact that this type of man was, at will, capable of seeing without being seen; that it was possible for those who found it to be in their interest to travel freely to all the innermost recesses of a residence after entering it; that they could be located, without being seen, next to guests in the process of dining, near people conversing with their dear ones, near sick people. Among Chaume's companions,[57] three or four were in the habit of seeing him regularly. In the Latin tongue, he produced posters exposing these chimeras and this empty balderdash, and the idea came to these men of exposing the posters on the street corners, the church doors and the busiest places. Those who had the desire to live well and blessedly were invited to this brotherhood and fortunate if they were initiated into these mysteries. The place, the time, the day and the hour were indicated on the posters.

The common people of France are rather credulous, but those of Paris are more so than all the others. There are no other people on earth who allow themselves to be fooled more easily, and who accept having been deceived more calmly once the deception is known.[58] Once these placards were posted, surprise seized the Parisians at first, followed by the most intense emotion. Fear, distress, indignation invaded nearly every house. People did not feel safe under their own roofs. People therefore dared neither to speak freely to their own families, nor to devote themselves to their affairs according to their desire. This panic of the citizens could not fail to attract the attention of the magistrate. The latter therefore demanded inquiries as to the authors, and even, if those who were sought were taken, that they be dealt with ruthlessly.[59]

Chaume, at first, was very astonished that what he had dreamed up would upset minds in such a general way. He laughed about it. But when he saw that very scrupulous inquiries of the incident were undertaken by people to whom power had been given to do so, he attended to his reputation and to his salvation by fleeing; prudently eluding danger, he returned to Vienne. He hid himself entirely in the breast of the Muses, and in a short number of years, having placed neither limits nor an end to his studies, he brought from his sleepless nights and his uninterrupted work those fruits of erudition which are amply sufficient for the honor, benefit, and usefulness of any man who is very absorbed by life. What he did not know, no other doctor would have known . . . He excelled in the Latin tongue . . . ; you would have said that he spoke it as his native language. . . . He died, having passed his fiftieth year, to the great grief of all and still more to my own. When, thanks to his care, I was restored from my troubles, with my strength having returned, he accepted for himself no payment from me, except (as he said) my good graces, which he requested of me. Any other payment was, in his eyes, not only contemptible to scholars and unworthy of a well-read doctor, but even, according to him, something absolutely dishonorable.[60]

What must be retained from this beautiful portrait concerning the incident of the placards of 1623? Let us begin with some observations. Pierre de Boissat having died in 1662, Étienne Chaume, who supposedly preceded him to the tomb by a few years at fifty-plus years of age, must have been born between 1603 and 1606, which would place his death between 1656 and 1659, at between fifty and fifty-three years of age. If he were born later, he would have been rather young to play, in 1623, the role described by Chorier; if he were born earlier, his life would have ended nearly ten years before Boissat's, which would no longer agree with Chorier's story. He was therefore between seventeen and twenty years old in 1623.[61]

But given these conditions, how could he have studied medicine beforehand in Montpellier and practiced it in Paris for some years? There is no choice but to suppose that on this point his biographer is mistaken. The archives of the University of Montpellier confirm this hypothesis: Only after the Parisian adventure, in December 1623, did Chaume enrol in the Faculty of Medicine in Montpellier.[62] It is therefore highly possible that he never practiced in Paris, unless one accepts that he went back there later, upon the completion of his studies and before settling definitively in Vienne.

This is not the only point on which Chorier's story can be seen as deficient. Chorier in fact left handwritten memoirs, addressed to his son, that, although they were interrupted after 1681, nonetheless cover the entire duration of Chaume's life and beyond. These memoirs being nothing other than an autobiography, one cannot expect to discover there a new telling of the Rosicrucian episode; Chaume himself only appears in it one time in 1646 or 1647, in reference to Chorier's illness.[63] Chaume, here, is far from having the role that Chorier attributed to him in the *De Petri Boessatii vita*. Far from saving Chorier's life, he saw no danger in his illness, which another physician considered very serious. It is readily conceivable, in these conditions, that he refused any salary from Chorier, having probably cared for him, but not cured him. The portrait of Chaume Chorier paints in the context of *De Petri Boessatii vita*, necessarily related to the praise genre, must therefore be viewed with caution.

In the Rosicrucian incident, two details in particular contradict the most known facts: Naudé's account seems to exclude the composition of the posters in Latin, which is quite natural according to Chorier; moreover, would not the Latin language have created an obstacle to the scope of circulation of the posters among the Parisians? Furthermore, these placards, as they are transmitted by Naudé, do not contain—far from it—all the details Chorier notes on the supposed invisibility of the Rose-Croix, to the point that one must wonder if Chorier did not assist his memory, after so many years, by using the text of the *Effroyables Pactions*. In this pamphlet, the necromancer Respuch promises the imprudent people who have agreed to sign his pact "also to render them invisible, not only in private but also in public, and to enter and exit Palaces and Houses, Rooms and Closets although all may be closed and locked with one hundred locks."[64]

Writing sixty years after the fact, Nicolas Chorier could have confused the various facts of the incident and constructed his account, as Baillet did ten years later in his *Vie de Monsieur Des-Cartes,* by substituting assertions with an authentic appearance for his doubts or for unknown elements.

The question emerges again with another document recently brought to light, for this one happens to agree, in its main points, with Chorier's story: an autograph note by John Locke (1632–1704), discovered by William Newman in one of the philosopher's numerous notebooks kept at Oxford.[65] Classified under the heading *Rosicrucians,* this note reproduces an anecdote told to Locke by Nicolas Thoynard (1628–1706), his friend and most assiduous correspondent, whose acquaintance the philosopher had made in Paris in 1677 in one of the capital's learned circles.[66] Thoynard himself heard this anecdote, according to his note, from the chief protagonist, a certain Mr. "Pallieure" or "Pailleure," none other than the mathematician Jacques Le Pailleur (?–1654), a member of the Mersenne Academy since 1637, a great friend of Blaise Pascal's father and of Gédéon Tallemant Des Réaux, a great lover of pranks, rather a skeptic and a supporter of "learned ignorance" as opposed to the credulity of excessively pious and superstitious people; the Mersenne Academy continued at his home after the death of the Minim scientist (1648).[67] One must therefore admit that the young Thoynard necessarily collected this story before Le Pailleur's death (1654), when Thoynard was 26, at most. The extent of time between this *terminus a quo* and the rather belated date when he shared it with Locke (not before 1677) perhaps explains the fact that some inaccuracies, and even some peculiarities, are to be found in it. But here is the anecdote:

> About the year 1618 or 20 Mr Pallieure who is mentioned in the preface of Mr Pascal's Hydrostaticks going to see one of his friends at his lodging in Paris found him & another of his acquaintance writeing but so as at his entrance they shufled aside the paper as unwilling to have known what they were about. After a litle while one of them says why should not Mr Pallieure make one. Agreed replys the other & upon that they produce their paper wch was the draught of a program to

be posted up in Paris to tell the people that there were certain persons of a brotherhood come to town to cure all diseases & doe other rare things amongst the rest make them that desired it invisible. And that these Adepti were to be found such a day every week in such a place, nameing a street in the Fauxbourg St Marceau where indeed there was no such street. And there lying on the table the Theatrum honoris of Rodolphus Conradus they subscribed their affiche R.C. These bills were posted up & downe all about Paris & had the [des *deleted*] effect they designed wch was to make themselves sport by seeing people & amongst them some of their acquaintance flock to the fauxbourg St Marceau in quest of these brothers RC.⁶⁸

That wch put them first upon doing this was to play a trick to Lullyists who were then in vogue at Paris & cried up as men that had more then ordinary skill in the secrets of nature. & therefor they presumed that this bill would be taken to be theirs. But that wch was a consequence of it w^{ch} [they did *deleted*] the contrivers did not designe nor forsee was that there spread thereupon an opinion that there was a brotherhood or society of men that could be when they pleased invisible & had other great secrets wch they called brothers of the Rosie crosse. who first gave this imaginare society y^t name Mr Pallieure (who counted this story to Mr Toynard) could not tell but concluded it to have been an interpretation of the R.C. wch was subscribed.

And thus began the invisible society of the Rosicrucians wch made such a noise at Paris & through all France that Naudaeus a little after writ a book ag^t them. And the world have soe much talked of since without knowing any thing of them or their original.

Mr Pailleure going not long after into Brittany at Vannes met with a young gent. with whome talking of News was told thus, that he ye young gent had receivd news from Paris of the Rosycrucians of whom all ye world talk^d strange things, espetially of their being invisible. wch the hostesse hearkening to as they were discoursing of it at table told them that there had been two of them lately at her house, who having dined vanishd soe that she & her people could not perceive wch way they were gon. A pleasant instance how easily people resigne up their beliefe to common fame & adde each one their testimony to a Lie that is once current.⁶⁹

We must begin with certain surprising details in this story. From a purely chronological point of view, the date 1618 or 1620 is of course erroneous: there were not two sessions of Rosicrucian posters in the capital at an interval of a few years, but only one in 1623: Otherwise, the previous posting could not have failed to be mentioned in 1623. What are we to think, furthermore, of the episode of the signature "R.C.," which was inspired, if we are to believe Le Pailleur's story, by the simple initials of "Rodolphus Conradus," an individual said to be the author of a *Theatrum Honoris?* It may first be noted that neither Thoynard nor Locke seem to have been aware that the Brotherhood of the Rose-Cross was famous in Germany as early as 1614, which would have been sufficient to ruin his explanation of the initials "R.C." and the rather belated origins of the Brotherhood. But if one researches more deeply, one quickly discovers that this very name, "Rodolphus Conradus," or whatever variants are given for it (Cunrad, Cunrat, Konrad, Konraedt, etc.) appears in no biographical dictionary, from Jöcher to the *Index notorum hominum*. On the other hand, a famous work exists with nearly the same title: the *Amphitheatrum Honoris* by Father Charles Scribani (1561–1629). But this anti-Protestant pamphlet, published on Plantin's printing press in Anvers in 1605 under a false address, appeared under the pseudonym "Clarius Bonarscius," which is very different from "Roldolphus Conradus."[70] This episode was therefore quite probably embellished and deformed after the event by Jacques Le Pailleur, whose memory was perhaps at fault and whose facetious nature would have enjoyed elaborating these fictitious origins of the Rose-Cross. But if we let the context of the time guide us, it is not impossible that Heinrich Khunrath's alchemical work *Amphitheatrum Sapientiae Aeternae* (1609, first ed. 1595), very much in fashion in France at that time and whose author was often spelled Conrad, especially by Naudé,[71] was hidden behind the vague memory of a Rodolphus Conrad and a *Theatrum Honoris*.

One detail, on the other hand, rings true in this version of the origins of the Rose-Cross: the fact that the posters had a great deal more success than they had anticipated. This unexpected success is also found in Chorier's story. In fact, the significance of Le Pailleur's recollection resides in its confirmation—or rather completion—of several aspects of Chorier's version. The latter showed Étienne Chaume making the posters himself, in the company of some friends, with the sole goal of laughing at the effect of these placards on the Parisians. Le Pailleur presents, without giving any names, the same version of the facts. Everything therefore leads to the belief that it was indeed Chaume whom he visited that day, which permits us to identify Le Pailleur himself as one of the friends of Chaume Chorier mentions.

The stories have other points in common: the idea of invisibility and that of the fictitious meeting set at the bottom of the poster. Here is the opportunity to raise the question of the contents of these placards. Since Le Pailleur's story confirms that they developed the theme of invisibility more than is seen in the version Naudé kept, and since this story adds the idea of treating all illnesses—a central theme, as we know, of the Rosicrucian myth but absent from Naudé's version—must we not think that there was not one version, but instead several separate posters? This would confirm the sentence uttered by Chaume or his friend in Le Pailleur's story: "Why should not Mr. Pallieure make one?" although it is possible to understand "one" as a copy of the poster and not a poster of his own invention. This point will therefore call for some clarification below.

By the same token, what are we to think of the Lullists Le Pailleur mentions to justify the fabrication of the posters? This detail seems parallel to the one in Chorier's story, according to which the idea for these posters was allegedly engendered in Chaume by the fact that the Rosicrucians were in fashion then. In Le Pailleur's story, the Lullist vogue inspired the idea for them, with the ulterior motive that the posters would be attributed to them (the Lullists). Therefore the two stories contradict each other on this point. It seems to me, however, that far from contradicting each other, they could well be complementary: If, for example, Chaume and his friends generally shared some skepticism similar to Naudé's concerning the "sciences curieuses,"[72] they could well have intended to mock the Lullists and the Rosicrucian supporters at once. Is this to say that Lullism was very much in fashion in 1623? We cannot doubt it: In a general way, the beginning of the seventeenth century marks a high point in the destiny of this movement of ideas: a publisher from Strasbourg, Lazarus Zetzner, re-edited Ramon Lull's works simultaneously with the enormous task of publishing the *Theatrum Chemicum,* whereas Giordano Bruno's philosophy, as well as alchemy and the Christian cabala, welcomed Lull's doctrines.[73]

More precisely, we know that at the beginning of the seventeenth century, there existed in Paris several groups of Lullists who were often concerned with alchemy. One of these groups left its memories in a vast handwritten collection—the Caprara collection in Bologna—assembled in Paris and its surroundings between 1617 and 1645. J. N. Hillgarth turned up the traces of another group in his synthetic study on Lullism in France.[74] Do we wish to have names? There is Pierre Morestel, who published in Paris in 1621 an *Artis Kabbalisticae Academia* that was quite a lot closer to Lull than to the cabala.[75] There is Jean Belot, author in 1623 of *L'Oeuvre des Oeuvres ou le plus parfaict des sciences paulines, armadelles et lullistes,* pinpointed by Naudé two years later in his *Apologie.*[76] There is Robert Le Foul, sieur de Vassy, "General Secretary and Lullist Doctor, member of the Order, Militia and Religion of the Holy Spirit," who most notably translated in 1634 *Le grand et dernier art de M. Raymond Lulle.*[77] Thus we should scarcely be astonished to see Gisbert

Voet, in the *Admiranda Methodus* of 1643 directed against Descartes, denounce the entire group of anti-Aristotelian innovators by naming them in the same breath—a rather revealing combination: "the group of Lullists and Paracelsians and the imaginary troop of the Brothers of the Rose-Cross."[78]

It is therefore rather plausible that Chaume and his friends aimed their posters at both the Parisian Lullists and the followers of the Rose-Cross at the same time, which tends to substantiate Le Pailleur's story. Another point is worth noting: the prudence that Chaume and his friend seem to have shown upon Le Pailleur's entrance, which led them momentarily to hide the poster they were composing. Indeed Chorier, for his part, insists upon the proceedings to which Étienne Chaume then exposed himself, and that led him to flee the capital. It seems, if this detail of Le Pailleur's story is reliable, that Chaume was aware of this danger right from the start of the venture.

Finally, Locke, following Thoynard, has noted that Le Pailleur, shortly afer the posting of the placards, went to Brittany; this voyage itself is attested to by Tallemant Des Réaux: "He led a life of debauchery in Paris for a rather long time. Weary of this life, he went to Brittany with the Count of Saint-Brisse, First Cousin of the Duke of Retz."[79] We can therefore situate this voyage in the second half of 1623, and we see that Le Pailleur's story finds a number of confirmations in its main points. It agrees on the whole, in spite of some divergences, with Chorier's story.

From these two accounts, if their questionable aspects are eliminated for the time being, it emerges that the principal author of the posters was Étienne Chaume, a future student of medicine. It was a matter of a simple student's hoax, inspired by the sensation that the Lullists and Rosicrucians were then arousing in France (this last point calls for clarification below). One text was written, or maybe several, which Chaume and some friends, including Jacques Le Pailleur, copied over many times; these friends helped him post them in the busiest places in the city, and the agitation that resulted, attested to by all the sources of the time, was so intense and so widespread that the Parlement ordered an investigation.[80] Chaume, worried by the turn the events had taken, then prudently left Paris.

Now that the author has been identified, there still remains the task of determining the posters' exact content. Must we content ourselves with the version indicated by Naudé, or should we take into account the different versions passed on by other sources and trust the stories of Chorier and Le Pailleur?

The Content of the Posters

If we return to Peiresc's account, which is the earliest of all those that can be dated precisely and therefore one of the most reliable, we will observe that Peiresc summarizes the poster in a way that already diverges from Naudé's text: The Rosicrucians, according to Peiresc, "in four lines promised the general public the knowledge of divinity and personal invisibility." Nothing of this is seen in Naudé's version, unless one extrapolates from his text; rather than a summary, then, Peiresc must have given an interpretation, quite removed from the poster's literal meaning. On the other hand, Peiresc's summary agrees extremely well with another version, that passed on by the *Effroyables Pactions*:

> Nous Deputez du College de Roze-croix donnons advis à tous ceux qui desireront entrer en nostre société & congregation, de les enseigner en la parfaicte cognoissance du Tres-hault, de la part duquel nous ferons ce jourd'huy assemblée, & les rendrons comme nous de visibles invisibles, & d'invisibles visibles, & seront transportez par tous les païs estrangers où leur desir les portera.[81] Mais pour parvenir à la cognoissance de ces merveilles, nous advertissons le lecteur que nous cognoissons ses pensées, que si la volonté le prend de nous voir par curiosité seulement, il ne communiquera jamais avec nous, mais si la volonté le porte reellement & de fait de s'inscrire sur le registre de nostre confraternité nous qui jugeons des pensees, nous luy ferons voir la verité de noz promesses, tellement que nous ne mettons point le lieu de nostre demeure puisque les pensees jointes à la volonté reelle du lecteur seront capables de nous faire cognoistre à luy & luy à nous.[82]

The poster summarized by Peiresc, if we are to believe him, only contained four lines. In fact, that is exactly what we get if we limit the text of the *Effroyables Pactions* to its first sentence, and if we attribute the following lines, which are rather awkward moreover, only to the creativity of this pamphlet's author. Thus the accounts of Peiresc, Chorier, and Le Pailleur could be connected: far from it being a case of Chorier's embellishing on the text of the *Effroyables Pactions*, it would be on the contrary a case of this pamphlet's using the authentic posters. From this we would have to conclude that Chaume and his friends indeed composed several posters, and not only one as Naudé leads us to believe. Considered from the angle of perpetrating a hoax, multiple versions were certainly more apt to hold the Parisians' curiosity.

Following this hypothesis, Le Pailleur's story would then attest to the existence of one poster of which there is no other evidence, that which announced a meeting in a fake street of the Faubourg Saint-Marceau. But even if this story remains the only trace of such a poster, other texts, on the other hand, have come to us in handwritten form. We must now examine these.

Two manuscripts, one kept in Paris, the other in London, present other versions of the posters. The Paris manuscript, which dates from the first half of the seventeenth century, is found in a collection of pieces from the former collection of Philippe de Béthune (1561–1649), the brother of Sully and the governor and first Gentleman of the Bedchamber of the king's second son, Gaston d'Orléans. Philippe De Béthune, whose doctor was Étienne de Clave, the chief protagonist in the incident of the Parisian theses of 1624,[83] was also the owner (and possibly the writer) of a group of more than 75 manuscripts that form a series of encyclopedic miscellanies including rather numerous notes on natural philosophy and alchemy, taken particularly from Heinrich Khunrath and Jean d'Espagnet.[84] In the manuscript that concerns us, one page alone is of interest. First of all, we find a copy of the version of the poster communicated by Naudé, with several variants:

Affiches des Chevalliers de la Croix rose

Par permission de nostre College des freres Europeans nous faisons nostre sesjour visible et invisible en cette ville. Nous scavons parler les langues des pays ou nous voullons estre. Et par la grace du Tres hault qui attire a soy les coeurs des justes sans marques, figures ny billetz nous enseignons aux hommes noz semblables a se retirer d'erreur & de mort.[85]

If we compare this text with Naudé's, we see that it is made of nearly the same elements, but that they are arranged in a completely different order, without, however, lacking coherence. Below that text, this one is found:

Autre des mesmes

Habitans depuis peu nostre Palais de France nous faisons estat d'enseigner les sages & les combler de felicitez pour plusieurs siecles sans qu'ilz soient obligez a la necessité de mourir. Nous donnons les biens & la vye sans fantosmes & illusions de la seulle grace de celluy qui conserve l'Estre des choses.[86]

As for this version, it is entirely different from the others. It reminds us, however, of a part of what Chorier wrote describing Chaume's posters: "those who had the desire to live well and blessedly [*bene beateque vivendi*] were invited into this brotherhood," a theme that does not appear in the other versions. It is therefore highly possible that these texts circulated, too, in the same way as those of Naudé and the *Effroyables Pactions*.

The other manuscript likely to interest us, that in London, contains correspondence of the Paracelsian physician and alchemist Théodore Turquet de Mayerne (1573–1655). Established in London since 1611 as the first physician of James I of England and his wife, Mayerne had kept his French title of ordinary physician of the king after the death of Henry IV, and his correspondence shows that he continued to maintain close ties with France while in London, and occasionally, with Louis XIII himself.[87] His only well-documented trips to France during these years date from 1618 and 1625, but one of his best biographers suggests the possibility of other trips carried out in 1621 and 1623.[88] It is therefore possible that Mayerne, too, saw the posters with his own eyes. If we bear in mind his keen interest for these types of subjects and his vast circle of connections—which, moreover, included Rubens himself, who at least was attested as such in 1630—Mayerne could also have received the documents presented below from one of his correspondents as well.

These documents bear no date, and the fact that the manuscript containing his correspondence does not follow a chronological order hardly allows us to remedy this. We are forced to confine ourselves to the texts as they stand in the manuscript. The whole of what interests us fits on a single page.[89] Under the general heading of "Fratres Societatis Roseae Crucis," Mayerne first notes a version of the poster close to that kept by Naudé: "Nous deputez de nostre College principal des freres de la Croix Rosée faisons sejour visible & invisible en cette ville par la grace du Tres hault, vers qui se trouve le coeur des justes. Nous enseignons sans livres, sans marques ny signes & parlons les langues des paÿs, où nous voulons estre, pour tirer d'erreurs & de mort les hommes nos semblables." This text, perhaps noted from memory rather than copied,[90] is followed by an astonishing list of ten questions bearing essentially on cabalistic themes that is highly problematic. Its source remains to be determined.[91] But its relation to the Parisian placards of 1623 is certain. On August 10, 1623, Peiresc indeed writes to Rubens: "I am sending you the little memoir of Mr. Pignorius on the hieroglyphic hand and at the same time some writings concerning the sect of the Rose-Cross, which were found in the room of a fugitive, & in which the rubbish and the astonishing stupidity of their cabalistic and alchemical mysteries is shown."[92]

Let us admit that it is very tempting to identify this "fugitive" as Étienne Chaume, although nothing allows us to be certain of this.[93] Still, a copy of these cabalistic questions, almost identical to that kept by Mayerne, is found among Peiresc's papers kept in Carpentras, in a group of letters and miscellanies Peiresc himself collected.[94] Here we find first, a few variants aside, the text of Naudé's poster (the brothers of the Rose-Cross are called *freres Roses*).[95] This is the very manuscript the editors of the Mersenne correspondence allude to in these terms: "Peiresc was able to obtain, perhaps before his departure from Paris, the copy of another document containing *"Certains articles des propositions des Frères-Roses,"* and dealing with the doctrine of the *Sephirots*, necessary for understanding the *Zohar* and for being enlightened by divine wisdom."[96]

All of the details gathered here lead one to ask—though it is scarcely possible to acquire the certainty of it—if Mayerne did not become aware of this document through Rubens himself. However that may be, the document focuses on the existence, the origins, the nature and the role of the ten *sephiroth,* those divine attributes Gershom Scholem has defined as "the ten spheres of divine manifestation in which God leaves His secret dwelling," the ensemble of which form "the unified universe of God's life" and upon which a large part of the cabala's speculations lie.[97] Here is this document, in the version kept by Mayerne:[98]

Articles de propositions faites par les freres de la Croix Rosée[99]

1. Si le monde a un, ou plusieurs coadjuteurs.[100]
2. Par quelle necessité on peut dire, qu'il y a des Zephirots, car l'on n'eu peu soustenir, qu'il n'y a[101] que l'infini.
3. S'il est necessaire qu'il y ait des Zephirots, comment il sera necessaire, qu'ils soyent dix, & que neantmoins les dix ne soyent, qu'une seule puissance.
4. D'ou se justifiera que les Zephirots sont emanez, & non creez, comme les creatures.
5. Comment on pourroit dire qu'il y en a eut[102] jusques à dix, & qu'ils s'immiscent; car en cecy est le secret des secrets.
6. Attendu qu'il y a indubitablement dix Zephirots, & qu'ils sont emanez ou influez, & non creez, & que le nombre n'est qu'une seule faculté, l'on[103] demande, pourquoy leur a esté donné une borne, mesure & definition.
7. Si les Zephirots sont irradiez de tout temps en quelque maintien[104] proche de la creation du monde.
8. Quelle est leur substance.
9. Quelle est la raison de leurs noms, & de leurs lieux.
10. Que chacun des dix a soubs soy cent, & c'est le sens de Moyse de cent bases, qu'il avoit faict, & le sens[105] du verset qui dit, que toutes rivieres retournent en la mer.[106]

In the absence of any additional explanation, it is quite difficult to grasp what link could attach these "propositions" to the brothers of the Rose-

Cross, unless it has to do with an extract from one of the countless treatises of Rosicrucian literature published before 1623. In Mayerne's manuscript, these ten articles are all the more astonishing in that they are followed by this note, written in English, probably a commentary by Mayerne himself, who in his manuscripts freely uses Latin, French, English, and sometimes even German (then substituting *Fraktur* for his usual handwriting): "This beinge fastned in diverse parts of Paris, there is strigt order given, to inquire, after these pretended brothers; but yf they keepe them selves upon the invisible in their propositions, I doubt not, but they will bee thougt as well, imperceptible in their Cabalisticall propositions."[107]

The irony of this commentary cannot make us lose sight of the questions it raises: Must we understand that not only was the well-known broadside posted, but also the cabalistic propositions that follow it in the manuscript? If we are to establish a connection between Mayerne's testimony and Peiresc's concerning the documents discovered in Paris at a fugitive's home relating to "cabalistic and alchemical mysteries" of the Rose-Croix, must we admit instead that these propositions, if they were not actually posted, were at least intended to be? But their rather technical nature, the desire displayed therein to examine thoroughly very precise points of the cabala, separate them quite a bit from the facetious and superficial character of the other posters encountered until now. In another respect, the detail noted by Mayerne concerning the investigation ordered by the Parlement lends a certain authenticity to his testimony and leads to the belief that it was contemporary to the facts.

The question resurges again with the last document contained in Mayerne's manuscript. Following the lines we have just read there is indeed found a new text, this time written in Latin, of which the translation follows:

May what follows be told to the most penetrating professor of the most excellent and most difficult science, to the most ardent defender of the Society of the Rose-Cross, to the brother worthy of being honored with all sorts of good offices.

Your great devotion to our society, the noble and numerous sleepless nights you have devoted to upholding its grandeur, are well-enough known to our primates, thanks to the distinguished benevolence of which, remarkable testimonies under the form of serious recommendations of your virtues, have very often reached us, the members of this noble mystical body. There are posted by our people for your aid, in the street corners of the very famous city of Paris, theses and problems filled with a more-than-Platonic spirit; the ability to penetrate their hidden meaning to the marrow is given only to an Oedipus, that of extracting their core is given only to a disciple of the rosy Society. Our invisible college scorns the vile minds of crawling men, disdains the crooked minds of the Beotians. The doors of the sacred sanctuary only open for the lynxes, among whose number we acknowledge you and congratulate you upon your election and send you—who are now our cross-bearing brother of a crimson brighter than the bright red color of the physical salamander—secrets to elucidate, and just as you are expected to reveal your feeling on them, by the same token it is right and appropriate for you to send us your principles in return. Farewell, and may the sevenfold ennead of the Sephiroth, an assistant in your efforts with the cabalistic decade of prime spirits in its concentric homogeneity, intervene in your plans with good fortune by the powerful rays of its influence. Be well and live. Done at our place, June 13, 1623.

Your P. ♂ N. ♃ S. ♄.
Your brothers by the mandate of W. ☉. "[108]

Even if the *Fama Fraternitatis* obviously inspired the play on initials that signs this text, it seems vain to wish to find in it a meaning that escapes us for want of being better informed. Beyond the enigmas it presents to the reader, the chief question that this document raises is whether it is an authentic poster, a plan for a poster, or perhaps a new hoax, possibly thought up by Mayerne himself, who, as a good alchemist, left in his papers proof of his mastery in this type of language and indications of his taste for such enigmas.[109] Moreover, one must acknowledge that this text is not found among the previously mentioned papers of Peiresc. Nevertheless, two details stand out: This text, intended to accompany the cabalistic theses that it follows, stands either as a message destined to the most interested onlookers—but must we emphasize the danger involved in having such notes circulate from one hand to another?—or as a new poster; what is more, it seems to attest that the cabalistic theses were indeed posted in Paris—and before June 13th—as problems submitted to the "initiables" by those signing the note. A third detail could furnish a response to the questions raised: This document is the only text written in Latin among those of the 1623 posters. Now, we remember that Nicolas Chorier affirmed that Étienne Chaume composed his posters in Latin. This document could be a remnant of this, either a plan for posters or posters actually put up on the walls. The hypothesis seems all the more tempting since the style is somewhat mannered, which would go along with the talents Chorier attributed to Chaume as a Latinist. Mayerne enjoyed the same gifts, but it is difficult to see why and with what goals he would have taken the trouble to elaborate such a hoax, for the well-established prudence of this quinquagenarian could hardly have led him to distribute such texts by a means other than private correspondence, and the indications that have come to us on his type of spirituality betray no penchant for the cabala.[110] A last detail argues for the authenticity of this text. Chorier summarized the posters' content in these terms: "Those who desired to live well and blessedly were invited to this brotherhood & were fortunate if they were initiated into these mysteries." The second clause of this sentence corresponds quite well to the above note, which is the only one to express this idea of fortunate election. This common point, combined with the preceding one, inclines us to take Mayerne's manuscript seriously.

At the conclusion of this investigation concerning the text of the posters, some certainties at last seem to be acquired. Étienne Chaume must have created several distinct posters:

1. The one we have from Gabriel Naudé; this version, also mentioned implicitly in the *Effroyables Pactions*,[111] was reprinted in the "Preface to the Reader" in the French version of Neuhus's work published in the autumn of 1623[112] and taken up again in the *Examen sur l'Inconnue et Nouvelle Caballe des Freres de la Croix Rosee*,[113] in the *Mercure François*,[114] then in the *Tresor chronologique* by Pierre Guillebaud,[115] and later still by Adrien Baillet;[116] it was also copied over in a number of manuscripts: those of Peiresc, Turquet de Mayerne, and Philippe de Béthune, to which we will add a manuscript coming from President Achille III of Harlay (1639–1712).[117]

2. The poster transcribed in the *Effroyables Pactions*, which agrees with the remarks Peiresc made to Rubens in his August 3, 1623, letter, as well as with Chorier's and Le Pailleur's stories.

3. The poster in Philippe de Béthune's manuscript, which, in accordance with Chorier's story, promises bliss to the wise.

4. Probably the ten cabalistic propositions kept by Turquet de Mayerne and Peiresc, if we are to believe Mayerne on this score and if this is the sense in which we are to interpret Peiresc's account in his August 10, 1623, letter to Rubens.

5. Finally, probably the long text in Latin kept by Mayerne, which is the only trace left of the use of Latin by Étienne Chaume, whose content agrees with that of part of the summary of the posters' text as given by Chorier.

From this we see that the accounts of Chorier and Le Pailleur, albeit rather divergent from the version of the facts presented by Naudé, are compatible with the facts gathered through the other accounts of the time, which argues still more in favor of their authenticity.

The First Reactions to the Posters: Hasty Associations?

How did contemporaries react to the posters? We have already seen this in part. Their discovery clearly provoked intense emotion in the capital. We also know from certain accounts that various swindlers quickly profited from rumors caused by the event, immediately passing themselves off as Rosicrucians in the opinion of credulous people to squeeze money out of them.[118] Their victims seem often to have been men of the law, which is indeed plausible if one considers that news circulated a great deal at the Palace of Justice Exchange.[119] Finally, we know that men like Peiresc echoed the matter in their correspondence. But the posters evoked different reactions by means of printed matter as well: The publication of several disturbing pamphlets was one of the causes that ultimately pushed Naudé to react. Furthermore, the coincidences of current events brought about confusion, and even associations, between the Rosicrucians and the Enlightened Ones of Spain, and even between the Rosicrucians and the libertine circle of Théophile de Viau.

An Irrational Rash of Pamphlets

One of the first reactions by means of printed matter seems to have been the translation of Heinrich Neuhus's work, through the initiative of an anonymous individual eager to "serve the public." This anonymous individual affirmed only that his "profession had always been to handle weapons more than books," although he knew Latin quite well and, if he is to be believed, he would have preferred to compose "some new work in accordance with his temperament" rather "than to subjugate himself to the forced translation of others' ideas."[120] The work must have been published in the autumn, since the translator hoped to satisfy the people enough not to have to regret "the small amount of time that I will have wasted in this harvest season telling you news of Germany and making you aware of those who in Paris are held to be Invisible."[121] His choice was clever, for although Neuhus, in 1618, scarcely sought more than to show the Rosicrucians to be "Anabaptists, or Socinians,"[122] the title of his book, accurately translated into French (*Advertissement Pieux & tres utile, Des Freres de la Rosee-Croix: A sçavoir, S'il y en a? Quels ils sont? D'où ils ont prins ce nom? Et à quelle fin ils ont espandu leur renommée?*), was quite apt for making it sell in Paris scarcely a few weeks after Chaume's posters.

Even if the translation of Neuhus's work had commercial motivations, such was not the case for other reactions. Two venomous pamphlets followed each other in the month of November or December:[123] the *Effroyables Pactions* and the *Examen sur l'Inconnue et Nouvelle Caballe des Freres de la Croix Rosee*. Both, related to the "rag" genre,[124] exploited the event in its most sensational dimensions by diabolizing it to the extreme, as we will see below.

Very shortly thereafter, toward mid-December, Gabriel Naudé published his *Instruction à la France sur la verité de l'histoire des Freres de la Roze-Croix*.[125] The year was then nearing its end, but the public's curiosity was not, which explains why each of these works was put on the market again in 1624 with a new title page. Neuhus's *Advertissement* and the *Effroyables Pactions* were thus again put on sale, to which must be added another edition without title of the latter, as well as a pirated edition of Naudé's work.[126] As for the *Examen sur l'Inconnue et Nouvelle Caballe*, three separate editions of it, all from Paris, were already in existence in 1623. One was said to be printed "for Pierre de La Fosse. With Permission";[127] another was published by David Ferrand, without permission, but in every way identical to the preceding one;[128] the last was published without an editor's name, without permission, with a different pagination, and bearing everywhere "Rozee-Croix" instead of "Croix Rosee;"[129] this one was put on sale again in 1624.[130]

Finally, Wallace Kirsop gave in his 1960 dissertation the title of another pamphlet published by Pierre Chevallier in 1624: the *Traicté des Atheistes, Deistes, Illuminez d'Espagne, et nouveaux pretendus Invisibles, dits de la Confrairie de la Croix-Rosaire. Elevez depuis quelques annees dans le Christianisme*.[131] This quarto pamphlet of fifty-six pages is not easy to identify, for far from figuring today among the anonymous works of the largest Parisian libraries, it is found only at the conclusion of Claude Malingre's continuation of Florimond de Raemond's *Histoire generale du progrez et decadence de l'Heresie moderne*. Announced on the title page, it can be attributed to Malingre himself.[132] After a preface by the latter, the brochure offers three successive chapters devoted to atheists, deists, and the Enlightened Ones of Seville; the fourth chapter, devoted to the Rosicrucians, is nothing other than Naudé's work.[133]

The privilege of Malingre's work dates from March 14, 1624: This shows that the matter of the Rosicrucians created an uproar for nearly nine months. But in exactly what senses were Étienne Chaume's posters interpreted?

In *The Rosicrucian Enlightenment,* Frances Yates considered quite rightly that the stir created by the Rosicrucians, particularly Father Garasse's work (*La Doctrine curieuse,* of which I have not yet said anything) and the *Effroyables Pactions,* were an attempt to launch a witch-hunt against the fictitious Brotherhood in France.[134] Naturally, Yates, with as much caution as insistence, connected this repressive fervor with her own interpretation of the end of the movement in Germany.[135] She then strayed into entirely erroneous views concerning Gabriel Naudé, misinterpreting his text as well as his thought,[136] for, by interpreting his *Instruction à la France* in the retrospective light of the *Apologie pour tous les grands personnages qui ont esté faussement soupçonnez de magie* (1625), a text whose true intention she did not grasp, perhaps because she did not know René Pintard's book, *Le Libertinage érudit,* she saw in Naudé a secret defender of the Rosicrucians, particularly well-informed concerning them but constrained by the repressive tendency of those years to hide his true opinions.[137] Nothing could be further from the position taken by Naudé, who, even if he was surely one of the best-informed Frenchmen of his time concerning the Rosicrucians, obviously did not know any more than what was contained in the various treatises that he had been able to read;[138] he was particularly unaware of the reality of Adam Haslmayr's destiny[139] and seems not to have been acquainted with Johann Valentin Andreae's work, certain aspects of which would not have failed to strike him. As an example, Andreae, in a work published in 1617, denied that magic was an art that came from the devil and could be learned in a very brief amount of time, declaring that on the contrary, "there is no magic other than the assiduous study of numerous and varied arts":[140] Without going into detail concerning the latent conceptions underlying this phrase, we must admit that a more likely position to attract the future author of the *Apologie* of 1625 could not have been invented.

But this is not yet the moment to study Naudé's position. To understand properly the reactions provoked by Chaume's posters, we should point out four repressive currents in France at the time that often combined with each other but that nevertheless did not aim at the same targets: far from it.[141] The French hostility to the Rosicrucians first fed on hasty associations. Even before it was a question of posters, a Jesuit apologetic treatise, that of

Father Gaultier, had already denounced the German Rosicrucians as a new Protestant sect, designating them for Catholic hatred. Then the Rosicrucians of Chaume's posters had scarcely appeared when they were assimilated in the public's mind with the Alumbrados of Seville and considered to be a sect of heretics. As for Father Garasse, solely for the need of his polemic against Théophile de Viau and his libertine circle, he had established a parallel between the Rosicrucians and those "atheists"; the image of the Rosicrucians then suffered, in the public's mind, the consequences of an association with Theophile's trial. But most often, the diverse comments on the fictitious Brotherhood especially fed the anti-Paracelsian polemic that had been latent in France uninterrupted since 1578. These different grievances, leveled against the Rosicrucians as Protestants, Alumbrados, atheists, and Paracelsians, most often combined with each other, but we see clearly that they did not answer to the same motives.

Protestants and Rosicrucians

As early as 1621, a great Jesuit controversialist, Father Jacques Gaultier (1562–1636), at the time a professor of theology at the College of the Trinity in Lyon, mentioned the German Rosicrucians in his *Table chronographique*. Contrary to what this shortened title might lead us to assume, it is in no way a historic work, but rather an apologetic book comparing all heresies to Protestantism, from the beginning of the Christian era to the seventeenth century, and showing the continuity of Catholicism through the centuries.[142] Presenting the Rosicrucians in the light of Michael Maier's *Themis aurea* (1618), Gaultier concludes: "Tous ces propos, partie enigmatiques, partie temeraires, partie heretiques, partie suspects de magie, nous donnent occasion de conjecturer que ceste Pretenduë Fraternité n'est pas si ancienne qu'elle se faict, ains que c'est un rejetton du Lutheranisme, meslangé par Satan d'empirisme & de Magie pour mieux decevoir les esprits volages & curieux."[143]

Lutherans, empirics, and witches: According to Father Gaultier, the Rosicrucians indeed deserved the pyre's flame. Even without exaggerating the importance of this passage, which is buried in a controversial work of enormous size, we still must realize that Father Garasse would soon make good use of it, and that Naudé, and then the *Mercure François,* would in turn cite it in their bibliographies.[144] Let us now tackle the very period of Étienne Chaume's posters.

Enlightened Brothers of the Rose-Cross and the Alumbrados of Seville

In 1691, in his *Vie de Monsieur Des-Cartes*, Adrien Baillet agilely depicted the reception of the Rosicrucians in France in 1623:

> Il s'étoit fait un changement considérable depuis l'Allemagne jusqu'à Paris sur les sentiments que le Public avoit des Rose-croix. On peut dire qu'à la réserve de M. Descartes & d'un trés-petit nombre d'esprits choisis, l'on étoit en 1619 assez favorablement prévenu pour les Rose-croix par toute l'Allemagne. Mais ayant eu le malheur de s'être fait connoître à Paris dans le même têms que les *Alumbrados*, ou les Illuminez d'Espagne, leur réputation échoüa dés l'entrée. On les tourna en ridicule, & on les qualifia du nom d'*Invisibles;* on mit leur histoire en romans; on en fit des farces à l'hôtel de Bourgogne; & on en chantoit déja les chansons sur le Pont-neuf, quand M. Descartes arriva à Paris.[145]

It is only too easy to criticize this description, which is of a more lighthearted nature than its reality: Naudé was in fact the only one who publicly ridiculed the Rosicrucians of Paris; even if it is true that they were called "invisible," this was not always in an ironic manner, far from it; the only "novels" concerning them were two pamphlets that were as violent as they were vehement, the *Effroyables Pactions* and the *Examen;* as for the farces of the Hôtel de Bourgogne and the songs sold on the Pont-Neuf, they stem directly from Baillet's imagination.[146] But there is, on the other hand, one point that Baillet touched upon correctly: the association made early on between the French Rosicrucians and the Alumbrados.

The incident of the Alumbrados of Seville seems to have exploded in the spring of 1623. It involved a sect of spiritualists whom the Spanish Inquisition immediately repressed.[147] The news appeared in Paris a short time later: The edict pronounced by the Inquisition was translated and printed there,[148] and the majority of the contemporary sources do not fail to mention the matter. The connection with the Rosicrucians—at least with the German ones—did not boil down solely to the notion of clandestine meetings. The Alumbrados, like most of the German authors involved in the Rosicrucian movement, did not acknowledge any church other than an internal one, claiming themselves to draw from the Holy Spirit. Indeed, Baillet did not exaggerate. From August 3, 1623, in the earliest of the accounts that we have at our disposition concerning the reception of Chaume's posters, Peiresc responds to Rubens in these words:

> The news of the Basilidians of Seville had not yet reached me and we would have looked willingly at the edict concerning them. But on the other hand we have had here certain other new Rosicrucian sectarians, who are rather famous in Germany

and who are perhaps the same as those of Seville. They thought up the idea of posting certain papers or notices on the streetwalls, in which they promised the general public, in four lines, the knowledge of divinity and personal invisibility and this in the name of the brothers of the Rose-Cross.[149]

We see here that even in the eyes of the most cultivated public, Rosicrucians and Alumbrados of Seville were easily confused with each other in the tangle of sects. Rubens himself who, since he lived in Antwerp, had heard of the Alumbrados—here jokingly called "Basilidians" in reference to the second-century Gnostic sect whose gems both men studied—before Peiresc had, answered his correspondent in the following week:

> For the time-being, it will be difficult to obtain the decree promulgated by the Inquisition against the Basilidians in Seville. As far as I know, we have received only one copy here. We will nevertheless do everything in our power to get it.
>
> The Rosicrucian sect is already old in Amsterdam, and I remember having read, three years ago, a little book published by their society, in which we found the glorious and mysterious life and death of their first founder, as well as all of their statutes and orders. I saw in all this nothing more than alchemists, pretending to possess the philosophers' stone, and it is indeed a pure imposture.[150]

This time, Rubens distinguished clearly between the Rosicrucians and the Alumbrados. Nonetheless, these two matters would still appear in his correspondence with Peiresc up to 1624, and in a systematically contiguous manner, as if one necessarily led to the other.[151] Such a connection is observed almost uniformly in the majority of contemporary accounts. In the autumn of 1623, here is how the translator of Neuhus's work painted the picture of the religious context of his time:

> au lieu de nous humilier devant Dieu & le servir selon ses commandemens, paroissent une infinité de nouvelles gens, les uns desquels souz pretexte, de je ne sçay quelle reforme, convertissent l'ancienne religion de nos Peres en un certain cult exterieur, auquel ils attachent entierement nostre salut: Les autres qui par une extreme impieté ont contraint les Magistrats d'uzer du glaive que Dieu leur a mis en main pour venger l'injure faicte à Dieu & à ses saincts bien-heureux, à la terreur de ceux qui desja par un desir de vivre licentieusement se laissoient emporter à l'Atheisme. En mesme temps on nous a apporté d'Espagne les nouvelles d'une secte Epicurienne de gens qui neantmoins se qualifient, Los Alombrados, ou Illuminez, que la Sacrosaincte Inquisition a bien de la peine d'exterminer. Puis d'une mesme volee se sont mis sur le tapis ces Freres de la Rozee-Croix, que l'on dit estre venus d'Allemagne.[152]

Garasse in turn painted a picture closely inspired by this text in his *Apologie* at the very beginning of 1624.[153] Shortly thereafter, an author who was himself inspired by Garasse, Claude Malingre, exercised the same association in volume 3 of his *Histoire de nostre temps*. After reproducing the May 29, 1623, edict against the Alumbrados, Malingre continued as follows: "Or puis que nous sommes icy sur le traict de ces nouvelles sectes qui se sont ainsi elevees dans le Christianisme depuis quelques annees, il est bien seant de rapporter en ce lieu, la naissance, le progrez, la doctrine, & facons de faire d'une autre secte endiablée qui paroist en nos jours en plusieurs lieux de la Chrestienté, particulierement en France, c'est de certains personnages originaires d'Allemagne qui se disent Confraires de la Confrairie & College de la Croix Rosaire."[154] And Malingre did the same in his brochure that was published simultaneously with the eloquent title: *Traicté des Atheistes, Deistes, Illuminez d'Espagne, et nouveaux pretendus Invisibles, dits de la Confrairie de la Croix-Rosaire*. Chapter 3 of this pamphlet, "Des Freres de la Confrairie de los Alumbrados ou Illuminez d'Espagne, & de leur profession empeschee par l'Inquisition Apostolique," is again only a reproduction of the edict of May 29, 1623, and immediately precedes the chapter devoted to the Rosicrucians.[155] A few months later, the *Mercure François* in turn dealt with the Alumbrados and the Rosicrucians successively, going on about the latter in these words (I emphasize the first words): "*En France aussi* il se veit plusieurs livrets contre une Fraternité pretenduë, appellee de la Rose-Croix, que l'Allemagne a produitte depuis quelques annees, les Confreres de laquelle furent appellez les *Invisibles,* par les faiseurs de nouvelles qui se vendent sur le pont neuf à Paris."[156]

The same remark is valid for the 1626 edition of the *Table chronographique* by Father Gaultier: inside of the passage on the Rosicrucians in the 1621 edition, Gaultier added fifteen lines on the episode of the Parisian placards, referring notably to the *Mercure François* and to Father Garasse; then, scarcely two pages later, he came to the "Enlightened Ones [of Spain]."[157] There is even a fictitious collection in the Mazarine Library in Paris, composed in the seventeenth century, that bears witness to this association: The collection contains successively the *Effroyables Pactions,* the *Edict d'Espagne contre la detestable Secte des Illuminez,* the *Examen sur l'Inconnue et Nouvelle Caballe des Freres de la Croix Rosee,* and Neuhus's *Advertissement Pieux & tres-utile*.[158] Such an association made the Rosicrucians themselves seem a sect deserving of the Inquisition's lightning bolts. The height of the confusion was reached, a quarter of a century later, by the historian and philosopher Scipion Dupleix (1569–1661), the king's historiographer.

In his *Continuation de l'histoire du regne de Louys le Juste* (1648), Dupleix constantly confused the Rosicrucians with the Alumbrados, and in the exposition of their doctrines he borrowed nearly verbatim from the *Mercure François* five of the articles of the edict pronounced by the Inquisition against the Alumbrados, while at the same time condensing several other articles; he even extended his confusion to the Enlightened Ones of Picardy, or "Guérinets," who appeared between 1620 and 1630, whose doctrines, it seems, consisted of "distorted simplifications of themes developed by the mystics of the North: Ruysbroeck, Tauler, Benoît de Canfield."[159] Here is Dupleix's story:

> During this pontificate [*that of Urban VIII (1568/1623–1644)*] a new sect appeared in France, that was named *Enlightened Ones* or *Invisible Ones*, some called it *Rose-Cross;* it had caused a stir in Spain as early as the year 1623, at which time the seven original authors of this accursed Heresy were condemned by an Edict, with an injunction to all those who might be affected to purge themselves to the Inquisition in the thirty days following the publication of the Edict, and this with grievous penalties for the offenders.
>
> This Edict reports seventy errors that this immodest Cabala received as principles of its doctrine; but our *Enlightened Ones* of France had reduced them to eight or ten which were neither to obey nor acknowledge any Secular or Ecclesiastic Superior at the expense of contemplation. That one can see the divine Essence during this life and know the secrets of the Holy Trinity. That the Holy Spirit directly governs those in their sect. That meditation on the sufferings of Jesus Christ was useless to them. That their perfect ones did not need to do good deeds, and that they were no longer obliged to go to Mass, nor to fast, nor to fulfill any practice that may exist, these precepts of God or the Church being for those who are not yet raised to this sublime state. That being taken over by the spirit and the love of God, they were no longer subject to the Laws: upon which they founded the basic article of their sect, which is to let the body give in to all its desires, even the most filthy contacts and the most abominable excesses of the flesh, because, they said, since they were in the perfect state of children, they should not be restrained to the servitudes of slaves, and because the operations of the body, whatever they may be, are rendered holy by the application of the spirit of man to the spirit of God, and of his own will to His love.
>
> This sect, which favored the senses, made a great deal of progress in a short amount of time, particularly towards Picardy and Beauce, where some weak and sensuous minds had so entirely given themselves over to these infamous practices, that it was difficult to turn them away from them, because the flesh prevailed over the mind, under the handsome pretext of perfection and refined spirituality; in such a way that whatever haste was brought to them through preaching, and through lectures, and through assault and corporal punishment, this evil fire was not so well extinguished that some sparks do not remain even yet: for, when I was in Paris in 1643, I saw two of them at the prison of the Conciergerie, and conversed for more than an hour with them, in the presence of the Bishop of Grenoble,[160] who had brought me there because one was the brother of his Chaplain; but I found that

they were more Jewish than Christian, and considered one to be suspect of Magic. Anyway, as they were sought by the Law, they dissipated without a sound, vanished and rendered themselves truly invisible.

This last detail consummates the tenuous connection between the Enlightened Ones of Spain or Picardy and the Rosicrucians: in Scipion Dupleix's work, which thus crowns an entire historiographical tradition, the association between the two is absolute. But this is only one aspect of the reception of Etienne Chaume's posters.

The Rosicrucians and Théophile's Trial

Among the 1623 works mentioning the Rosicrucians, we find that of Jesuit Father François Garasse, *La Doctrine curieuse des beaux Esprits de ce temps,* directed primarily against the libertine poet Théophile de Viau, his friends and (if possible) his protectors.[161] In a recent article, Louise Godard de Donville brought to the fore the ties insidiously woven by Garasse in this book between the "clever minds of this time" and the Rosicrucians, with the obvious goal of discrediting Théophile by placing him parallel to the diabolized characters of the brothers of the Rose-Cross.[162] It must be realized that by doing so, Garasse inscribed himself in an anti-Rosicrucian campaign undertaken in 1618 by another member of his order, Father Jean Roberti, a campaign that I will discuss below. Moreover, he did not restrict his association of Théophile with the Rosicrucians to *La Doctrine curieuse;* in January 1624, he also linked Théophile, the Rosicrucians, and the Alumbrados in his *Apologie,* his attempt to reply to the attacks in the *Jugement et Censure du Livre de la Doctrine Curieuse* published by François Ogier three months prior.[163]

Now Théophile de Viau, when he was at last interrogated in 1624 after a long period of detention, was briefly suspected of having written or having had printed a book entitled *Les Enffans de la Croix Roze* "which is full of impiety," found in his trunk at the moment of his arrest, in September of the preceding year, and to which it was believed he had alluded based on some words written in his hand "mentioning the children of the Rose-Cross." Théophile responded that he had never seen this book, nor had he himself placed it in his trunk, and that he had never written anything on the Rose-Cross except according to what he had "heard said without however having seen it."[164] It is indeed conceivable that someone—by Garasse's order?—had placed the book, whatever book it may have been, in Théophile's possessions, in the hope of compromising the poet.¹⁶⁵ We also know how Théophile himself reacted to the posters: finding himself charged, shortly before the incident with the book, with lines written in his hand in the margin of a handwritten piece in which he mocked some Parlement members' belief in the existence of the Rosicrucians, he answered that his mockery only aimed at discrediting the Rosicrucians, "whose posters he [called] 'game-flushers,' that is to say, traps for intercepting the impious curiosity of those who enjoy them."¹⁶⁶ Théophile, in other words, perhaps aware of Garasse's maneuver, thought that the posters were perhaps an ambush the authorities had set to trap the unwise who might have declared their sympathy toward the Rosicrucians.

Would this suspicion have been justified? In his letter to Rubens on August 3, 1623, Peiresc, after having put forward the hypothesis of an identity between the Alumbrados and the Rosicrucians of Paris, continued in these terms: "Someone inquired about them [i.e., the Rosicrucians] and a certain Théophile, a courtier suspected of atheism, was put in prison, and will be put on trial for another charge, since no proof can be found that he belongs to the sect of the Rosicrucians."¹⁶⁷ From this surprising account, it seems that the suspicions of Rosicrucian sympathy that weighed upon Théophile de Viau in 1624 could have been born as early as the period of the Parisian posters one year earlier, and directly in connection to these posters. If one takes into account the exaggeration in Peiresc's story—we know well that it is not this suspicion that incited the Parlement to order the poet's arrest on July 11, 1623¹⁶⁸—the fact remains that the pieces concerning the Rosicrucians about which Théophile was interrogated in June 1624 had been seized, or perhaps planted in his belongings, as early as his arrest in 1623. Of course one must not exclude the possibility that Peiresc, confronted with the influx of news coming from all sides, mixed up distinct incidents and melded them into one. Nevertheless, he insisted on this point in the continuation of his correspondence, establishing a direct correlation between *La Doctrine curieuse* and the Rosicrucian matter. Thus on September 17, 1623, one month after his departure from Paris, here is what he wrote to Rubens:

> I saw a very meticulous account of the interrogations performed by the Inquisitors of the Adombrados [*sic*] or Denudos of Lisbon, who, under the guise of congregation and piety committed the most abominable obscenities [*sporchezze*

nefandissime]. If you do not have it, I will have it sent to you from Provence along with the commission of the Inquisitor Bishop.

I am very obliged to you for your opinion on the Rosicrucians, against whom Father Garasse is having a very handsome book printed, which has not yet been published, at least it was not during my stay in Paris.[169]

Peiresc therefore made Garasse's work (at least such as he had heard of it) into a treatise directed against the Rosicrucians, which proves that section 14 of *La Doctrine curieuse,* the only one devoted to the Rosicrucians, must have caused a certain stir even before its publication. The Rosicrucians were thus implicitly assimilated with the libertines even beyond Garasse's book. Moreover, this association had almost started with Father Gaultier's *Table chronographique,* in which the chapter on the "Fraternity of R.C." already appeared, rather significantly, between "Marc Antoine de Dominis Apostat" and "Lucilius the Atheist," that is, the famous Vanini tortured and executed in Toulouse in 1619, whom Garasse later presented as Théophile's primary inspirer.

But the most obvious tie between the matter of the Parisian posters and *La Doctrine curieuse* is found precisely in one of the pamphlets these posters brought about: We have only to reread the *Effroyables Pactions* to recognize indeed the very essence of Garasse's thought, pushed here to a paroxysm that only the anonymous genre of rags could unleash.[170]

Taking over from Gaultier and Garasse, the *Effroyables Pactions,* challenging the common opinion that tended to assimilate the Rosicrucians with the Alumbrados, on the contrary distinguished carefully between the two sects, which instantly permitted the author to situate the Rosicrucians directly in the tradition of Vanini: "On tient que les Illuminez d'Espagne, & les Invisibles de France n'ont rien de commun en leur croyance, ains qu'elle est differente grandement de l'un à l'autre: les Illuminez croyent l'immortalité de l'ame, & noz Invisibles n'en croyent point: Toute leur croyance n'est qu'Epicurienne, enseignent la mesme leçon & la mesme methode que ce Philosophe Italien qui fut brulé à Thoulouze en la place de Salin par Arrest du Parlement dudit lieu, en l'année 1619."[171] The term "Epicurean" is in no way innocent: It is not only the chief complaint of the sixth book of *La Doctrine curieuse* and one of the significant traits of the sect of "clever minds" Garasse constantly brought up, here fallaciously attributed to the Rosicrucians; it is also, let us not forget, the theme of one of Chaume's posters, preserved in one of the manuscripts of Philippe de

Béthune—a prince who had become Théophile's friend just at the start of the years 1616–17.[172] The *Effroyables Pactions* did not strike at random, and here Étienne Chaume made himself, without being aware of it, the objective ally of Garasse's side, providing supplementary material for the "insidious parallel" between the Rosicrucians and the libertines, which perhaps justified Théophile de Viau's mistrust toward these posters he called "game-flushers."

The *Effroyables Pactions* does indeed develop the theme of the Rosicrucians' Epicureanism. But beforehand, the author takes care to identify his target by name. According to the fanciful story implemented in this pamphlet, six Rosicrucians had been sent to France by the Devil to preach a new religion there; they found accommodations in various places on the edges of the capital: "Là proposoient les leçons qu'ils devoient faire en particulier avant de les rendre publiques, & de la difficulté qu'il y avoit d'enseigner une nouvelle religion à Pairis [sic], tant à cause des livres Theophiliques, que de tant de Predicateurs qui ne demandent autre chose que d'entrer dans le combat de la verité pour confondre les ennemis de la Religion & les fleaux ou plustost les bourreaux de la vertu."[173] Thus two types of obstacles conflicted with the teaching of the new Rosicrucian religion: works of Christian apologists, and the "Theophilic books," that is, those that exposed the "curious doctrine" of the libertines. In other words, the Devil had to confront two distinct adversaries: the theologians, who were his natural enemies, and the followers of Théophile, who found himself set up as Satan's rival in the work of the corruption of souls.

A little later in the pamphlet, the Rosicrucians had finally succeeded in trapping "a Lawyer of the Parlement of Paris"—which raises some thorny questions for us;[174] they taught him, made him take an oath of fidelity and undertook with him a rite of sorcery: "Toutes ces ceremonies faictes, on commence à boire & manger à l'Epicurienne aux despens de l'Advocat qui n'espargnoit rien de ce qu'il possedoit pour traicter ses compagnons."[175] The repetition of the term "Epicurean" clearly marks the desire to insist upon the Rosicrucians' Epicureanism, so that a reader of Garasse could not fail to establish a connection with that of the "beaux Esprits."

And what are we to say about these words concerning the powers conferred upon the Rosicrucians by Astaroth, which seem to echo the very title of *La Doctrine curieuse*: "Item je leur donne parole qu'ils seront admirez des *Doctes*, & recherchez des *Curieux* en telle sorte que l'on les recognoistra pour estre plus que les Prophetes Antiens qui n'ont enseigné que des fadaises."[176] There are yet other similarities between these two texts. In *La Doctrine curieuse*, Garasse entered into a polemic against the mockeries of the libertines concerning the resurrection of the dead; the polemic prepared vengeful sections against Vanini's ideas on this theme. Moreover Garasse attacked, in passing, alchemists who claimed "to show that Resurrection is an act as natural as the birth of a chicken."[177] This theme is found again in the *Effroyables Pactions*, applied to the Rosicrucians: "Puis-je passer soubz silence ceste abjuration qu'ils font de la Resurrection de la chair, veu que les plus infidelles, les plus Payens, & les plus incredules y ont aucunement adjousté foy."[178] Here the author writes exactly as if the denial of the resurrection was a commonplace of the Fraternity, which is of course not the case. It is, on the other hand, a reproach that goes without saying in *La Doctrine curieuse*.

A. E. Waite saw, quite rightly, "the work of a venal pamphleteer" in these *Effroyable Pactions*;[179] it now seems obvious that this pamphleteer could only have been in the pay of Garasse, whose chapter on the Rosicrucians in *La Doctrine curieuse* found in the *Effroyables Pactions* its most convenient and most natural continuation. We can ask ourselves if the same pen did not compose the *Examen sur l'Inconnue et Nouvelle Caballe des Freres de la Croix Rosee*. This pamphlet was by far the most violent of those directed against the Rosicrucians. The author first attacked those whom he named the "Cabalists," in whom he saw only "Satan's Vanguard":[180] by the same token, the Rosicrucians became witches. The author called upon the king himself to dispense justice concerning these:

> Et vous Louys le Juste, sera t'il dit qu'en la Metropolitaine de vostre Royaume, à la barbe du plus Auguste de vos Parlements, sejour ordinaire de vostre Sacrée Majesté tels endiablés ozent jetter leurs envenimées racines pour y commencer le Regne des fils de perdition? Est point parvenu jusqu'en vostre Louvre le bruit commun *Des Freres de la Croix Rosée*, Bande infernalle, mortes-payes de Sathan, brigade abandonnée, sortie ces derniers temps des Manoirs Plutoniques pour achever de corrompre un tas de desbauchés qui courent le grand galop aux Enfers, & dont les brutalles actions font voir combien peu ils estiment le salut de leurs ames.[181]

Even if it is perhaps too risky to identify this "heap of debauched persons" with the libertines Garasse denounced, we cannot, on the other hand, avoid recognizing in the expression "these handsome Dogmatists," used later on to designate the Rosicrucians,[182] the very term Garasse used constantly in *La Doctrine curieuse* to designate the libertine followers of Théophile.[183] And yet another significant detail is that Théophile himself took up the term "dogmatists" in 1624 to speak of the Rosicrucians.[184]

Having called upon the king in this way, the author of the *Examen* denounced the turpitude that he attributed to the brothers of the Rose-Cross, namely the various crimes attributed to witches, and he concluded with a direct attack against Théophile, whom he avoided calling by name:

> Voyla les fruicts plus suaves de ceste abominable Magie, puis les bons compagnons demandent s'il est loisible de les faire mourir, si l'on doit proceder judiciairement contre eux, & s'il n'est pas plus a propos de les renvoyer a leurs Pasteurs & Curez, comme gens estropiez de cervelle, que regler leur procez à l'extra-ordinaire.
>
> O Ames peu zelées de l'honneur de Dieu sçachez que l'heresie & la Sorcellerie sont deux monstres qu'on doit estouffer au berçeau. . . . C'est pourquoy les SS. Cayers en conseillent l'extirpation en ces termes expres, *Maleficos non patieris vivere.*[185]

These "good companions" were none other than Théophile and his comrades. Théophile had indeed given his verdict in 1620 about a case of demonic possession in Agen, in which he concluded the absence of all things supernatural, and he had cited it in his published works, a fact that Garasse had not failed to bring to the foreground.[186]

I will finish my observations of the supreme confusion between the Rosicrucian issue and that of Théophile by pointing out the surprising way several historians of the time presented the events of 1623 to posterity. In his *Thresor de l'histoire generale de nostre temps,* published in the spring of 1624, the historian M. Gaspard expressed himself thus: "En ce temps icy [*by August 6, 1623*] courut le bruit de diverses sortes de sectes incogneuës, nouvellement souslevees dans le Christianisme, principalement en France, Espagne & Allemagne: En Espagne le seminaire des Illuminez: En Allemagne des Deistes ou des croix Rosaires: En France des mesmes croix Rosaires, Atheistes & esprits curieux du temps."[187] Here Gaspard confused purely and simply the Rosicrucians of Étienne Chaume and the "beaux Esprits" of *La Doctrine curieuse*. His confusion is explained somewhat when we observe that with regard to the Rosicrucians, Gaspard limited himself to copying a highly significant paragraph of the *Effroyables Pactions*.[188] Gaspard seems in fact to have been above all directly inspired by the much more complete books by Claude Malingre, published almost at the same time.

In his *Histoire de nostre temps,* Malingre, as we have seen, reproduced the edict of May 29, 1623, against the Enlightened Ones of Spain and continued with another *"secte endiablée:"* the Rosicrucians. The latter were presented as "personnages originaires d'Allemagne qui se disent Confraires de la Confrairie & College de la Croix Rosaire, dont le pere Garasse Jesuite a parlé en sa doctrine curieuse, personnages qui se nomment invisibles, mais plustost Magiciens, Necromanciens & endiablez, ainsi que le discours suivant fera cognoistre." Malingre next reproduced the text of the *Effroyables Pactions,* then added a commentary conspicuously directed against the *curieux* and the *doctes:* "Je concluray donc en Chrestien par les regrets que je reçois en l'ame, de voir tant de pauvres esprits curieux se precipiter d'eux mesmes dans le gouffre de l'enfer. . . . Bien heureux sont les pauvres d'esprit, puisque le plus souvent nous voyons abysmer dans les ondes infernales les doctes & les plus relevez en doctrine."[189]

The report of Théophile's trial, however, came only much later in the work,[190] and the lines we have just read did not designate Théophile by name. The association between the Rosicrucians and the targets of *La Doctrine curieuse,* however, is evident in them. Moreover, it appears better still in the *Traicté des Atheistes, Deistes, Illuminez d'Espagne, et nouveaux pretendus Invisibles,* published by Malingre at the same time,[191] as we will see below when we study Gabriel Naudé's position.

Another revealing connection can be inferred from the texts accompanying the pages that preserve, in a manuscript of literary and historical miscellanies, a copy of the *Recherches sur les Rose-Croix* hastily attributed by some historians to Jacques Dupuy (1586–1656). We find there successively the interrogation of a magician in June 1623, led by Gilbert Gaulmin, then the criminal lieutenant of Moulins;[192] then two letters of Théophile de Viau;[193] then a translation of the theses of Jean Bitaut, Antoine de Villon and Étienne de Clave, announced but banned on August 24, 1624;[194] and finally the pages on the Rosicrucians.[195] Thus one finds reunited in a fairly cohesive ensemble documents concerning magic and sorcery, libertinism, alchemy, anti-Aristotelianism, materialism, and atomism as well as the Rosicrucian Fraternity, the whole forming an association characteristic of the confusion of the time.

We see then how the incident of Étienne Chaume's placards is inextricably intertwined with other events of the time. Théophile's trial, or rather the crusade Garasse undertook against him, seems particularly to have contaminated the incident of the brothers of the Rose-Cross, to such a point that everything happens as though the two violent pamphlets written about the latter had only aimed, under the appearance of aiming at the Rosicrucians, at Théophile alone, thus dangerously prolonging "the insidious parallel" of Father Garasse. Nevertheless, we would be mistaken if we attributed the task of diabolizing the Rosicrucians, at work in these pamphlets, to the strategies of *La Doctrine curieuse* alone. This task of diabolizing comes from further back: Its origins are found in a quarrel that, from 1615, set the Jesuit Father Jean Roberti against the Paracelsian Rudolph Goclenius and later Jean-Baptiste van Helmont. Thus we can also regard the episode of Chaume's posters as a specific moment in the reception of Paracelsus in France.

An Episode and a Caricature of the Reception of Paracelsus in France

From 1578, Paracelsianism was in France the occasional object of violent polemics capable of exceeding by far its medical context. Far from being limited to the quarrel over antimony, as modern historiography is still pleased to think,[196] Paracelsus's reception in France, as in Germany, touched upon the latent quarrel of the ancients and the moderns as well as the problem of religious orthodoxy. In 1578, the proceedings started by the Faculty of Medicine in Paris against the Norman Paracelsian Roch Le Baillif led the Sorbonne itself to censor some sixty propositions of Paracelsus as impious, and Étienne Pasquier accepted the role of defending Roch Le Baillif before the Parlement in large part because this cause was also that of the moderns of the time.[197] This abscess opened in the Faculty's side by the ghost of Paracelsus, temporarily closed after this case in 1581, was opened again in 1603, and, although more or less scarred over in 1607 still suppurated until 1610, inciting an extremely prolix battle—at the cost of thousands of pages, several solemn censures, and resounding exclusions—between the Faculty of Medicine and various doctors (including members of its own body), of whom the best known were two of the king's doctors, Joseph Du Chesne and Théodore Turquet de Mayerne.[198] Paracelsianism in France had thus become a sensitive subject if ever there was one in the medical circle, and we must now situate the episode of the placards of 1623 in this extremely controversial context.

A Favorable Context for an Anti-Paracelsian Reaction

When Étienne Chaume and his friends came up with the idea for their trick, the Rosicrucians were already newsworthy, even in France, because of a quarrel of an entirely Paracelsian nature. Since 1615, the Jesuit Jean Roberti had been confronting a professor from Marburg, Rudolph Goclenius the younger (1572–1621), on the theme of the "magnetic" curing of wounds.[199] The therapy advocated by Goclenius, stemming from a pseudo-Paracelsian text, the *Archidoxes magicae,* consisted of preparing an ointment (the *unguentum armarium*) with the help of the victim's blood and applying it not to the wound, but to the weapon that had caused it. The idea governing this curious therapy was that of a "magnetic" or "sympathetic" force, present in all of nature, and capable of uniting and connecting the most distant objects to each other; the cure therefore operated, according to Goclenius, in a natural way. This idea was the very object of the controversy.

Roberti was fiercely opposed to the idea that inert material could be invested with spiritual qualities, except through demonic activity. The anonymous author of the *Archidoxes magicae* had foreseen this objection, maintaining, as Paracelsus himself would have done, that the therapeutic powers of this type of remedy were proof enough of their divine, not diabolical, origins.[200] But in Roberti's eyes, attributing such an act to natural causes denied the presence of the supernatural. The polemic thus raised the question of naturalism, one of the major problems Paracelsianism posed in the eyes of the theologians, on the basis of which the Sorbonne had pronounced most of its censures in 1578, anticipating by several years the placing of Paracelsus's works on the Index.[201] Either Goclenius was making himself guilty of demonic magic, practicing cures that could only be explained by the activity of the Devil, or his therapy was ineffective; but in either case, he incurred an accusation of atheism, for he upheld theories that denied the supernatural. When Jean Baptiste van Helmont in turn took part in the polemic with the publication—voluntary or not—of his *De Magnetica vulnerum curatione* in 1621, his views, which were scarcely different from those of Goclenius, as well as his sarcastic remarks against the Jesuits, earned for him, thanks to Roberti, a long Inquisition trial that could have ended very badly.

This quarrel became associated with the Rosicrucian Fraternity in 1618, when Roberti, in his *Goclenius Heautontimorumenos,* indulged himself in a digression in which he fulminated against the Rosicrucians, who he claimed were guilty of magic just as Goclenius was. The success of this diatribe in France was not immediate,[202] but it was undeniable. In the course of the first six months of 1623, no fewer than three works referred to it. Not surprisingly, their authors were all theologians.

The first of these authors was Mersenne in his *Quaestiones celeberrimae in Genesim*, whose printing was completed February 1, 1623. *Quaestio* 53 of this work consisted of finding out "if the blood from Abel's cadaver had gushed forth onto Cain, and if this happens in the murderer's presence, as it is commonly said."[203] This very banal question was constantly debated at the time, notably in the context of the explanations the Paracelsians provided, for it too came down to the dilemma between natural and supernatural causes. Andreas Libavius had devoted a treatise to it in 1594,[204] and as late as 1640 the Bureau d'Adresse of Théophraste Renaudot made it again the subject of one of its lectures.[205]

In article 5 of his *quaestio*, Mersenne, as expected, examined the hypothesis according to which "this gushing of blood must be attributed to a natural cause," and in the context of the impious theme of the soul of the world, brought about by the Paracelsian explanations of Johann Ernst Burggraf, Goclenius, and others, Mersenne, relying especially on what Roberti had said in his *Goclenius Heautontimorumenos,* launched invectives against the heretical and impious Rosicrucians, "who at nearly every one of the fairs of Frankfurt introduce into the Christian world pamphlets smelling of impiety."[206]

A few months later, toward June 1623, it was the turn of a fanatical former member of the Catholic League, Jean Boucher, previously a Parisian priest at Saint-Benoît and a refugee at Tournai since Henry IV's abjuration, to mention the Rosicrucians in the course of his *Couronne mystique*.[207] In every place where heresy—that is, Protestantism—had gotten a foothold, atheism and magic, if we are to believe Boucher, were multiplying. Witness "two new monsters of magic": Goclenius with his magnetic cure—Boucher added to it the "perpetual lamp of Paracelsus"—and the "Fraternité des Rosecrucians."[208] Boucher's admitted source here was Roberti's *Goclenius Heautontimorumenos*. The Rosicrucians, he continued, blasphemed in addressing the promise of the Evil One—*Eritis sicut Dei*—to princes and kings, and the idea of attributing so much power to remedies like the *unguentum armarium* amounted to clearing a path for atheism

(Boucher here confused Goclenius and the Rosicrucians in the same anathema).[209] Furthermore, the "Rosecruceans," like Goclenius himself, were heretics, which was the "source of this evil."[210]

Here Father Gaultier's tradition meets that of Father Roberti. The latter had already denounced the underlying Protestantism in Goclenius's naturalism, recognizing its mark in his obstinacy in denying miracles. But Gaultier, by qualifying the Rosicrucian movement in 1621 as "the child of Lutheranism, mixed with empiricism and Magic by Satan," gave primacy to the reformed religion, in accordance with the overall intention of his book. In the *Couronne mystique,* Boucher offered a synthesis of these two attitudes: The accusation of magic, developed through Roberti, brought Goclenius and the Rosicrucians to the edge of atheism, but in the final analysis Boucher attributed the origin of their "monstrosity" to their "heresy": Protestantism.

For his part, Father Garasse scarcely insisted upon the Protestant religion, yet one of his targets of choice, in the chapter of *La Doctrine curieuse* devoted to the Rosicrucians. But his sources were nonetheless (besides Michael Maier) Goclenius, Roberti, and Gaultier himself.[211]

Let us add in a more general way that the Rosicrucians smacked of heresy in a large part of Catholic Europe. Not only had Adam Haslmayr and Benedictus Figulus found themselves condemned in 1612 by a power entirely devoted to the Jesuits of Tyrol and Austria, but from the beginning of the 1620s, one of the inspirers of the Rosicrucian manifestos, Christoph Besold himself, barely converted to Catholicism, began to revise his works in a Catholic sense, notably by hiding as well as he could his own participation in Rosicrucian matters; and although Naudé believed that he had reproduced Campanella's opinion on the fictitious Fraternity as an appendix to his *Instruction à la France,* we have recently learned from Michel-Pierre Lerner that this negative judgment reproduced by Naudé should be attributed, according to all probability, not to Campanella, whose real opinion on the subject is unknown, but to Besold himself.[212]

In the years preceding the incident of Chaume's placards, a number of works likely to feed the anti-Paracelsian hostility had been issued successively in Paris itself. The year 1621 saw the appearance of the *De Magnetica vulnerum curatione* by Jean Baptiste van Helmont and, under an identical title, a new edition of the treatise by Goclenius (who died that year).[213] The books of two of Mersenne's future targets were also published in 1620 and 1621: the *Traictez du vray Sel secret des Philosophes, et de l'Esprit universel du Monde* by Clovis Hesteau de Nuysement, and the last volume of the *Utriusque Cosmi Historia* by Robert Fludd.[214] Although the following year was marked by the deaths of Saint François de Sales and Jacques-Auguste de Thou—and of Michael Maier—it saw above all, in the domain of alchemy, a new edition of the *Traicté du feu et du sel* by Vigenère, quoted in 1623 by Gabriel Naudé in his *Instruction à la France*,[215] to say nothing of the *Problemata* by Georgius Venetus, which Mersenne would attack as well. Let us add that in that year in Strasbourg the fifth volume of the *Theatrum Chemicum* was published, and that among the minor Parisian publications, which were quite likely to irritate a man like Naudé, was found a curious *Interpretation des secrets Hebrieux, Chaldées & Rabins, du prince Dorcas, Philosophe Ethiopien, pour augmenter l'or & l'argent à dix pour cent de profit chaque semaine*.[216] Furthermore, we know that the plague was raging in Paris in 1622; there was a fresh outbreak in August of 1623, and probably for this reason a fragmentary new edition of a treatise on the plague, more than forty years old, by Roch Le Baillif himself was published.[217]

There is still more: on April 8, 1623, that is, between the *Quaestiones in Genesim* by Mersenne and the *Couronne mystique* by Jean Boucher, the printing was completed in Paris, of *La petite Chirurgie, autrement ditte la Bertheonee, de Philippe Aoreole* [sic] *Theophraste Paracelse grand Medecin & Philosophe entre les Allemans*, translated by a certain Daniel Du Vivier, "Surgeon and Barber of the King," who dedicated this thick volume of 750 pages to the young Charles of Schomberg (1601–56), the future duke of Halluyn, cared for by Du Vivier during the siege of Sommières (1622). An interesting detail: The editor himself emphasized that this translation, produced "in order to give this pleasure to the importunity of those who for twenty years have done nothing but ask for it in our Shops," had been undertaken several times already and abandoned because of the difficulties of Paracelsus's language.[218] The book, apparently very long awaited, opened with "the Basel program," the June 1527 manifesto by which Paracelsus had announced to the students of Basel the orientation of the courses that he was preparing to give. Printed here in large italic characters under the title *Paracelse aux Estudians, Salut,* this text contained most notably these words: "car ce n'est pas le titre, les ornemens du langage, la cognoissance des langues, ou la lecture de plusieurs Livres [quoy que ces choses soient à priser] qui rendent un Medecin capable en son art: mais la profonde cognoissance de la nature des choses, & des mysteres plus cachez qui suppleent au defaut de tout le reste."[219]

We can easily imagine the Faculty of Medicine's reaction when confronted with such a publication, especially as the latter clearly involved the king's immediate entourage.[220] The translator, fully aware of the work's subversive potential, had moreover warned Schomberg that this book might be subject to a rather hostile reception, "because of the novelty of its doctrine, known by few people, which however has always allowed me to succeed when other remedies could not do anything, or seemed to me more delayed than they should have been."[221] And Naudé may have been thinking precisely of these lines of the "Basel program" when he listed ironically in his *Instruction à la France* "la quint'essence, Medecine universelle, pierre des Philosophes, signature des choses, thresors, planettes, intelligences, Magie, Cabale, Chymie, & *mysteres les plus cachez*."[222] We see at least that the currency of the Parisian book scarcely allowed one to forget the existence of Paracelsianism and the quarrels it raised.

As for the works by Mersenne, Boucher, and Garasse, although all three had been published in 1623, none of them mentioned Chaume's posters, nor could they have done so: the latest of them, Boucher's work, must have appeared at the latest in June, just before the Parisian placards or perhaps just at the same time.[223] Thus we see the exact context that must have given Chaume the idea for his posters, which also explains to a large extent the reactions that they elicited. Let us recall indeed that the theories of Goclenius referred all of the polemicists to one and the same source: Paracelsus himself.

The Anti-Paracelsianism of Garasse and Mersenne
In fact, Garasse and Mersenne had indeed identified Paracelsus as the main person responsible for the naturalism of the doctors and alchemists, and several passages of *La Doctrine curieuse* and the *Quaestiones in Genesim* indeed fulminated against him. In the eyes of Garasse, the atheist Vanini, Théophile's intellectual master, was only a summary of Paracelsus, Pomponazzi, Cardano, and Agrippa. This curious shrinkage is found in an attack by Garasse on the critics of demonic possession: "je trouve qu'il y a cinq meschans & pernicieux Escrivains, qui ont tasché de rendre cette verité mesprisable & profane, par leurs mal-heureuses inventions, sça-

voir, Paracelse, Pomponace, Cardan, Agrippa, & Lucilio Vanino. Paracelse estoit un resveur hypocondriaque, Pomponace un Atheiste parfaict, Cardan un profane, Agrippa un endiablé, Lucilio Vanino l'abbregé des autres quatre."[224]

It is a most striking thing to see Paracelsus associated here with the most notorious and the most frequently denounced representatives of libertinism at the time (only Charron is missing). Paracelsus is labeled only a "delusional dreamer," which remains far short of the invectives usually launched against the Swiss doctor. But later on, Garasse clarifies his thoughts. In the *Section dixiesme. Quels sont les Livres Cabalistiques de nos nouveaux Dogmatizans,* he compares Theophile's friends' imaginary library to that of the Rosicrucians, and he situates Paracelsus on the first shelf, this time between Pomponazzi and Machiavelli:

> Il n'est pas jusques aux Boëmiens, aux Gueux & aux Couppeurs de bourses qui n'ayent leurs livres confidens & cabalistiques. Les Freres de la Croix de Roses, qui sont de pauvres gueux, ont leur Bibliotheque au rapport de Goclenius, & là dedans ils gardent quatre ou cinq volumes de grande recommandation: Le premier s'appelle, *Fama:* Le second, *Axiomata:* Le troisiesme, *Proteus:* Le quatriesme, *Rotae.* . . . Cela supposé je voy cinq ou six especes de livres, qui font comme la Bibliotheque des Libertins, laquelle je desire parcourir pour en dire mon advis.
> Le premier rang contient le Pomponace, le Paracelse, & Machiavel. . . .
> Paracelse est plustost un resveur, & Alchymiste dangereux, qu'un Atheiste ou Libertin, il est neantmoins tres-defendu, d'autant que ses curiositez, & douces resveries l'ont porté hors des bornes de la science naturelle, quoy qu'à mon advis il fust comme Cosme Ruggeri, plus ignorant qu'on ne pense, nommement au faict de la Magie, ainsi qu'il se void par quelques lambeaux rapportez dans les Disquisitions de Martin Delrio; car il semble en les lisant que la teste de cet homme fust comme une vive lanterne remplie d'estranges fantaisies, & d'imaginations frenetiques.[225]

This opinion, moderate in appearance, has a high probability of being that of Garasse himself. The Jesuit priest, by his very anti-Protestant fanaticism, had excluded himself from having recourse to a prime source: the *Disputationes de Medicina nova Philippi Paracelsi* by Thomas Erastus (published in 1571–73), used all along by the Sorbonne in 1578, then by Mersenne himself in his *Quaestiones in Genesim.*[226] This gap in Garasse's sources probably explains the verbal indulgence he displays toward the Swiss doctor, although in his eyes Paracelsus, when all was said and done, was no less deserving of the stake than Pomponazzi or Machiavelli. Paracelsus's writings seem particularly to have disconcerted Garasse. But it is quite possible, on the other hand, that the work of another Jesuit Father, Martin Del Rio, had suggested to him his way of associating Paracelsus with Pomponazzi, Agrippa, Cardano, and Vanini. When examining the illicit remedies, Del Rio gave indeed examples of superstitious therapies and magic remedies taken from pagan authors, then added this: "Those who have been baptized are not any more exempt from these superstitions. Indeed, we meet many of them in Pomponazzi's works, many in Heinrich Cornelius Agrippa's works, but even more still in Philip Aureolus', also known as Bombast Paracelsus, in a number of his works, the reading of all of which has been prohibited by the Church."[227]

Erastus had already named Pomponazzi beside Paracelsus in several places in his *Disputationes*.[228] Del Rio remembered this and added a companion to them who was nearly unavoidable in this context: the author of *De occulta philosophia*. Pomponazzi, Agrippa, Paracelsus: We have here something like a sketch of the association Garasse later effectuated.

Del Rio then reproached Paracelsus for his recourse to demons by the application of seals, images, and monstrous characters, quoted his exhortations to frequent old women of the countryside rather than university professors, and finally denounced the naturalism the Swiss doctor professed through his magic; and, citing the apocryphal objection of the *Archidoxes magicae* according to which an effective remedy cannot come from the Devil, he commented on it in these words: "Impure mouth! As if it were not up to theologians alone to judge what is against God and what is not against God! It is in this way that the vile charlatan claims the court of Faith for himself."[229]

We see that Father Garasse, in his verbal moderation, certainly remained very far short of his sources. Father Mersenne, on the other hand, had read with care Erastus's *Disputationes,* citing them by name in two separate places. First taking up the attacks of the German doctor against the naturalistic explanations of Paracelsus concerning the resurrection of the dead, he followed this diatribe as he raged against the alchemization of the mysteries of religion:[230] This passage, well known in Mersenne's work but rarely placed into its context,[231] is directly connected, in fact, to the Minim's anti-Paracelsian polemic, of which it is only the continuation on another level. The common target of these two attacks is nothing other than naturalism scandalously applied to sacred mysteries.

Later on, Mersenne again took up Erastus's attacks, on a theme every bit as important: the creation of the world. Now all of Paracelsus's conceptions found themselves questioned, and the *mysterium magnum* was in the front line.[232] In the course of these attacks in which, as we know, Mersenne advocated that the alchemists be thrown into the sea with a millstone attached to their necks, the Minim scarcely restrained himself from using violent invectives against Paracelsus, calling him, on the occasion of a passage on the *homunculus,* "the Germanic monster."[233]

The novelty of these anti-Paracelsian attacks by Mersenne and Garasse must be emphasized above all. Before 1623, such offensives are scarcely found in France outside of the medical domain. With Fathers Garasse and Mersenne, the polemic against Paracelsus suddenly crosses beyond the walls of the Faculty of Medicine and finds itself projected into the public sphere, exposed to the full light of day by the theologians' pens. Only the trial of Roch Le Baillif, followed by hundreds of people, had previously attained this degree of diffusion; but at that time the matter had remained innocuous: It was only a matter of whether or not to ban an empirical doctor, of defending or condemning the new medicine, of whether or not to pronounce themselves in favor of the moderns, and the Sorbonne's censure had then not carried weight in the eyes of the Parlement. In 1623, on the other hand, the theologians were the ones to take the matter in hand, and this was not the least of the factors that would push Naudé to react.

Gabriel Naudé's Position

As Louise Godard de Donville has recently remarked, it seems legitimate to consider Naudé's *Instruction à la France* a reply to Garasse's attacks.[234] Still we must not lose sight of the other dimensions of this work, which, in spite of its conciseness, is as dense as we have the right to expect from such a sedulous disciple of Seneca. Lorenzo Bianchi has characterized the *Instruction à la France* as one of the first manifestations of Naudé's political thought and has emphasized the importance of Naudé's medical training in his attitude toward alchemy and Paracelsianism, even if this attitude corresponded as well and above all to choices of a philosophical nature.[235] If therefore we wish to appreciate Naudé's position from all its angles, nothing remains but to let each of these different facets shine in its turn.

A Reaction against "La Doctrine curieuse" One of the best arguments for establishing the connection between Naudé's *Instruction à la France* and *La Doctrine curieuse* is the scarcely disguised comment on the poet Guillaume Colletet with which the *Instruction à la France* concluded: Naudé vigorously took up the defense of the poet ("le sieur C.") against Garasse, testifying to Colletet's orthodoxy and making himself the guarantor of it.[236] Let us recall, indeed, that the order of the Parlement of July 11, 1623, decreed the arrest not only of Théophile, but also of Colletet, Frenicle and Berthelot, considered to be the primary authors of the *Parnasse des Poëtes Satyriques*. In a similar fashion, on August 18, the order that condemned Théophile to the stake also banished Colletet from the kingdom for nine years (though he exiled himself only to Saint-Denis, just outside of Paris, where several members of his family occupied positions of some importance in the famous abbey).[237] To defend Colletet was thus to align oneself with the party opposed to Garasse. In fact, the presence of Colletet took up so to speak the whole of the *Instruction à la France:* Not content with providing his conclusion to Naudé's work, Colletet also gave him a prefatory sonnet as an opening, signed with his initials, in which are found—thumbing his nose at Garasse's attacks against the "beaux Esprits"—these verses in carefully weighted terms:

Continue then, *bel Esprit,* purge this universe,
As Hercules did formerly, of these varied monsters,
Who produce everywhere such harmful effects

(Poursuy donc, *bel Esprit,* purge cet univers,
Comme Hercule jadis, de ces monstres divers,
Qui produisent par tout des effects si nuisibles)[238]

When we consider that Colletet himself wrote—but published a long time after—an epigram against "Ragasse" (Garasse) and that it was one of his friends, Francois Ogier, who, in October 1623 struck the Jesuit priest with the blows that were to ensure the decline of *La Doctrine curieuse* in public opinion,[239] we are no longer in doubt concerning the motivations that pushed Naudé to publish the *Instruction à la France*. Indeed, Naudé was situated on the side of the scholars, traditionally opposed to the Jesuits,[240] and he made this quite clear, for example, by including in his book an emphatic praise of Guillaume Postel, whose follies he nonetheless denounced, but whom he revered as an individual who was "eminent in knowledge, elevated in doctrine and admirable in reputation" if ever there was one, "the great doctrine of whom is worthy of every kind of excuse."[241] Thus we discover in his book a number of veiled attacks against Garasse and the Jesuits.

Naudé's irony exerts itself against the Jesuits in disguised terms when he benignly embarks upon clearing, with as much zeal as calculated weightiness, "three Holy Fathers of the Company of Jesus"—Fathers Gaultier, Roberti, and Garasse—of having believed the ridiculous fable of the Rosicrucians and "followed in this the stupidity of the populace."[242] Naudé finishes this passage by discreetly taking a dig at Garasse's bloodthirsty zeal by means of a Latin quotation from Tertullian on "the wholly merciful Church," which should seek less to make the blood of heretics flow than to spread shame in their souls.[243] Other cutting remarks respond with exactitude to *La Doctrine curieuse*. Garasse, for example, had assumed that the libertines, incapable of perceiving their crimes, judged that they had been punished for nothing, and he was ironical concerning this Nothing, making a show of speaking knowledgeably about it. Naudé seems to have this in mind when in one breath he cites, among the examples of man's folly, "[Jean] Demons [amusing himself] by philosophizing on the quarter of Nothing," and a little below Charles de Bovelles for "the excellent treatise he composed *de plusquam nihilo*."[244] Elsewhere, Garasse had praised the Jesuits of Tyrol for having condemned Adam Haslmayr: Therefore Naudé, in his desire to reduce the Rosicrucians to a ridiculous fable, hastens to deny—quite wrongly, moreover—that Haslmayr's zeal for the supposed Fraternity had been the true cause of this condemnation.[245] At the end of the *Instruction*, the attack becomes sharper. Delicately tossing a bit of oil onto the fire, Naudé obligingly emphasizes the major defect of *La Doctrine curieuse*: He pretends to get indignant that the libertines of the time, "with an unequaled rashness and impudence," describe Garasse's work "with the very pernicious title of Atheism reduced to art." Thus discrediting the treatise of the Jesuit priest, Naudé works in the very direction of the *Jugement et Censure* published by François Ogier at the same time; but by choosing dissimulation, he adopts, yet without spelling it out, the attitude Ogier displayed in broad daylight, making himself also appear—but implicitly—"more pious than the spokesperson for the pious" and stabbing Garasse while seeming to defend him, which allows him to support Guillaume Colletet all the better.[246]

The *Instruction à la France* met with a most unexpected destiny. Partly directed against Garasse, the book was carelessly re-published by a member of the devout party in a brochure entirely devoted to upholding and prolonging the crusade of *La Doctrine curieuse*'s author. Ogier's *Jugement et Censure*, not content with provoking a reply from Garasse, had indeed stimulated the zeal of the Jesuit priest's partisans, among whom Claude Malingre himself was not the least active. Not content with reactivating in his *Histoire de nostre temps* the parallel Garasse had established between atheists and Rosicrucian sorcerers, Malingre wanted to intensify the effect by having his *Traicté des Atheistes, Deistes, Illuminez d'Espagne, et nouveaux pretendus Invisibles* published at the same time. This brochure opened with an anonymous "Preface," probably attributable to Malingre himself, in which the latter, taking up Garasse's quarrel as his own, deplored the number of his enemies with loud cries:

> mais quoy, le monde ne peut supporter ces lumieres, les medisans politiques deployent tous leurs efforts contre ces veritez: mais sur tous les Athees, les Deistes & autres sectes infernales, comme particulierement interessez en la descouverte de leurs impietez, s'esmeuvent en ce temps, fremissent d'horreur, & se lancent aujourd'huy contre ceux qui esclairent de trop prez leurs Atheismes & horribles deffauts. . . . C'est pourquoy . . . Je veux en ce Traicté . . . ruyner les desseins du Diable, les Athees & Heretiques modernes ses suposts, qui luy [*i.e.*, God] veulent non seulement ravir sa toute-puissance, sa providence, sa bonté, sa misericorde, son eternité, voire jusques a son essence que les Athees luy ostent, que les Deistes partagent, que les Illuminez mesprisent, & les Croix Rosaires supposent faire servir en l'execution de leurs prestiges. C'est de ces quatre sortes de nouvelles sectes eslevees en ces derniers temps dans l'Empire Chrestien, dont je traicteray icy fort succinctement.[247]

Malingre was obviously reacting in this case to the attacks against Garasse by Ogier and, perhaps, by Guy de La Brosse;[248] perhaps, too, without being aware of it, he was reacting to Naudé's work: the term "political slanderers" can quite easily be applied to each of these authors. As in the *Histoire de nostre temps,* Malingre abstained at any rate from naming Théophile explicitly. But the first chapter of his brochure, "On the Atheists of Our Time," was nothing other than an apologetic and edifying assembly of extracts from *La Doctrine curieuse,* perhaps mixed in with passages attributable to Malingre himself, in any case devoid of any ambiguity concerning their intent. Thus the final lines read: "Dieu soit loüé qu'en ce temps les Cours de Parlements de France & Messieurs les Magistrats, conformement aux loix divines & les commandemens de sa Majesté tres Chrestienne & tres-pieuse, se sont saisis d'aucuns Atheistes recens, ont faict brusler leurs livres & travaillent journellement à faire recherche des lieux où s'enseignoit cette doctrine impie, & des Autheurs d'icelle qui desbauchoient sa Noblesse Catholique par la publication de leurs detestables maximes."[249]

Chapter 2, "On Deists or Trinitarians, Also Called New Arians," although borrowed from Father Gaultier's *Table chronographique* and referring to the years 1564–66, was an extremely clear allusion to the crusade Mersenne led against the "deists" at the time.[250] Chapter 3 reproduced the edict of May 29, 1623 against the Alumbrados, and chapter 4, under the misleading title, "On the Brothers Who Claim to Be Invisible & Enlightened of the *Croix-Rosaire*, on Their College, Meetings and Doctrine," reproduced Naudé's *Instruction à la France* without the author's name and without its true title, an incredibly stupid blunder if Malingre's intent with this publication was in fact, as we have good reason to believe, to take up Garasse's defense.[251]

A Discreet Manifesto of Libertinism   Naudé's attitude in the *Instruction à la France,* so prudently hostile to the Jesuit priest and his order that Malingre did not understand an ounce of it, revealed itself as well by some typically libertine features here and there. We have just seen how Naudé opposed the merciful attitude of the Church Fathers to the murderous zeal of Garasse, advocating persuasion rather than the stake. Elsewhere, Naudé dared to take up as his own some lines of Lucretius against religion itself:

> n'est-ce pas une chose surpassante la portee de nostre esprit, qu'un Arius, un Luther, un Calvin, . . . ayent armé le fils contre le pere, bouleversé les plus grandes Monarchies, & pensé esteindre la race du genre humain, & occasionné de si grands malheurs & calamitez, que je suis contraint de dire avec Lucrece,
>
> *Tantum relligio potuit suadere malorum.*
> ou plus veritablement,
> *Relligio peperit scelerata & impia facta.*[252]

Whether they signify a sign of recognition or a discreet profession of libertine doubt, these cleverly executed lines can be added to other evidence. Thus we can see a frank irony in the use Naudé made of two authorities: Pierre Charron and Pierre de Lancre, mentioned to justify a depiction of the diversity of beliefs in the domain of religion: "Charon ayant recueilli les diverses & monstrueuses opinions des hommes és loix & statuts, & de l'Ancre vous ayant faict voir diverses extravagances de leurs actions; il ne me resteroit qu'à vous representer comme en deux belles pieces de tapisserie les diversitez qui se rencontrent tous les jours és deux dernieres sources de tant de caprices, sçavoir la profession du culte divin, & le cercle de toutes les sciences."[253]

Without lingering to identify more precisely the sources of this new veiled attack on religion, let us note that Pierre de Lancre, the bloodthirsty magistrate who specialized in witchcraft trials, seems to have had a highly amicable relationship with Garasse, whereas Charron with his *Sagesse*, "bible of the libertines," was nothing other than one of the chief bêtes noires of *La Doctrine curieuse* (and also of Mersenne).[254] By juxtaposing them, Naudé armed himself with the poison and the antidote at the same time, neutralizing the sulfurous exhalations of the one, *pace* Garasse, with the orthodoxy of the other. It is moreover not a coincidence if, at the same time, François Ogier took up Charron's defense against Garasse, to the great indignation of the author of *La Doctrine curieuse*;[255] and furthermore, we know that Naudé, in 1625, composed his *Apologie* against Pierre de Lancre himself, whom he designated by name.[256]

We also know that Naudé intended, in his *Instruction*, "to set all things evenly at the level of reason"; the sentence has been rightfully pointed out as an indication of a "rationalistic and skeptic attitude."[257] Now one passage from the *Instruction* has preserved a remnant of an opinion on the true nature of the Rosicrucians that must, from this point of view, have especially attracted Naudé. This passage is found at the end of a whirlwind of contradictory opinions concerning the Fraternity:

> je conjecturay incontinent que suivant cette opinion le venerable Pere illuminé premier autheur de la Congregation . . . devoit estre principalement redevable au Reverend Pere endiablé Picatrix. . . . Mais la verité de cette conjecture & interpretation estant difficile à persuader à la trop grande incredulité de quelques uns qui estans accoustumez à telles fictions & narrations fabuleuses, *bustorum formidamina, noctium occursacula, larvarum terriculamenta, nocturnos lemures, portentaque Thessala risu excipiunt.* . . .[258] Je me suis persuadé que l'opinion de ceux-là estoit plus recevable qui ont estimé que c'estoit une Compagnie de gens doctes & curieux, lesquels desirans par la communication qu'ils avoient ensemble parvenir à la cognoissance des secrets les plus cachez de la nature, . . . nous ont plustost par leur Manifeste & Confession representé le modele des choses qu'ils devoient rechercher, que non pas le catalogue de celles qui estoient en leur puissance, & lesquels ne se vouloient embarquer à la conqueste de cette toison d'or qu'apres avoir authorisé, comme d'autres Argonautes, le dessein de leurs voyages du favorable tiltre de quelque Compagnie ou Congregation . . . : laquelle opinion

est authorisee par le consentement du Sieur Adami Gentilhomme Allemand, auquel nous serons perpetuellement obligez pour les oeuvres de ce phoenix de tous les Philosophes & Politiques Thomas Campanella, ausquelles il sert tous les jours de sage femme . . . ; car en l'une de ses Epistres manuscrites, de laquelle j'ay la copie, addressee au defunct Pere Baranzani, il parle expressement d'eux en ces termes: *Votum forte fuit hominum bonorum, qui communicationem in literis desiderabant.*[259]

But Naudé immediately rejected this hypothesis in the name of an argument that was for him irrevocable, inspired by right reason: "Neantmoins si cette Compagnie estoit telle que porte cette conjecture, il faudroit dire . . . que ne respirant rien autre chose que le bien, & instruction d'une fourmiliere d'esprits qui perdent tous les jours leur temps à la recherche de ce qui leur est impossible de conduire à perfection, . . . elle les auroit voulu retirer de cette queste, les engageant à une autre de bien plus grand merite & consequence, & par mesme moyen exercer nostre jugement à descouvrir sans aucunes conjectures le lieu de leur demeure & invisible Congregation."[260] The goal of a company desirous of establishing more communication in the scholarly world obviously not being to mislead the learned, Naudé could only reject the attractive advice of Tobias Adami, "set evenly at the level of reason."

There is another indication of this attitude: Naudé, as aware as he was of living in a time rich in discoveries and upheavals, nevertheless displayed a resolute hostility toward millenarianism, whose inner workings he patiently dismantled.[261] And the lesson of the *Instruction à la France* can be summarized by the following absolute rebuttal, more clearly expressed still in the *Mascurat* in 1650, but concerning which Naudé had not changed his mind since 1623: "This is why I affirm to you again that Magic is an empty science, & incapable of producing any effect, the same as Alchemy."[262]

This position of Naudé as a philosopher is also that of Naudé as a scholar with a medical education. A reaction against the dangerous zeal of Father Garasse, a sign of recognition and a discreet manifesto of libertinism, the *Instruction à la France* for all that still bears, in more than one place, the anti-Paracelsian trademark of an author educated at the Faculty of Medicine in Paris.

Naudé's Anti-Paracelsianism: A Typical Feature from the Faculty of Medicine Naudé's teacher René Moreau, who crossed the pages of the *Instruction* more than once, was one of the regents of the Faculty. He most likely shared his colleagues' fierce opposition to Paracelsus, without necessarily lapsing into caricature, as Guy Patin did. He had at the very least demonstrated in his *École de Salerne* the impossibility of extending life, and Naudé, in 1625, willingly referred the *Apologie*'s readers to that work.[263] Highly knowledgeable in the topic of alchemy, Naudé himself was not to be outdone in producing an entire arsenal of anti-alchemical sources.[264] In fact, in his *Instruction,* he identified the origins of the entire Rosicrucian movement with a primary cause that he denounced without delay: Paracelsus himself. Indeed Naudé related the millenarianism of the Rosicrucian movement first of all to Paracelsus' Helias Artista prophecy, and this was the first target he chose in dismantling the myth of the Fraternity, distorting the Swiss doctor's name (Hohenheim) in a ridiculous onomatopoeia:

> C'a esté une des principales resveries de cet Hermite Philippe Bombast, Aureole, Theophraste, Paracelse, de Hohenhehin, lequel sorti d'un des Cantons de Suisse, s'est voulu signaler par la multitude de ses noms, aussi bien que par la pernicieuse nouveauté de sa doctrine, de nous vouloir persuader qu'il n'estoit que le precurseur d'un certain Helie Artiste, lequel devoit venir apres luy & esclater au monde fourny du secret veritable de la transmutation des metaux, & de toutes les sciences, pour reformer la corruption qui par laps de temps s'estoit glissee en icelles. . . . Voila, Messieurs, la base de cette Confrairie, l'origine de ce Manifeste, la cause de tant de chimeres, & le gond sur lequel tournent tant de fantaisies. Cette source estant tarie les ruisseaux se secheront: cette racine coupee les branches se fanneront: ce fondement sappé adieu tout l'edifice, *Hic Rhodus, hic Saltus, hoc opus hic labor est.*[265]

Here we see Naudé's skeptical and rationalist tendencies, which made him the enemy of every millenarianism, united with his nearly professional hostility to the doctrine of Paracelsus. We could rightly be astonished that a skeptic like Naudé, having grasped extremely well how far the polemic of Roberti went along with that of Garasse,[266] did not have more sympathy for the Paracelsians' naturalism, which, after all, amounted to denying the reality of miracles. This naturalism however, rested on foundations that Naudé could only reject as being superstitious; moreover, this libertine never did risk upholding naturalistic theories openly, even if they were of Aristotelian inspiration, preferring the historical method to them by far and, as René Pintard said, "dissolving philosophy in history."[267]

Later Naudé, so that no one might remain unaware of it, again designated Paracelsus as "the fundamental stone of all this Congregation" and launched a classic invective against him: "Voire mesme cette marque de division est si essentielle au mensonge, que Paracelse, le Luther de la

Medecine, a plustost esté diversifié par Crollius, du Chesne, Hartman, & une infinité d'autres, que nous n'avons recognu par l'intelligence de ses oeuvres, les blasphemes & absurditez de sa nouvelle doctrine."[268] "The Luther of Medicine"—this epithet was at the time frequently used for Paracelsus, and the usage had started in his lifetime. In 1600, the reformed alchemist Bernard G. Penot had used it in praise, arguing against Andreas Libavius that Paracelsus had been to medicine what Luther was to theology, impelled by God to bring new remedies to new ills.[269] But in France, Henri de Monantheuil, dean of the Paris Faculty of Medicine at the time of the proceedings instituted against Roch Le Baillif, had, completely on the contrary, employed it as an invective at the heart of two speeches, one at the University and the other in the open Parlement.[270] It was a cutting invective indeed, for Paracelsus himself, in the *Paragranum,* had risen up against the epithet "Luther of the doctors" that his enemies let fly at him, replying to them with haughtiness: "I will make life difficult, for him and for you!"[271] In fact, at the beginning of the seventeenth century, the Lutherans themselves considered the Swiss doctor an atheist.[272] To reactivate this epithet in Paris in 1623 with regard to the Rosicrucians was not only to link oneself to Father Gaultier's anti-Protestant movement: It was above all to reveal a resurgence of the anti-Paracelsianism characteristic of the Paris Faculty of Medicine, a bastion of orthodoxy as much in the domain of medicine as in that of Catholic faith—although this last, as we know, mattered little to Naudé in and of itself.[273]

A Reaction against Popular Agitation   But if the opposition to Paracelsus is definitely one of the various motives that pushed Naudé to denounce the Rosicrucian myth, at the very heart of which he had seen the shade of the Swiss doctor in the position of a protective spirit, this fact does not exhaust the richness of the *Instruction à la France*. Warned of the danger of seditious pamphlets and aware of the perils that they were likely to produce to the security of the State—we know that Naudé himself had criticized in *Le Marfore* all those pamphlets who were aiming at Luynes in 1620—the author of the *Instruction* reacted in 1623 in an apparently similar manner: He rose up against the madness that seemed to have seized the people of Paris after the posting of Étienne Chaume's placards and endeavored to fight against a disorderly production of pamphlets and opinions that threatened the social order, going as far as to print in his book the text of the poster "to relieve the infinite number of people who have never seen it, from scribbling it in their tablets," that is, to attempt to contain the uncontrollable proliferation of manuscripts, in which seditious texts risked being still more dangerously distorted.[274]

Lorenzo Bianchi has explained extremely well how certain passages of the *Instruction à la France* anticipated Naudé's *Considerations politiques sur les coups d'estat* of 1639, showing the people as a weak and idiotic mass lost in the "orchestra pit of lies," in great need of being enlightened by the learned, whose role was to denounce the superstitions and prophecies that were likely to attract the people and thereby bring trouble into the kingdom. Naudé's position can then be stated in the following manner: Hostile to Paracelsus in the name of reason and not that of religion, Naudé could not tolerate that the denunciation of the Swiss doctor and his followers who were disguised as Rosicrucians—a denunciation that was in principle salutary—had turned, as was the case with the pamphlets that were circulating, into a witch-hunt, and he rose up precisely against this caricature of anti-Paracelsianism. From this point of view, his *Instruction à la France* was simultaneously aimed at three targets: Paracelsus and his sectarians, whose absurdities must be denounced; those who condemned them in the name of religion, for they were the primary dupes of these absurdities and made others (such as Théophile, Colletet, and their friends) suffer the consequences of them; and those who exploited the situation in an extremely dangerous manner by publishing pamphlets like the *Effroyables Pactions,* for these pamphlets risked frightening the people to the point of inciting riots.

And if the *Instruction à la France* did not yet offer views on the interest and the necessity of absolutism and religion in the preservation of states, at least Naudé knew where to find gripping words for endeavoring to send the supposed Rosicrucian Fraternity back to the void, and with it not only all the Étienne Chaumes, capable of troubling the social order, but also all the Garasses and hired pamphleteers who, pandering to the obsessions of the people, dangerously misjudged the people's propensity for mounting Pegasus, or who on the contrary exploited it without restraint at the risk of bringing about "the complete ruin" of the State:

> Genereux esprits transcendans & eslevez par les aisles de vostre jugement au dessus du commun d'une populace, & qui comme du theatre de la verité contemplez une infinité d'esprits qui perdent leur credit dedans le parterre du mensonge,

c'est à vous à qui il appartient de leur donner à cognoistre comme tous ces faux bruits, nouveautez, propheties & opinions anticipees, ont tousjours esté cause de la subversion des Estats & entiere ruine des plus grandes Monarchies. . . . Et moy j'adjousteray, [que les nouveautés] ayans esté cause en ce Royaume de quatre batailles donnees, un million d'hommes occis, trois cens villes surprises, cent cinquante millions despensez pour le payement seul de la gendarmerie, neuf villes, quatre cens villages, & dix mille maisons tout à faict bruslees ou rasees; le ressouvenir d'une calamité si estrange nous devroit faire dresser les cheveux à la teste aux premiers bruits de telles superstitions & nouveautez, lesquelles comme trespernicieuses ont tousjours esté defenduës par les loix Imperiales, qui ont prefix certaines peines à ceux qui s'efforcent d'estonner les autres par quelque vaine superstition.[275]

By giving this horrific account of the wars of religion in the declamatory conclusion of the *Instruction à la France,* Naudé might hope to produce some effect on a generation who, in the beginning of the 1620s, had not finished warding off its obsessive fear of an Iron Age that was still quite close in its collective memory.[276] The outer fringes, at least, of the opinions closest to power seem to have heard him, since in 1624, having given a summary of Father Gaultier, Naudé, and Garasse, the *Mercure François* turned to the *Instruction à la France* to conclude its article devoted to Chaume's placards, precisely with Naudé's idea that the Rose-Cross was only an imposture, which, for this reason, "had been unable to find a foothold for establishing itself in France."[277]

This research into the incident of the Rosicrucian placards considered as an episode in the reception of Paracelsus in France has thus revealed several important points. In France, as soon as 1621, there existed a trend hostile to the Rosicrucians within the Jesuit milieu. This hostility, which could be called endemic within the Company of Jesus if we only recall the condemnation of Adam Haslmayr in 1612, generated in France a quarrel which had begun in Germany between Father Roberti and the Paracelsian Goclenius, and it instantly adopted that quarrel's characteristic marks, denouncing the Rosicrucians as guilty of diabolic magic and atheism. In the French religious milieu, Father Gaultier made himself the first echo of this in 1621, followed in the first part of 1623 by three theologians of whom the first was Mersenne, eager to fight against naturalism in all its forms and especially against that of the Paracelsians; the second, Jean Boucher, a former member of the Holy League and prior of the Sorbonne with extreme Roman tendencies, who took refuge in Tournai for having refused to acknowledge the legitimacy of Henry IV after the latter's conversion;

and the third, Father Garasse, scarcely less hostile toward naturalism than Mersenne, who saw in the Rosicrucian matter an additional means of discrediting Théophile by placing him in parallel with the diabolized threat of the brothers of the Rose-Cross.

Most likely this very context inspired in the young Étienne Chaume the idea for his placards, for we do not see that Rosicrucians were in any other way an issue in France until shortly before the spring of 1623. And this context again certainly explains to a great extent the violent reactions the posters provoked. The coincidences of current events, about which Chaume had probably scarcely worried, led to a double association: on the one hand between the Rosicrucians and the heretical sect of the Alumbrados of Seville, on the other hand between the former and Théophile's trial. The first of these associations resulted in the assimilation of the Rosicrucians in public opinion with a sect of heretics; Garasse (or pamphleteers in his command) immediately exploited the second association in a veritable enterprise to demonize the Rosicrucians by means of two widely diffused pamphlets, the *Effroyables Pactions* and the *Examen sur l'Inconnue et Nouvelle Caballe des Freres de la Croix Rosee*. Whereas Théophile thought he saw in the Rosicrucian placards yet another trap set by Garasse's party, Naudé, watching the matter take on disturbing proportions, finally decided to take up his pen shortly after his friend Francois Ogier to fight Garasse in his turn, by casting disrepute upon *La Doctrine curieuse,* but also to stop a proliferation of pamphlets that threatened the social order, by denouncing the follies characteristic of occult sciences, prophecies, and superstitions. Naudé sought at the same time to dispossess the theologians of a quarrel, that of Paracelsianism, that was extremely dangerous in their hands, for by adopting it as his own, with an almost professional hostility inherited from his medical education, he meant to settle it in the name of reason and not religion.

The Impact of the Incident of the Rosicrucian Placards

Étienne Chaume's placards doubtless produced still other reactions in France after 1624. Not only did they belatedly give birth to a myth, or rather to a pair of myths, those of Descartes and Mersenne as Rosicrucian brothers, but we can even follow their impact in various forms in France until the eighteenth century.

Twin Myths: Descartes and Mersenne as Rosicrucian Brothers

We know the anecdote, reported by Adrien Baillet and repeated a hundred times, according to which René Descartes's stay in Germany allegedly earned for him, upon his return to Paris in 1623, the suspicion of belonging to the brothers of the Rose-Cross. Henri Gouhier carried out a critical analysis of Baillet's story, accompanied by an equally critical inspection of his sources, in 1958, and subsequent research has not brought any new information concerning the incident to light.[278] Here we will limit ourselves to comparing Baillet's account, clarified by Henri Gouhier, with the various known elements of the matter of the Parisian placards to determine whether Baillet's story has any degree of plausibility. While doing this, we will guard against intervening on the no less controversial question of a possible influence of the Rosicrucian movement on Descartes's thought, which extends far beyond the subject of the present chapter. Let us observe only that from a methodological point of view, in light of Carlos Gilly's research, it appears extremely difficult to bring out any so-called "Rosicrucian" thought from a text other than the *Fama* and the *Confessio*, for how is one to favor, among the some 400 writings provoked by these two manifestos, this one or that one as representative of the thought of a Brotherhood that, as we well know, never really existed? The same remark applies to the *Chymische Hochzeit* by Andreae, still cited rather recently, for example, by Paul Arnold:[279] Whatever the undeniable literary value of this text may be, it is difficult to see why a dominating role of importance should be attributed to it since at the time, it scarcely exercised more than a very modest influence within German Rosicrucian literature and had absolutely no influence in France and England until quite a long time later. As for the dreams of Descartes, is there really any need to seek a precise source for them when the seventeenth century abounds in dreams of this type, as Sylvain Matton has well demonstrated, and as is exemplarily shown in a dream, well studied by Robert Halleux, that came to the young van Helmont, of which Descartes could not have been aware?[280]

To return to the anecdote Baillet reported, we know that according to him, Descartes, coming back from Germany, left the Spanish Low Countries for France at the beginning of February 1622. Toward mid-March he was in Rennes, having decided to avoid Paris, since the city was not yet "free of the contagion with which it had been infected for two years." Only at the end of February in 1623 did he go there "to see his friends again, and to learn the news of the State and of literature." Well informed concerning Germany's affairs, Descartes was able to satisfy his friends on this point.[281]

In Baillet's mind, these friends could hardly have been other than Mydorge and Mersenne. According to Baillet, in 1613, when Descartes went to Paris for the first time, he made the acquaintance of Claude Mydorge, ten years his senior, and struck up a friendship with Mersenne, nearly eight years his senior, whom he had met at the school of La Flèche and who was called to Nevers in 1614. They apparently saw each other again only in 1623.[282] This chronology, rendered doubtful by the difference in age between Descartes and the other two, has been unanimously refuted by Charles Adam, Gustave Cohen, Cornelis de Waard, Robert Lenoble, Henri Gouhier, and Geneviève Rodis-Lewis: The relationship between Mersenne and Descartes in fact began only after his return from Germany, when he came to Paris in 1623.[283] Baillet continues his story:

> En revanche ils luy firent part d'une nouvelle qui leur causoit quelque chagrin, toute incroyable qu'elle leur parût. Ce n'étoit que depuis trés-peu de jours qu'on parloit à Paris des Confréres de la Rose-croix, dont il avoit fait des recherches inutilement en Allemagne durant l'hiver de l'an 1619: & l'on commençoit à faire courir le bruit qu'il s'étoit enrollé dans la confrérie. M. Descartes fut d'autant plus surpris de cette nouvelle, que la chose avoit peu de rapport au caractére de son esprit, & à l'inclination qu'il avoit toûjours euë, de considérer les Rose-croix comme des imposteurs ou des visionnaires. Il jugea aisément que ce bruit desavantageux ne pouvoit être que de l'invention de quelque esprit mal intentionné, qui auroit forgé cette fiction sur quelque-une des lettres qu'il en avoit écrites à Paris trois ans auparavant, pour informer ses amis de l'opinion qu'on avoit des Rose-croix en Allemagne, & des peines qu'il avoit perduës à chercher quelqu'un de cette secte qu'il pût connoitre.[284]
>
> Il s'étoit fait un changement considérable depuis l'Allemagne jusqu'à Paris sur les sentimens que le Public avoit des Rose-croix. On peut dire qu'à réserve de M. Descartes & d'un trés-petit nombre d'esprits choisis, l'on étoit en 1619 assez favorablement prévenu pour les Rose-croix par toute l'Allemagne. Mais ayant eu le malheur de s'être fait connoître à Paris dans le même têms que les *Alumbrados*, ou les Illuminez d'Espagne, leur réputation échoüa dés l'entrée. On les tourna en ridicule, & on les qualifia du nom d'*Invisibles;* on mit leur histoire en romans; on en fit des farces à l'hôtel de Bourgogne; & on en chantoit déja les chansons sur le Pont-neuf, quand M. Descartes arriva à Paris. Il en avoit reçu la premiére nouvelle par une affiche qu'il en avoit lûë aux coins des ruës & aux édifices publics, dés son arrivée. L'affiche étoit de l'imagination de quelque bouffon, & elle étoit conçuë en ces termes. *Nous Députez du collège principal des Fréres de la Rose-croix, faisons séjour visible & invisible en cette ville. . . Nous montrons & enseignons sans livres*

ni marques à parler toutes sortes de Langues des pays où nous habitons. Sur la foy de cette affiche, plusieurs personnes sérieuses eurent la facilité de croire qu'il étoit venu une troupe de ces Invisibles s'établir à Paris. On publioit que de 36 députez que le chef de leur société avoit envoyez par toute l'Europe, il en étoit venu six en France; qu'aprés avoir donné avis de leur arrivée par l'affiche que nous venons de rapporter, ils s'étoient logez au Marais du Temple; qu'ils avoient ensuite fait afficher un second placart portant ces termes. . . .[285]

Le hazard qui avoit fait concourir leur prétenduë arrivée à Paris avec celle de M. Descartes, auroit produit de fâcheux effets pour sa réputation, s'il eût cherché à se cacher, ou s'il se fût retiré en solitude au milieu de la ville, comme il avoit fait avant ses voyages. Mais il confondit avantageusement ceux qui vouloient se servir de cette conjoncture pour établir leur calomnie. Il se rendit visible à tout le monde, & principalement à ses amis, qui ne voulurent point d'autre argument pour se persuader qu'il n'étoit pas des Confréres de la Rose-croix ou des Invisibles: & il se servit de la même raison de leur *invisibilité*, pour s'excuser auprés des curieux, de n'en avoir pû découvrir aucun en Allemagne.

Sa présence servit sur tout à calmer l'agitation où étoit l'esprit du Pére Mersenne Minime son intime ami, que ce faux bruit avoit chagriné d'autant plus facilement, qu'il étoit moins disposé à croire que les Rose-croix fussent des *Invisibles*, ou des fruits de la chimére, aprés ce que plusieurs Allemands & Robert Fludd Anglois avoient écrit en leur faveur. Ce Pére ne put tenir secréte la joye qu'il avoit de revoir & d'embrasser M. Descartes.[286]

This is, it must be acknowledged, a case *par excellence* for applying the historical method specially implemented by Gabriel Naudé in his *Apologie pour tous les grands personnages qui ont esté faussement soupçonnez de magie.*

If we establish a connection between Baillet's story and the currently known facts concerning the incident of the Rosicrucian placards, the story gains elements of plausibility that it did not possess before. For example, it might initially appear astonishing that the suspicion of belonging to the Rosicrucian Brotherhood appeared to be so serious to the friends of Descartes. But in the climate of widespread mistrust that successively motivated Étienne Chaume's flight, Théophile de Viau's mistrust and Gabriel Naudé's reaction, we scarcely have any difficulty in envisioning the danger of such a suspicion and its seriousness, for example, in the eyes of a man such as Mersenne.

It is unfortunate that a large part of Baillet's story relies precisely upon the reaction of the "friends" of Descartes; yet we have just seen that these friends, in Baillet's mind, could only have been Mydorge and Mersenne, and that this hypothesis proves to be inadmissible in the light of modern criticism. Thus a part of Baillet's story crumbles.

Nevertheless Baillet relies upon one source that he names: the work of Father Poisson, published in 1670. Henri Gouhier has checked this source and remarked that Poisson in no way mentioned the 1623 Parisian episode. He concluded with a great deal of plausibility that Baillet, "a scrupulous scholar and a suspect historian," "does not know how to remain silent when his documentation leaves lacunae"; as he desires to write "a biography without gaps," he gathers information about the time and "shows, according to plausibility, what could have happened."[287] Let us add to this that Baillet treated his information in a curious manner: Basing a part of his story on the 1624 article in the *Mercure François*,[288] he distorted it by embellishing it with several details so as to ridicule the Parisian episode, by which he produced without realizing it a serious misinterpretation; in a similar fashion, he retained from the *Effroyables Pactions* only certain colorful elements, eliminating any trace of witchcraft found in this pamphlet. In a word, Baillet handled his sources with the greatest of liberties.

With Mersenne's trail having stopped short and the verification of Baillet's sources having ended only by making him highly suspect of fabrication, very little remains of Baillet's story, as we can see. One last question, however, remains: When exactly did Descartes come to Paris in 1623?

Baillet is reasonably precise on this point. It appears to be established that the philosopher went to Paris a first time from the end of February to the beginning of May in 1623, that is to say, some time before the posting of Chaume's placards.[289] Leaving Paris in early May, then passing through Rennes, Descartes stayed in Poitou from May to July. We know that on July 8, he signed there a sales contract with a gentleman of the region.[290] Baillet asserts, or at the very least he supposes, that Descartes stayed in Poitou during all of July, for in his summary of the philosopher's life, he writes: "Having returned to Paris in the month of August. . . ."[291] Descartes left for Italy in September 1623.[292] It therefore seems certain, as far as we can trust Baillet on this point—and he is unfortunately the only known source—that Descartes was in Paris again in August of 1623, at the very heart of what Frances Yates has called "the Rosicrucian scare." In this case, however, we have the right to wonder what Baillet's remark is worth concerning the plague that dissuaded Descartes from going to the capital the year before. In effect, on August 18, 1623, the epidemic was raging to the point that only a few magistrates had remained at their post to proceed with the expeditious judgment of Théophile. In September it had scarcely calmed down: On the advice of the civil lieutenant, the start of the school year was delayed until November 5.[293] If Descartes was in Paris at that time (did he have to pass through there to go to Italy?), we know nothing of his reaction to Chaume's posters, for the story of it Baillet gives, in spite of its praiseworthy attention to plausibility, appears to be based on no authentic elements and, because of this, is not admissible. We thus see with what prudence it should be considered, which unfortunately is not always the case.[294]

The most curious thing is that after this story by Baillet, used toward satirical ends starting the following year by Huet in the *Nouveaux mémoires pour servir à l'histoire du cartésianisme*,[295] another serious author invented a similar anecdote about one of Descartes's contemporaries. In 1709, Father Thuillier, in his biography of Mersenne, related what follows with regard to a stay that Mersenne had made in the countryside shortly after the printing of the *Quaestiones celeberrimae in Genesim* (February 1, 1623):

> Meanwhile, the rumor of the brothers of the Rose-Cross being spread in the city gave certain idle and turbulent individuals the opportunity to denigrate Mersenne, whose writings had stung when he had drawn his pen against the atheists. They slandered him by presenting him as one of the members of this brotherhood, arguing above all from the fact that he had not been seen for a few days; as nothing about these *Roses* was in effect so notorious as their aptitude for becoming invisible at will, they persuaded themselves that the absence of Mersenne was a solid enough argument to convince people to pay attention to it. This tale, when Mersenne heard it upon his return, inflamed him and not without reason, and so that the authors might not peddle it with impunity, he came to avenge himself in a noble manner by the French work which he called *L'Impiété des déistes, des athées et des plus subtils libertins de ce temps, combattue et renversée point par point par des arguments tirés de la philosophie et de la théologie*.[296]

Thuillier thus attributed this highly improbable accusation, obviously inspired by Baillet, to Théophile's clan. And under the influence of Baillet's authority, not yet refuted by Henri Gouhier, Robert Lenoble in turn was not afraid to echo the story in 1943.[297] Yet Lenoble had just shown what he thought of another fanciful explanation by Thuillier, put forward with regards to the composition of the *Quaestiones celeberrimae in Genesim*: According to Thuillier, this vast treatise could have had only one goal at the outset: the refutation of the arts of magic and divination. But Father Rangueil allegedly made Mersenne observe that this intent would force him to make these doctrines known, at the risk of propagating the taste for them; Mersenne then supposedly decided to wrap them in "the scholarly apparel" of a biblical commentary. A simple comparison of dates made short work of this story, as Lenoble has shown.[298] Yet Thuillier's intent is more or less the same concerning *L'Impieté des Deistes:* If he is to be believed, Mersenne supposedly composed this new book to exact vengeance on the libertines who had tried to pass him off as a Rosicrucian. Unfortunately for Thuillier's story, Mersenne obtained the permit to print the work as soon as June 12, 1623, which puts the date of its composition well before the posting of Chaume's placards.[299] Furthermore, if I am not mistaken, *L'Impieté des Deistes* contains not a single word about the Rosicrucians.

This example, as well as that of Baillet's untrue anecdote about Descartes, both products of a somewhat excessive rationalistic zeal, are still striking testimonies of the destiny of the 1623 episode in France. There were many others up to the time of the Enlightenment.

Some Aspects of the Fortunes of the Rose-Cross in France in the Seventeenth and Eighteenth Centuries

Is it possible that Étienne Chaume's hoax contaminated Holland as well? According to the editors of Mersenne's *Correspondence,* on June 19, 1625, the counsels of Holland, Zealand, and Frisia invited the magistrate from Haarlem to take precautions against the brothers of the Rose-Cross "who, having stayed in Paris, have now come to Holland."[300] This information, cited without reference but borrowed from Gustave Cohen, came from a book by Willem Meijer that referred to the *Historisch Verhael* by Nicolaes Janszoon van Wassenaer, for the years 1624–25.[301] Although a brief examination of this book, analogous to the *Mercure François,* did not allow for the verification of Meijer's assertion, the link established between these Rosicrucians of Holland and those of Paris could nevertheless reveal a certain diffusion of the rumor.

Numerous manuscripts and various printed materials testify to the fortunes of Chaume's posters, almost to the end of the Ancien Régime. Among the manuscripts, a copy of Van Helmont's *De Magnetica vulnerum curatione,* unfortunately difficult to date accurately, nevertheless attests to the interest aroused in France by the debate over the *unguentum armarium.*[302] A collection of handwritten extracts of German Rosicrucian treatises that presents the same difficulties in dating proves at least, if proof is needed, that these texts of German origin circulated without difficulty in seventeenth-century France;[303] furthermore we are aware of the existence of a French version of the *Speculum Rhodostauroticum* by Theophilus Schweighart [Daniel Mögling], shown to Gabriel Naudé in 1623 by a Parisian bookseller of his acquaintance who had had it made for his own use.[304] Let us add to these points of interest the numerous copies of the anonymous *Recherches sur les Rose-Croix,* rashly attributed to Jacques Dupuy by Marion Kuntz following the catalogues of the Dupuy collection of the BNF, on the authority of one of these copies written in Jacques Dupuy's own hand.[305] These *Investigations,* whose guiding principle seems to have been to accumulate as many data as possible (even if they were contradictory), are either contemporary with or come soon after the posters of 1623. Indeed, they contain the following: "Ils ont promis par leurs affiches de se faire voir dans peu de jours, lors que le periode du temps qui leur a esté prefix par leurs restaurateurs, seroit escheu; cependant ils s'assemblent visiblement ou invisiblement par toute la terre, attirans a leur opinion les esprits les plus credules flattez par les promesses qu'ils font de l'immortalité, laquelle fait acquerir une perseverance aus labeurs, & une agitation continuelle cent ans durant, au bout desquels il est permis de se reveler."[306]

The accusation of naturalism, characteristic of the quarrels of the time against Paracelsianism and of Mersenne's attacks against the alchemists is also found therein:

> Les auteurs qui ont escrit de cette société disent qu'elle est aussi antienne que le monde. Qu'Adam, Seth, Noë, Abraham, Moyse, David, Salomon ont receu les admirables secrets de leurs sciences par tradition de l'un a l'autre. Que c'est par une occulte intelligence des merveille[s] de la nature qu'Elie a esté ravi; que saint Jean est immortel, & que Moyse parut a la transfiguration du Christ, porté dans le vuide de l'air, soustenu de son propre poids, quoi qu'il y parust revestu d'un corps visible. Que depuis l'absence de ce Christ plusieurs grands personages illuminez ont operé divers miracles parmi les hommes, qui pourtant n'estoient que purs effets de la nature bien comprise. Que les Arabbes sont demeurez possesseurs de ces souveraines vertus jusques en l'an 1413, auquel temps leur doctrine fut communiquee aus estrangers par ce moien [*here follows a brief summary of the myth of Christian Rosenkreutz, taken from the* Fama Fraternitatis].
>
> . . . Quand a la religion, il y a apparence qu'ils pretendent a un changement ou a la nullité: ils tirent a leur sens [*that is, toward alchemy*] les passages de la Genese, de la Sapience & des Pseaumes de David, avec des conceptions si formelles qu'il semble que ces grands personnages n'ayent escrit que pour auctoriser leur croiance, en quoi ils s'aident fort de la racine des langues.

En fin ce sont gens tres dangereux, ingenieux en meschanceté, grands seducteurs de peuples, perturbateurs d'Estat, & precurseurs d'une abominable secte; ce qui se voit par quelques articles de leur foy, & par leur serment de fidelité.[307]

One last manuscript, this one from the eighteenth century, is worth our attention.[308] It is by the philologist Jean Boivin de Villeneuve, keeper of the manuscripts of the King's Library (1663–1726). Containing only thirteen pages, it is entitled *Des Freres de la Roze Croix* and bears at the end the date of April 24, 1717. "Ce sont ici," Boivin writes, "les premieres reflexions qui me sont venuës sur ces écrits, a mesure que je les transcrivois. On en peut corriger quelques unes, on y en peut ajouter d'autres."[309] After giving the text of the poster as presented by Gabriel Naudé, Boivin offers his opinion as follows (1): "Les freres de la Roze Croix ne sont autres que tous les corps des Prétendus Réformez. . . . Calvinistes, Lutheriens, Zvingliens &c. Toutes ces especes conviennent a briser les croix, & les images. En cela elles se ressemblent: ils sont tous freres. Tous sont Brise-Croix. C'est ce que signifie Roze en hebreu רצץ: & au feminin, a cause de *croix*, רעצה. Ainsi *Roze-Croix,* ou *Rosea Crux,* c'est *Brise-Croix.* C'est pour cela qu'il est écrit par un z qui répond a la lettre hebraïque γ. Mr Naudé a ainsi écrit *Roze,* sur l'affiche qu'il a copiée." Here we recognize the tradition of the *Table chronographique* by Father Gaultier, which reduced the Rosicrucians to a Protestant sect. Boivin, commenting sentence by sentence upon the summary of the *Fama* provided by Naudé, gives according to this surprising key an interpretation of the origins of the Fraternity that is not devoid of ingenuity, by applying his theory in a systematic manner to the smallest details of the myth. His exegesis contains some gems. Here is his delectable manner of explaining the role of Paracelsus in the Rosicrucian legend: "*Il y avoit* [dans la grotte abritant le corps de Christian Rosenkreutz] *des livres de diverses sortes, & entre autres le Dictionnaire des mots de Paracelse,* c'est a dire, des mots pour rire: c'est ce que signifie ici Paracelse, mot hybride, moitié Grec & moitié Hebreu: παρα קלס."[310] Erudition and fantasy are here in the service of apologetics and rationalism: The supposed mysteries of the Rosicrucians, all related to the reformed religion, are explained in a manner that deprives them of any magical or supernatural dimension.

Among printed texts, we find traces of the episode of Chaume's posters in all sorts of works published or composed between 1623 and 1771. At least one of Théophile's friends, the poet Saint-Amant—who was moreover close to Mr. de Saint-Brice, the duke of Retz's cousin, whom Jacques Le Pailleur himself had followed to Brittany toward the autumn of 1623— left in his works a memory of the incident. He seems only to have retained the colorful aspect of the association between Rosicrucians and witches, within a satire composed between 1623 and 1629, aimed at an old woman:

> Mais qu'est-ce-cy mes camarades?
> Voicy d'estranges algarades!
> On nous en baille, on nous en vent,
> Nous ne bernons plus que du vent,
> Et le Demon qui la possede
> Mieux qu'il ne fit jamais Salcede,
> La rendant ainsi que vous trois
> De l'ordre de la Rose-crois,
> Droit aux Enfers l'a transportée,
> Pour estre si bien tourmentée,
> Qu'au prix d'elle les Gaufridis
> Penseront estre en Paradis.[311]

Although in 1627 Garasse's friend Pierre de Lancre limited himself to reproducing the article of the *Mercure François* in his work *Du Sortilège, où il est traité, s'il est plus expédient de supprimer et tenir sous silence les abominations et maléfices des sorciers, que de les publier et manifester*,[312] other authors betrayed here and there some more expressive recollections of the poster incident. Thus in 1629, obviously with reference to the Parisian incident, Pierre Gassendi ironically mentioned "these fortunate crucirose brothers who, as if protected by the ring of Gyges, fly around us, always invisible," with an untranslatable play on words involving the two possible meanings of *invisi*: "invisible" or "hateful."[313]

A few years later, the 199th conference of Théophraste Renaudot's Bureau d'Adresse (May 16, 1639) dealt with *The Brothers of the Rose-Cross*.[314] Renaudot began the conference by justifying the fact that he displayed here "un titre dont la profession estoit criminelle en cette ville il n'y a pas long temps: me souvenant que quelques-uns furent emprisonnez pour s'estre vantez d'estre de cette societé, & dits invisibles." Then he listed, according to some of the books published by the Rosicrucians, all of the secret societies that had preceded them since Antiquity, taking up in this context the entire tradition of the *prisca theologia*. The second speaker summarized the Rosicrucian myth according to the *Symbola Aureae Mensae* by Maier. The third—probably a member of the Faculty of Medicine in Paris—mainly gave his verdict against Paracelsus's medicine, which he considered to be the Rosicrucians' most noticeable characteristic, and particularly against antimony, which he designated in an allusive yet clear manner.[315] Here is the speech of the fourth speaker:

> Le 4e, qui se professoit ouvertement estre l'un des freres de cette societé dist, Que le Chevalier Anglois Flud n'a pas mauvaise grace d'interpreter ces trois lettres F *fide*, R *religione*, C *charitate*: la commune opinion toutesfois a prévalu, qui veut qu'elle[s] signifient, *Fratres Roseae Crucis*. Mais aucune de ces interpretations ne peut passer pour grand secret. C'est pourquoi il en faut cercher un autre: dans lequel examen nous trouverons que la croix est veritablement de la partie, mais en un autre sens; qui est que dans cette + le mot LUX se trouve: d'où l'on croid que ces freres ont pris en Espagne le mot d'Illuminez. Mais ozeroy-je passer outre sans rompre le seau céleste, mettre Diane tout à nud, & la clef du cabinet de la nature entre les mains du vulgaire? Oüy, puisque les sages nous l'ont promis au declin des siécles où nous sommes. La rosée (qui est le plus puissant dissolvant de l'or entre les corps naturels & non corrosifs) n'est autre chose que cette lumiere espoissie & renduë corporelle: laquelle estant cüite & digerée artistement par un temps convenable, en son propre vaisseau, est le vray menstruë du dragon roux; c'est à dire de l'or, veritable matiére des Philosophes. Duquel secret cette societé ayant voulu laisser à la posterité dans son nom des marques qui ne peussent estre effacées par le temps, a retenu celui des freres de la Rozée cüite. C'est pourquoi la benediction de Jacob à Esaü ne contenoit que ces deux matiéres, *de Rore coeli & pinguedine terrae det tibi Deus*. Au reste ce qu'on impute à cette compagnie d'estre invisible, s'entend de ce qu'elle n'a pas des marques visibles, & qui la distinguent des autres, comme le reste des societez, telles que sont les diverses couleurs & façons d'habits; mais n'est connüe & visible qu'à ceux de la societé mesmes.[316]

The fifth lecturer responded to this discourse by arguing for a moderate position, open to new ideas, most notably in the subject of medicine, but resolutely hostile to an excessive enthusiasm that would make the Rosicrucians "the only wise men & the darlings of nature": "De fait, quelle apparence qu'une matiére si volage que la rozée fust la médecine de ces trois corps si differens [*vegetables, animals, and minerals*]? & pourquoi elle plustost que la pluye; voire plustost que la manne, qui est cette rozee toute cüite par la nature? Ces freres n'ont donc rien de recommandable en eux que l'amour qu'ils professent de sagesse, & la recherche des secrets de la nature, que nous recherchons à la verité trop négligemment." But having produced nothing until present except their promises, the brothers of the Rose-Cross had scarcely given any reasons for being taken seriously, the same as the "neant de la pierre Philosophale."[317]

One month later (June 20, 1639), the 203rd conference had as its theme: *Qu'est-ce qu'a voulu entendre Paracelse par le livre M?*[318] This title betrayed an error on the part of Théophraste Renaudot: In fact Paracelsus was not the one who mentioned the *Liber M.*, but rather it was the text of the *Fama Fraternitatis*, which portrayed Christian Rosenkreutz as so rapid in his progress in Arabic at Damcar that after one year he even found himself able to translate into Latin the *Liber M*. As for this *Liber M.* (habitually understood as the *Liber Mundi*, "Book of the World"), the *Fama* specified however that in it Paracelsus "was able to inflame and sharpen his genius."[319] Renaudot was therefore not entirely mistaken; his error showed above all to what degree the Rosicrucians found themselves, in the minds of Renaudot and his contemporaries, bound up with and inseparable from Paracelsianism.

Among the speakers of this session, we find the alleged Rosicrucian of the May 16 meeting. Once again he was the fourth speaker:

> Le 4ᵉ dist, Que ce livre M ne sçauroit estre celui du Monde, puisque le Monde ne peut estre tourné d'Arabe en Latin, estant leu également de tous peuples.... Que s'il est permis de parler en figure: je trouve bien plus d'apparence à dire, que ce livre n'est autre chose qu'une figure ou charactére talismanique gravé en un cachet, dont les freres de la Rose-croix se servent pour s'entreconnoistre, appellée le livre M; pource qu'elle represente une M croizée avec quelques autres lettres, de la combination desquelles resulte le mistere du grand oeuvre: qui désigne sa matiére, son vaisseau, son feu, & ses autres circonstances: La premiere desquelles est la rosee, le vrai menstrüe, ou dissolvant du dragon roux, qui est l'or. Bref, dans cette figure sont comprises tant de choses, par la diverse combination des lettres qui y sont representées, qu'elle merite bien d'estre appellée un livre.[320]

But the following speaker mocked him:

> Le cinquiesme dist, Que si c'estoit là le secret des freres de la Rose-croix, qu'ils estoient invisibles en toutes leurs procedeures: pource qu'il ne se void point là de secret, mais force absurditez ... puis que dans ces trois lettres *sic*, diversement couppées & appliquées les unes aux autres, vous y rencontrez non seulement toutes les lettres mais encor par leur combination, tous les livres & toutes les choses qui sont au monde. Où il n'y a pas plus d'industrie qu'à faire entonner toutes sortes de notes à un flageolet.... Disons donc plustost ... qu'au lieu que nous avons estimé que cette *M* signifiast *Mons*, nous voyons maintenant qu'elle ne signifie rien davantage que *Mus*.[321]

Of course, these two conferences must be put into context, as they are quantitatively negligible in the midst of all those held in the setting of the Bureau d'Adresse between 1633 and 1642; Howard Solomon indicates, however, that among the 460 subjects debated therein, the session devoted to the Rosicrucians was among only three to which eight pages were devoted, instead of the four pages usually reserved for each theme.[322]

One would scarcely have to look very long before discovering in France all sorts of allusions to the Rosicrucians in the years 1640 to 1660. We know, for example, that the libertine Nicolas Le Gras cherished the dream, very close to the ideal set out in the *Fama Fraternitatis,* of "founding a sect or an order composed only of doctors and which, by distributing its care to all the nations of the world, would restore the state of nature to entire humankind."[323] And, without saying anything about the attacks coming from La Mothe Le Vayer, who was visibly inspired by Naudé and Gassendi, we find precise allusions to the episode of Chaume's placards in Father Le Moyne, who, in 1640, in his *Peintures morales,* betrayed clear recollections of the quarrels of 1623–25, detecting in the "Secte des Esprits Forts" the worthy successors of the "Cabala of those of the Rose."[324] Various works of ecclesiastical history from the same time preserve the memory of the Parisian incident of 1623. Those of the Feuillant monk Pierre Guillebaud (1647) and of Scipion Dupleix (1648) have already been mentioned: They contributed, probably along with others that a new investigation would easily unearth, to saving the event from oblivion (while distorting it) for it is clear that the most read pages of these books were those that dealt with the seventeenth century.

In a manner that is equally close to that of the Parisian posters, here is how, in *Les Estats et Empires de la Lune,* the Genius of Socrates recounted his travels in Europe to Cyrano de Bergerac: "J'y vis Agrippa, l'abbé Triteme, le docteur Fauste, La Brosse, Coesar, et une certeine caballe de jeunes gens que le vulgaire a connus sous le nom de chevaliers de la Rose Croix, à qui j'enseigné quantité de soupplesses et de secrets naturels, qui sans doute les auront faict passer chez le peuple pour de grans magiciens."[325] In this typically libertine passage that strips magic of any supernatural traces, the association between the Rosicrucians and magic, as well as the mention of Guy de La Brosse, a friend of Théophile, refer directly to the Parisian episode of 1623. Let us note in passing that a close association between alchemy and witchcraft is found in Cyrano in his letter "On Behalf of Witches."[326] But the context, which is entirely literary, is that of a brilliant stylistic exercise: Cyrano moreover knew full well how to establish the division between these two disciplines, as is shown not only in the other instances of alchemical themes in his work,[327] but also in the letter "Against Witches," which is symmetrical to the preceding one but of an entirely different nature. In that letter, the author categorically denies the supernatural effects attributed to the devil and to the supposed witches and the possessed.[328]

The quarrel between Fludd, Mersenne, and Gassendi equally contributed to the fortunes of the Brothers of the Rose-Cross in France. In 1662, the authors of the *Logique de Port-Royal,* above all by basing themselves on Gassendi's examination of Fludd's philosophy, in a long tirade against unintelligible style became indignant at seeing "appliquer ce que l'Ecriture dit des vrais Chrétiens, qu'ils sont la race choisie, ... à la chimerique confrerie des Rosecrois, qui sont selon eux, des Sages qui sont parvenus à l'immortalité bienheureuse, ayant trouvé le moyen par la pierre philosophale de fixer leur ame dans leurs corps, d'autant, disent-ils, qu'il n'y a point de corps plus fixe & plus incorruptible que l'or."[329] Arnauld and Nicole thus took up as their own the type of criticism formulated by Mersenne against certain alchemists, starting with the *Quaestiones celeberrimae in Genesim*. But whereas the context of the *Quaestiones* situated this attack in the wake of Mersenne's anti-Paracelsian polemic, Arnauld and Nicole came, by means of Gassendi, to apply it more specifically to the brothers of the Rose-Cross.

This constant, albeit diffuse, presence of the Rosicrucian myth in French literature throughout the seventeenth century explains why in 1681, Thomas Corneille, Donneau de Visé, and perhaps Fontenelle based their comedy *The Philosophers' Stone* in part on the Rosicrucian Fraternity. This play was inscribed in the sinister context of the "Affaire des Poisons" (1679–82), in which more than 400 people from every level of society, from the lowest to the highest, were implicated in all sorts of criminal acts: counterfeiting, fraud, abortion, infanticide, poisoning, and possible plots against the king's life, as well as kidnappings and sacrifices of children, trade in human blood, Satanism, and celebrations of black mass. The large majority of these acts were hidden behind practices of a considerably more innocent appearance: particularly divination, alchemy, and searches for hidden treasures. The principal defendants who had to answer to the interrogations cited these very pretexts, for as we can imagine, the creation of poisons and abortifacient powders scarcely necessitated equipment different from that of an alchemist.[330] In November 1679, the police lieutenant La Reynie, wishing to ruin the Parisians' belief in fortune-tellers, or, more probably, concerned with interposing "between the public and the sordid reality" "the painted canvases, the palettes, the gold and the lights of enchantment," ordered two authors who were in vogue, Thomas Corneille and Donneau de Visé, to create a spectacular play entitled *La Devineresse ou les Faux Enchantements*. De Visé probably accepted all the more easily since his own brother had lost two successive wives in the Affaire des Poisons.[331] This play had a run of five months. Thus the same team was ready to have another go in February 1681, perhaps assisted by Fontenelle on this occasion. This play was *La Pierre philosophale*, which met with resounding failure.[332] This failure can be explained in various ways: Perhaps the play was too scholarly, for had it not incorporated too many allusions that were inaccessible to the general public, leaving them cold while amusing only the connoisseurs? After the failure of the first two performances, its costly luxuriousness probably precluded the possibility of cherishing any hope that the public might be awakened, and thus the play had to be abandoned immediately. Finally, perhaps the play came at a time when the Affaire des Poisons, then stagnant but having already lasted for two years, had exhausted and disgusted an audience that was henceforth overwhelmed by horrors.[333] Yet the authors had neglected nothing: machines, a score by Marc-Antoine Charpentier; all the enchantments of the *Comte de Gabalis* (a great success at the bookstores), which was adapted to the theater for the first time; and the colorful aspect of the brothers of the Rose-Cross.[334]

The authors of *La Pierre philosophale* referred directly to the 1623 episode in the play. In the preface to the reader, they cited in turn Naudé, then Garasse, then Scipion Dupleix.[335] On the other hand, they were careful not to exploit the association between Rosicrucians and witchcraft presented by the *Effroyables Pactions*, probably trying to create above all an effect of ridicule and not to feed the audience's fears, which current events already provoked and justified only too much. Thus, at the very time in which this association was taking shape in a sinister way, the authors, who were in charge of amusing the general public, eliminated it entirely and substituted in its place the marvelous enchantment of elementary spirits borrowed from the *Comte de Gabalis*, although the source of this fantastic element—the *De Nymphis* by Paracelsus—would have possessed enough subversive energy in the beginning of the seventeenth century to provoke accusations of demonic magic against those who made use of it.

It would be wrong to think that 1681 brought an end to this subversive potential thanks to the triumph of reason. To do so we would have to forget the fact that individuals of every social class ran the risk, at exactly the same time, of being burned alive for having taken part in bloody black masses, to which reason had failed to put a stop. If the enchantments of elementary spirits could take on a fairylike character at this point in time, it was more probably due to their erudite nature. This type of magic, which had become too scholarly to cause worry, was the concern of another time, of the dusty world of learned men that *Les Provinciales* had, not so long before, contributed to discrediting with so much brilliance, imposing on the public the taste for common sense against an erudition that was accused of pedantry and left to the specialists alone: The black masses did not demand so much knowledge. Paradoxically, this scholarly dimension both permitted the plot of the *Comte de Gabalis* to be brought to the stage and at the same time contributed to the failure of the play.

Ten years later, in 1691, an anonymous collection on alchemy published in Paris, the *Traitez du Cosmopolite Nouvellement découverts*, announced with a great uproar the existence of a supposed society of "unknown Philosophers," obviously inspired by the Rosicrucian myth.[336] This is probably the society that the physician François Alary was thinking of some time later, when in the preface of one of his anonymous works, *Le Texte d'Alchymie, et le Songe-verd*, he openly displayed his submission to the "Brothers of the true Rose-Cross": "Mais comme ce n'est que par la Croix que doivent être éprouvez les veritables Fidéles, c'est à vous, Freres de la vraye Rose-Croix, qui possedez tous les trésors du Monde, c'est à vous à qui j'ai recours. Je me soûmets entierement à vos pieux & sages conseils.... Tout est à vous, tout vient de vous, tout retournera donc à vous. Recevez (Messieurs) cét Acte de soûmission que je vous fais aujourd'hui."[337] Furthermore, it is this author's double fascination with Paracelsus and with the Rosicrucian myth that allowed for his recent identification.[338] Not very common in his time, at least in France, this fascination is in fact as noticeable in another one of his anonymous works as in a prophecy published under his own name in Paris in 1701, which moreover earned him some difficulties with the government: the *Prophetie du Comte Bombast Chevalier de la Rose-Croix, Neveu de Theophraste Paracelse, Publiée en l'année 1609. Sur la Naissance miraculeuse de Louis Le Grand*. Here the connection to the incident of Chaume's placards is entirely lost: There remains only the original solidarity between the figure of Paracelsus and the Rosicrucian myth.

The other anonymous work by Alary was none other than *Le Parnasse assiegé ou La Guerre declarée entre les Philosophes Anciens & Modernes,* published in Lyon in 1697. The goal of this curious alchemical contribution to the quarrel of the ancients and the moderns was to "demonstrate the reality of the science of Hermes, & the truth of the medicine of Paracelsus."[339] Alary imagined that after the death of Apollo, the master of Parnassus, "ce fut à qui de toutes les sectes des Philosophes, y établiroit la premiere le trône de sa réputation." But since Parnassus, surrounded by thick clouds, was turning out to be impregnable, the siege was quickly degenerating into a civil war in which the majority of schools were confronting one another. Only the Gassendists and the Cartesians succeeded in advancing. Aristotle then sent the greatest orators of Antiquity to exhort the "peuple Philosophique" to make peace. Once this peace was obtained, a scout who had been sent to Parnassus on reconnaissance came back in the company of four spies, who informed the attackers about the mysterious enemies who were occupying and defending Parnassus. Having explained that only the philosophers from the school of Hermes knew how to dissipate the fog and reach the summit, the spies were questioned concerning the fortifications: one of them "responded wisely that the true mark of the perspicacity of Hermes consisted only of the three letters F.R.C. painted on the banners that his officers bore":[340] Thus the disciples of Hermes who were occupying and defending the summit of Parnassus were none other than the brothers of the Rose-Cross. Here again, we are quite far from the 1623 episode.

A last survival of this period is found considerably later and in a rather unexpected place: in a work directly inspired by the Faculty of Medicine of Paris, the *Dictionnaire historique de la médecine ancienne et moderne* by Éloy (1778), under the entry "CROIX (Les Freres de la ROSE)."[341] Éloy gave an extremely critical brief review of the fraternity, including it only because of its links with medicine, which shows that the Faculty, even at the very end of the eighteenth century, had not renounced its hostile position toward any manifestation of Paracelsian doctrines.

As we can see from the number and variety of the works discussed in this section, the impact of the incident of 1623 was on the whole rather lasting and contributed to the ongoing presence of the Rosicrucian myth in French literature in a more or less continuous manner until the end of the eighteenth century, in forms both favorable and unfavorable to the Rosicrucian Fraternity, in recollections of Théophile's trial, in Mersenne's or the Faculty of Medicine's anti-Paracelsianism, or even in the libertine and naturalistic forms that Mersenne himself had fought. As for the anecdote invented by Baillet, it has probably been, from 1623 on, one of the most definite means of transmitting the recollection of Chaume's posters to the scholars up to our time.

This hoax involving the Rosicrucian placards posted on the streets of Paris in the summer of 1623, conceived merely as a piece of adolescent humor, but taking place in a context loaded with polemics, must be seen as the very first manifestation of the association, typical of those years, between Paracelsianism and libertinism. The hunt for libertines that began when Vanini was burned at the stake (1619) did not intensify until Théophile de Viau's trial: Only then did religious authorities in France notice to what extent Paracelsianism was an objective ally of irreligion and attack it as such. These attacks reached their highest point in 1625: Alchemy and Paracelsianism—refuted by Naudé in his *Apologie,* partially condemned by Father Mersenne in *La Verité des sciences* for their impiety and the obscurity of their language, censured by the Sorbonne in a ruling pronounced against Khunrath's *Amphitheatrum sapientiae aeternae,* and attacked by the Inquisition in the proceedings started against Jean Baptiste van Helmont—then encountered a crisis that has not yet been analyzed in depth, the stakes and consequences of which it would be beneficial to measure.[342] The religious issue, deferred until then for all sorts of reasons, played a dominating role in this, and the Rosicrucian incident of 1623 was its first symptom.

Acknowledgment

This chapter synthesizes several portions of my doctoral dissertation, "Paracelsisme et alchimie en France à la fin de la Renaissance (1567–1625)," Université de Paris IV, 1998, to be published in a revised form by Droz Editions (Geneva) in 2006. My sincere thanks go to the editors of this volume, who had the present chapter translated, and to the copyeditor for his skill and effort in preparing the final version.

Notes

This chapter was translated by Laura Dennis-Bay.

1. See Carlos Gilly, *Cimelia Rhodostaurotica: Die Rosenkreuzer im Spiegel der zwischen 1610 und 1660 entstandenen Handschriften und Drucke. 2. verbesserte Auflage* (Amsterdam: In de Pelikaan, 1995), 7. See also Gilly and F. Niewöhner

(eds.), *Rosenkreuz als europäisches Phänomen im 17. Jahrhundert* (Amsterdam: In de Pelikaan, 2002) and P. van der Kooij and Gilly (eds.), *Fama Fraternitatis. Das Urmanifest der Rosenkreuzer Bruderschaft zum ersten Mal nach den Manuskripten bearbeitet* (Haarlem: Rozekruis Pers, 1998). Other bibliographic indications will be found below.

2. I quote the summary of Roland Edighoffer, *Rose-Croix et société idéale selon Johann Valentin Andreae*, 2 vols. (Neuilly-sur-Seine: Arma Artis, 1982–87), 1:221.

3. Frances A. Yates, *The Rosicrucian Enlightenment* (London: Routledge and Kegan Paul, 1972; reissued, London-New York: ARK Paperbacks, 1986), 221 (French translation, *La Lumière des Rose-Croix: L'illuminisme rosicrucien* [Paris: CELT, 1978; re-ed. Paris: Retz, 1985], 252). The French translation is weighed down by mistranslations and approximations, each one more absurd than the next.

4. Carlos Gilly has recently observed that in 1786, J. S. Semler had also proposed the idea, sustained by Yates and already shared by certain of Andreae's contemporaries, Lutheran as well as Catholic, of complicity between the Calvinists and the Rosicrucian movement. Cf. Gilly, *Cimelia Rhodostaurotica*, 75–76; also "*Theophrastia Sancta*: Der Paracelsismus als Religion im Streit mit den offiziellen Kirchen," in J. Telle, ed., *Analecta Paracelsica: Studien zum Nachleben Theophrast von Hohenheims im deutschen Kulturgebiet der frühen Neuzeit*, Heidelberger Studien zur Naturkunde der frühen Neuzeit, no. 4 (Stuttgart: Franz Steiner Verlag, 1994), 425–88, at 469 and n. 97.

5. Edighoffer, *Rose-Croix et société idéale*, 1:222.

6. On the thesis of John Warwick Montgomery (*Cross and Crucible: Johann Valentin Andreae (1585–1654), Phoenix of the Theologians*, 2 vols. [La Haye: Martin Nijhoff, 1973]), stained by an "apologetic zeal" that compels its author to "put on Manichean glasses," see the criticism of R. Edighoffer, *Rose-Croix et société idéale*, 1:223 (see also, for example, 190). It is better to say nothing about a more recent article of Montgomery's ("The World-View of Johann Valentin Andreae," in *Das Erbe des Christian Rosenkreuz: Vorträge gehalten anläßlich des Amsterdamer Symposiums 18–20 November 1986* [Amsterdam: In de Pelikaan, 1988], 152–69).

7. Gilly, *Cimelia Rhodostaurotica*, 22 and 74.

8. Ibid., 76–77. Gilly makes clear (77) that the success of the Rosicrucian manifestos was due not only to their mythical attire (without which they could have provoked only scant interest) but also and above all to the fact that they had presented the Fraternity as already formed and that they had invited scholars and princes of the time to respond through the channels of printed matter.

9. On the exact content of this philosophy, see the quotations from the original text of the *Fama Fraternatis* as Gilly has collected them, ibid., 77.

10. Ibid., 78–79 and 45–47 on Tobias Hess. On the evolution of Christoph Besold, cf. Carlos Gilly, "*Iter Rosicrucianum*: Auf der Suche nach unbekannten Quellen der frühen Rosenkreuzer," in *Die Erbe des Christian Rosenkreuz*, 63–89, at 72.

11. Gilly, *Cimelia Rhodostaurotica*, 82; cf. Edighoffer, *Rose-Croix et société idéale*, 1:230–31. Let us remember that the *Chymische Hochzeit*, composed in

German, never had a Latin translation. It was translated only into English in 1690 and into French not before 1928.

12. Gilly, *Cimelia Rhodostaurotica*, 1.

13. Ibid., 29.

14. Ibid., 29–31. On the Tyrolian Adam Haslmayr (1562–1630), organist, schoolmaster, and an enthusiastic Paracelsian, see Carlos Gilly, *Adam Haslmayr: Der erste Verkünder der Manifeste der Rosenkreuzer*, Texts and Studies published by the Bibliotheca Philosophica Hermetica, no. 5 (Amsterdam: In de Pelikaan, 1994) to be corrected by Gilly, *Cimelia Rhodostaurotica*, 30, as to his date of birth.

15. Gilly, *Cimelia Rhodostaurotica*, 29.

16. Ibid., 40–41.

17. Gilly, *Adam Haslmayr*, 18, 48, and 187–201; see also his *"Theophrastia Sancta."*

18. Gilly, *Adam Haslmayr*, 89; *Cimelia Rhodostaurotica*, 36. On August von Anhalt, see also Julian Paulus, "Alchemie und Paracelsismus um 1600: Siebzig Porträts," in Telle, *Analecta Paracelsica*, 335–406, at 343–44.

19. Gilly, *Adam Haslmayr*, 125–26.

20. Ibid., 127–28, 123, 134–36. Gilly found Haslmayr's text, which previously could not be located, in Weimar and reproduced it in facsimile; ibid., 69–80 (A. Haslmayr, *Antwort An die lobwürdige Bruderschafft der Theosophen von RosenCreutz N. N. vom Adam Haselmayr Archiducalem Alumnum, Notarium seu Judicem ordinarium Caesareum, der zeyten zum heyligen Creutz Dörflein bey Hall in Tyroll wohnende. Ad Famam Fraternitatis Einfeltigist geantwortet. Anno 1612* [s.l. [Halle ?], s.n.e. [Joachim Krusicke ?], 1612]); cf. Gilly, *Adam Haslmayr*, 22.

21. Gilly, *Adam Haslmayr*, 32–67; *Cimelia Rhodostaurotica*, 32–33.

22. Quoted by Gilly, *Cimelia Rhodostaurotica*, 34: "Was aber für ein Ruehloß, whuestes, ungeheurigs, verzweifeltes, Sodomitisches Leben auf den Galern, von den Welschen Gefangnen sonderlich, ist gefhiert worden, ist daher wegen der keuschen ohren nicht zu schreiben. . . . und ist All mein Schrifft, darumb ihr mich verdambt, nichts anders, auf kein andern weg nicht gemeint, als daß ich mich einen armen Theophrastischen Christen bekenne."

23. Gilly, *Adam Haslmayr*, 59–60.

24. Ibid., 98–99; *Cimelia Rhodostaurotica*, 38. On Figulus (Benedikt Töpfer), see also Joachim Telle, "Benedictus Figulus: Zu Leben und Werk eines deutschen Paracelsisten," *Medizinhistorisches Journal*, 22 (1987): 303–26; Paulus, "Alchemie und Paracelsismus um 1600," 352–54.

25. Gilly, *Adam Haslmayr*, 142–43; *Cimelia Rhodostaurotica*, 70.

26. Gilly, *Adam Haslmayr*, 84 (facsimile of the title page of the *editio princeps* of the *Fama*): "*Auch einer kurtzen Responsion, von dem Herrn Haselmeyer gestellet / welcher deßwegen von den Jesuitern ist gefänglich eingezogen / und auff eine Galleren geschmiedet.*"

27. Gilly, *Cimelia Rhodostaurotica*, 70.

28. Gilly, *"Iter Rosicrucianum,"* 82; *Adam Haslmayr,* 99; *Cimelia Rhodostaurotica,* 38 and 70.

29. Gilly, *Adam Haslmayr,* 143–45 and 150, n. 32.

30. Ibid., 85–90; *Cimelia Rhodostaurotica,* 43 and 75–76.

31. Gilly, *"Theophrastia Sancta,"* 469–72. There is insufficient space here to summarize the contributions of other recent works on some of the figures traditionally attached to the Rosicrucian movement. See among others Karin Figala and Ulrich Neumann, "A propos de Michel Maier: quelques découvertes biobibliographiques," in D. Kahn and S. Matton, eds., *Alchimie: art, histoire et mythes: Actes du Ier colloque international de la Société d'Étude de l'Histoire de l'Alchimie* (Paris-Milan: S.É.H.A.-ARCHÉ, 1995), 651–64; Ulrich Neumann, "Mögling" in Traugott Bautz, ed., *Biographisch-Bibliographisches Kirchenlexikon,* vol. 5 (Herzberg: 1993), col. 1582–84, and in the *Neue Deutsche Biographie,* vol. 17 (Munich: C. H. Beck, 1994), 613–14; idem, *"Olim, da die Rosen Creutzerey noch florirt, Theophilus Schweighart genant:* Wilhelm Schickards Freund und Briefpartner Daniel Mögling (1596–1635)," in Friedrich Seck, ed., *Zum 400. Geburtstag von Wilhelm Schickard. Zweites Tübinger Schickard-Symposion 25. bis 27. Juni 1992,* Contubernium: Tübinger Beiträge zur Universitäts- und Wissenschaftsgeschichte, no. 41 (Sigmaringen: Jan Thorbecke Verlag, 1995), 93–115.

32. On this point see also Gilly, *Adam Haslmayr,* 89.

33. Cf. Gilly, *Cimelia Rhodostaurotica,* 156–157, no. 278.

34. Quoted by Gilly, *"Theophrastia Sancta,"* 471, n. 99. See also his *Adam Haslmayr,* 109 and 114–15, notes 14–19. Gilly emphasizes in passing that this trial tarnishes somewhat the brilliant picture of the "hermetic patronage" of Moritz von Hessen-Kassel, far more interested in the possible practical alchemical success of the Rosicrucians than in their spiritual aspirations. See Bruce T. Moran, *The Alchemical World of the German Court: Occult Philosophy and Chemical Medicine in the Circle of Moritz of Hessen (1572–1632),* Sudhoffs Archiv, Beiheft 29 (Stuttgart: Franz Steiner, 1991).

35. Quoted by Gabriel Naudé, *Instruction à la France sur la verité de l'histoire des Freres de la Roze-Croix* (Paris: François Julliot, 1623; reprinted, Paris: Gutenberg Reprint, 1979), 27. Naudé's text bears "d'erreur de mort"; I am correcting it: see below, note 114. Nearly all the sources studied here have already been inventoried by René Pintard, *Le Libertinage érudit dans la première moitié du XVIIe siècle* (1943, re-ed. and extended, Geneva-Paris: Slatkine, 1983), 584–85 (n. 4 on 48).

36. Naudé, *Instruction à la France,* 26.

37. Peter Paul Rubens, *Correspondance de Rubens et documents épistolaires concernant sa vie et ses oeuvres,* ed. Max Rooses and Ch. Ruelens, vol. 3, *Du 27 juillet 1622 au 22 octobre 1626* (Anvers: Jos. Maes, 1900), 224. The original text is in Italian. The correspondence between Peiresc and Rubens concerning the Rosicrucians has already been put to good use by Robert Halleux, "Helmontiana," *Academiae Analecta: Mededelingen van de Koninklijke Academie voor Wetenschappen, Letteren en Schone Kunsten van België. Klasse der Wetenschappen,* 45, no. 3 (1983), 35–63, esp. 60–61.

38. Mrs. P. Tannery, C. de Waard and R. Pintard, eds., *Correspondance du P. Marin Mersenne, religieux Minime,* vol. 1 (1617–27) (Paris: Gabriel Beauchesne & fils, 1932), 154; *Effroyables Pactions Faictes entre le diable & les pretendus Invisibles: Avec leurs damnables Instructions, perte deplorable de leurs Escoliers, & leur miserable fin* (s.l., s.n.e., 1623; reprinted in G. Naudé, *Instruction à la France,* 1979, 3e pagination), 15 (see infra for the dating of this pamphlet). Basing his argument on the *Instruction* by Naudé, Henri Gouhier situates the placards in the month of August, but the letter of Peiresc positively contradicts this dating; cf. *Les Premières pensées de Descartes: Contribution à l'histoire de l'Anti-Renaissance,* De Pétrarque à Descartes, no. 2 (Paris: Vrin, 1958), 125–26.

39. Carpentras, Bibliothèque Inguimbertine, MS 1777, fol. 475r: "Coppie d'une affiche mise par les carrefours de Paris en juillet 1623." See the analysis of this document below.

40. See below the analysis of this other document.

41. The Exchange ("La place au Change"), equivalent to the Stock Exchange ("La Bourse"), was located "in the courtyard of the Palace, beneath the 'salle Dauphine'" (that is, the Palace of Justice ["Palais de Justice"]). It was until 1700 "the meeting place of the merchants 'to deal in trade and bills of exchange, every day at noon'" (René Pillorget, *Nouvelle histoire de Paris: Paris sous les premiers Bourbons, 1594–1661* [Paris: Hachette, 1988], 185).

42. *Mercure François,* vol. 9 (1622–24) (Paris: Jean & Étienne Richer, 1624), 371.

43. Naudé, *Instruction à la France,* 26; *Effroyables Pactions,* 17. There were four lines according to Peiresc, who does not specify whether the posters were handwritten or printed.

44. The fact that they trotted "from hand to hand" (to which no other source has attested) could seem quite plausible; upon further consideration, however, it was quite risky to have notes circulated in such an ill-concealed manner, notes that we know were very quickly considered by the authorities to be seditious. This detail could therefore be nothing but an embellishment, added after the fact by the writer from the *Mercure* and brought about by the mention of the "Change du Palais" which stems from Naudé's story, as we will see.

45. "Rather than uncovering immediately in the city the things buried in the shadowy depths of the forest" (Virgil, *Aeneid,* VI, 267).

46. Count Ernst von Mansfeld (c. 1580–1626), a mercenary at the beginning of the Thirty Years War, had become the military leader of Protestantism in Germany. In 1622, during a campaign of Louis XIII against the Protestants of the South of France, Mansfeld, in order to relieve the latter, had created a diversion by having his troops enter Champagne. It was necessary to negotiate with him in order to turn him away from the kingdom. A short time after, by the Treaty of Paris (February 1623), Mansfeld had passed into France's service and from then on fought in Flanders against the Spanish. See Victor L. Tapié, *La France de Louis XIII et de Richelieu* (Paris: Flammarion, 1967; reissue 1980, coll. "Champs,") 129–34; *The New Cambridge Modern History,* vol. 4, J. P. Cooper (ed.), *The Decline of Spain and the Thirty Years War, 1609–48/59* (Cambridge: Cambridge University Press,

1970), 309–24; and the article by Reinhard R. Heinisch, "Mansfeld," in the *Neue Deutsche Biographie*, vol. 16 (Munich, 1990), 80–81.

47. See the text of the *Mercure François* quoted above: "Ceux-cy voyant le change du Palais de Paris sans nouvelles . . ."

48. Naudé, *Instruction à la France*, 25–26 (follows the text of the poster).

49. [Adrien Baillet], *La Vie de Monsieur Des-Cartes*, 2 vols. (Paris: Daniel Horthemels, 1691), 1:107.

50. François Secret, "Notes sur quelques alchimistes de la Renaissance. I: Un témoignage oublié sur l'épisode des placards des Frères de la Rose-Croix," *Bibliothèque d'Humanisme et Renaissance*, 33 (1971): 625–26.

51. Cf. Gilly, *"Iter Rosicrucianum,"* 83 and 89, n. 48; Lorenzo Bianchi, "Gabriel Naudé critique des alchimistes," in J.-C. Margolin and S. Matton, eds., *Alchimie et philosophie à la Renaissance: Actes du colloque international de Tours (4–7 déc. 1991)*, De Pétrarque à Descartes, no. 57 (Paris: Vrin, 1993), 405–21, at 411, n. 20; Bianchi, *Rinascimento e libertinismo: Studi su Gabriel Naudé* (Naples: Bibliopolis, 1996), 185, n. 23 (I thank Alain Mothu for having provided me with this work).

52. Nicolas Chorier, *De Petri Boessatii, Equitis et Comitis Palatini viri clarissimi, Vita amicisque litteratis. Libri duo* (Grenoble: Fr. Provensal, 1680). On Chorier (1612–92), see Jean Serroy's notice in G. Grente, dir., *Dictionnaire des Lettres françaises: Le XVIIe siècle* (1954), new ed., dir. P. Dandrey, in series Encyclopédies d'aujourd'hui (Paris: La Pochothèque, 1996), 281; see also Philippe Tamizey de Larroque, *Quatre lettres inédites de Jacques Gaffarel*, extract from *Annales des Basses-Alpes, Bulletin de la Société scientifique et littéraire de Digne* (Digne, 1886) 22. On the erotic poem Aloysia Sigea (ca. 1660), see Alain Mothu, "À propos du 'philosophe soldat' Antoine de Villon," La Lettre Clandestine, 13 (2004): 179–91. On Pierre de Boissat, a colorful individual and one of the first members of the Académie Française, see the bibliography given by Secret, "Un témoignage oublié sur l'épisode des placards," 625. A handwritten French translation exists of the section of Chorier's book concerning the friends of Pierre de Boissat (Grenoble, Bibliothèque municipale, MS 2357, XIXth century). I did not have the opportunity to consult it.

53. See, for example, Eucharius Cygnaeus (cited infra), 5: "Verum vos, Illuminatissimi Fratres, . . ."; Naudé, *Instruction à la France*, 40 and passim.

54. *Pauli di Didis A. M. A. Σοφια Παναρετος Qua spiritu & ritu, norma & forma, filo & stylo, aura & auro Gratiosi Ordinis Fratrum Rosatae Crucis ad Beatam & Bonam vitam itur & pervenitur, Dei & hominum gratia conciliatur & impetratur* (s.l.n.d. [1618]). Cf. Gilly, *Cimelia Rhodostaurotica*, 110–11, no. 150.

55. It should be the following work: Theophilus Schweighart [Daniel Mögling], *Speculum Sophicum Rhodo-Stauroticum, Das ist: Weitläuffige Entdeckung deß Collegii unnd axiomatum von der sondern erleuchten Fraternitet Christ-RosenCreuz: allen der wahren Weißheit Begirigen Expectanten zu fernerer Nachrichtung, den unverständigen Zoilis aber zur unaußlöschlicher Schandt und Spott. Durch Theophilum Schweighardt Constantiensem* (s.l., s.n.e., 1618). Cf.

Gilly, *Cimelia Rhodostaurotica*, 131, no. 204; but see also 130–31, no. 202, another work by Schweighart/Mögling: *Sub umbra alarum tuarum Jehovah. Pandora Sextae aetatis, Sive Speculum gratiae* [. . .] (s.l.n.d. [1617]), as well as a pseudo-Schweighart who is no other than his adversary Friedrich Grick (Gilly, *Cimelia Rhodostaurotica*, 129, no. 201). On Daniel Mögling (1596–1635), doctor and astronomer, friend of J. V. Andreae, see Ulrich Neumann, "Mögling" and "*Olim, da die Rosen Creutzerey noch florirt.*"

56. Heinrich Neuhus, *Pia & utilissima admonitio, De Fratribus Rosae-Crucis, Nimirum, An sint? Quales sint? Unde nomen illud sibi asciverint? & quo fine ejusmodi famam sparserint? Conscripta, & Publicae utilitatis causa in lucem emissa Ab Henrico Neuhusio Dantiscano, Medicinae & Philosophiae Magistro, P. in Mörbisch H.* (s.l. [Dantzig]: Christoph Vetter, 1618). The work of Cygnaeus is a reply to the work of Neuhus: Eucharius Cygnaeus, *Conspicilium notitiae, Inserviens oculis aegris, qui lumen veritatis ratione subjecti, objecti, medii & finis ferre recusant. Oppositum Admonitioni futili, Henrici Neuhusii De Fratribus R.C. An sint? Quales sint? Unde nomen illud sibi asciverint? Et quo fine ejusmodi Famam sparserint? Et ex Fama, Confessione, & veritatis fonte filiis doctrinae exhibitum, ab Euchario Cygnaeo Philadelpho & Philalitheo* (s.l.n.d. [1619]). Cf. Gilly, *Cimelia Rhodostaurotica*, 164–65, nos. 296 and 297.

57. One could also read "Among the affiliate members," but that reading would go against what precedes it.

58. Naudé does not say anything different (*Instruction à la France*, 5): "C'est toutesfois ce que vous faittes & pratiquez journellement, sans que vostre trop grande credulité, estant tous les jours accusee & convaincuë de faux par le temps, grand maistre & censeur de la verité, vous ayez le regret en l'ame & la honte sur le front."

59. Chorier's text shows *certus*. I am correcting this by *certius*.

60. Chorier, *De Petri Boessatii vita*, 248–52:

STEPHANUS CHAUMEUS, aliquot ante annos, vita exierat, quam Boëssatius; qui, si ille vixisset, etiam viveret. Nam medicam artem mirabiliter edoctus leto eripiebat, quos vis morbi ad mortem damnabat. Viennae natus medendi artem Monpelii didicerat, quam Parisiis, nonnullos per annos, exercuerat: In ea Imperii Gallici Metropoli, stabiles sibi sedes, ponere, cum animo suo, constituerat. Fabula obstitit. Varia de Roseae Crucis illuminatissimis Fratribus (sic loquebantur) per id tempus, jactabantur: Homines esse qui quaecunque scire vellent, seu eruditas vulgaresque linguas, seu artes & scientias, nullo labore, quam vellent, paucos post dies, quam concupiissent, notitiam haberent; ex vilioribus meliora, mirum in modum, metalla conflarent, atque adeo in aurum mutarent: Demum, nihil negante aeterna numinis mente, quae vel sibi, vel aliis cuperent illico consequerentur. Merae nugae. At commentitiae fabulae scripti a nonnullis, de ea re, tractatus, quorum in numero Paulus Didisus, Theophilus Schweighartius, Euchariusque Cygnaeus (ficta nomina) ac Henricus Neuhusius Dantiscanus fuerunt, magnum authoritatis pondus afferebant. De suo Chaumeus, per ludum & jocum, juvenilemque lasciviam, multa adjecit; hocque maxime, id hominum genus, cum vellent, videre, nec videri; quorum, si e re sua esset, ingressi essent domos, omnes pervagari libere per angulos; coenantibus, colloquentibus cum domesticis, cubantibus adesse nec videri. Ex sodalibus, tres quatuorve convenire Chaumeum soliti erant. Latina lingua, qui commenta haec nugasque frivolas enarrarent, libellos confecit. Illis vero ad compita, aedium sacrarum fores, & loca quae plurimum frequentarentur, proponere venit in mentem. Quibus bene beateque vivendi libido esset ad id sodalitium invitabantur; felices his si sacris initiarentur futuri: Locus, tempus, dies, hora indicebatur. Credulae admodum in Galliis sunt plebes, at omnium plurimum, Parisiensis. In Terris sane nulla est, quae verba dari sibi facilius patiatur, cognitaque fraude, aequiori se esse illusam animo ferat. Proscriptis his tabulis admiratio primum ac dein ingens perturbatio Parisienses cepit. Metus, luctus, indignatio omnes plerasque domos invasit. Suis se in tectis securos non putabant. Igitur nec libere cum domesticis loqui, nec rebus suis operam dare, pro arbitrio, audebant. Consternationem civium fieri non potuit, quin Magistratus animadverteret. Igitur in auctores inquiri jubet, etiam, investigati si deprehendantur, saevire cert[i]us. Principio Chaumeus, quae excogitasset, tanto opinionum progressu omnium fere animos perculisse, mirabatur. Ridebat. Verum, ab iis, quibus potestas data esset, cum diligentissimas iniri quaestiones de re videret, famae salutique fuga consuluit; periculo se prudens subduxit; Viennam rediit. Totum se in Musarum sinum abdidit; & paucos intra annos, cum nullus studiorum modus finisque esset, eos ex vigiliis, non intermissisque laboribus, eruditionis fructus retulit, qui omni abunde vitae occupatissimo, ad laudem, compendiumque & utilitatem, sufficerent. Quod nesciret, nullus medicorum noverat. . . . Latinam linguam. . . . optime callebat. Vernacula & propria lingua, cum latino sermone uteretur, loqui dixisses. . . . Quinquagesimum aetatis annum praetergressus, magno omnium, meoque maximo dolore, mortem obiit. Honorarium sibi a me nullum, postquam refectis viribus, e morbis, sua ope, convaluissem, meam praeter benevolentiam, quam optabat, ut loquebatur, volebat. Aliam a litteratis viris mercedem, non inhonestum modo & turpe litterato medico, sed & prorsus flagitiosum quid arbitrabatur.

61. The same type of reasoning, founded exclusively on the facts furnished by Chorier, caused Adolphe Rochas to say: "[Chaume] died toward 1660, at the age of fifty or sixty years." (*Biographie du Dauphiné, contenant l'histoire des hommes nés dans cette province qui se sont fait remarquer dans les lettres, les sciences, les arts, etc., avec le catalogue de leurs ouvrages et la description de leurs portraits*, 2 vols. [Paris, 1856–60; reprinted, Geneva: Slatkine, 1971], 1:153).

62. Montpellier, Bibliothèque interuniversitaire de médecine, Archives de la Faculté de médecine, S 20, fol. 192v: "Ego Stephanus Chaume Viennensis examinatus fui a reverendis dominis procuratoribus et ab amplissimo domino Ranchino cancellario relatus in numerum studiosorum soluto prius universitatis jure et praestito juramento de observandis statutis. Actum in Collegio papae die octava decembris 1623. S. Chaume."

63. *Mémoires de Nicolas Chorier, de Vienne, sur sa vie et ses affaires*, trans. F. Crozet (extract from the *Bulletin de l'Académie Delphinale*, sessions of April 12 and December 20, 1867) (Grenoble: Prudhomme, 1868), 36–37. We learn in this book that Chorier met Molière in Vienne and in Lyon (ibid., 150).

64. *Effroyables Pactions*, 10.

65. I am indebted to William Newman, whom I thank very warmly here, for all the information concerning this manuscript and its origins.

66. It was Henri Justel's circle. On Thoynard, see Antoine Adam, *Histoire de la littérature française au XVIIe siècle*, 3 vols. (1948, new ed. 1962), *Bibliothèque de l'Évolution de l'Humanité*, nos. 24–26 (Paris: Albin Michel, 1997), 3:402, n. 2; Alexandre Cioranescu, *Bibliographie de la littérature française du dix-septième siècle*, 2d ed. (Paris: Éditions du Centre National de la Recherche Scientifique, 1969), 3:1914; Maurice Cranston, *John Locke: A Biography* (London: Longmans, 1957; new ed. Oxford: Oxford University Press, 1985), 174 and passim; and especially E. S. De Beer, ed., *The Correspondence of John Locke*, 8 vols. (Oxford: Clarendon Press, 1976–89), 1:579–82.

67. On Jacques Le Pailleur, see Gédéon Tallemant Des Réaux, *Historiettes*, ed. Antoine Adam (Paris: Bibliothèque de la Pléiade, 1960–61), passim, esp. 2:99–101 and 1000; Adam, *Histoire de la littérature française*, 1:384, n. 3; Robert Lenoble, *Mersenne ou la naissance du mécanisme* (1943; re-ed., Paris: Vrin, 1971, 591); René Pintard, *Le Libertinage érudit*, 91 and 349–50; and especially Jean Mesnard, "Pascal à l'Académie Le Pailleur," in *L'Oeuvre scientifique de Pascal*, Centre international de Synthèse, section d'histoire des sciences (Paris: Presses Universitaire de France, 1964), 7–16, essential for this highly interesting but unfortunately poorly documented individual.

68. Cf. Louis Sébastien Mercier, *Le Tableau de Paris* (1st ed. 1781), 1, chap. 85: "Le Faubourg Saint-Marcel", ed. J.-C. Bonnet et al., 2 vols. (Paris: Mercure de France, 1994), 1:217 (and 1:1554 *sq.*, n. 4): "C'est dans ces habitations éloignées du mouvement central de la ville, que se cachent les hommes ruinés, les misanthropes, les alchimistes, les maniaques, les rentiers bornés, et aussi quelques sages studieux, qui cherchent réellement la solitude, et qui veulent vivre absolument ignorés et séparés des quartiers bruyants des spectacles." This is the neighborhood where Jacques Gohory held his "Lycium philosophal" in 1572.

69. Oxford, Bodleian Library, MS Locke c. 42, 70 of the second pagination. I reproduce here the transcription done by William Newman, to whom I again extend all my gratitude for having so generously passed this text on to me.

70. Clarius Bonarscius [Charles Scribani], *Amphitheatrum Honoris in quo Calvinistarum in Societatem Jesu criminationes jugulatae* (Palaeopoli Aduaticorum [Anvers]: Alexander Verheyden [ex officina Plantiniana], 1605). Cf. Carlos Sommervogel, *Bibliothèque de la Compagnie de Jésus*, new ed., vol. 7 (Brussels-Paris: O. Schepens–A. Picard, 1896), cols. 982–83.

71. Naudé, *Instruction à la France*, 97–105. See the chapter devoted to Khunrath in my dissertation ("Paracelsisme et alchimic en France à la fin de la Renaissance (1567–1625)," Université de Paris IV, 1998), to be published by Droz Editions (Geneva) in 2002.

72. See Gabriel Naudé, *Advis pour dresser une bibliotheque Présenté à Monseigneur le President de Mesme* (1627) (Paris: Rolet Le Duc, 1644; reprinted, Paris: Klincksieck, 1990 and 1994), chap. 4, 51, classifying pell-mell "the Cabala, artificial Memory, Art of Lull, Philosophers' Stone, Divinations, & other similar subjects"

among those that "teach nothing but vain and useless things" but that nonetheless must be collected in every scholarly library.

73. Rita Sturlese, "Lazar Zetzner, 'Bibliopola Argentinensis': Alchimie und Lullismus in Straßburg an den Anfängen der Moderne," *Sudhoffs Archiv,* 75 (1991): 140–62; François Secret, *Les Kabbalistes chrétiens de la Renaissance* (1965), ed. extend. (Milan–Neuilly-sur-Seine: Archè–Arma Artis, 1985), 288–91.

74. Michela Pereira, *The Alchemical Corpus Attributed to Raymond Lull,* Warburg Institute Surveys and Texts, no. 18 (London: The Warburg Institute, 1989), 30; D. Kahn, "Le fonds Caprara de manuscrits alchimiques de la Bibliothèque Universitaire de Bologne," *Scriptorium,* 48 (1994): 62–110; J. N. Hillgarth, *Ramon Lull and Lullism in Fourteenth-Century France,* Oxford-Warburg Studies (Oxford: Clarendon Press, 1971), chap. 7 ("Epilogue"), 294–311.

75. Secret, *Les Kabbalistes chrétiens de la Renaissance,* 288; [idem], *Kabbale et philosophie hermétique. Exposition à l'occasion du Festival International de l'Ésotérisme, Carcassonne, 17–19 novembre 1989* (Amsterdam: Bibliotheca Philosophica Hermetica, s.d. [1989]), 64–67.

76. Secret, *Les Kabbalistes chrétiens de la Renaissance,* 290–91.

77. Robert Le Foul, sieur de Vassy, *Le Grand et Dernier Art de M. Raymond Lulle, M^e. es Arts liberaux, et tres Illustre Professeur dans la sacree Theologie* (Paris: Louis Boulanger, 1634). On this individual, see Hillgarth, *Ramon Lull and Lullism in Fourteenth-Century France,* 294–311, as well as the chapter of my dissertation, "Paracelsisme et alchimie en France," devoted to Khunrath.

78. René Descartes, *Oeuvres,* ed. C. Adam and P. Tannery, 13 vols. (1897–1914; rev. ed., ed. B. Rochot & P. Costabel, 9 vols., 1964–74) (reprinted, Paris: Vrin, 1996) 8/2: 27–28, note a.

79. Tallemant Des Réaux, *Historiettes,* 2:99. This Saint-Brice also numbered in his circle the poet Saint-Amant; cf. Saint-Amant, *Oeuvres,* ed. Jacques Bailbé, in series Société des Textes Français Modernes (Paris: Marcel Didier, 1967–71), 1:252 and n. 57.

80. This investigation could be probably documented by means of a research into the judicial archives of the time. But confirmation of it will be found below.

81. The text gives here not a period, but rather a question mark. I made the correction.

82. *Effroyables Pactions,* 16–17. Let us remember that this pamphlet could only have come into being in November or December 1623: Peiresc's letter, dated August 3, is clearly earlier.

83. I owe this information to Antonio Clericuzio, who is preparing a study on Étienne de Clave for his book *Elements, Particles, and Atoms: A Study of Seventeenth-Century Chemistry and Corpuscular Philosophy* (Dordrecht: Kluwer, 2001, forthcoming).

84. More details on this ensemble will be found in my doctoral dissertation, "Paracelsisme et alchimie en France." Let me only point out here that one of these manuscripts, MS fr. 2528 in the Bibliothèque Nationale de France (BNF) in Paris,

entitled "Effetz de nature, Ier volume" (the second volume is MS fr. 2529), contains in fol. 141r–143r, under the title of "Rose Croix," some brief and hasty reading notes on the Rosicrucian movement in general.

85. BNF, MS fr. 3653 (anc. 10344), fol. 46r. On the origins of this manuscript, cf. Henri Omont, *Anciens inventaires et catalogues de la Bibliothèque nationale*, vol. 4: *La Bibliothèque royale à Paris au XVIIe siècle* (Paris: Ernest Leroux, 1911), 166, no. 10344, and 238.

86. BNF, MS fr. 3653, fol. 46r.

87. Here I can only refer to my dissertation ("Paracelsisme et alchimie en France"), in which (based upon a suggestion from Antonio Clericuzio, for which I am very grateful) I studied in detail the circle of Turquet de Mayerne through his many manuscripts.

88. Irene Scouloudi, "Sir Theodore Turquet de Mayerne, Royal Physician and Writer, 1573–1655," *Proceedings of the Huguenot Society of London*, 16 (1940): 301–37, at 304, n. 4. I warmly thank Julian Paulus for this precious reference.

89. London, British Library, Add. MS 20921, fol. 58r.

90. As is suggested by the few variants it presents from that of Naudé: *vers qui se trouve* instead of *vers qui se tourne*, which is much better; *d'erreurs & de mort les hommes nos semblables* instead of the inverse in Naudé's case (*les hommes nos semblables d'erreur [&] de mort*).

91. I have not had enough time to peruse the 440 columns published by Mersenne following his *Quaestiones in Genesim* under the heading *Observationes, et emendationes ad Francisci Georgii Veneti Problemata. In hoc opere Cabala evertitur; editio vulgata, et inquisitores Sanctae fidei catholicae ab Haereticorum, atque Politicorum calumniis accurate vindicantur* (Paris: Sébastien Cramoisy, 1623). It would probably be especially necessary to consult the German Rosicrucian literature. The ten questions asked, however, do not betray a very advanced knowledge of the cabala and could have arisen from themselves—except the last one—on the occasion of a simple meeting with the doctrines of the *Zohar*. On the vast diffusion of the cabala in France at the beginning of the seventeenth century, see Secret, *Les Kabbalistes chrétiens de la Renaissance*, esp. 333–52. See also BNF, MS fr. 19948 (anc. St. Germain, 1909): *Double du livre de la Caballe des Hebrieux. Livre entien et trouvé depuis peu de temps, contenant plusieurs grands et admirables secretz de la theologie et philosophie* (*inc.*: "Quand j'eu bien profondement pansé en moy mesme la briesveté, temperie et inconstance de tous hommes . . ."). This manuscript was copied between September 23 and December 14, 1623, by an individual whose name has been completely scratched out and replaced by the name "Hermes" (fol. 111v). The reverse side of the last page bears the name and the signature "Rossignol." This could be Nicolas Rossignol, procurer at the Grand Châtelet in Paris, the copyist of several manuscripts on alchemy, on which see my article "Notre-Dame de Paris et l'alchimie: un traité inédit du début du XVIIe siècle, le *Discours des visions sur l'oeuvre*," *Chrysopoeia*, 5 (1992–96): 443–52.

92. M. Rooses and C. Ruelens, eds., *Correspondance de Rubens et documents épistolaires concernant sa vie et ses oeuvres*, 3:239.

322   Didier Kahn

93. An investigation into the judicial archives of the time would perhaps allow it.

94. Carpentras, Bibliothèque Inguimbertine, MS 1777, fols. 475r–477r (see the description in M. Duhamel & M. Liabastres, *Catalogue général des manuscrits des bibliothèques publiques de France. Départements,* vol. 35 [Paris: Plon, 1899], 225–34, esp. 232). A microfilm of this manuscript, which can be consulted at the Institut de Recherche et d'Histoire des Textes in Paris, fortunately helped me to avoid submitting to the exorbitant prices of reproduction imposed by the current management of the Bibliothèque Inguimbertine, which seems to do its best to increase the difficulties inherent in this type of research.

95. Carpentras, Bibliothèque Inguimbertine, MS 1777, fol. 475r: "Coppie d'une affiche mise par les carrefours de Paris en juillet 1623. Nous deputez de nostre College principal des freres Roses faisons sejour visible & invisible en cette ville par la grace du Tres haut, vers qui se tourne le coeur des justes. Nous enseignons sans livres, marques ny signes & parlons les langues des pays, ou nous voulons estre, pour tirer les hommes nos semblables d'erreur & de mort."

96. Tannery, de Waard, and Pintard, *Correspondance du P. Marin Mersenne,* 1:154. See also Pintard, *Le Libertinage érudit,* 48.

97. Gershom G. Scholem, *Major Trends in Jewish Mysticism* (1941), French trans. *Les Grands courants de la mystique juive,* in series Bibliothèque scientifique (Paris: Payot, 1968), chap. 6, esp. 229.

98. London, British Library, Add. MS 20921, fol. 58r. I give here the variations between this manuscript and Bibliothèque Inguimbertine, MS 1777, fol. 477r. (In the following notes, the text on the left side of the colon gives the reading from the manuscript in the British Library, and the text on the right side presents the corresponding variant from the manuscript in the Bibliothèque Inguimbertine.)

99. "Articles . . . Rosée": "Certains articles des propositions faites par les freres Roses."

100. "coadjuteurs": "conducteurs"; the ten propositions are not numbered in the Bibliothèque Inguimbertine manuscript.

101. "l'on n'eu": "on eust eu"; "n'y a": "n'y a rien."

102. "on pourroit": "ne pourront"; "eut" is omitted in the Bibliothèque Inguimbertine manuscript.

103. "l'on": "je."

104. "maintien": "maintenant."

105. "sens": "secret" (in both occurrences in item 10).

106. Cf. Exodus 38:27: "The 100 talents of silver were used to cast the bases for the sanctuary and for the curtain—100 bases from the 100 talents, one talent for each base." Ecclesiastes 1:7: "All streams flow into the sea, yet the sea is never full. To the place the streams come from, there they return again."

107. London, British Library, Add. MS 20921, fol. 58r.

108. Ibid.: "LECTISSIMAE ET ABSTRUSISSIMAE SCIENTIAE ACUTISSIMO professori, Societatis Roseae Crucis defensori acerrimo, fratri omni officiorum genere colendo tradantur.

Quam sis addictus societati nostrae, quot & quantis majestatem ipsius fulcire lucubrationibus conatus sis, primatibus nostris notum satis, a quibus benevolentia insignis eximia per serias virtutum tuarum commendationes, testimonia, ad nos corporis illius mystici membra saepissime pervenerunt. En tibi a nostris Lutetiae Parisiorum per celeberrimae urbis compita theses affixas, & plus quam Platonico genio turgida problemata, quorum medullitus reconditos sensus soli Oedypo penetrare, soli roseae societatis alumno enucleare datum. Viles reptilium hominum mentes despicit invisibile nostrum collegium & Boeotica ingenia naso adunco suspendit. Sacrarii augusti solis lynceis valvae patent. In quorum numerum ascitum te ut agnoscimus, sic gratulamur inaugurationi tuae, tibique tanquam fratri crucigero & murice nostro supra Salamandrae physicae colorem roseum fulgenti, enodanda arcana mittimus, de quibus ut tuam aperire teneris sententiam, sic te ad nos tua remittere placita aequum & decorum est. Salve. Et Zephyrotarum septupla enneas cum concentrica in homogeneitate sua spirituum principum Cabalistica decade conatuum tuorum coadjutrix, per energeticos influentiarum suarum radios sese tuis coeptis foeliciter immisceat. Vale & vive. Datum apud nos, Eid. Junii 1623.

Tui P. ♂ N. ♃ S. ♄.
Confratros tui
Ex mando [sic] W. ☉."

109. See, for example, the same manuscript (Add. MS 20921), fols. 17r–18r, containing a letter from 1632 (it can be found transcribed as an appendix to my "Paracelsisme et alchimie en France.")

110. Among the manuscripts of Mayerne there is a text in Dutch entitled *De Cabala van de philosophe Isaac den Hollander int cort*. (London, British Library, MS Sloane 2097, fols. 56v–78v), but we know that only treatises of alchemy circulated with the name of Isaac Hollandus, and it is doubtful that this text would be an exception to the rule. Another manuscript (MS Sloane 2100, fols. 117r–141v) contains a *Tabula revelationis majestatis divinae comprehensae capite primo Geneseos. In qua indicatur; quomodo Deus in principio semetipsum omnibus creaturis suis patefecerit re et verbis, et qua ratione omnia opera sua, eorumque naturas, proprietates, virtutes, atque operationes in brevem scripturam compendiose redegerit, atque haec omnia primo homini quem ipse ad imaginem suam condidit, tradiderit, ita ut ad nos etiam hucusque illa dimanarint*. This text, preserved in a manuscript supposed to have belonged to Mayerne himself, is composed only of questions, as if it were the table of contents for a considerably larger book (like Mersenne's *Quaestiones in Genesim*) that was probably never written. Each series of questions is organized successively around each word of the first chapter of Genesis. In it are found many metaphysical questions and also numerous questions of an alchemical nature: nothing, in any case, that relates to the cabala.

111. This pamphlet does not cite the text of the poster; but when it goes on to comment on the one he is recommunicating, the author writes (17): "everyone is astonished by this invisibility and by the perfect ability in speaking all kinds of languages." Now the theme of the knowledge of languages only appears in the poster preserved by Naudé.

112. Heinrich Neuhus, *Advertissement Pieux & tres utile, Des Freres de la Rosee-Croix: A sçavoir, S'il y en a? Quels ils sont? D'où ils ont prins ce nom? Et à quelle

fin ils ont espandu leur renommée? Escrit, & mis en lumiere pour le bien public (Paris: s.n.e. ["Et se vendent au Palais"], 1623; reprint, Paris: Gutenberg Reprint, 1979, following the *Instruction* by Naudé), "Adresse au lecteur," 3. See below my remarks on the dating of this book.

113. *Examen sur l'Inconnue et Nouvelle Caballe des Freres de la Croix Rosee, habituez depuis peu de temps en la Ville de Paris. Ensemble l'Histoire des Moeurs, Coustumes, Prodiges, & particularitez d'iceux* (Paris: for Pierre de La Fosse, 1623), 14.

114. *Mercure François*, vol. 9 (1622–24), 371. The only variant in comparison with Naudé's version is the reading "où nous habitons" instead of "où voulons estre." The *Mercure* then (371–72) commented upon the reading "d'erreur de mort," which is that of Naudé and which the *Mercure* kept: "Aucuns en les transcrivant [*i.e., the posters*] y meirent *d'erreur, & de mort*: Tellement que les Curieux de nouvelles eurent de quoy discourir de ces nouveaux *Illuminez & Immortels*." In fact, all the other printed and handwritten sources give the reading "d'erreur & de mort."

115. Pierre Guillebaud (in religion, Pierre de Saint-Romuald), *Tresor chronologique et historique contenant ce qui s'est passé de plus remarquable & curieux dans l'Estat, tant Civil qu'Ecclesiastique, depuis l'an de Jesus-Christ 1200 jusqu'à l'an 1647*, 3 vols. (Paris: Antoine de Sommaville, 1647), 3:831. On the incident of the placards of 1623 (which Guillebaud placed in 1622), see ibid., 858–59. An error by Guillebaud curiously transformed the name of Michael Maier into "Majerne" (859). All of his information is drawn from Naudé. Guillebaud (1585–1667) was a Feuillant monk who strove to publish historical works compiling the most diverse of sources; see the article "Guillebaud" by M. Standaert in the *Dictionnaire d'histoire et de géographie ecclésiastiques*, vol. 22 (Paris: Letouzey & Ané, 1988), cols. 1053–54.

116. [Baillet], *La Vie de Monsieur Des-Cartes*, 1691, 1:107–8.

117. Paris, BNF, MS fr. 15777 (17th c.; papers concerning various religious orders), fols. 83r–84v: *Extraictz tirez des livres qui traictent des freres de la Rose Croix*, here fol. 84v. Cf. Léopold Delisle, *Le Cabinet des Manuscrits de la Bibliothèque Nationale*, 3 vols. (Paris: Imprimerie Nationale, 1868–81), 2:100–1.

118. Neuhus, *Advertissement Pieux & tres utile*, "Adresse au lecteur," 5–6. See also the end of Jacques Le Pailleur's story, as well as the account of Théophraste Renaudot in 1639, reproduced below in the last part of this chapter.

119. Neuhus, *Advertissement Pieux & tres utile*, 5: "a lawyer;" cf. *Effroyables Pactions*, 17–20: "a Lawyer of the Parlement of Paris."

120. Neuhus, *Advertissement Pieux & tres utile*, "Adresse au lecteur," 7–8.

121. Ibid., 8; see also 6: "Et de fait m'en allant aux champs en la saison des vendanges, je l'emportay [*i.e., the book by Neuhus*] avec moy [. . .]"

122. Ibid., 56.

123. The author of the *Effroyables Pactions* mentions, 21–22, two adventures that happened "in the month of last October."

124. Cf. Alain Mothu, "Une petite satire de l'alchimie sous Louis XIII: *Le vray secret et invention de la Pierre Philosophale, trouvee dans l'escuelle de Bois,*" *Chrysopoeia*, 5 (1992–96): 669–73.

125. We read in fol. [i4]r that Naudé, on December 1, 1623, accorded the benefit of his privilege to the bookseller François Julliot.

126. H. Neuhus, *Advertissement Pieux & tres utile*, 1624; shelfmark BNF: Rés. H. 2134 (2); another copy: Bibliothèque de l'Arsenal, shelfmark: 8° S. 12952 (1). *Effroyables Pactions*, 1624; shelfmark Bibliothèque de l'Arsenal: 8° S. 12952 (2). The *Effroyables Pactions* are also found, without their title, in [Claude Malingre], *Troisiesme Tome de l'Histoire de nostre temps, ou Suitte de l'Histoire des guerres contre les Rebelles de France, ez années 1623. & 1624.* (Paris: Jean Petit-Pas, 1624), 99–124 (I will return to this below). The 1624 edition of Naudé's work, which is impossible to find using the habitual methods as it is anonymous and without its title, is attested to by the *Catalogus omnium operum Gabrielis Naudaei*, of which Lorenzo Bianchi has edited some extracts (*Rinascimento e libertinismo. Studi su Gabriel Naudé*, 276), thanks to which I succeeded in locating this edition in the *Traicté des Atheistes* pointed out by Wallace Kirsop (see below, note 131).

127. Shelfmark BNF: Hp. 681. Another copy: Paris, Bibliothèque Mazarine, shelfmark: 36683, no. 5. This is the edition I am citing throughout the chapter.

128. Bibliothèque Mazarine, shelfmark: 37231, no. 13.

129. Bibliothèque Mazarine, shelfmark: Rés. 37206, no. 5. Another copy: BNF, Hp. 1134.

130. There are two copies of the 1624 issue at the BNF (shelfmarks Y^2 75469 and Hp. 1135).

131. Wallace Kirsop, "Clovis Hesteau, sieur de Nuysement, et la littérature alchimique en France à la fin du XVIe et au début du XVIIe siècle," 2 vols. (Ph.D. diss., Université de Paris, 1960), 2: 153, n. 109.

132. [Claude Malingre], *Traicté des Atheistes, Deistes, Illuminez d'Espagne, et nouveaux pretendus Invisibles, dits de la Confrairie de la Croix-Rosaire. Elevez depuis quelques annees dans le Christianisme* (s.l., s.n.e. [Paris: Pierre Chevalier], 1624), in [Claude Malingre], *Histoire generale du progrez et decadence de l'Heresie moderne. Tome second. A la suite du premier de M. Florimond de Raemond. Conseiller du Roy en sa Cour de Parlement de Bordeaux . . . Plus un traicté des Atheistes, Deistes, Illuminez d'Espagne, et nouveaux pretendus de la Croix-Rosaire* (Paris: Pierre Chevalier, 1624). On Malingre, novelist and historian (c. 1580–c. 1653), see the entry in the *Dictionnaire des Lettres françaises. Le XVIIe siècle*, 805. Only the *Catalogus omnium operum Gabrielis Naudaei* edited by Lorenzo Bianchi allowed me, thanks to a reference to Malingre, to locate this brochure.

133. "Des Freres pretendus Invisibles & Illuminez de la Croix-Rosaire, de leur College, assemblees & doctrine," in [Malingre], *Traicté des Atheistes, Deistes, Illuminez d'Espagne, et nouveaux pretendus Invisibles*, 21–55.

134. Yates, *The Rosicrucian Enlightenment*, ed. 1986, 103 and 104–5 (French trans., ed. 1985, 135 and 136–37).

135. Ibid., 106 (French trans., 138).

136. Ibid., for example, 107–8 (French trans., 139–40), misunderstanding of 15–16 of Naudé, to whose irony Yates has remained hopelessly deaf.

137. Ibid., 107–11 (French trans., 139–42).

138. In his *Instruction à la France,* Naudé shows that he has at least read treatises by F. G. Menapius [Friedrich Grick], by Michael Maier (*Verum Inventum, Silentium post clamores, Themis aurea, Symbola aureae mensae, Arcana arcanissima* and *De volucri arborea*) and by Heinrich Neuhus, Eucharius Cygnaeus, Ludwig Combach, Michael Potier, Theophilus Schweighart [Daniel Mögling], Rudolph Goclenius, Georg Molther, and Andreas Libavius. See the epigram by Menapius reproduced after the privilege of the *Instruction,* fol. [i4]r–v; for the other authors, cf. ibid., 54–59, 64, 90, and 96. See also the "Catalogue des Livres qui sont en l'estude de G. Naudé à Paris" (before 1642), partially edited by Bianchi, *Rinascimento e libertinismo. Studi su Gabriel Naudé,* 253–270, esp. 264–65. Naudé also possessed information passed on to Father Baranzano by Tobias Adami: I will come back to this. On the works by Friedrich Grick, see Gilly, *Cimelia Rhodostaurotica,* 78–79; 129, no. 201; 131, no. 203; 132–37, nos. 205–18; and 140–41, no. 229.

139. Naudé, *Instruction à la France,* 62.

140. Johann Valentin Andreae, *Institutio magica pro curiosis* (c. 1609–11), published in his *Menippus Sive Dialogorum Satyricorum centuria inanitatum nostratium Speculum* (s.l., s.n.e. [Strasbourg: heirs of Lazarus Zetzner], 1617), 200 and following (quoted by Gilly, *Cimelia Rhodostaurotica,* 51).

141. In what follows I will go more deeply into the connections already established by Halleux, "Helmontiana," 58–61.

142. Jacques Gaultier, *Table chronographique de l'estat du Christianisme, Depuis la naissance de Jesus-Christ, jusques à l'année M. DCXX. Contenant en douze colomnes les Papes, & Antipapes: les Conciles & Patriarches des quatre Eglises Patriarchales: les Escrivains sacrez, & autres Saincts & Illustres personnages: les Empereurs & Roys, tant de nostre France, qu'Estrangers: les Autheurs Profanes, les Heretiques, & les Evenemens remarquables de chasque Siecle, ou Centurie. Ensemble le rapport des vieilles Heresies aux modernes de la Pretenduë Reformation: Et douze des principales Veritez Catholiques attestées contre le Calvinisme, par l'Escriture Saincte, & de Siecle en Siecle par les Saincts Peres & Docteurs de ce temps-là* (Lyon: Pierre Rigaud, 1621). On Father Gaultier, see A. Boland, "Gaulthier," in *Dictionnaire d'histoire et de géographie ecclésiastiques,* vol. 20 (Paris: Letouzey and Ané, 1984), cols. 59–60.

143. Gaultier, *Table chronographique,* 875.

144. On Garasse, see below, note 161; Naudé, *Instruction à la France,* 59–60; *Mercure François,* vol. 9 (1622–24), 374 ff. Gaultier's article was again summarized by Henry de Sponde in his continuation of the *Annales* of Baronius: *Annalium Emin^{mi} Cardinalis Caes. Baronii Continuatio, ab anno M.C. XCVII. quo is desiit, ad finem M. DC. XL.,* vol. 3 (Paris: Denis de La Noue, 1641), 871.

145. [Baillet], *La Vie de Monsieur Des-Cartes,* 1691, 1:107.

146. Baillet must have embellished upon the text of the *Mercure François*. The latter, after discussing the Parisian placards, added that other people "pour en faire encor plus accroire," had written the *Effroyables Pactions* and the *Examen*; the *Mercure* then gave a summary of these two pamphlets, then concluded thus: "C'est ce qui s'est escrit boufonnesquement de ces Invisibles, ou Freres de la Rose-Croix, pour donner de l'esbat aux esprits curieux" (*Mercure François*, vol. 9 [1622–24], 372–74). See below as well the mention of Pont-Neuf in the *Mercure François*.

147. Alvaro Huerga, *Historia de los Alumbrados (1570–1630)*, vol. 4: *Los Alumbrados de Sevilla (1605–1630)* (Madrid: Fundacion Universitaria Española, 1988), 238 ff. Cf. the pamphlet by Antonio Farfán de las Godos, *Discurso . . . en defensa de la religión católica contra la secta de los Alumbrados* (Seville, 1623).

148. *Edict d'Espagne contre la detestable Secte des Illuminez. Eslevez és Archevesché de Seville & Evesché de Cadiz. Traduict sur la coppie Espagnole imprimée en Espagne* (s.l., s.n.e., 1623) (Bibliothèque Mazarine, shelfmark: 36683, no. 4). This is the text that the *Mercure François* took up again the following year.

149. Rooses and Ruelens, *Correspondance de Rubens et documents épistolaires concernant sa vie et ses oeuvres*, 3:224; quoted and commented upon by Halleux, "Helmontiana," 60–61.

150. Rooses & Ruelens, *Correspondance de Rubens et documents épistolaires concernant sa vie et ses oeuvres*, 3:230 (August 10, 1623).

151. See below the quotes concerning Théophile.

152. Neuhus, *Advertissement Pieux & tres utile*, "Adresse au lecteur," 4.

153. I will come back to this point below.

154. [Malingre], *Troisiesme Tome de l'Histoire de nostre temps*, 82–98 (see also 779–81) and 98–99 (quote). The privilege of this book bears the date March 7, 1624.

155. [Malingre], *Traicté des Atheistes, Deistes, Illuminez d'Espagne, et nouveaux pretendus Invisibles*, 13–20. The privilege for Malingre's continuation of Florimond de Raemond's *Histoire generale du progrez et decadence de l'Heresie moderne*, in which this brochure was inserted, dates from March 14, 1624.

156. *Mercure François*, vol. 9 (1622–24), 354–70 (Alumbrados) and 371–87 (Rosicrucians), here 371.

157. Jacques Gaultier, *Table chronographique . . . jusques à l'année M. DC. XXV.* (Lyon: Pierre Rigaud et associés, 1626), 877–78 and 880 ff.

158. Bibliothèque Mazarine, shelfmark 36683, nos. 3, 4, 5 and 6.

159. Scipion Dupleix, *Continuation de l'histoire du regne de Louys le Juste, treiziesme du nom* (Paris: Claude Sonnius & Denis Béchet, 1648), 406 (heading "Estat de l'Eglise", XXXIIX–XL). Dupleix has taken up the text of articles 4, 9, 10, 17, and 18 of the edict, which consists of seventy-six articles. On the Enlightened Ones of Picardy or Guérinets, see Pillorget, *Nouvelle histoire de Paris: Paris sous les premiers Bourbons*, 608–10.

160. Pierre Scarron (1584–1668).

161. François Garasse, *La Doctrine curieuse des beaux Esprits de ce temps, ou pretendus tels. Contenant plusieurs maximes pernicieuses à la Religion, à l'Estat, & aux bonnes Moeurs. Combattue et renversee par le P. François Garassus, de la Compagnie de Jesus* (Paris: Sébastien Chappelet, 1623; reissue, Paris: Sébastien Chappelet, 1624). On Father Garasse (1585–1631) and his *La Doctrine curieuse*, see François Berriot, *Athéismes et athéistes au XVIe siècle en France*, 2 vols. (Paris: Éditions du Cerf, s.d. [1977]), 2:753–74; Marc Fumaroli, *L'Age de l'éloquence: Rhétorique et* res literaria *de la Renaissance au seuil de l'époque classique*, Bibliothèque de l'Évolution de l'Humanité, no. 4 (Genève: Droz, 1980; reprint, Paris: Albin Michel, 1994), esp. 326–34; Louise Godard de Donville, *Le Libertin des origines à 1665: un produit des apologètes* (Paris-Seattle-Tübingen: Papers on French Seventeenth Century Literature, 1989) (*Biblio 17*, no. 51), and the review of Godard de Donville's book by Jean Jehasse in *XVIIe siècle*, no. 172 (1991): 300–2.

162. Louise Godard de Donville, "Théophile, les 'Beaux Esprits' et les Rose-Croix: un insidieux 'parallèle' du père Garasse," in Wolfgang Leiner and Pierre Ronzeaud, eds., *Correspondances. Mélanges offerts à Roger Duchêne*, Etudes littéraires françaises, no. 51 (Tübingen/Aix-en-Provence: Gunter Narr Verlag/Publications de l'Université de Provence, 1992), 143–54. I warmly thank Michel-Pierre Lerner for having pointed this article out to me.

163. François Garasse, *Apologie du Pere François Garassus de la Compagnie de Jesus, pour son Livre contre les Atheistes & Libertins de nostre siecle. Et Response aux censures et calomnies de l'Autheur Anonyme* (Paris: Sébastien Chappelet, 1624) (privilege dating January 10, 1624), 5–6.

164. Frédéric Lachèvre, ed., *Le Libertinage devant le Parlement de Paris*, Part 1: *Le Procès de Théophile de Viau (11 juillet 1623–1er septembre 1625). Publication intégrale des pièces inédites des Archives nationales* (Paris: Honoré Champion, 1909), vol. I, 453 and 500. See also Antoine Adam, *Théophile de Viau et la libre pensée française en 1620* (Paris: E. Droz, 1935), 389–90. I do not understand how Godard de Donville, "Théophile, les 'Beaux Esprits' et les Rose-Croix," 144, can deny the existence of this book and its supposed discovery in Théophile's belongings, which have been clearly established by the terms of the interrogation edited by Lachèvre.

165. Garasse, *La Doctrine curieuse*, 83–91, calls the Brotherhood of the Rose-Cross "La Confrerie de la Croix de Roses." He is about the only one who translated in this manner the words *Fraternitas Rosae* [or *Roseae*] *Crucis*: His contemporaries used the terms "Roze-Croix" (Naudé, the *Effroyables Pactions* and *Mercure François*), "Rosee-Croix" (the translator of Neuhus, and one of the editions of the *Examen sur l'Inconnue et Nouvelle Caballe*), "Croix Rosée" (the manuscript of Turquet de Mayerne, the *Examen sur l'Inconnue et Nouvelle Caballe*), "croix Rosaires" (Gaspard: see below, note 187), "Rosecrucians" or "Rosecruceans" (Boucher: cf. infra, note 208), "Freres Roses" (the manuscript of Peiresc in the Bibliothèque Inguimbertine in Carpentras). We find, however, "Chevalliers de la Croix rose" in the manuscript of Philippe de Béthune (see above, note 85), which is closer to the terms figuring in Théophile's interrogation. See below, note 172.

166. Lachèvre, *Le Libertinage devant le Parlement de Paris,* Part 1: *Le Procès de Théophile de Viau,* vol. 1, 436 and 444; vol. 2, 397. The handwritten piece in question began with these words: "Quoy qu'on me puisse veoyr accablé de malheurs . . ." It has not been located.

167. Rooses and Ruelens, *Correspondance de Rubens et documents épistolaires concernant sa vie et ses oeuvres,* 3:224. This testimony, not taken into account by the historians of Théophile de Viau, even the most recent ones, has on the other hand been well brought out and commented on by Halleux, "Helmontiana," 60–61, in which we already find all the imperative connections between Théophile and the Rosicrucians.

168. The pretext the Parlement put forward was the publication of the *Parnasse des Poëtes Satyriques,* of which Théophile was considered to be one of the principal authors (Lachèvre, *Le Libertinage devant le Parlement de Paris,* Part 1: *Le Procès de Théophile de Viau,* vol. 1, 131–32).

169. Rooses & Ruelens, *Correspondance de Rubens et documents épistolaires concernant sa vie et ses oeuvres,* 3:247 ("de la barque sur la Garonne, près de Cadillac"). The printing of Garasse's work was completed on August 18, the exact day of Peiresc's departure and of the first ruling of the Parlement that condemned Théophile. Peiresc's term "Adombrados" is not a slip of the tongue: It is found, for example, under Naudé's pen, *Instruction à la France,* 113. In February 1624, Peiresc still had not been able to read *La Doctrine curieuse;* he wrote to Rubens on February 12 (Rooses & Ruelens, *Correspondance,* 3:287, from Aix-en-Provence): "I am sending to you the copy of the poster of the Adombrados and of a letter that I recently wrote to a friend in Italy concerning a fragment of Egyptian antiquity along with the drawing of this piece. Father Garasse's book has still not reached me from Paris."

170. Frances Yates, who had already hinted at it, was nonetheless not able of establishing the link with the polemic directed against Théophile and limited herself to associating the author of the *Effroyables Pactions* with Garasse in their common effort to launch a witch-hunt against the Rosicrucians (*The Rosicrucian Enlightenment,* ed. 1986, 105; French trans., ed. 1985, 137).

171. *Effroyables Pactions,* 6.

172. Adam, *Théophile de Viau et la libre pensée française en 1620,* 71, 103, and 408–9. Théophile and Béthune were close enough that in 1626, on the path to exile, the poet enjoyed the hospitality of the latter's son, Hippolyte de Béthune, at his château in Selles. Furthermore, the term "Croix Roze" used in the questionings of Théophile is only found once, to my knowledge, in the very manuscript that comes from Philippe de Béthune (*"Affiches des Chevalliers de la Croix rose"*). Could it be that these posters were collected by Théophile himself?

173. *Effroyables Pactions,* 15.

174. Théophile reproached the members of the Parlement for believing in the Rosicrucians. But this could not explain the implication, here, of the Parlement of Paris, expressly designated. Could it be a matter of settling a score that escapes us, or of an individual scandal that was widely but only briefly discussed? The lawyer is indeed presented as being pursued by the police, hence his desire to make him-

self invisible, in the same way as in the *Examen sur l'Inconnue et Nouvelle Caballe des Freres de la Croix Rosee,* in which a similar adventure takes place with a lawyer in distress over debts.

175. *Effroyables Pactions,* 17–19, at 19.

176. Ibid., 10 (emphasis added). Let us recall, after Louise Godard de Donville, that "in the title of *La Doctrine curieuse,* 'curieuse' means (among other things) 'magic.'" The repetition of this word is therefore less innocent than it seems. Cf. Godard de Donville, "L'oeuvre de Théophile de Viau aux feux croisés du 'libertinage'," *Oeuvres et Critiques,* 20, no. 3 (1995): 185–204, at 201. On the notion of "curiosité," "already laden with a long history of reprobation in secular and Christian literature," cf. Jean Dupèbe, "Curiosité et magie chez Johannes Trithemius," in Jean Céard, ed., *La Curiosité à la Renaissance* (Paris: Société d'Édition de l'Enseignement Supérieur, 1986), 71–97, as well as the thoughts of Jean Céard at the beginning of the same volume, 14–18 (cf. 16 his quote from Martin Del Rio: sometimes "*curiosus* pro malefico seu mago accipitur").

177. Garasse, *La Doctrine curieuse,* 296–99.

178. *Effroyables Pactions,* 9.

179. Arthur Edward Waite, *The Brotherhood of the Rosy Cross* (1924; reprinted, Secaucus, N.J.: University Books, 1973), 360: "the work of a venal pamphleteer who saw money in the madness and fed it with incredible stories of drowning and suicide which followed the experiences of initiation."

180. *Examen sur l'Inconnue et Nouvelle Caballe des Freres de la Croix Rosee,* 8–10: "Fourriers de Sathan."

181. Ibid., 10–11.

182. Ibid., 14: "ces beaux Dogmatiseurs."

183. See, for example, the very title of the chapter on the Rosicrucians (Garasse, *La Doctrine curieuse,* 83–91): "Section Quatorziesme. La Secte de nos beaux Esprits dogmatisans en cachettes, est semblable à la faction de ces gens qui s'appellent, *La Confrerie de la Croix de Roses.*"

184. Lachèvre, *Le Libertinage devant le Parlement de Paris,* Part 1: *Le Procès de Théophile de Viau,* vol. 1, 444.

185. *Examen sur l'Inconnue et Nouvelle Caballe des Freres de la Croix Rosee,* 16. The pamphlet ends with this recommendation. The "SS. Cayers" are the Holy Scriptures.

186. On this matter see Lachèvre, *Le Libertinage devant le Parlement de Paris,* Part 1: *Le Procès de Théophile de Viau,* vol. 1, 49–51, and Garasse's allusion pointed out below, note 224.

187. M. Gaspard, *Thresor de l'histoire generale de nostre temps. De tout ce qui s'est fait & passé en France soubs le regne de Louis le Juste. Depuis la mort déplorable de Henry le Grand jusques à present. Contenant les troubles arrivez au Royaume durant la Regence de la Royne sa Mere, dans la Majorité du Roy, & pendant les guerres de la Rebellion, jusques apres la Paix donnée par sa Majesté à ses subjects de la Religion pretenduë reformée. Par M. Gaspard, N. Historien.*

Troisiesme edition, reveuë & augmentée par l'Autheur (Paris: Joseph Bouillerot, 1624) (Bibliothèque de l'Arsenal, shelfmark: 8° H. 6965), 671. This edition compiles a registry of current events until March 10, 1624. The 1623 edition, that can be consulted at the BNF, contains nothing on the Rosicrucians.

188. Ibid., 692–93 – *Effroyables Pactions,* 6: "It is held that the Enlightened Ones of Spain, & the Invisible Ones of France . . ." (text already quoted in part above, 270). The reader will not be surprised to learn that Gaspard's passage on the Rosicrucians is framed by those that concern the Enlightened Ones of Spain (671–92) and Théophile's trial (693–96). Gaspard was to be followed by the *Table chronographique* of Gaultier, who was no less confused in the 1626 edition; cf. Godard de Donville, "Théophile, les 'Beaux Esprits' et les Rose-Croix," 148.

189. [Malingre], *Troisiesme Tome de l'Histoire de nostre temps,* 98–99, 99–124 (*Effroyables Pactions*) and 124–25. Let us remember that the privilege of this book is dated March 7, 1624.

190. Ibid., 330–47: report of Théophile's trial (August 1623) and his arrest (September 1623).

191. Let us remember that the privilege for Malingre's continuation of Florimond de Raemond's *Histoire generale du progrez et decadence de l'Heresie moderne,* in which this brochure is located, dates from March 14, 1624.

192. BNF, MS fr. 19574 (XVIIth c.), fols. 28r–61v. On this questioning, see among others Robert Mandrou, *Magistrats et sorciers en France au XVIIe siècle. Une analyse de psychologie historique,* in series L'Univers historique (Paris: Seuil, 1980), 105; 27, no. 19; and passim.

193. BNF, MS fr. 19574, fols. 61r–62v.

194. Ibid., fols. 65r–68v. On this incident, I refer once more to my dissertation, "Paracelsisme et alchimie en France." See also D. Kahn, "Entre atomisme, alchimie et théologie: la réception des thèses d'Antoine de Villon et Étienne de Clave contre Aristote, Paracelse et les 'cabalistes,'" *Annals of Science,* 58 (2001), forthcoming.

195. BNF, MS fr. 19574, fols. 69r–72v.

196. A characteristic example of this shortcoming is provided by Allen G. Debus's *The French Paracelsians: The Chemical Challenge to Medical and Scientific Tradition in Early Modern France* (Cambridge–New York–Melbourne: Cambridge University Press, 1991). The author, in spite of constant efforts to widen his perspectives in the manner of his work *The Chemical Philosophy* (New York: Science History Publications, 1977), comes back steadily, as if it were a leitmotiv, to a mere account of the stages of the too-famous quarrel.

197. I discuss this matter at length in "Paracelsisme et alchimie en France." See also Kahn, "La Faculté de médecine de Paris en échec face au paracelsisme: enjeux et dénouement réels du procès de Roch Le Baillif," in Heinz Schott and Ilana Zinguer, eds., *Paracelsus und seine internationale Rezeption in der frühen Neuzeit. Beiträge zur Geschichte des Paracelsismus,* Brill's Studies in Intellectual History, vol. 86 (Leiden: E. J. Brill, 1998), 146–221.

198. These various episodes are also discussed in my dissertation, "Paracelsisme et alchimie en France."

199. On this quarrel, cf. Walter Pagel, *Joan Baptista Van Helmont, Reformer of Science and Medicine*, Cambridge Monographs on the History of Medicine (Cambridge: Cambridge University Press, 1982), 8–12; Halleux, "Helmontiana," 51–57 and 61; Moran, *The Alchemical World of the German Court*, 36–39; Wolf-Dieter Müller-Jahncke, *Astrologisch-magische Theorie und Praxis in der Heilkunde der frühen Neuzeit, Sudhoffs Archiv*, Beiheft 25 (Stuttgart: Franz Steiner, 1985), and his "Magische Medizin bei Paracelsus und den Paracelsisten: Die Waffensalbe," in P. Dilg and H. Rudolph, eds., *Resultate und Desiderate der Paracelsus-Forschung, Sudhoffs Archiv*, Beiheft 30 (Stuttgart: Franz Steiner, 1993), 43–55, at 47–54; and my dissertation, "Paracelsisme et alchimie en France."

200. See the beginning of the *Archidoxes magicae* in Theophrast von Hohenheim, gen. Paracelsus, *Sämtliche Werke*, ed. Karl Sudhoff (Munich-Berlin: R. Oldenbourg, 1933), I, 14: 448; on this text published for the first time in 1570 in Gérard Dorn's Latin translation, cf. ibid., xxv–xxviii.

201. See on this topic Ralf Georg Bogner, "Paracelsus auf dem Index. Zur kirchlichen Kommunikationskontrolle in der frühen Neuzeit," in Telle, *Analecta Paracelsica*, 489–530; D. Kahn, "59 thèses de Paracelse censurées par la Faculté de théologie de Paris le 9 octobre 1578," in S. Matton, ed., *Documents oubliés sur l'alchimie, la kabbale et Guillaume Postel offerts, pour son 90e anniversaire, à François Secret par ses éleves et amis* (Geneva: Droz, 2001), forthcoming.

202. Is this what incited an Italian prelate, Agesilao Marescotti, to ask Peiresc about the Rosicrucians on June 13, 1619? See BNF, MS fr. 9540, fol. 184r: "Mi resta da supplicar' il signor di Peiresc ... di raccogliermi tutto ciò che appartiene alla società della Rosea Croce" (pointed out by Pintard, *Le Libertinage érudit*, 48 n. 4, and 584).

203. Marin Mersenne, *Quaestiones celeberrimae in Genesim, cum accurata textus explicatione. In hoc volumine Athei, et Deistae impugnantur, & expugnantur, & Vulgata editio ab haereticorum calumniis vindicatur. Graecorum, & Hebraeorum Musica instauratur. Francisci Georgii Veneti cabalistica dogmata fuse refelluntur, quae passim in illius problematibus habentur. Opus Theologis, Philosophis, Medicis, Jurisconsultis, Mathematicis, Musicis vero, & Catoptricis praesertim utile* (Paris: Sébastien Cramoisy, 1623), *quaestio* LIII (*An e cadavere Abelis sanguis adversus Cain proruperit, & an id fiat homicida praesente, ut vulgo affirmant*).

204. Andreas Libavius, *Tractatus duo physici; prior de impostoria vulnerum per unguentum armarium sanatione Paracelsicis usitata commendataque. Posterior de cruentatione cadaverum in justa caede factorum praesente, qui occidisse creditur. ... His accessit epistola de examine Panaceae Amwaldinae...* (Frankfurt: Johann Saur for Peter Kopff, 1594), 140–392.

205. It is the 241st conference of the Bureau d'Adresse, on March 19, 1640 (*Quatriesme Centurie des questions traitees aux conferences du Bureau d'Adresse*, 1641, 209 [*recte*: 213]–216: "Why dead bodies bleed in the presence of their murderers"). See on this subject Alain Boureau, "La preuve par le cadavre qui saigne au XIIIe siècle: entre expérience commune et savoir scolastique," *Micrologus*, 7 (1999), 24–81.

206. Marin Mersenne, *Quaestiones celeberrimae in Genesim*, quaestio LIII, art. 5 (*Si illa sanguinis ebullitio naturali tribuenda est, cui potissimum, ac probabilius*), col. 1452: "qui fere quibuslibet nundinis Francofurtensibus libellos impietatem redolentes in orbem Christianum inducunt."

207. Jean Boucher, *Couronne mystique ou Armes de pieté, contre toute sorte d'impieté, heresie, atheisme, schisme, magie, & Mahometisme. Par un signe ou hieroglyphique mysterieux, fait en forme de Couronne, autant rare & ancien, que divinement descouvert en nos jours. Avec dessein sur ce sujet, de milice ou chevallerie Chrestienne, contre tous mescreans. Specialement contre le Turc. Oeuvre plein de varieté et meslange, tant de doctrine divine & humaine, que d'histoire sacrée & prophane, & remarque de choses rares. Le tout divisé en V. Livres. Aux Rois & Princes souverains. Et specialement aux deux freres doublement alliez, les deux plus grands Rois de Chrestienté* (Tournai: Adrien Quinqué, 1623; reissue, 1624). Boucher transferred the privilege, which bears the date April 24, 1623, to the publisher on May 1: The work was therefore published at the end of May or the beginning of June. On Jean Boucher (ca 1548–1644 or 1646), see the article "Boucher" by J. Dedieu in the *Dictionnaire d'histoire et de géographie ecclésiastiques*, vol. 9 (Paris: Letouzey & Ané, 1937), cols. 1457–60, and the notice in the *Dictionnaire des Lettres françaises. Le XVIIe siècle*, 191.

208. Boucher, *Couronne mystique*, reissue 1624, 551, 552, 553–54.

209. Ibid., 553.

210. Ibid., 554.

211. Garasse, *La Doctrine curieuse*, 83–91.

212. On Besold's attitude toward the Rosicrucian movement, cf. Gilly, "*Iter Rosicrucianum*," 72. On Besold as the translator into German of Campanella's *Monarchia di Spagna* (translation published in 1620 and reissued in 1623), and on the addition to this text reproduced by Gabriel Naudé and attributable to Besold himself, see Michel-Pierre Lerner, *Tommaso Campanella en France au XVIIe siècle*, Lezioni della Scuola di Studi Superiori in Napoli, no. 17 (Naples: Bibliopolis, 1995), 34–36.

213. Jean Baptiste van Helmont, *De Magnetica vulnerum curatione. Disputatio, Contra opinionem D. Joan. Roberti, Presbyteri de Societate Jesu, Doctoris Theologi, in brevi sua anatome sub censurae specie exaratam* (Paris: Victor Le Roy, 1621) (the printer's address to the reader is dated from Lyon, "*ex typis nostris,*" in the year 1620). Rudolph Goclenius, *De magnetica vulnerum curatione* (Paris: Samuel Celerius, 1621) (pointed out by Robert Halleux, "Helmontiana," 53 and n. 122).

214. Cf. Lenoble, *Mersenne ou la naissance du mécanisme*, reissue 1971, XLVI. The alleged "Jacques de Nuysement" found in nearly every recent work on seventeenth-century alchemy never existed: This author was no other than the poet and alchemist Clovis Hesteau de Nuysement, as Wallace Kirsop showed forty years ago. "Jacques" is a modern bibliographic error.

215. Naudé, *Instruction à la France*, 14. Cf. Blaise de Vigenère, *Traicté du feu et du sel* ("A Paris, En la boutique de L'Angelier. Chez Claude Cramoisy au premier

pilier de la grand'Salle du Palais. 1622. Avec privilege du Roy") (copy in London, British Library, shelfmark 1033.i.8). This edition is but the posthumous edition of 1618 being put back on sale; the privilege is the same, and with good reason: It was accorded for ten years. Only the title page was modified, with an omission: "Blaise Vigenere" (the particle "de" is missing). Cf. [Secret], *Kabbale et philosophie hermétique. Exposition à l'occasion du Festival International de l'Ésotérisme*, 66, ill. 25.

216. This book by a certain de Mérac (?), published by Pierre Ramier, is pointed out by Pierre Borel, *Bibliotheca Chimica, seu Catalogus librorum philosophicorum hermeticorum* (Paris: Charles Du Mesnil & Thomas Jolly, 1654; 2nd ed., Heidelberg: Samuel Brown, 1656; reprint, Hildesheim: Georg Olms, 1969), 246.

217. *Le veritable et souverain remede pour la maladie Pestilentieuse. Extraict des oeuvres de M. Roch le Baillif, Sieur de la Riviere Conseiller & Medecin ordinaire du Roy* (Paris: Gervais Aliot, 1623). This is a simple little book of eight pages that reproduces a fragment of the *Traicté du remede à la peste* by Le Baillif (Paris, 1580) starting with fol. 20r ("Faut prendre une once poisant de vieil theriaque de Venise..."). On the plague in Paris in 1623, see the passage on Descartes below.

218. Paracelsus, *La petite Chirurgie, autrement ditte la Bertheonee, de Philippe Aoreole Theophraste Paracelse grand Medecin & Philosophe entre les Allemans. Plus les traittez du mesme Autheur, des Apostemes [,] syrons ou noeuds, des ouvertures du cuir, des ulceres, des vers, serpens, taches ou marques qui viennent de naissance, & des contractures. Avec notes & explications des termes & mots plus difficiles, Table des Chapitres, & matieres* (Paris: Olivier de Varennes, 1623) (Paris, Bibliothèque Ste-Geneviève, shelfmark: T. 8° 1751 inv. 4549 FA), fol. ã2r–v. This work is described by Karl Sudhoff, *Versuch einer Kritik der Echtheit der Paracelsischen Schriften,* vol. 2: *Paracelsus-Handschriften* (Berlin: G. Reimer, 1899), "Nachträge zum ersten Bande," 803–4 = no. 323.

219. Paracelsus, *La petite Chirurgie,* fol. ã3v–ã4r. Here is the original in Latin: "Non titulus, non eloquentia, non linguarum peritia, nec multorum librorum lectio, etsi haec non parum exornent, in medico desideranda, sed summa rerum ac mysteriorum cognitio, quae una facile aliorum omnium vices agit" (Th. von Hohenheim, gen. Paracelsus, *Sämtliche Werke,* ed. Sudhoff, I, 4 [Munich-Berlin: R. Oldenbourg, 1931], 3).

220. The publisher, Olivier de Varennes, had just published at Louis XIII's expense Marino's *L'Adone,* adorned with the king's coat of arms (cf. M. Fumaroli, *L'Inspiration du poète de Poussin. Essai sur l'allégorie du Parnasse*, Les Dossiers du Département des Peintures, no. 36 [Paris: Réunion des Musées Nationaux, 1989], 52). As for the dedicator, Schomberg, he was at the time very close to Louis XIII.

221. Paracelsus, *La petite Chirurgie,* fol. [†4]r.

222. Naudé, *Instruction à la France,* 64 (emphasis added).

223. As for the first book of *La Doctrine curieuse,* Louise Godard de Donville reminds us that it was composed at the beginning of March at the latest ("Théophile, les 'Beaux Esprits' et les Rose-Croix," 143).

224. Garasse, *La Doctrine curieuse,* 849. This should be connected to the attack by the *Examen sur l'Inconnue et Nouvelle Caballe des Freres de la Croix Rosee,* studied above, 273, against the "good companions" who doubt the reality of demonic possession.

225. Garasse, *La Doctrine curieuse,* 1010–17, here 1012–13. Cf. Martin Del Rio, *Disquisitionum magicarum Libri sex: Quibus continetur accurata curiosarum artium, & vanarum superstitionum confutatio; utilis Theologis, Jurisconsultis, Medicis, Philologis* (1599), book VI, chap. II, sect. I, quaest. 1 (ed. Lyon: Jean Pillehotte, 1612, 405–6). On Del Rio and Paracelsus, see the analysis by Halleux, "Helmontiana," 54–56.

226. We will see this below. Let us point out that the Sorbonne's censure against Paracelsus had remained very little known outside of the Faculty of Medicine of Paris. See Kahn, "59 thèses de Paracelse censurées par la Faculté."

227. Del Rio, *Disquisitionum magicarum Libri sex,* ed. cited, 405: "Nam nec ii, qui Baptismate tincti; satis horum immunes. Nam multa occurrunt apud Pomponatium, multa apud Henricum Cornelium Agrippam; sed plura apud Philippum Aureolum, alias Bombastum Paracelsum variis ejus operibus, quorum omnium libri prohibitae per Ecclesiam sunt lectionis." Banned for the first time in 1580 in the Parme *Index,* Paracelsus had particularly appeared in the Tridentine *Index* of Pope Clement VIII, published in Rome in 1596. See Bogner, "Paracelsus auf dem Index" and Kahn, "59 thèses de Paracelse censurées par la Faculté."

228. Thomas Erastus, *Disputationum de Medicina nova Philippi Paracelsi Pars prima . . . Pars quarta et ultima,* 4 vols. (Basle: Pietro Perna, 1571–73); see, for example, 1: 54.

229. Del Rio, *Disquisitionum magicarum Libri sex,* ed. cited, 405: "Os impurum! quasi non sit Theologorum judicare; quid contra Deum, quid non contra Deum sit. Sic agyrta nequissimus Fidei tribunal sibi vendicat."

230. Mersenne, *Quaestiones celeberrimae in Genesim,* cols. 576–77 (start of the polemic) and cols. 649–54 (text of the actual polemic).

231. See especially Sylvain Matton, Introduction to: Dom Belin, *Les Aventures du philosophe inconnu* (1646; Paris: Retz, 1976) (coll. "Bibliotheca Hermetica"), 13–69, at 31–32; Matton, "Créations microcosmique et macrocosmique. La *Cabala Mineralis* et l'interprétation alchimique de la Genèse," in Siméon Ben Cantara, *Cabala Mineralis* (Paris: J.-C. Bailly Éditeur, 1986), 25–33, esp. 26–29, which does not, however, cite the passage of the *Quaestiones celeberrimae in Genesim* under discussion here; W. L. Hine, "Mersenne and Alchemy," in Z. R. W. M. von Martels, ed., *Alchemy Revisited: Proceedings of the International Conference on the History of Alchemy at the University of Groningen, 17–19 April 1989,* Collection de travaux de l'Académie Internationale d'Histoire des Sciences, no. 33 (Leiden: E. J. Brill, 1990), 188–91, esp. 191, also does not cite the *Quaestiones celeberrimae in Genesim;* Armand Beaulieu, "L'attitude nuancée de Mersenne envers la chymie," in Margolin and Matton, *Alchimie et philosophie à la Renaissance,* 395–403, at 396–97.

232. Mersenne, *Quaestiones celeberrimae in Genesim,* cols. 711–18 (esp. cols. 711–12). Sylvain Matton has pointed out that Mersenne at first criticized, without

naming his source, a text that is in fact by Robert Fludd ("Créations microcosmique et macrocosmique: La *Cabala Mineralis* et l'interprétation alchimique de la Genèse," 26–27).

233. Mersenne, *Quaestiones celeberrimae in Genesim*, col. 651.

234. Godard de Donville, "Théophile, les 'Beaux Esprits' et les Rose-Croix," esp. 148–54.

235. Lorenzo Bianchi, "Gabriel Naudé critique des alchimistes;" idem, *Rinascimento e libertinismo. Studi su Gabriel Naudé*, passim.

236. Naudé, *Instruction à la France*, 114–15; already pointed out by Godard de Donville, "Théophile, les 'Beaux Esprits' et les Rose-Croix," 152–53. For a commentary on this passage of Naudé's *Instruction*, cf. Pintard, *Le Libertinage érudit*, 161; 166, n. 5; and 601.

237. Lachèvre, *Le Libertinage devant le Parlement de Paris*, Part 1: *Le Procès de Théophile de Viau*, vol. 1, 131–32 and 141–43; Josephine de Boer, "Colletet's Exile after His Condemnation in 1623," *Modern Language Notes*, 47 (1932): 159–62; Pasquale Aniel Jannini, *Verso il tempo della ragione. Studi e ricerche su Guillaume Colletet* (1965), vol. 28 of Biblioteca della Ricerca, Cultura Straniera (reprinted, Fasano: Schena Editore, 1989), 74–80.

238. Emphasis added. See Naudé, *Instruction à la France*, fol. [e3]v, sonnet signed "G. C. P. A.," that is, "Guillaume Colletet, Parisien, Advocat." Josephine de Boer cites four of Colletet's poems signed, in 1619, in the same way but with the signature written out in full ("Colletet's Exile After His Condemnation in 1623," 159, n. 5). Furthermore, Colletet repeated this sonnet in his *Divertissemens* of 1631. In 1625, he honored in the same way the *Apologie* by Naudé. See Jannini, *Verso il tempo della ragione. Studi e ricerche su Guillaume Colletet*, 139–40 and 155, nn. 33 and 35.

239. See Jannini, *Verso il tempo della ragione. Studi e ricerche su Guillaume Colletet*, 77–78, and, on the quarrel between Ogier and Garasse, Fumaroli, *L'Age de l'éloquence*, 329–34. Louise Godard de Donville rightly reminds us that Guy de La Brosse joined forces with Naudé and Ogier in the attack directed against Garasse ("Théophile, les 'Beaux Esprits' et les Rose-Croix," 153–54).

240. Godard de Donville, "Théophile, les 'Beaux Esprits' et les Rose-Croix," 152; Fumaroli, *L'Age de l'éloquence*, for example, 252–56.

241. Naudé, *Instruction à la France*, 49–52.

242. Ibid., 59–60; already pointed out by Godard de Donville, "Théophile, les 'Beaux Esprits' et les Rose-Croix," 149.

243. Naudé, *Instruction à la France*, 60; already pointed out by Godard de Donville, "Théophile, les 'Beaux Esprits' et les Rose-Croix," 149, n. 18.

244. Naudé, *Instruction à la France*, 14–15. On Garasse and Nothing, see Lachèvre, *Le Libertinage devant le Parlement de Paris*, Part 1: *Le Procès de Théophile de Viau*, vol. 1, 173–74.

245. Garasse, *La Doctrine curieuse*, 90–91. Naudé, *Instruction à la France*, 62: "voire mesme l'industrie de l'architecte de tous ces mensonges [*i.e., the author of*

the Fama] a esté si perçante en la Preface de son Manifeste, que de nous vouloir persuader qu'un *Adamus Haselmeyer* (condamné pour ses malefices) avoit esté envoyé aux galeres à cause que par un zele & trop grande devotion à cette compagnie, il avoit avancé quelques paroles à sa loüange, esperant par cette terreur de supplice avoir une excuse legitime de ne se manifester plus à descouvert."

246. Naudé, *Instruction à la France*, 114–15. See Marc Fumaroli's analysis of Ogier's reply to Garasse (*L'Age de l'éloquence*, 329–32).

247. [Malingre], *Traicté des Atheistes, Deistes, Illuminez d'Espagne, et nouveaux pretendus Invisibles, dits de la Confrairie de la Croix-Rosaire*, 3.

248. The privilege of Guy de La Brosse's Traicté contre la mesdisance bears the date of Nov. 18, 1623. Cf. Adam, *Théophile de Viau et la libre pensée française en 1620*, 379 and 413–14.

249. [Malingre], *Traicté des Atheistes, Deistes, Illuminez d'Espagne, et nouveaux pretendus Invisibles, dits de la Confrairie de la Croix-Rosaire*, 4–11, here 11.

250. Ibid., 11–13. The dedication to Richelieu of the first part of *L'Impieté des Deistes* is dated June 8, 1624, but the approval of the Minims, Fathers Claude Rangueil and Jacques Bremant, dates from January 25.

251. Ibid., 21–55.

252. Naudé, *Instruction à la France*, 30. The verses from Lucretius (I, 83, and 101) mean: "Religion has been able to counsel so many crimes" and "Religion is what gave birth to criminal and impious actions."

253. Ibid., 10–11.

254. On Garasse and de Lancre, see Godard de Donville, *Le Libertin des origines à 1665*, 153 and n. 3; Pintard, *Le Libertinage érudit*, 442. For some violent passages directed against Charron, see Garasse, *La Doctrine curieuse*, for example, 27–29 or 1015; Marin Mersenne, *L'Impieté des Deistes, Athees, et Libertins de ce temps, combatuë, & renversee de point en point par raisons tirees de la Philosophie, & de la Theologie. Ensemble la refutation du Poëme des Deistes* (Paris: Pierre Billaine, 1624), chap. 9 (180–210); Lenoble, *Mersenne ou la naissance du mécanisme*, 64 and 204.

255. See Garasse's reply to Ogier: *Apologie du Pere François Garassus*, chaps. 21 and 22 (259–85).

256. Gabriel Naudé, *Apologie pour tous les grands personnages qui ont esté faussement soupçonnez de magie* (Paris: François Targa, 1625), Préface, fol. [ã5]r–v. On the *Apologie*, besides René Pintard (*Le Libertinage érudit*) and Lorenzo Bianchi (*Rinascimento e libertinismo*), see Anna Lisa Schino, *Tradizione ermetica e astrologia giudiziaria in Gabriel Naudé* (Florence: Olschki, 1992), esp. 133–227 (I thank Robert Halleux for this reference).

257. Naudé, *Instruction à la France*, 64. Cf. Pintard, *Le Libertinage érudit*, 446; Bianchi, "Gabriel Naudé critique des alchimistes," 412; idem, *Rinascimento e libertinismo. Studi su Gabriel Naudé*, 186–87.

258. "Welcome with laughter the cadavers in frightening shapes, the apparitions of the night, the spectral bogeymen, the nocturnal ghosts and the wonders of Thessaly." (Apuleius, *Apologia sive de Magia*, 64).

259. "It was perhaps a wish emanating from good men who wished for communication in literature." Cf. Naudé, *Instruction à la France*, 77–79. On Naudé's relations with Father Redento Baranzano (1590–1622) and Tobias Adami, see Pintard, *Le Libertinage érudit*, 133–34, and especially Lerner, *Tommaso Campanella en France au XVIIe siècle*, 33–34 and n. 53 (on Baranzano, add the bibliography of Kurd Lasswitz, *Geschichte der Atomistik vom Mittelalter bis Newton*, 2 vols. [Hamburg-Leipzig: Leopold Voss, 1890], 1:330–31, n. 12). Thanks to Adami, Naudé had fairly good information at his disposal: see the *Instruction à la France*, 88, in which Naudé cited a letter from Adami on October 15, 1622, that declared with regards to the Rosy-Cross: *fabula ista jam fere peracta est* ("This fable is almost over already").

260. Naudé, *Instruction à la France*, 79.

261. Ibid., 22–24 and 42–48. On Naudé's complete historical awareness, see Bianchi, *Rinascimento e libertinismo. Studi su Gabriel Naudé*, 96–97.

262. [Gabriel Naudé], *Jugement de tout ce qui a esté imprimé contre le Cardinal Mazarin, Depuis le sixiéme Janvier, jusques à la Declaration du premier Avril mil six cens quarante-neuf* [also called the *Mascurat*] (s.l.n.d. [Paris, 1650]), 469.

263. Naudé, *Instruction à la France*, for example, 51 and 84; *Apologie pour tous les grands personnages*, 365.

264. Naudé, *Instruction à la France*, 49.

265. Ibid., 42–48, at 42 and 44. On the prophecy of Helias Artista, see Eugène Olivier: "Bernard G[illes] Penot (Du Port), médecin et alchimiste (1519–1617)," *Chrysopoeia*, 5 (1992–96): 571–668, at 611–12 (with bibliography); William R. Newman, *Gehennical Fire: The Lives of George Starkey, an American Alchemist in the Scientific Revolution* (Cambridge: Harvard University Press, 1994), 3–4.

266. Naudé, *Instruction à la France*, 59–60.

267. Pintard, *Le Libertinage érudit*, 449 (quotation) and 443–44, 447, 467–68.

268. Naudé, *Instruction à la France*, 59–60, 105, and 75.

269. See Olivier, "Bernard G[illes] Penot (Du Port), médecin et alchimiste (1519–1617)," 620.

270. See Kahn, "La Faculté de médecine de Paris en échec face au paracelsisme," 160.

271. Quoted by Gilly, *"Theophrastia Sancta,"* 426.

272. On the hostility toward Paracelsus in the Lutheran circles, see the significant reactions of the Faculty of Medicine of Tubingen between 1599 and 1609 against Tobias Hess, accused of having made himself "the disciple of the atheist Paracelsus" (Gilly, *"Iter Rosicrucianum,"* 70–71; *Cimelia Rhodostaurotica*, 45–47).

273. On the importance of the Faculty of Medicine of Paris in Naudé's intellectual development, see Pintard, *Le Libertinage érudit*, 158–59. Elsewhere Naudé describes the Paracelsian physicians and the iatrochemists as *fumivenduli Paracelsistae, Hermetici medicastri, Basilicae Crollianae fautores, & id genus omne Medicorum;* cf. Lorenzo Bianchi, "Renaissance et libertinage chez Gabriel Naudé," in Antony McKenna and Pierre-François Moreau, eds., *Libertinage et*

philosophie au XVIIe siècle, Part 2: *La Mothe Le Vayer et Naudé* (Saint-Étienne: Publications de l'Université de Saint-Étienne, 1997), 75–90, at 79.

274. Naudé, *Instruction à la France,* 26. See the judicious parallel established in this respect by Lorenzo Bianchi between *Le Marfore* and the *Instruction à la France* (*Rinascimento e libertinismo. Studi su Gabriel Naudé,* 89–92).

275. Naudé, *Instruction à la France,* 108 and 110. Cf. Pintard, *Le Libertinage érudit,* 542–43; Bianchi, "Gabriel Naudé critique des alchimistes," 414–15; idem, *Rinascimento e libertinismo. Studi su Gabriel Naudé,* 190–92.

276. See Marc Fumaroli, "Sous le signe de Protée, 1594–1630," chap. 1 ("Saturne et les remèdes à la mélancolie") in Jean Mesnard, dir., *Précis de littérature française du XVIIe siècle* (Paris: Presses Universitaires de France, 1990), esp. 41–42.

277. *Mercure François,* vol. 9 (1622–24), 374–87. I do not know how Godard de Donville can write that the *Mercure,* in this article, "gives Garasse the last word" ("Théophile, les 'Beaux Esprits' et les Rose-Croix," 150).

278. Gouhier, *Les Premières pensées de Descartes,* esp. 121–27.

279. Arnold's *Histoire des Rose-Croix* had a second edition in 1990. On the supposed analogies between the dreams of Descartes and the *Chymische Hochzeit,* see Fernand Hallyn's brief refutation of Arnold: "Une 'feintise,'" in F. Hallyn, ed., *Les "Olympiques" de Descartes,* Romanica Gandensia, no. 25 (Geneva: Droz, 1995), 91–111, at 101–3.

280. Sylvain Matton, "Le rêve dans les 'secrètes sciences': spirituels, kabbalistes chrétiens et alchimistes," *Revue des Sciences Humaines,* no. 211 (July–September 1988): *Rêver en France au XVIIe siècle,* 153–80 (179–80 on Descartes); Robert Halleux, "Helmontiana II: le prologue de l'*Eisagoge,* la conversion de Van Helmont au paracelsisme, et les songes de Descartes," *Academiae Analecta. Mededelingen van de Koninklijke Academie voor Wetenschappen, Letteren en Schone Kunsten van België. Klasse der Wetenschappen,* 49, no. 2 (1987): 18–36.

281. [Baillet], *La Vie de Monsieur Des-Cartes,* 1691, 1:105–6 and 107.

282. Ibid., 1:36–37.

283. Gouhier, *Les Premières pensées de Descartes,* 126, n. 35, taking sides with Cornelis de Waard's opinion in Tannery, de Waard, and Pintard, *Correspondance du P. Marin Mersenne,* 1:149; Geneviève Rodis-Lewis, *Descartes. Biographie* (Paris: Calmann-Lévy, 1995), 87.

284. Baillet indicates here: "Le P. Poiss. *Rem. sur la Méth. de Desc.* part. 2, pag. 30. 31. 32." This passage from Poisson is reproduced in the *Oeuvres* of Descartes (ed. Adam and Tannery), 10:197–98, note a.

285. Baillet cites here the text of the poster transmitted in the *Effroyables Pactions.*

286. [Baillet], *La Vie de Monsieur Des-Cartes,* 1691, 1:107–8. Baillet indicates here: "Hil. de Cost. vie de Mers. pag. 15. 16.," but there is nothing about Descartes in these pages of the book of [Hilarion de Coste], *La Vie du R. P. Marin Mersenne theologien, philosophe et mathematicien de l'Ordre des Peres Minimes* (Paris: Sébastien & Gabriel Cramoisy, 1649).

287. Gouhier, *Les Premières pensées de Descartes*, 20 and 124; also Rodis-Lewis, *Descartes. Biographie*, 13.

288. We become certain of this thanks to a variant in the text of the first Rosicrucian poster: Only the *Mercure François* gives, as Baillet does, the reading "in which we live" instead of "in which <we> wish to be." See above, note 146, for the analysis of Baillet's use of the *Mercure*.

289. [Baillet], *La Vie de Monsieur Des-Cartes*, 1691, 1:106 and 116; Descartes, *Oeuvres* (ed. Adam and Tannery), 1:2–4; Mrs. Tannery, de Waard, and Pintard, *Correspondance du P. Marin Mersenne*, 1:149. In his abridged version of the life of Descartes (1693), Baillet specifies: "M. Descartes fut deux mois et quelques jours à Paris," which agrees with these dates (Adrien Baillet, *Vie* [abridged] *de Monsieur Descartes* [Paris: La Table Ronde, 1946, reissued 1992], 55).

290. [Baillet], *La Vie de Monsieur Des-Cartes*, 1691, 1:116; Descartes, *Oeuvres* (ed. Adam and Tannery), 1:2; Baillet, *Vie* [abridged] *de Monsieur Descartes*, 55: "De là il alla en Poitou, et pendant les mois de juin et de juillet qu'il y demeura, il vendit, du consentement de son père, la plus grande partie des biens qui lui étaient venus du côté de sa mère."

291. Baillet, *Vie* [abridged] *de Monsieur Descartes*, 55–56.

292. [Baillet], *La Vie de Monsieur Des-Cartes*, 1691, 1:118.

293. Besides Frédéric Lachèvre's *Le Libertinage devant le Parlement de Paris*, part 1: *Le Procès de Théophile de Viau*, 1:141, cf. Arlette Lebigre, "Les procès de Théophile de Viau," in Jean Imbert, dir., *Quelques procès criminels des XVIIe et XVIIIe siècles*, Travaux et recherches de la Faculté de droit et des sciences économiques de Paris, Série "Sciences historiques," no. 2 (Paris: Presses Universitaires de France, 1964), 29–43, here 37; Charles Jourdain, *Histoire de l'Université de Paris, au XVIIe et au XVIIIe siècle*, 2 vols. (Paris: Hachette, 1888), 1:190.

294. A number of scholars accept this story in spite of or in complete ignorance of Gouhier's serious objections. See, for example, William R. Shea, *The Magic of Numbers and Motion: The Scientific Career of René Descartes* (New York: Science History Publications, 1991), 95–96 and 109–15; Debus, *The French Paracelsians*, 67, and even Rodis-Lewis, *Descartes. Biographie*, 59. For her part, Louise Godard de Donville goes further still, claiming that "if Naudé affects [in his *Instruction à la France*] to have no tie with Germany, . . . it is for fear of meeting with the misfortune of a Descartes" ("Théophile, les 'Beaux Esprits' et les Rose-Croix," 152).

295. See Gouhier, *Les Premières pensées de Descartes*, 127–29.

296. René Thuillier, *Diarium Patrum, Fratrum et Sororum Ordinis Minimorum Provinciae Franciae sive Parisiensis Qui religiose obierunt Ab Anno 1506 ad Annum 1700* (Paris: Pierre Giffart, 1709), 2:96: "Haud paulo post cum Rangolio in Merianum Domini Peschei viri bello egregii Arcis & Urbis Guisiacae Praefecti praedium abiit; interim vulgata per urbem Fratrum Roseae Crucis fabula, locum dedit obtrectandi Mersenno quibusdam otiosis & inquietis hominibus quos sua dum in Athaeos stilum stringit, scripta pupugerant. Diffamatus scilicet ab eis est, tanquam unus aliquis ex ea sodalitate, vel hoc maxime argumento quod ab aliquibus diebus visus non fuisset; cum enim nihil aeque in Rosanis illis notatum

sit, quam quod pro suo nutu fierent inaspectabiles, Mersenni absentiam satis firmum sibi esse argumentum rati sunt, ut factum esse in conspicuum persuaderent. Hoc ludibrium, ut in reditu audiit Mersennus, non immerito exarsit, & ne id impune ferrent autores, liberali ultionis genere vindicatum ivit, scripto gallice libro quem inscribit Impietas Dcistarum & Atheorum ac maxime subtilium hujus aevi libertinorum articulatim expugnata, & eversa ductis ex Philosophia & Theologia argumentis."

297. Lenoble, *Mersenne ou la naissance du mécanisme,* 31 (referring to Thuillier): "Descartes was accused of being one of them [i.e., the Rosicrucians]. There were even hoaxers who launched suspicions against Father Mersene, the sworn enemy of the sect. Upon his return from a short sojourn that he had made in the country with Father Rangueil, he learned that someone had made the rumor circulate that he had fled with the mysterious Brothers. It was obviously a farce, but it seemed to him to be in poor taste."

298. Ibid., 25.

299. Mrs. Tannery, de Waard, and Pintard, *Correspondance du P. Marin Mersenne,* 1:148–49.

300. Ibid., 1:155.

301. Gustave Cohen, *Écrivains français en Hollande dans la première moitié du XVIIe siècle* (Paris: Édouard Champion, 1920), 388. Willem Meijer, *De Rozekruisers of de Vrijdenkers der XVIIde eeuw* (Haarlem, the Netherlands: F. Bohn, 1916), 66. I was unable to consult this latter work, which is not found in the BNF, and I admit to having retreated when faced with a complete perusal of Van Wassenaer's text, which is prolix and written in a language of which I have hardly any mastery.

302. Paris, Bibliothèque Sainte-Geneviève, MS 2237 (XVIIth c.), fols. 1r–56r (without any mention of the author). It is apparently a transcription of the printed version.

303. Paris, Bibliothèque Sainte-Geneviève, MS 2240 (XVIIth c.). This volume contains inter alia extracts of the *Nucleus Sophicus sive Explanatio in Tincturam Physicorum Theophrasti Paracelsi* of Liberius Benedictus (Frankfurt: Lucas Jennis, 1623) and even a transcription of Haslmayr's "Charakter Cabalisticus oder paracelsischer Signatstern" (cf. Gilly, *Adam Haslmayr,* 18 and 203; idem, *Cimelia Rhodostaurotica,* 31, ill. 29). It also contains extracts from the *Cabala artis & naturae* (Augsburg, 1615), wrongly attributed to Stephan Michelspacher, who is only its editor (cf. Gilly, *Cimelia Rhodostaurotica,* 107). See the description of the whole manuscript by Marie-Thérèse d'Alverny and Françoise Hudry, "Al-Kindi, De radiis," *Archives d'histoire doctrinale et littéraire du Moyen-Age,* 4 (1974): 139–260, at 203–4.

304. Naudé, *Instruction à la France,* 58. This translation could be one and the same as the French version of the *Speculum Rhodostauroticum* that is found in BNF, MS fr. 17154 (first half of the XVIIth c.), fols. 33r–44r.

305. BNF, MS Dupuy 550, fols. 70r–73r, edited only in part in Descartes, *Oeuvres* (ed. Adam and Tannery), 10:199n., and in full by Marion Kuntz, "The Rose-Croix

of Jacques Dupuy (1585–1656)," *Nouvelles de la République des Lettres,* 2 (1981): 91–103; see on this subject Gilly, *"Iter Rosicrucianum,"* 89, n. 48. Other copies can be found in the BNF: MS fr. 15777, fols. 83r–84v (this is the manuscript coming from President Achille III de Harlay); MS fr. 19574, fols. 69r–72v (in this manuscript these *Recherches* come after a few pages on an incident of witchcraft, two letters from Théophile, and the 1624 theses of Étienne de Clave and Antoine de Villon). Gilly (*"Iter Rosicrucianum"*) points out another copy in the manuscripts of Jean Vauquelin des Yveteaux (microfilms can be consulted at the Institut de Recherche et d'Histoire des Textes [Paris], Section latine).

306. BNF, MS Dupuy 550, fols. 70v–71r. In my transcription I do not reproduce the underlining and the punctuation added on the manuscript in black ink.

307. Ibid., fol. 70r and fols. 72v–73r.

308. We can disregard BNF, MS fr. 14778 (XVIIIth c.), entitled in the BNF's catalogue *Dictionnaire de Géomancie et des Rosecroix:* Actually the last three words reproduced in the catalogue title were added later and seem only to attest to a great confusion, for the manuscript does not seem to contain anything other than geomancy.

309. BNF, MS fr. 15250 (II), 12. René Pintard had given the following reference: BNF, MS lat. 5506 A (*Le Libertinage érudit,* 585, line 1), but this manuscript has changed shelfmark in the meantime.

310. BNF, MS fr. 15250 (II), 9. The *Fama*'s commented sentences are in italics. Let us observe that the Hebrew roots as indicated by Boivin do possess indeed the meaning that he attributes to them. I warmly thank Jean-Pierre Brach for having checked them for me.

311. Antoine Girard, sieur de Saint-Amant, "La Berne," in *Les Oeuvres du sieur de Saint-Amant* (1629). Cf. Saint-Amant, *Oeuvres,* 1:240.

312. Pierre de Lancre, *Du Sortilège,* 25, as cited in Godard de Donville, "Théophile, les 'Beaux Esprits' et les Rose-Croix," 150 and n. 19.

313. Pierre Gassendi, *Epistolica Exercitatio* (against Fludd) (Paris: Sébastien Cramoisy, 1630), cited in Mrs. Tannery, de Waard, and Pintard, *Correspondance du P. Marin Mersenne,* 2:183: "beatissimi isti fratres Crucirosei, qui ut Gygis annulo tuti nos semper invisi circumvolitant." On the Rosicrucians, see especially this work by Gassendi, 188–274.

314. [Théophraste Renaudot], *Quatriesme Centurie des questions traitees aux conferences du Bureau d'Adresse, depuis le 24ᵉ Janvier 1639. jusques au 10ᵉ Juin 1641* (Paris: Au Bureau d'Adresse, 1641), 53–60.

315. Ibid., 53–55, 55–57, 57–58.

316. Ibid., 59.

317. Ibid., 59–60.

318. Ibid., 73–76.

319. Cf. Bernard Gorceix, *La Bible des Rose-Croix. Traduction et commentaire des trois premiers écrits rosicruciens (1614–1615–1616),* Collection Hier (Paris: Presses Universitaires de France, 1970), 5 and 7. The beautiful translation by Gor-

ceix, which is quite sufficient here, is unfortunately not literal enough. On Paracelsus and the *Liber ·M.*, see Roland Edighoffer, "Le *Liber M.*," *ARIES*, no. 15 (1993): *Magie du livre, livres de magie*, 74–84, esp. 78–83.

320. [Renaudot], *Quatriesme Centurie des questions traitees aux conferences du Bureau d'Adresse*, 75.

321. Ibid., 75–76 (*Mons:* "mountain"; *Mus:* "mouse").

322. Howard M. Solomon, *Public Welfare, Science, and Propaganda in Seventeenth Century France: The Innovations of Théophraste Renaudot* (Princeton: Princeton University Press, 1972), 93. On the Bureau d'Adresse, see also Lerner, *Tommaso Campanella en France au XVIIe siècle*, 57, n. 24.

323. René Pintard, "Les problèmes de l'histoire du libertinage. Notes et réflexions" (1980), republished in *Le Libertinage érudit*, xiii–xliii, here xxii.

324. Cited in ibid., xxxvi–xxvii; ibid., 175 and 653 (note 6 of 543) on La Mothe Le Vayer.

325. Cyrano de Bergerac, *L'Autre monde ou les Estats et Empires de la Lune* (composed in 1650, 1st ed. 1657), ed. Madeleine Alcover, Société des Textes Français Modernes (Paris: Honoré Champion, 1977), 66–67; about these pages, see my article "Les apparitions du Démon de Socrate parmi les hommes," in P. Harry, A. Mothu, Ph. Sellier (eds.), Dissidents, excentriques et marginaux de l'Age classique. Mélanges en l'honneur de Madeleine Alcover (Paris: Champion, 2005), p. 483–550.

326. Cyrano de Bergerac, "Pour les sorciers" (1654), in *Lettres satiriques et amoureuses*, eds. J.-C. Darmon & A. Mothu (Paris: Desjonquères, 1999), 79–84.

327. For example, Cyrano de Bergerac, *Oeuvres complètes*, 115, 378, 498.

328. *Lettres satiriques et amoureuses*, 85–93.

329. Antoine Arnauld and Pierre Nicole, *La Logique ou l'art de penser*, critical ed. by Pierre Clair and François Girbal, Le mouvement des idées au XVIIe siècle, no. 3 (Paris: Presses Universitaires de France, 1965), 93; ed. Charles Jourdain (1865), reprinted in series Tel (Paris: Gallimard, 1992), 85.

330. See Arlette Lebigre, *L'Affaire des Poisons, 1679–1682*, in series La Mémoire des Siècles (Paris: Éditions Complexe, 1989).

331. Ibid., 94.

332. On *La Pierre philosophale*, see my article "L'alchimie sur la scène française aux XVIe et XVIIe siècles," *Chrysopoeia*, 2 (1988): 62–96, here 66–76; on this play and on *La Devineresse*, cf. Lebigre, *L'Affaire des Poisons*, 82 and 125 (quotation 82); on the possibility of Fontenelle's collaboration, cf. Alain Niderst, *Fontenelle à la recherche de lui-même (1657–1702)* (Paris: Nizet, 1972), 24–25 and 104–9.

333. This is the opinion of Lebigre, *L'Affaire des Poisons*, 125: "Occultism was no longer a big success at the box office, the 'Affaire' had made the public sick of it, the magicians were in prison and no one knew when and how the interminable drama would be resolved." When we know what sinister dramas were concealed behind the theme, harmless in appearance, of the search for treasures, we can understand that the audience was scarcely disposed to laugh at it: see *L'Affaire des Poisons*, esp. 117 and 113.

334. Nicolas Montfaucon de Villars, *Le Comte de Gabalis,* ed. Roger Laufer (Paris: Nizet, 1963). On Charpentier's collaboration, noted also by Lebigre (*L'Affaire des Poisons,* 125), see Catherine Cessac, "Éléments pour une biographie," *L'Avant-Scène Opéra,* 68 (October 1984): 4–22, at 12 (I warmly thank Jean-Noël Laurenti for this reference).

335. See my "L'alchimie sur la scène française aux XVIe et XVIIe siècles," 69–71.

336. *Traitez du Cosmopolite Nouvellement découverts. Où aprés avoir donné une idée d'une Société de Philosophes, on explique dans plusieurs Lettres de cét Autheur la Theorie & la Pratique des Veritez Hermetiques* (Paris: Laurent d'Houry, 1691). On this work, see Robert Amadou, "Le 'Philosophe Inconnu' et les 'Philosophes Inconnus.' Etude historique et critique," *Les Cahiers de La Tour Saint-Jacques,* 7 (1961): 65–138, at 90–103.

337. [François Alary], *Le Texte d'Alchymie, et le Songe-verd* (Paris: Laurent d'Houry, 1695), Préface, 25–28.

338. See my article "Littérature ou alchimie? A la recherche de l'authentique *Songe vert,*" in D. Garrioch, H. Love, B. McMullin, I. Morrison, and M. Sherlock, eds., *The Culture of the Book. Essays from Two Hemispheres in honour of Wallace Kirsop* (Melbourne: Bibliographical Society of Australia and New Zealand, 1999), 218–31.

339. [François Alary], *Le Parnasse assiegé ou La Guerre declarée entre les Philosophes Anciens & Modernes* (Lyon: Antoine Boudet, 1697), Préface, fol. A2r. On this text, see Allen G. Debus (who was unaware of the author's name), "The Paracelsians in Eighteenth Century France: A Renaissance Tradition in the Age of the Enlightenment," *Ambix,* 28 (1981): 36–54, here 38–39, repeated in Debus, *The French Paracelsians,* 164–65.

340. [Alary], *Le Parnasse assiegé,* 1, 3–5, 6, 9–10, and 14–15.

341. N. F. J. Éloy, *Dictionnaire historique de la médecine ancienne et moderne* (Mons: H. Hoyois, 1778), 1:734–35.

342. More facts will be found about the years 1623–1625 in my doctoral dissertation, "Paracelsisme et alchimie en France."

7

"The Food of Angels": Simon Forman's Alchemical Medicine

Lauren Kassell

Simon Forman is infamous for his astrology, notorious for his magic, and legendary for his sexual exploits.[1] He was born in Wiltshire in 1552, received little education, and spent his early life acting as a tutor and schoolteacher, occasionally practicing medicine, and studying astronomy, astrology, medicine, magic, and alchemy. He moved to London in 1591, within a few years established an immensely popular astrological practice, and became the self-appointed nemesis of the London College of Physicians. Forman styled himself as an astrologer-physician, but until Cambridge granted him a license to practice physic and astronomy in 1603, he had no official credentials to practice any sort of medicine.[2] He argued that astrology, which was conventionally one of the tools of physicians, should determine all medical diagnoses and therapies.[3] He had arrived at this position not through formal study, but from eclectic reading. His knowledge, he thought, was divinely ordained, and for him astrology was one of a handful of hermetic arts that he aspired, and strove, to achieve.[4] On one occasion he described himself as "a god among men" and "borne to find out arte and to make yt perfecte."[5] Accordingly, he wrote numerous, lengthy astrological treatises.[6] Around 1606 he began a treatise on the philosophers' stone, but after writing that this was intended to preserve knowledge for posterity, "wherin they shall see the course of natur and facility of things done in tyme with discretion," he left off.[7] Forman had begun drafting this tract on pages that contained an unfinished transcript of a treatise attributed to Hermes Trismegistus, and this likewise remained incomplete. However inspired Forman might have been, he failed to articulate in a coherent text his ideas on the philosophers' stone and on the relationship between alchemy and medicine.

Forman's industrious pursuit of the secrets of the philosophers' stone falls into two phases.[8] During the first phase, which was underway by the mid-1580s and lasted a decade, Forman transcribed more than a hundred alchemical texts.[9] As with many Elizabethans, Forman's pursuit of alchemy did not depend solely on printed texts, of which there were very few in the vernacular, but on the circulation of manuscripts.[10] The second phase in Forman's study of alchemy was the compilation of three commonplace books.[11] Forman's first volume dates from around 1597, when he began reorganizing and synthesizing his alchemical notes under subject headings covering alchemical materials, preparations, and principles, arranged loosely alphabetically in a volume on the "Principles of Philosofi."[12] This enormous volume was at some point conceived of as a whole. Between c.1607–9 Forman compiled an even larger alchemical volume, "Of Appoticarie Druges," in which he incorporated most of the information from the earlier volume.[13] A third, shorter commonplace book has no title and no date.[14] The entries in this volume were organized according to preparations and principles, and, unlike in the other volumes, specific chemical and other substances were not discussed under their own headings.

Why did Forman engage in this extensive project of compiling alchemical manuscripts? He was not a gentleman or an academic like John Dee, Robert Fludd, or Girolamo Cardano; Forman did not draft a text for the printing presses or even to be circulated in manuscript.[15] While studying these texts he established an astrological practice in which he was consulted many times a day. Most of the questions that he was asked were medical. His copying, annotating and recopying of these alchemical manuscripts document an association between medicine and alchemy, and reveal a definition of hermetic knowledge in which alchemy, astrology, and magic were integral to medical diagnoses and therapies. His alchemical notes were more than idle doodlings, and his medicine more than quackery.[16] Forman saw himself as a magus, divinely chosen to possess knowledge that others could not. However hubristic Forman's self-portrayal, his manuscripts reveal a culture of the occult in Elizabethan London that centered on medicine.[17]

"Of Cako," or Alexander von Suchten's "Second Treatise on Alchemy"

The numerous copies of alchemical treatises that Forman made during the 1580s do not constitute evidence that he conducted alchemical experiments at that time. An alchemist needed time, money, and space, and Forman was not born with any of these. The earliest evidence of Forman's chemical pursuits dates from after the time he moved into his own rooms in London in 1592. In 1593 he recorded that he earned much money by distilling strong waters.[18] When he was first interviewed before the London College of Physicians, he claimed that he had cured twenty-three people of a fever by giving them an electuary of syrup and roses with wormwood water. In December 1594 he was involved in an elaborate procedure for distilling *argentum vivum*.[19] In April 1595 Forman noted making syrup of violets, and at Lent he wrote, "I began ye philosophers ston & befor mad my furnes and all for yt, as in my other bock yt aperes. I mad many sirups & drugs & distilled many waters & bought stills."[20] The following September Forman noted, "I drempt of 3 black cats, and of my philosophical powder which I was distiling."[21] Like many magi Forman frequently recorded his dreams, and in October 1595 he recorded one in which a man wrote some words about the philosophers' stone on Forman's coat and gave him two kinds of white powder.[22] The following December he noted that he had been very unlucky, broken "two glasses and lost the water."[23] Three months later, in April 1596, Forman recorded another dream in which a friend gave him some of the philosophers' stone in liquid form. Forman took it in his hand, and before he could find a glass to hold it, it ran through his fingers.[24] His pursuit of the philosophers' stone continued, and in March of 1596 he noted that "[i]n subliming of ☉ & ☿ my pot & glasse brokee & all my labour was lost per lapidem."[25] In November 1598 Forman was more successful. He and an unnamed associate made an amalgam of gold and silver, then philosophical mercury, iron and "cako." This went through several operations, and a few days later was put into "our ege."[26] These disparate notes illustrate that Forman was engaged in distilling herbal preparations and conducting alchemical experiments, though they contain no evidence that Forman's alchemy had any bearing on his medicine. For this we must turn to Forman's copying of alchemical texts.

The same month as Forman had included "cako" in an alchemical preparation, November 1598, he had also transcribed a short treatise "On Cako" and interjected his alchemical experiences and observations throughout the text.[27] Although Forman did not know who the author of this text was, it was to become well known in the next century as

Alexander von Suchten's "Second Treatise on Antimony."[28] Suchten was a Prussian follower of Paracelsus.[29] Although Suchten's treatises on alchemy were not published in English until 1670, manuscript translations of the second treatise had been circulating since the 1570s.[30] Forman's text differed from other English versions of Suchten's text in two ways. First, Forman's text was not attributed to Suchten or anyone else; Forman simply noted "authoris incogniti." Second, in Forman's text antimony was referred to throughout as "cako," and the word antimony was omitted.[31] This word, which was of Hebrew origin, did not appear in any of the other versions of Suchten's text or in other alchemical works.[32]

In addition to these differences, Forman's copy of Suchten's text (which we will call "Text A") diverged in content from the other English versions. First, Forman's text was framed by a different beginning and a different ending. Forman's copy began with a description of cako in the florid language of medieval alchemy, which might have been drawn from an older text. With a series of tropes it outlined the preparation of cako in the language of life and death. In contrast, the 1575 translation had an epistolary opening in which Suchten stressed that he would write about common antimony in a plain style. Once he had described the manual operations for preparing antimony, he would leave it up to the reader to decide if this was the medicinal arcanum of which the old magi and Paracelsus had written.[33] Furthermore, Forman's text continued with a lengthy section headed "Opus Magnum" that contained more preparations for cako, most of which employed florid alchemical language. Second, most of the references to Paracelsus had been omitted from the text that Forman had copied, though the attacks on "Galenists" were preserved. It appears that Suchten's text had been subsumed within an older alchemical tradition, and Forman had copied it in this form.

Unlike most of the alchemical tracts that Forman had copied, Suchten's "Second Treatise on Antimony" was explicit about the medicinal virtues of alchemical preparations. He described how antimony came out of the mine in a crude form and had to be purged and cleansed in a series of four operations in order to reduce or mature it into a series of purer substances and ultimately into gold. Three quarters of the way through this tract there was a section on "what medicine ther is in vulgar Cako."[34] Suchten stated that many men had prepared antimony and used it in medicines, but none had perceived the medicinal secret because the antimony of physicians was not the same thing as the antimony of philosophers. In Forman's text Suchten's explanation of the medicinal virtues of antimony read as follows: "For in the cako of philosophers ar all the medicins potentiall and for that cause it is called quintesscence. But in the Cako vulgar is not the quintessence medicinalle but the elemente and the matter only of the Quintessence. The which essence is a medison againste all diseases which proceed of the fier of the lyttle wordle [sic]."[35] This was elucidated by an analogy: "Ther is a fier within the wood, which we must have in our kytchens for to dress our meate. Soe ther is a fier in Cako, by which we dresse our medison, the which by it receyveth the essence, and by the same essence extinguishethe the elementalle heate in our diseases." The cako of physicians was an ordinary and crude substance; philosophical cako rendered substances medicinal by purifying them and making them essential, and an essential substance restored health. This, according to Suchten, was contrary to the practices of Galenic physicians, who did not believe in the virtue of purity and were forever combining substances in juleps and masking their natures with honey. Furthermore, the alchemist differed from the Galenist in his theory of medicine: "The Gallienestes doe boaste them selves to doe awai the heate [of diseases] with endive and poppie and nightshade and with other cold simples, which they cannot doe, unlesse the heat doe naturally cease of it self."[36] The virtue of cako was that it caused the body to purify itself by the natural means of sweating, not by altering the balance of the humors with purges or emetics.[37] In these passages Suchten outlined some of the tenets of Paracelsian medicine.[38]

Forman copied most of Text A into "Principles of Philosofi" under the heading "Cako," which will henceforth be referred to as "Text B."[39] He signaled the importance of cako to the alchemist in a verse preface to this commonplace book. The verses of this preface were written in the language of marriage commonly used by alchemists, and Cako, Mars, Mercury, and Luna played the leading roles.[40] In incorporating the text on cako into this commonplace book, Forman significantly altered it in two ways.

First, he replaced the opening section of Text A with a more lucid definition: "Of this Cako coms the great secreate of philosophers. For from Cako commeth Regulus, and well yt may be called soe for yt ruleth and governeth all the reste. For by this regulus is drawen forth the Sulfur of all mettals apt to the philosophers stone, and with out the which yt cannot be done." Forman then, perhaps referring to the opaque description at the beginning of his initial copy of this text (Text A), proceeded to explain that cako was a philosophers' secret that was not written down in books but transmitted by tradition from generation to generation; when it had finally been written about, endless tropes had been used to obscure it from posterity. Forman drew on his experience to support this:

> I cam to the knowledge therof and with my owne handes and eyes. I sawe and proved the experience ther of at my owne coste and charges secretly for yt was the will of god yt should be soe. For in all my practizes and and [sic] workings I never came to any knowledge, but only by the will of god and by my owne industry & coste. Neither had I any frinds that ever gave me 40s. in all my life. Not that they could not—but because they wold not, but put me to live and shifte for my selfe when I was but 12 years old. Yet when they all forsoke me for that I was soe moch bent to my bocke—god toke me to his grace and delivered me from hunger penury and mysery, and from imprisonments sclanders death & sicknes, and from infinit of other trobles and from many enimies which wer very mightie and strong as from Lawiars Councellors Justics Judges bishops and many others. And this had I by the will and grace of god. . . .[41]

Forman signed his name at the bottom of this passage. In addition, he removed the final section on "Opus Magnum." He also removed the evidence that he engaged with this text as a reader by omitting the passages that he had interjected into Text A. In other words, Text B is almost the same as Text A, except that it appears to be written by Forman.

Second, in copying Text A to Text B Forman introduced a section of "notes." He sympathized with the anti-Galenic sentiment Suchten expressed and took this further by adding a section on philosophical medicine. Forman described how the life of a substance could be revived "of and in the multiplication of the forme and not of the matter of mettalls:"[42] "for all the mistery of nature doth issue out of on fountaine & ar on essence, but miraculosly severed according to the will of god, the which is a specifica of all his creatures which is not comprehensible more then god is."[43] No dead thing could be raised without the addition of this soul, and any who taught to the contrary was not a complete philosopher:[44] "And yt followeth that in a lyving thinge ther be a nature & fashion of the thinge which should be raised again. For yt is the will of god that all things shall dye and that is the specifick of nature, the which after the death is multiplied infinitly. For his mortalytie doth put one ymmortallity and his corruption doth put one incorruption and his mortall body after his resurrection, is becom a glorified body able to give life to ded things."[45] He elucidated this with an organic metaphor: "As youe see that a grain caste into the earth is made quicke by the water, vz that in the grain is a ded water by which the water becommeth again a lyvinge water and it is a ferment of waters. Vz. yt giveth to the water his nature specifick, thus of on grain then groweth other infinite. So moste youe understand in this worke ☿ of cako in ♄ was his death."[46] With this section, Forman added a hermetic component to this text, a component previously effaced by the removal of references to Paracelsus. Thus Forman reconstructed this text to make it appear that he was its author, while at the same time strengthening the anti-Galenic, even Paracelsian, sentiments that it expressed.

Forman transformed this text further. Although in Forman's copy of Suchten's treatise cako directly replaced antimony, there is no evidence in Texts A or B that Forman equated these substances. For instance, the 1575 version of this text explained that crude antimony had to be purified into the regulus of antimony; Text A described the same sequence in terms of changing cako into regulus.[47] Then, in Text B, Forman attempted to relate cako and antimony by replacing this passage with an account of changing cako into antimony, then regulus.[48] Text B was recorded in "Principles of Philosofi," which has a separate entry for antimony that does not mention cako.[49] In "Of Appoticarie Druges" there is no entry for cako, and the entry for antimony notes that from antimony was made cako, and from cako, regulus, thus reversing the relationship between antimony and cako depicted in Text B.[50]

The progression from "Of Cako" to the entry of antimony in "Of Appoticarie Druges" is clear: In the first stage, Forman copied another author's text, adding notes of his experiments; in the second stage, Forman recopied the text and made several changes, including adding a section on philosophical medicine with hermetic elements; in the third stage, Forman reclassified cako under antimony, though he continued to define them as different substances. This is only one example of Forman's processes of compiling alchemical commonplace books. In this case he appropriated a text that, though he did not know it, was authored by a prominent Paracelsian physician, and despite Suchten's suggestions about the medicinal virtues of antimony and Forman's experiments with what he thought was cako and with antimony, there is no evidence that Forman used remedies made from antimony or other metals in his medical practice. In what follows first I will look more closely at Forman's engagement with hermeticism and Paracelsianism and the status of these subjects in England, then I will assess the role of alchemy in Forman's medical practice.

In the Beginning

In 1599 Forman transcribed the "Life of Adam and Eve" (hereafter "the Life"),[51] which, unlike most of the texts he copied, was not alchemical. It was written around the second century A.D. and was part of the tradition of the apocryphal texts known as the Books of Adam and Eve. It recounted the legend of Adam and Eve from their expulsion from the garden to their deaths.[52] Although many variants of the Life circulated throughout medieval Europe, there has so far been no study of the circulation of this text in Latin or in English in sixteenth-century England.[53] Most extant manuscripts of the Life produced in England date from the fifteenth century.[54] I have identified four versions of the Life in English, all dating from the late fourteenth or fifteenth centuries.[55] Forman's copy of the Life seems to be the only extant sixteenth-century copy of English origin in English or Latin.[56]

For our purposes, a structural comparison between Forman's copy of the Life and the Latin and English versions is instructive.[57] Forman's copy is in three parts. The first part (A) recounts the creation of Adam, then that of Eve, and their expulsion from paradise. This section is not present in the standard Latin version of the Life. The fifteenth-century versions of the Life of English provenance begin with the creation of Adam and Eve but do not contain as much detail as Forman's text.[58] The remaining sections are consistent with the other versions. The second part (B) "showeth what became of Adam after he was caste out of paradice." The third and final part (C) narrates "[h]owe Adam calleth together all his children and enformeth them of many things, and also telleth them that he is nere his death."[59] Here Forman's version contains elaborate details of the afflictions of the Fall and provides an extensive lists of diseases. In all cases except one, Forman's text has more detail than either the English or Latin versions.[60]

Forman not only copied the Life, he read it closely,[61] as is evident from his marginal notes. Marginalia are often intractable and difficult to discuss because they are not simply linear sequences; as in the following cases, a passage was often extracted from one context and aligned with another. The majority of Forman's annotations to the Life concern either the genealogy of knowledge or the nature of Adam and Eve before the Fall. We will look at each of these in turn. Forman wrote these marginalia by leafing through, juxtaposing and physically manipulating the Life and numerous other texts, many of which were alchemical.[62] The result was a conglomeration of hermetic ideas about the creation of the world and the macrocosm-microcosm analogy with a Paracelsian emphasis on the causes of disease.

To begin with, Forman annotated the margins of the Life with mythical details about the genealogy of knowledge. For instance, Forman annotated the passages describing Adam's knowledge with more particulars. Next to the passage in the Life describing how "god did replenishe him [Adam] with all kinds of wisdom Arte and Conninge and in the Science of Astrologie and knoweledg of the stars," Forman noted that after the Fall the angel Raziel gave Adam a book of astronomy and magic.[63] Later in the text, next to passages about the trials of Adam and Eve after the Fall, Forman gave further details about these books. Adam and Eve were discovered by Solomon, and

[w]hen Cabrymael the Angell byd him loke secretly in the Arke of the testament of god in the which he found all the boockes of Moyeses and Aron, and the bocks of Noah and of Jerimy and of the other profets the which Sallomon had long tyme sought for, and therin alsoe he found the boock which was called Raziel, the which god gave unto Adam, by the Angell Raziell, when Adam was dryven out of paradice. And he found also therin on another bocke named the Semiphoras which god alsoe gave unto Adam in paradice. And also he found therin another bocke that god gave unto Moyses in the Mounte Synay after Moises had fasted 40 dais & 40 nights. Therin did he find alsoe the rod of Moyses which was changed into a serpente and from a serpent again to a rod.[64]

This time capsule also contained the tablets on which the commandments were written, a square, golden table inlaid with fourteen precious stones, and a box inscribed with the seven great names of God. Forman had collected this information from a variety of sources, including Josephus, Augustine, and a popular fifteenth-century world chronicle, Werner Rolewinck's *Fasciculus Temporum*.[65]

The genealogy of knowledge was a common theme among alchemists. For instance, in a text that Forman transcribed and claimed to have corrected, Bernard of Treves, the late-fourteenth-century physician, gave the following history of alchemy: "The firste inventers of this arte as youe shall reed in the artes of memory: in the aunciente gests of the romains, in the imperiall boockes, and in the exposition of Alevetus upon the tables, and in many other boocks, was Hermes Trismegistus: for he made and composed the bocks of the 3 kinds of naturall philosophie, that is to say, vegitable, minerall & animall."[66] Another text that Forman copied conceded that God made "the first mane Adam perfecte in all naturall things, and didest endue him with sufficient knowledge." The text then outlined how this knowledge was imparted to the rest of mankind throughout the ages, beginning with "Beraliel" and "Abholiab" receiving the knowledge to invent metalwork.[67]

Whereas the philosophers' stone was conventionally divided into three types, animal, vegetable, and mineral, in "Of Appoticarie Druges" Forman recorded a four-part scheme according to which each type had a different history. Hermes had the animal, or angelic stone; Moses had the magical, or prospective stone; Solomon had the vegetable, or growing stone; and Lull, Ripley, and others had the mineral stone. Forman concluded, "The angellical stone is true medison to mans bodie against all infirmities and makes a man live longe and by that stone he obteined wisdom and knowledge of thinges in dreams & otherwise."[68] We will return to this below.

In collecting these details Forman was engaging with debates about the genealogy of alchemical knowledge. All accounts agreed that this knowledge was divinely imparted, but when and to whom was uncertain. One of the greatest points of contention was whether Adam had the knowledge of alchemy.[69] This was the subject of the earliest theoretical alchemical text printed in English, Richard Bostocke's *The Difference Betwene the Auncient Phisicke . . . and the Latter Phisicke* (1585).[70] Bostocke constructed an iatrochemical genealogy beginning with Adam. He challenged the assertion that Paracelsus promoted a new physic. In his scheme, medicine had become corrupt since the Fall, and Galenic medicine perpetuated this corruption. Paracelsus had restored knowledge of the original, true, and ancient physic.[71] The year after Bostocke's text was published, a similar text appeared, likewise printed in English: *A Coppie of a Letter . . . by a Learned Physician* (1586), by I. W., who, like Bostocke, to whom he referred, argued that Paracelsianism was not "a new sect." Rather, "*it had his beginning with our first father Adam,* and so from that time to time hath continued untill this day: but indeed so amplified and enlarged of late, and brought unto every mans sight (that hath both his eies) by the long labour and infinite paines of *Paracelsus*, that it seemeth to be borne a new with, him."[72] These treatises were printed during the period when Forman was pursuing alchemy, and he may have read them.[73]

The issue of whether Adam had scientific knowledge (which was conventionally expressed as astronomy and natural history, not alchemy) had theological importance.[74] When knowledge was imparted to Adam and what this knowledge constituted (the fall of angels, the Fall and redemption of man) had ramifications for the interpretation of free will. The significance of Adam's knowledge for alchemists was that it potentially contained the secrets of the relationship between the microcosm (man's body) and the macrocosm (the universe). The Fall resulted in the corruption of life, in disease, and in the loss of the prime substance. For Paracelsus and his followers, the pursuit of the philosophers' stone was the pursuit of the prime substance, the material from which the universe was made. Accordingly, the theme of creation, which is prominent in hermetic texts, became a constant point of reference for alchemists, especially Paracelsus.[75] The operations for creating the philosophers' stone were analogous to those undertaken by God in the creation of the world.

Before returning to Forman's version of the Life, we should note that although some alchemists discussed creation, few English alchemical writers seem to have turned from the alchemical analogy with creation to general expositions on creation. Forman did this in "Upon the firste of Genesis."[76] He began with definitions of God and creation, then discussed the creation of heaven and earth and when and of what they were created. He structured this tract like a conventional Biblical commentary, quoting a verse and then expanding on it. He drew on a range of authorities, including Augustine, Nicholas de Lyra, and the Picatrix. Forman narrated how God had created something out of nothing, which was God's prerogative.[77] According to Forman, God spoke the word "fiat" and "of this word fiat came the chaos. For as the breath in cold weather goinge out of a mans mouth becommeth thicke and is seen: and is condensate to a cloud or water which riseth like a miste from the mouth of a man and after dropeth downe and ys seen which before was nothinge: soe lykewise of that worde fiat beinge once pronounsed cam firste an Invisible Substance by power Imperiall of the Creator."[78] This was more than a simile. The second part began with a description of how and of what God had created the heavens and the earth: "The heavens firste as chife and principalle Agente and father of all thinges superioure, and the Earth as principall passive mother of all thinges inferioure."[79] The macrocosm-microcosm analogy was implicit in this statement. An explicit analogy between the creation of the world and the art of the alchemist followed:

Then was the sprite of the Lord borne upon the face of the waters which was the liquide forme of thinges and the moste apteste to make moste formes shapes and creatures of. As for example a man taketh a great pote & filles yt with water honni oylle wine verjuce milke and such lyke lyquid thinges, and he setes yt on fier to distill howe many sortes of water may he drawe out of this smalle Chaos, & every on better then other. And yet in thend [sic] ther is dregs lefte, which may be Congealed into a thicker or harder masse, out of which again also, a man may drawe or make divers other thinges, and formes.

Soe was yt with god in his huge Chaos, out of which by his worde he drewe all the formes in the wordle [sic].[80]

The text is rich with alchemical meaning in its description of the division of the waters, the creation of the eighth heaven, and further analogies to minerals.[81] This followed alchemical tradition, and Forman's attention to it was marked by his frequent note of "chaos" in the margins of the texts he had copied.[82] Under the headings for "Chaos" in his commonplace books Forman gave a very similar account of creation.[83] One of the definitions of Genesis was "The trewe knowledge, wherof to many is rare which moch may healpe thee in tyme of need."[84] The extent to which Genesis contained the secrets of alchemy, the secrets of life and death, was manifest in Forman's annotations to the Life on the subject of the nature of Adam and Eve at creation and the corruption that ensued with the Fall.

Forman's copy of the Life began with an account of the creation of Adam similar to that in his "Upon the firste of Genesis." The first sentence of Forman's copy of the Life stated that when God had made heaven and its ornaments he saw that they were good. The he made man out of nothing, in his own image and likeness. This fell within the boundaries of the debates about whether man had one spirit or three.[85] In this version Man was like God in three things: "The first like unto his ymage: for at that tyme he had put him selfe in the same ymage wherin he made Adam, that ys to say in ye ymage of a man: The second was, he made him to his liknes: That is to saie In Righteousnes and puer hollynes. For Adam was righteouse Innocente and holy. The Thirde, he made him like unto himselfe in

Eternity, for he breathed into his nostrels the breath of life, and made him a lyvinge soule, to live for evermore."[86] The similarities between God and Adam in substance, appearance, and eternity corresponded to the three parts of the soul: animate, sensitive, and rational. Adam, furthermore, was created out of red earth, the slime of the earth, and the quintessential substance.[87] In the margin, next to this description, Forman added a passage defining the relation of the body, soul, and spirit and the composition of each.[88] At the bottom of the page he added a description of the three parts of the body of man—natural, animal and vital—and the same parts of the soul of man. Thus, Adam shared being with inanimate things, life with plants, sense with animals, and mind and intelligence with angels.

This annotation was very similar to passages in Forman's "Of Appoticarie Druges" under the heading "Anima & Spiritus."[89] The entry began with a discussion of whether man had only a soul or a soul and a spirit. In a passage almost identical to the one that flanks the composition of Adam in the Life, Forman stated that every man was composed of three things: body, soul and "sprite."[90] Next Forman outlined the relationships between these three things. Forman concluded "Anima & Spiritus" with a direct citation of almost the entire chapter from Cornelius Agrippa von Nettesheim's *De occulta philosophia* on the joining of man's soul with his body (book 3, chapter 37).[91] Agrippa described how, according to the Platonists, the soul proceeded immediately from God and was linked to the body by a "celestial vehicle of the soul": the *spiritus mundi*, "the quasi-material vehicle for the *anima*." The soul was thereby infused into the middle of the heart, from which it was diffused throughout the body. If disease (or "mischief") impeded the mediating force, the soul retracted to the heart. When the heart failed, the soul left the body. Thus an ethereal body was joined to a gross body. It seems that Forman, like Thomas Vaughan, read Agrippa's *De occulta* as a text sympathetic to and informative about alchemy.[92] These passages served as the introduction to a number of alchemical procedures, and Forman effected an abrupt transition, noting, "Now we com to speak of the soulle and sprite in mineralles and metalles and metal, animals & vegitable philosophically."[93]

In the margins of the Life, Forman was silent about Agrippa and named Paracelsus. Although Forman referred to Paracelsus throughout his alchemical notes, this was the only instance in which he singled out Paracelsus from the herd of alchemical magi. In 1591 Forman had copied an

English translation of a pair of works by Paracelsus.⁹⁴ Although Forman had engaged with these texts by adding many of his experiences to them, he did not rely on them any more than on the other alchemical texts he had transcribed. His notes reveal that he had access to other texts attributed to Paracelsus, some genuine, others spurious. In "Of Appoticarie Druges" Forman cited Paracelsus more than in his earlier texts, though he did not give him any particular authority. Since many of these references are drawn from the eleventh volume of Zaccharias Palthen's edition of the writings of Paracelsus, printed in 1605, Forman may have owned a copy of this volume.⁹⁵

Forman read texts attributed to Paracelsus and Agrippa's *De occulta* as containing information about creation. He synthesized these ideas, it seems, in a treatise on the microcosm that is now missing. Forman referred to the chapters in this work on the eighth, ninth, and tenth heavens throughout "Anima & Spiritus." A section of notes headed "microcosmos" is preserved, and it begins with an account of creation and concludes with an exposition on natural magic, drawing on Hermes, Paracelsus, the Picatrix, and other sources.⁹⁶ In this fragment, as in his other alchemical and related notes, Forman gives Paracelsus no special authority. Forman's annotations to the Life on the subject of the nature of Adam, however, cite Paracelsus at length. Above the description of the creation of Adam, Forman noted: "Pa[ra]celsus: The materiall seed of microcosmus was taken out of all the elements from all the places of the whole wordle [sic] into on place, and created man out of it and yt was don upon the water (which was matrix majoris mundi) and out of all thes did god mak man even of the vertu of all things."⁹⁷ Next to the description of God breathing into Adam's nostril Forman added definitions of the divine and animal souls that he attributed to Paracelsus.⁹⁸ The divine soul was eternal, only the imagination of man could kill it; the animal or elemental soul died with the body. At the bottom of the page Forman added a description of Adam as born with eternal life, power over all creatures, and "altitude": "that is the heighte and glory of all thinges. For when he was created ther was noe creatur in beuty shape and wisdom like to Adam. For the upper parte of Adam from the girdle upwards was in heaven in respecte of his purity and beinge. And the lowar parte from the girdle downwards was on earth, till he had broken the commandments. And before his fall he had no genitors, but after he was put forth of paradice, his genitors began to growe forth of him."⁹⁹ In these annotations Forman was drawing on the pseudo-Paracelsian "Liber Azoth." This text appeared in Palthen's eleventh volume of the works of Paracelsus, which Forman seems to have owned, and if this is the version of the text that Forman used, then these annotations can be dated to sometime after 1605.[100]

The annotations continued. The naming and composition of Adam was flanked by a note that glossed the story of creation, beginning by describing Adam and Eve as created "Angelically with the necrocomish soulle, that is with a soulle puer righteouse & undefiled & ymortalle." The description continued with gnostic terminology, recording that Adam and Eve had "a peculiar and dyvine Lymbus which was separated and did differ from the Lymbus of the earthy [sic] Adam." Limbus, a margin between the eternal and the concrete, was a gnostic concept that pervaded Paracelsian texts.[101] The annotation continued: when Adam and Eve were in the Garden, "they did feed and eate angelically of divine food, wherin ther was noe corruption poison nore infection of mortallity nor eternall death. For they eat of all the trees of the garden that wer good, ^after their nature,^ and ther was noe tre evill nor infected with dedly poison, but the tree of good and evill. For Adam & Eve were made good and did knowe nothing but good." God had put a commandment on this tree:

youe shall eat of all the fruits of the garden but only of [excluding?] this tre. For in this tree is good and evill life and deth, honny and galle, good meat & poison that will infecte thee with a contynuall sicknes and diseas therfore take heed of yt. So Adam & Eva did eat good fruite in which was noe poison nor evil to troble their bodies, as men use to eat good and holsom meats & never feell sicknes nor diseas, nor distemper of their bodies. But yf they eat the appell colloquitida or som rubarb ellebor agarick or som such thinge wherin ther is a poisoned substance (although yt show well) then theyr stomakes, bowells and whole bodi is sick & sore trobled, by which they presently knowe they have eaten som poysoened & evill thinge, wherby they presently knowe that ther ar bad meats as well as good.

But one could not know that there were poisonous as well as wholesome meats by being told; this knowledge had to be learned through experience. Adam could not have understood the poison until he had tasted it:

But then yt is to late. The poison & venom hath taken hould and root in them and they most die or be deformed or become monsters, for then their bodies swell and becom full of sore botches and blains and soe they ar altered and presently from their firste form & shape, as Adam was, which was first dyvine and had a heavenly form, but after he had eaten he was poisoned with sinn and felt the operation of the apple in his hart & body wherby he knewe he had don evill. . . . For his bodie being poisoned wth sinne he becam monstrouse and lost his first form and shape divine & heavenly and becam earthy [sic] full of sores and sicknes for evermore. And soe as a leprose man is chased or expelld out of the company of good hole and sound men leste they should be infected by him, even soe was Adam cast out of paradise."[102]

In his reading of the Life Forman combined gnostic, hermetic, and Paracelsian ideas about the origins of disease.

"The Food of Angels"

Disease was a major theme within the text of the Life. From his deathbed Adam called his children together (chapter 30). They asked him why he was lying in bed, and Seth asked if he could get him anything. Adam replied: "Sonne I desier nothinge but I wax full sicke and have greate sorowe and penance in my bodie."[103] Seth said that he did not understand. So Adam recounted the story of the Fall and God's anger. God had said that for having forsaken his commandments, "I shall caste into thy bodie seventy wondes ^ & too ^ of divers sorowes, from the crowne of the head unto the soule of thy feete, and all in divers members of thy bodie, be they tormented with soe many sicknises thou and thyne offspringe forevermore."[104]

In addition to the description of disease as the effect of the Fall, the physical condition of Adam and Eve, especially with reference to food, was writ large.[105] When they were expelled from Paradise, Adam and Eve went west and built a tabernacle where they stayed for seven days lamenting their Fall, "for losinge and wantinge their naturalle foode."[106] They had nothing to eat and were very hungry, so Adam began to look for food but could find none. This caused discord between them, and they searched together but could find nothing except herbs and grass such as beasts ate. And Adam said to Eve: "Our lorde god delyvered meate to beastes but to us he delivered meate of angells, the which he hath nowe deprived us of: and given us over to feed with the beastes of the filde."[107] After the births of Cain and Abel, Adam led them into the east, and God sent Michael to teach Adam to "worke & to till ye lande, and to provide fruite to live by" (Life, chapter 22). Forman's text continued with further details that were in neither the Latin nor the English versions. Adam's descendants lived by tilling the ground until after the flood. They ate herbs and fruit and roots but not flesh. It never rained, and the ground was moistened with mist.[108]

"The Food of Angels" 361

Lastly, at the end of the Life, Adam told Seth that he had this information because "I had my knowinge and my understandinge of things that is to com, by eatinge, that I eate of the tree of understandinge."[109]

In the sixteenth century food was central to interpretations of the Fall. The eating of the apple and the necessity to work for bread were integral to theological definitions of free will and knowledge. The questions of what man ate before the Fall and when man began to eat meat and discussions of the meaning of the necessity to till the ground and the longevity of the patriarchs figured in Biblical commentaries of the time. These themes were prominent, for instance, in the printed English commentaries by two Elizabethan divines, Gervase Babington and Nicholas Gibbens.

In *Certain Plaine, Briefe, and Comfortable Notes, Upon Every Chapter of Genesis* (1596), Babington included a section entitled "How dooth God appoint man foode before his fall":

Man is appointed heere his foode of God that he should eate, and some moove the question how that shall be. For if man were created immortall if he sinned not, what needed he any meate to be appointed for him, since yet he had not sinned. Answer is made by some, that there be two kindes of Immortall, one that cannot die but ever live, an other that may may live for ever, a condition being observed, and die also if that condition be broken. One imortall after the first sort needeth no meate, but he that is immortall after the second sort dooth neede, and such was Adam: if he had not sinned he had not dyed, but sinning he was so made, that he might die, and therefore his flesh and nature not such that could live without meate. Others answer that this appointment of meate was made by God in respect of their fall, which he knew would bee. Howsoever it was, curiositie becometh us not: but this comfort we may rightlie take by it, that what the Lord hath made, he will maintaine and nourish, and casteth for them his providence ever to that end.[110]

Likewise, Gibbens's *Questions and Disputations Concerning the Holy Scripture* (1602) discussed the significance of food and the longevity of the patriarchs. Gibbens asked whether God showed his liberality as much in providing food for Adam and Eve as in creating them.[111] He argued that man did not have flesh to eat, "which while mens bodies were immortall, because they were void of sinne, was no convenient foode to nourish them."[112] He answered the question "wherein consisteth the punishment of Adam?" with an exposition of "in sorrow thou shalt eate thereof [the ground]" (Genesis 3.17), explaining that Adam became proud, and God, like a physician, administered "a potion of humilitie, wherby man being dailie emptied of his old corruptions, might with hunger and thirst, gaspe for the death of Christ, which is the fruit of life."[113] He then outlined

Adam's tripartite punishment: "the curse of the earth; the miserie of life; and the end therof by death." Whereas before the Fall the earth had brought forth fruit of its own accord, it had been corrupted by sin; where once wheat had grown now grew weeds. This curse was man's misery: Working the soil bridled him from waxing proud. Other afflictions accompanied work and included bodily diseases and mental vexations.[114]

Gibbens elaborated these final points in the answer to the question of why the patriarchs lived so long. The answer was that it was "of the wisedom of the Lord for the disposition of his counsailes, for our sinnes, and the weakenes of our bodies, that we cannot now live so long as they." He explored this point further, noting the argument that over time the human body has become increasingly corrupt and unable to resist disease. Moreover, he gave two further reasons for the longevity of the patriarchs. First, "because they were of temperate and sober diet, not given so much to fleshlie appetite, nor mixing their meat with such varieties, but content with simple food, which the aboundance of the earth brought forth unto them." Second, "because the fruits of the earth were much more nourishable and healthfull before the floud, then afterward they were, either thorough the waters of the sea, bringing barrennes and saltnes to the earth, and to the fruits therof; or for that the Lord had given unto man more libertie of food, the fruite of the field was not so necessarie."[115] Although Forman did not necessarily read these commentaries, they demonstrate that within Elizabethan theology food and disease were current and associated themes.[116]

The food of angels, as a phrase or a concept, appeared throughout medieval religious texts.[117] There was also a medieval tradition, prominent in the writings of Roger Bacon,[118] of associating the prolongation of life with Adam and Eve. Food, furthermore, was central to the ideas attributed to Paracelsus and his followers. According to his doctrine of "Tartar," in simplified terms, food consisted of parts that were pure and impure, and the body could use only those that were pure. The impurities remained in the body, causing obstructions and resulting in disease.[119] Similarly, the Life explained disease as originating with the Fall. In the Garden Adam and Eve had consumed the food (or meat) of angels, but once they had tasted the forbidden fruit, their bodies and souls were corrupted, and man had thereafter suffered disease. Such a Paracelsian history of disease was articulated by Robert Bostocke in *The Difference Betwene the Auncient*

Phisicke . . . and the Latter Phisicke and by I. W. in *A Coppie of a Letter . . . by a Learned Physician*. It should be remembered that while Bostocke and I. W. gave clear formulations of these ideas, Forman did not.

In his second chapter, "the originall causes of all diseases in the greate worlde, and the little worlde, which is man," Bostocke recounted the story of the Fall in order to demonstrate that the binary principle of Galenic medicine, which employed opposites, was corrupt. He named the serpent that engineered the Fall "Binarius" and explained how he had persuaded man to eat the apple.[120] Adam had eaten, "[w]hereupon by the curse of God impure Seedes were mingled with the perfect seedes, and did cleave fast to them, and doe cover them as a garment: and death was joyned to life." This impurity was in all things, depending on the nature of the soil in which they grew or the food on which they fed, as experience showed: "But the foode and nourishments for mans body, though they have in them mingled, venemous, sickly or medicinable properties, yet for all that, by reason of that mixture with their good seedes, as long as unitie and concord is kept betweene them, they be tempered, seperated, resolved and expelled out of mans body." If this did not work, "the seedes of diseases do then take roote in mans body." If this did not work, "the seedes of diseases do then take roote in mans body." Man knew by the ancient art and by experience how to separate the good from the bad and the life from the death in all things. Thus, diseases proceeded from the breach of unity, and only in unity could they be cured.[121]

I. W. explained, "Now we may see that before the fall of *Adam* all thinges were good, all things came unto him and were bred unto his hand without his labour. But afterward part of it was joyned to poison, part of it so fast lockt up, that without great sweate of browes he should not eat of it." He took this opportunity to accuse the humanist physicians of regressive conservatism and laziness: "And in these our latter dayes sloth is growen so strong & idlenes hath gotten such masterie, that there are very few which will let one drop fal from their browes to seeke this bread, but indevour by all methodicall meanes to maintaine this idlenes." They sat on cushions in their chambers and wrote prescriptions for apothecaries to fill.[122] The original food, or "bread," was the philosophers' stone, which medical men should be seeking. He continued "If you had bestowed but half of your study in the first booke of Moses, which youe spent in the foolish Philosophie of *Aristotle,* you had espied your errors long agoe . . . for the offence of our first parents, death was not onely laid upon them, but for the same transgression God planted a death in every thing he had made, in every thing he put a death able to destroy such a life."[123] Genesis, then, taught good medicine.

Aside from Forman's copy of the Life, I have located only two other sixteenth-century texts referring specifically to the food of angels. The first is an eclectic manuscript on physic owned by Forman that includes a chronology of the afflictions of man and the discoveries of the knowledge to remedy them. For instance, in the year 3616 God had given Moses the commandments and had struck those who would not obey them with numerous infirmities, none of which could be helped with medicine. Those who were obedient were fed with "mana or angells food."[124]

A second sixteenth-century English manuscript referred to the food of angels as a form of the philosophers' stone: "The Epitome of the Treasure of all Welth," written in 1562 by one "Edwardus Generosus Anglicus Innominatus."[125] According to this text, St. Dunstan had departed from the standard description of a tripartite philosophers' stone (animal, vegetable, mineral) by adding a fourth type and redefining the other three.[126] Edwardus described the angelic stone as "preservative to the state of mans body," "by this stone shall mans body be kept from corrupcion also he shalbe [sic] endued with divine giftes & foreknowledge of things by dreames and revelations."[127] It was invisible and aromatic; it could also be tasted: "& therefore in St Dunstans worke itt is said that Solomon King David's sonne did call itt the foode of Angell, because a man may live a long time without any food having som taste of this stone."[128] This is the same scheme that, as noted above, Forman recorded in "Of Appoticarie Druges." Although there is no evidence that Forman read Edwardus's treatise or St. Dunstan's work on the angelic stone, somewhere he encountered St. Dunstan's definition of it. He did have a copy of an alchemical text attributed to St. Dunstan, which he recopied in 1608.[129] This text had a long history, Forman's copy being one of the three earliest.[130] It may be related to the work that Edwardus described as St. Dunstan's. Its subject was the mineral stone, and it did not mention the angelic stone. Elias Ashmole used St. Dunstan's description of the philosophers' stone, as recounted by Edwardus, almost a century later in the prolegomenon to his collection of English alchemical poetry, *Theatrum Chemicum Britannicum*.[131]

In whatever text or in whatever form Forman encountered the description of the angelic stone as the food of angels, his reading of the Life is evidence that he engaged with an alchemical tradition concerned with health. In addition to annotating this text with alchemical notes, Forman expanded further on the theme of corruption by inserting a list of diseases in the section of the Life (chapter 34) that described Adam's deathbed scene. Adam had announced that God had inflicted seventy-two diseases on him and his offspring for the Fall. In the Latin and English versions the narrative continued with Adam's expression of sorrow and pain to his children (chapter 35). Forman interrupted the narrative and inserted a list of diseases divided according to physiology and sex.[132] Twenty-one of these diseases afflicted both men and women alike and might occur in all parts of the body; twelve occurred only in the head; three diseases affected the throat; four each affected the breast and stomach; two affected the left side; four were diseases of the heart; three were of the bowels; twelve were diseases of the reins (kidneys); and fourteen afflicted only women. That Forman inserted this list into his transcript of the Life is clear from his changes to the numbers of diseases and the shifts in format and script. The number of diseases listed, for instance, adds up to seventy-five, not seventy-two. The list began: "Of thes diseases ther be 21 that be generalle both to man and woman." The "1" of the "21" was crossed through and a "2" was inserted so that it read "22." Twenty-three diseases were then actually listed, though the twenty-second and twenty-third extended into the margin. Forman's hand became less neat as the list proceeded, and he left blank spaces for the numbers of diseases to be inserted.

Whether Forman devised this list or derived it from another text, it might not have been a coincidence that two similar lists appeared in tracts published in England around this time. The first was in *A Coppie of a Letter* by I. W., discussed above, which explained that causes of disease were not "humors intemperie & obstructions"; humors were "the fantasticall inventions of an idle head, having no foundation or ground in nature." A list of twenty-nine diseases followed, most of which also appeared in Forman's list. I. W. concluded: "This is the cause that man dieth such sundry deaths, because hee eateth in his bread the death of all other things, which when perfect separation is not made, bringeth foorth fruit according to his kinde. Over these deaths hath the Physician power, and not over that which was injoyned to the body of man particularly."[133] Therefore the above diseases listed were curable, because they were diseases of the fruits of the world and not of man: They did not grow naturally in man but came to him through transplantation. Again, whereas I. W. articulated a Paracelsian definition of disease, Forman did not.

The association between disease and the Fall in the late sixteenth century was evident in a work of another genre. It appeared in Guillaume Saluste Du Bartas's *Divine Weeks and Works,* in the third part of the first day of the second week, "The Furies." Josuah Sylvester's translation of this section into English was printed in 1599, the same year Forman copied his treatise on Adam and Eve. Du Bartas recounted how after the Fall man was beset by three furies, sickness, war, and dearth. Sickness was described as attacking Adam in a mock-heroic battle, beginning with diseases of the head and moving down the body.[134] Du Bartas did not, however, describe Adam's deathbed scene, though the "Handicrafts" concluded with Adam near death from the sadness of his vision of the destructions of the future.[135]

Whether or not Du Bartas or I. W. influenced Forman's taxonomy of disease, their texts demonstrate that there were precedents in Forman's time for cataloguing the diseases that afflicted man after the Fall. No precedent for these diseases being catalogued in Adam's deathbed scene, however, has been identified. There are alchemical elements to Du Bartas's text, and thus it is possible that Forman, I. W. and Du Bartas were all influenced by similar texts and traditions. Whether or not the associations are purely coincidental, the presence of these lists in this range of literature in late-sixteenth-century England indicates that Forman was not alone in associating the Fall, disease, and alchemy. When Adam and Eve ate the fruit of the tree of knowledge, they sowed the seeds of disease within their bodies. When God expelled them from paradise, Adam and Eve kept free will and the knowledge of good and evil, the two vehicles, it seems, by which man had thereafter tried to return to the tree of life, once again to eat the food of angels and to achieve eternity.

Forman's annotations to the Life, and to his alchemical notes, contained the same blend of alchemy and the Fall. It does not require a leap of the imagination to envisage Forman with his copy of the Life, perhaps the 1605 volume of works attributed to Paracelsus, his notes on creation, and "Of Appoticarie Druges" simultaneously opened in front of him, each influencing how he read and annotated the others.[136] For Forman, the philosophers' stone was the food of angels, and its secrets lay in Genesis.

Magic and Medicine

Forman's annotations to the Life and his copying and modifying of Suchten's treatise on antimony are evidence that he subscribed to a Paracelsian philosophy of medicine; they do not reveal whether these ideas went beyond the academic and had some bearing on his astrological physic. According to Forman, his therapies differed from those of humanist physicians in that he looked to the stars for the cause of the disease and for the timing of the treatment. As recorded in his casebooks, his prescriptions and therapies were conventional: Sometimes he evacuated superfluous humors by phlebotomy, purging, or vomiting; sometimes he recommended a fortifying drink.[137] There is, moreover, no evidence in Forman's notes that his alchemical pursuits produced chemical medicines; rather, alchemical philosophy informed, or even inspired, Forman's use of magic as a means of performing medical diagnoses and treatments.[138]

For Forman, alchemy and magic, along with astrology and geomancy, were kindred arts. In an outline of the types of knowledge, Forman defined what he termed "astromagic" and "alchemagic" as the operative components of astronomy.[139] Astromagic involved the use of amulets and other objects to harness the power of the stars. Alchemagic was defined baldly as the means of transmuting metals and making the philosophers' stone. This sort of operative magic was the culmination of hermetic philosophy. The magus could employ his knowledge of the cosmos, through rituals and amulets, to make interventions in the workings of the world.[140] Forman's papers provide an unusual example, shrouded in secrecy, of the actual uses of operative and revelatory magic.[141] His activities involved a combination of cabala, amulets, sympathetic magic, and angel calling. His use of magic linked his hermetic philosophy and his medical practices.

In Elizabethan England the question was debated as to whether diseases caused by witchcraft and possessions had natural or supernatural causes.[142] Some scholars had argued that supernatural diseases existed that had natural manifestations, though this position seems to have been uncommon in England.[143] Forman held a similar position. He divided the causes of diseases into three categories, natural, "unnatural," and supernatural. Unnatural diseases had supernatural causes but were manifested naturally. According to Forman, God allowed evil people to do evil deeds, and the devil to do his will, but they could only work by natural means.[144] Although the physician was not to intervene in supernatural diseases, he could do so for unnatural and natural diseases, provided that he consulted the stars to assess what sort of disease he was dealing with and to determine whether to use natural remedies, supernatural remedies, such as amulets and prayers, or a combination of the two.[145] Supernatural remedies relied on natural magic. They were used to influence nature, via the stars, and to effect things that might not have occurred naturally in a certain time or place but were within the scope of natural possibilities. For instance, in a section on sigils in "The Astrologicalle Judgmentes of Phisick and Other Questions" (c.1599), Forman, probably following Paracelsus, noted that oranges did not naturally grow in England, but, with the use of appropriate amulets, could be made to do so.[146]

In practice Forman did not draw a strict line between therapies for natural and unnatural diseases. Discovering the cause of a disease could be difficult, and there was always room for doubt about whether the causes were natural, unnatural, or supernatural, especially when the patient's disease did not respond to therapy. For instance, in December 1598 Nicholas Chapman consulted Forman a number of times. Initially Forman determined that Chapman's disease was natural, but he later decided that it was caused by witchcraft.[147] Another example is the case of Jackamyne Vampenathe (or Vampena), a 47-year-old Dutch woman who first came to Forman in June 1601 suffering from melancholy. Ten days later she consulted Forman again, and he noted "she despairs in god." Three weeks later she was no better, and her husband, John Stockbridge, a merchant, agreed to pay Forman £12 if he could cure her. Forman recorded an extensive pharmacy in his attempts to purge and to sedate this woman. He gave her five types of pills and seven types of strong water over the next two weeks. The "ingredients" of one of the waters included the immersion of a ring engraved with the symbol of Jupiter.[148]

Forman had designed numerous amulets and rings, which he also referred to as "sigils," "laminas," and "characts." He commissioned other people to make them and oversaw the process.[149] These objects often included astrological, cabalistic, and other magical symbols and had to be made at astrologically propitious moments. In Forman's words, as part of the definition of astromagia, these objects "enclosed som parte of the vertue of heaven and of the plannets according to the tyme that [they are] stamped caste or engraven or written in."[150] As already noted, Forman was familiar with numerous magical and cabalistic texts, including the Picatrix, Agrippa's *De occulta philosophia*, and texts attributed to Paracelsus, and he might also have been influenced by Trithemius's *Steganographia*.[151]

Forman often specified the medicinal powers of amulets. In 1611 he sent Richard Napier, the rector of Great Linford, Buckinghamshire, and Forman's astrological protégé, some molds and in the accompanying letter noted the following: "Yf youe have them, and can tell howe to use them youe have a good thinge aswelle for the cueringe of diseases as for divers other purposes."[152] In 1609, among his notes on cabala, Forman copied extensive passages from Agrippa's *De occulta philosophia*, added his revisions, and specified that these symbols could be used in cases of diseases.[153] In notes on "electrum," an amalgamation of two or more metals, Forman recorded Paracelsus's description of the virtues of rings made of electrum against poisons and witchcraft. If such a ring was worn on the "heart finger" it prevented cramps and the falling evil and would change color when sickness or evil was directed at the wearer.[154] Forman was also interested in more traditional sympathetic magic. He made notes describing how to use a homunculus, or a small clay or wax image of a person, for healing and other types of magic at a distance.[155] In 1603 he recorded the case of a man imprisoned for bewitching the son of Sir John Harris and found in possession of a seven-inch, hirsute mandrake that moved, spoke, and drank blood.[156]

As already noted, Forman's interest in amulets was more than academic. In 1583 he had a ring made with the "eagle stone," which may have had some magical properties.[157] In his November 1597 casebook he recorded a design for a sigil.[158] That year he had lost a gold lamina that he had worn on his chest.[159] In April 1598 he infused a ring with astral properties, and that summer he made several more.[160] In 1599 Forman had an amulet with a coral stone made, as well as rings for himself and for Mrs. Blague.[161] One of these rings had a piece of parchment inserted beneath the stone on which was written the names of the stars underneath which Forman was born. It was designed to be worn on the little finger of his left hand and would protect him against witchcraft and other ills as well as giving "favour & credit & to make on famousse in his profession & to overcom enimies."[162] A decade later, in March 1609, Forman drew a series of characters on his left arm and right breast in semipermanent ink to alter his destiny.[163] In April 1601 Forman designed an amulet for Martha Shackleton, perhaps one of his clients.[164]

Although Forman used operative magic for medical and other purposes, he also practiced revelatory magic by calling angels. In 1588 he first "began to practice foiygomercy [sic] and to calle angells & sprites," and a boy named Steven skryed for him.[165] In 1590 Forman noted that he "entred a cirkell for nicromanticall spells."[166] In notes on invoking apparitions and spirits Forman included many examples of questions about whether he would achieve the power of necromancy as well as records of his attempts. For instance, in September 1591, he recorded that a spirit did not appear.[167] Again Forman's dreams are revealing, and in August 1594 he noted that he dreamed a spirit appeared in three shapes in three different glasses and that his three companions could see it, but he could not.[168] Four days later Forman recorded, "I drempt I did see in a glas when I did call and that I did heare alsoe & that yt was the first tim that ever I did heare or see & I was annswered directly of all things."[169] The next month Forman dreamed that he was lost on a highway and the angel Michael appeared to him to show him the way.[170] In the late 1590s Forman used John Goodage, "a gelded fellowe," as a skryer.[171] On October 29, 1597, "the sprite came and shook the bed for or five times and cast out such a fire and brimstone that it stank mightily and that night he kept much adoe and rored mightily but I saw him not. but I sawe the fire & then sawe him in a kind of shape but not perfectly." Two days later it appeared again; this time it cast out much fire but could not be brought to a human form: Instead it took the form of a large black dog. In another account of the same session Forman recorded that he heard the spirit but could not see it.[172] In 1599 Forman simply noted "I had a sear sometimes to call," and in August of that year he dreamed about one John Ward calling angels in a church.[173]

These examples reveal that Forman tried to call angels but say little about how he did it, what he hoped to achieve, or if he succeeded; if Forman, like John Dee, recorded his conversations with angels, these notes are missing. A manuscript that Forman copied and recopied contains instructions for how to call angels for particular purposes, including medical diagnoses. In his diary entry for the year 1600, Forman noted, "This yere I wrote out the 2 boocks of de Arte memoratum of Appolonius Niger drawn with gould of the 7 liberal sciences."[174] This was the Solomonic text known as the Ars Memoriae, or Ars Notoria, with a commentary by Apollonius.[175] Forman copied this text at least three times and illuminated it at least twice.[176] It began with an account of the history of how God selected Solomon as a recipient of his wisdom, knowledge, and grace. He sent an angel, Phanphilius, to Solomon with some golden tablets on which were written orations containing the names of holy angels in Chaldean, Greek, and Hebrew. The orations were prayers invoking the names of angels and of God. They were inscribed in images of angels in various positions. Phanphilius taught Solomon how to use these tablets, and, following the angel's instructions, Solomon obtained all wisdom and knowledge. This art was then passed to Apollonius, a learned doctor and philosopher, who translated the orations into Latin and wrote a commentary on them. He who rehearsed these prayers in the appropriate sequence and at the appropriate times would achieve the understanding of all sciences, a perfect and enduring memory, and the eloquence with which to express such knowledge. This was the first step. The adept could then proceed to the knowledge of the seven liberal sciences, again by speaking the orations and observing the images at the correct moments over a period of time.[177] This was angel magic.

Although Forman's copies of many alchemical texts and of the Life are fair copies, it is unclear why he made multiple, illuminated copies of the Ars Notoria.[178] One copy, which is now in Jerusalem, included only a single interjection by Forman, an astrological note, and was probably the most similar to the parent text.[179] The colophon read, "This booke and al the figures and signs therin contained as youe here find yt was drawen out & written according to the old coppie by Simon Forman gentleman and d. of physick with his own hand 1600 Anno Eliz 42 June."[180] The other two copies were working texts. The copy that is now at Trinity College, Cambridge, had no illustrations and was dated June 28, 1600, the same month as the Jerusalem copy. It also contained a lengthy gloss, in English, on the text. At the end Forman added a number of prayers from other texts, including some from printed books, and others that "I toke out of the other bock that was writen in paper that Mr Conie brought me."[181] The copy of the manuscript that is now in the Jones collection in the Bodleian, Oxford, was in progress between 1600 and 1603, and in it Forman incorporated his notes among the text.

Unlike his annotated copies of the Life, Forman's copies of the Ars Notoria contain few clues as to how he read it. The text itself nonetheless contains evidence of a link between Forman's hermetic ideas and his astrological physic. Each of the angelic figures represented a different art, and one of them was devoted to physic. This figure was accompanied by the following instructions, with the precept that these operations were to be done only by someone who had achieved the preliminary knowledge. While standing at the sickbed, the oration was to be spoken with great reverence and in a low voice: "by and by it shalbe [sic] declared to thee and suggested in thy minde by angelical vertues wheather that sicke partie shall recover health or die of that same sickness."[182] Forman often sought an answer for the same question in the stars. This oration had further medical uses. To determine whether a woman was pregnant, the practitioner was to stand in front of her and to utter the prayer. The voices of angels would reveal whether she was with child, and if so, what sex it was. Likewise for the question of a woman's virginity. These were three questions an astrologer-physician frequently asked.[183] Whereas the astrologer mapped the heavens at the time of the question and judged the answer according to a set of rules, the Solomonic adept performed the required ritual and was inspired with knowledge about the patient. The ritual for an astrological interview and a magical action were almost the same.

To conclude that Forman called angels in his consulting room would be to overstate the case. There is no evidence that he, or any of the authors whom he followed, did this.[184] Forman's philosophy of medicine is most evident in how he responded to questions about disease. He calculated an astrological, or on occasion a geomantical, figure for the time at which he was consulted and read this for the cause of the disease or to foresee its outcome. He may then have negotiated his conclusions with the patient or the person who had asked the question. If he judged the disease to be natural or unnatural, he might treat it with herbal or magical remedies, all of which had to be administered at astrologically propitious moments. All of Forman's medical activities relied on an analogy between the microcosm and the macrocosm and on his role as a magus in possession of celestial and supercelestial knowledge. This knowledge was recorded in texts and revealed in dreams, and Forman thought it enabled him to read the stars and to hear the voices of angels. Forman's astrological expertise, pursuit of the philosophers' stone, and magic were components of hermeticism.

Through his copying and recopying of "Of Cako," annotations to the Life, and attention to the Ars Notoria, we have charted Forman's pursuit of the hermetic and Paracelsian secrets of medicine. Forman was neither an innovative alchemist nor a rigorous scholar; his study of alchemy, however, was inscribed in his medical practice. For Forman the spiritual pursuits of the magus were grounded in the mundane ambitions of making a living as an astrologer-physician.

Acknowledgments

I am indebted to David Colclough, Tony Grafton, Nick Jardine, William Newman, Margaret Pelling, Joad Raymond, and Charles Webster for comments on various versions of this chapter. Early versions received many helpful comments from audiences at seminars at All Souls' College, Oxford, and the Wellcome Unit, Manchester. This chapter was written before I was able to make full use of the work of Deborah Harkness, David Harley, Michela Pereira, and Lawrence Principe, citations of which can be found in the notes. Most of my research on Simon Forman has been funded by the Wellcome Trust.

Notes

1. William Lilly, *History of His Life and Times,* ed. Charles Burman (London, 1774), 15–23; Michael MacDonald, "The Career of Astrological Medicine in England," in Ole Grell and Andrew Cunningham, eds., *Religio Medici: Medicine and Religion in Seventeenth Century England* (Aldershot: Scolar, 1996), 62–90; A. L. Rowse, *Simon Forman: Sex and Society in Shakespeare's Age* (London: Weidenfeld and Nicolson, 1974); Keith Thomas, *Religion and the Decline of Magic,* 2nd ed. (Harmondsworth: Penguin, 1973), 356–82 passim. These accounts draw on Forman's autobiographical writings and casebooks. For a full study of Forman's manuscripts, see my "Simon Forman's Philosophy of Medicine: Medicine, Astrology and Alchemy in London, c.1580–1611" (Oxford University, D.Phil. diss., 1997).

2. Oxford, Bodleian Library, Ashmole MS 208, Item 13, fol. 225v; 802, Item 13, fol. 133 (hereafter Ashm.). The license is Ashm. 1301, and Ashm. 1763 is Ashmole's copy of this.

3. For astrological medicine, see Allan Chapman, "Astrological Medicine," in Charles Webster, ed., *Health, Medicine and Mortality in the Sixteenth Century* (Cambridge: Cambridge University Press, 1979), 275–300.

4. The word "hermetic" is used to denote Forman's engagement with the tradition of occult ideas that were attributed specifically to Hermes Trismegistus. For

discussions of this terminology, see Brian Copenhaver, "Natural Magic, Hermeticism, and Occultism in Early Modern Science," in David Lindberg and Robert Westman, eds., *Reappraisals of the Scientific Revolution* (Cambridge: Cambridge University Press, 1990), 261–301.

5. Ashm. 1491, fol. 1248.

6. "The Grounds of Arte Gathered Out of Diverse Authors," 1594–95 (Ashm. 1495); "Liber Juditiorum Morborum," 1600 (Ashm. 355; for an exact copy by Thomas Robson see Ashm. 1411). There are several versions of "The Astrologicalle Judgmentes of Phisick and Other Questions," c.1599 (Ashm. 403, Ashm. 389, Ashm. 363; London, British Library, Sloane MS 99 (hereafter Sloane)). Ashm. 363 is the most complete version.

7. Ashm. 1433, Item 2, fol. 23r–v.

8. This is based on the dated, extant manuscripts.

9. These are contained in twelve notebooks, now bound in Ashm. 208, 1423, 1433, and 1490.

10. Charles Webster, "Alchemical and Paracelsian Medicine," in Webster, *Health Medicine and Mortality in the Sixteenth Century,* 301–34. Contrast Allen Debus, *The English Paracelsians* (London: Oldbourne, 1965), and Paul Kocher, "Paracelsian Medicine in England: The First 30 Years," *Journal of the History of Medicine,* 2 (1947): 451–80. For recent works on manuscript culture, see David Carlson, *English Humanist Books: Writers and Patrons, Manuscript and Print, 1475–1525* (Toronto: University of Toronto Press, 1993); Harold Love, *Scribal Publication in Seventeenth-Century England* (Oxford: Clarendon Press, 1993); Arthur Marotti, *Manuscripts, Print, and the English Renaissance Lyric* (Ithaca: Cornell University Press, 1995); and Henry Woudhuysen, *Sir Philip Sidney and the Circulation of Manuscripts, 1558–1640* (Oxford: Clarendon Press, 1996).

11. For an introduction to commonplace books as kept by John Dee and his Cambridge contemporaries, see William Sherman, *John Dee: The Politics of Reading and Writing in the English Renaissance* (Amherst: University of Massachusetts Press, 1995), 59–65. See also Ann Moss, *Printed Common-Place Books and the Structuring of Renaissance Thought* (Oxford: Oxford University Press, 1996). Forman's commonplace books were paginated or foliated through at some point, and according to this pagination each has had a third of its pages extracted. This was probably done by Forman as part of the process of compiling and recopying.

12. Ashm. 1472. This is the heading of the first entry, and as the subsequent entries are alphabetically organized, it is probably a prefatory section. The title suits the volume and will be used to refer to it.

13. Now bound in two volumes: Ashm. 1494 and 1491.

14. Ashm. 1430.

15. None of his works, with the exception of *The Groundes of Longitude* (London, 1591), a pamphlet advertising a hermetic method for calculating longitude, was printed.

16. For the assertion that Forman's "escapades" might even have deterred other Elizabethans from an interest in alchemy, see Debus, *English Paracelsians,* 102.

17. Most histories of occult subjects in Elizabethan England focus on John Dee, and, following Frances Yates, subordinate issues of spiritualism and improvement to medicine. See esp. Yates's *Giordano Bruno and the Hermetic Tradition* (Chicago: Chicago University Press, 1964), 60–61. For studies of Dee see Nicholas Clulee, *John Dee's Natural Philosophy: Between Science and Religion* (London: Routledge, 1988); Peter French, *John Dee: The World of an Elizabethan Magus* (London: Routledge, 1972); Deborah E. Harkness, *John Dee's Conversations with Angels: Cabala, Alchemy and the End of Nature* (Cambridge: Cambridge University Press, 1999); and Sherman, *John Dee*.

18. Ashm. 208, Item 1, fol. 49r–v.

19. Ashm. 1472, fol. 56.

20. Ashm. 208, Item 1, fols. 53v, 54.

21. Ibid., fol. 54v.

22. Ashm. 1472, fol. 809. Cf. Nancy Siraisi, *The Clock and the Mirror: Girolamo Cardano and Renaissance Medicine* (Princeton: Princeton University Press, 1997), 175–76.

23. Ashm. 208, Item 1, fol. 55.

24. Ashm. 1472, fol. 809.

25. Ashm. 208, Item 1, fol. 57.

26. Ashm. 195, fol. 203v.

27. Ashm. 208, Item 2, fols. 78–93v. Parts of a rough copy and parts of another fair copy of this text in Forman's hand are bound in Ashm. 1486, Item 3, fols. 7–11, 12–20. Some of these pages are damaged.

28. Alexander von Suchten, *Of the Secrets of Antimony: In Two Treatises* (London, 1670). This treatise was printed in German in 1570 and in Latin in 1575.

29. J. R. Partington, *A History of Chemistry*, vol. 2 (London: Macmillan, 1961), 156. For a discussion of Suchten's influence on the tracts by George Starkey, alias Eirenaeus Philalethes, see William Newman, *Gehennical Fire: The Lives of George Starkey, an American Alchemist in the Scientific Revolution* (Cambridge: Harvard University Press, 1994), 135–41, and idem, "Prophecy and Alchemy: The Origin of Eirenaeus Philalethes," *Ambix*, 37 (1990): 102–6.

30. A text copied by Thomas Robson was dated 1575, the year of the Latin publication (Ashm. 1418, Item 3, fols. 17–30). Ashmole in turn copied Robson's text (Ashm. 1459, Item 2, fols. 136–61v). Another text, which has not been examined, is in the Hartlib Papers (University of Sheffield, 16/1/48–63). Robson also copied Forman's copy of this text, without noting the similarities between it and the Suchten text (Ashm. 1421, fols. 29–34v).

31. Though on fol. 84v it is noted along with other metals.

32. I am indebted to David Katz for explaining that the Hebrew word "kochav" means "star" and to William Newman for explaining that Forman probably used this word to indicate the star-regulus of antimony.

33. Ashm. 1818, fol. 17r–v.

34. Ashm. 208, Item 2, fol. 83v.

35. Ibid.

36. Ibid.

37. Ibid., fol. 83.

38. Massimo Bianchi, "The Visible and the Invisible: From Alchemy to Paracelsus," in Piyo Rattansi and Antonio Clericuzio, eds., *Alchemy and Chemistry in the Sixteenth and Seventeenth Centuries* (Dordrecht: Kluwer, 1994), 17–50.

39. Ashm. 1472, fols. 136–41.

40. Ibid., fol. 6v.

41. Ibid., fol. 200.

42. Ibid., fol. 206.

43. Ibid., fol. 205.

44. Ibid. In a later version Forman added divines, physicians, and astrologers to this (Ashm. 1494, fol. 56v).

45. Ashm. 1472, fol. 205. This is elaborated on in Ashm. 1494, fol. 59.

46. Ibid., fol. 206.

47. Ashm. 1418, Item 3, fol. 17v; 208, Item 2, fol. 78. Forman does not note that this is the regulus of antimony, and it is not clear whether he used the word "regulus" to mean pure antimony or something else.

48. Ashm. 1472, fol. 200.

49. Ibid., fol. 136.

50. Ashm. 1494, fol. 53. It is clear that the notes on cako in "Of Appoticarie Druges" were copied from Text B, not Text A, because the marginal notes in Text B were subsumed directly into "Of Appoticarie Druges."

51. Ashm. 802, Item 2. This is the title used by modern scholars of pseudo-epigraphical texts.

52. H. F. D. Sparks, ed., *The Apocryphal Old Testament* (Oxford: Oxford University Press, 1984), 141–67, and R. H. Charles, ed., *The Apocrypha and Pseudepigrapha of the Old Testament in English* (Oxford: Oxford University Press, 1913), 123–54. Recent scholarship on this text has been summarized by Michael E. Stone, *A History of the Literature of Adam and Eve* (Baltimore: Johns Hopkins University Press, 1994).

53. Sparks, *Apocryphal Old Testament*, 143. According to Charles, this text circulated widely in England, among other places, in the sixteenth century (*Apocrypha and Pseudepigrapha*, 124). Charles probably concludes this from Meyer, who drew attention to the vernacular versions (M. E. B. Halford, "The Apocryphal Vita Adae et Evae: Some Comments on the Manuscript Tradition," *Neuphilologische Mitteilungen*, 82 [1981]: 417).

54. Halford, "Apocryphal Vita Adae et Evae," J. H. Mozley, "The 'Vita Adae,'" *Journal of Theological Studies* 30 (1929): 121–49.

55. These are as follows: Oxford, Bodleian Library, Bodley MS 596 (hereafter Bodley) is a prose version, of the late fourteenth or early fifteenth century; Cam-

bridge, Trinity College MS R.3.21, Item 48, fols. 249–57, is virtually identical to Bodley MS 596 (and was owned by John Stowe); London, British Library, Additional MS 35,298, fols. 162–65 is also in prose; Oxford, Trinity College MS 57, fol. 157v is in verse.

56. This impression may be distorted by the failure of manuscripts to survive and the difficulty in identifying postmedieval as well as Biblical manuscripts in catalogues. There are numerous sixteenth-century copies of the Life on the Continent (Stone, *Literature of Adam and Eve*, 9). The Life was printed in Latin in the late fifteenth century and in 1518.

57. The following chapter designations are taken from Sparks, *Apocryphal Old Testament*.

58. Forman's text closely resembles two fifteenth-century manuscripts of the Life, Bodley 596 and Cambridge, Trinity College MS R.3.21, and he was probably copying a related manuscript. Moreover, the way in which Forman ruled his text with wide margins, rubricated it, and wrote in a upright hand might indicate that he was mimicking a fifteenth-century manuscript. For details on the creation of Adam and Eve in other versions of the Life, see Halford, "Apocryphal Vita Adae et Evae," 420, n. 11; S. Harrison Thomson, "A Fifth Recension of the Latin 'Vita Ade et Eve,'" *Studi Medievali*, N.S. 6 (1933): 271–78; Sparks, *Apocryphal Old Testament*, 161, n. 1. Mozley identified a tradition, which he called the Arundel class, of English manuscripts that have these details at the end. These include descriptions of the eight substances out of which Adam was made and his naming after the four cardinal points (Halford, "Apocryphal Vita Adae et Evae," 419; Stone, *Literature of Adam and Eve*, 17).

59. The sections are as follows: A (fols. 1–8v), B (fols. 9–20), C (fols. 20–30).

60. Forman's text does not contain Satan's description of his rebellion (Life, chaps. 14 and 15). This is present in both the English versions and the Latin Life.

61. This discussion is restricted to a textual analysis of the Life. This text demonstrates a recording and transforming of legendary elements, particularly aspects of the Holy Rood. There is some evidence that versions of the Life persisted in Cornish creation plays through the sixteenth century (Stone, *Literature of Adam and Eve*, 16). John Aubrey recorded the following note as potentially relevant to Forman: "My grandfather Lyte told me that at his Lord Maier's shew there was the representation of the creation of the world, and writ underneath, 'and all for man.'" (John Aubrey, *The Natural History of Wiltshire*, ed. John Britton [London, 1847] [written between 1656 and 1691], 79).

62. For other reading practices, see Anthony Grafton and Lisa Jardine, "'Studied for Action': How Gabriel Harvey Read His Livy," *Past and Present*, 129 (1990): 30–78; Sherman, *John Dee*, chaps. 3 and 4.

63. Ashm. 802, Item 2, fol. 3v.

64. Ibid., fol. 14r–v.

65. See also ibid., fols. 15v–16, 28v, 29.

66. Bernard of Treves, "The most excellent and true booke" (Ashm. 1490, Item 81, fol. 222v).

67. "A dialogue of Egidius de Vadius [sic]" (ibid., Item 1, fols. 28–36v).

68. Ashm. 1494, fol. 623.

69. Arnold Williams, *The Common Expositor: An Account of the Commentaries on Genesis, 1527–1633* (Chapel Hill: University of North Carolina Press, 1948), 82.

70. Sections of Bostocke's text are edited by Allen Debus in "An Elizabethan History of Medical Chemistry," *Annals of Science*, 18 (1962): 1–29.

71. Richard Bostocke, *The Difference Betwene the Aunctient Phisicke . . . and the Latter Phisicke* [London, 1585], sigs. Fiiii, viii. For arguments that Bostocke's text was an anomaly, see Debus, *English Paracelsians,* chap. 2; Kocher, "Paracelsian Medicine in England," 465. See also David Harley, "Rychard Bostok of Tanridge, Surry (c. 1530–1605), M. P., Paracelsian Propagandist and Friend of John Dee"; *Aubix,* 47 (2000), 29–36.

72. I. W., *A Coppie of a Letter Sent by a Learned Physician to His Friend* (London, 1586), sig. A1v.

73. For a mocking of the notion that Adam was the first alchemist, see Conrad Gesner, *The Treasure of Euonymus,* trans. Peter Morwyng (London, 1559), sig. A1v. In Ben Jonson's *The Alchemist* there is a joke about Adam's writing a treatise on the philosophers' stone in High Dutch (Act II, scene i, 84–86).

74. See Williams, *Common Expositor.*

75. Charles Webster, *From Paracelsus to Newton: Magic and the Making of Modern Science* (Cambridge: Cambridge University Press, 1982), 49. See also Norma Emerton, "Creation in the Thought of J. B. van Helmont and Robert Fludd," in Rattansi and Clericuzio, *Alchemy and Chemistry,* 85–101.

76. For instance, within the thousands of papers in the Ashmole collection there are only two manuscript commentaries on Genesis, one of which was by Forman (Ashm. 802, Item 1; 766, Item 6). The two parts of Forman's text are now inversely bound. The original foliation makes it clear that in its initial form this text began with "Upon the firste of Genesis" (now fols. 3–12) and was followed by "the heaven and the earth and what the heavens are that were first created" (now fols. 1–2). In the original foliation the first item is fols. 1–6 and the second fols. 9 and 10, which indicates that two pages (7 and 8) are missing. The hand and inks are consistent throughout, and the catchwords conform to this reading. It is unclear whether this is the introduction to what was intended as a long tract or is most of it. Some of Forman's related notes, in rough format, are now bound in his copy of *The Hystory of Kyng Boccus and Sydrack* (1537?) (St. John's College, Oxford). These notes are on the birth of Lucifer, which figures in Forman's creation tract as the beginning of supercelestial time (fols. 5–6). For a printed version of these notes see Philip Bliss, "Extracts from a Manuscript of Dr. Simon Forman," *Censura Literae,* 8 (1807): 409–13.

77. See Williams, *Common Expositor,* 44–45.

78. Ashm. 802, Item 1, fol. 4v; see also fol. 8.

79. Ibid., fol. 1.

80. Ibid., fol. 2.

81. For conflict between theologians and alchemists on the subject of creations, see Williams, *Common Expositor,* 46.

82. See his transcripts, in chronological order from 1585: Ripley (Ashm. 1490, Item 44), Bernard of Treves (Ashm. 1490, Item 81), J. J. (Ashm. 1490, Item 83), Henry Lock (Ashm. 1490, Item 84), Egidius Devadis (Ashm. 1490, Item 1), and Blomfild (Ashm. 1490, Items 67 and 68).

83. Ashm. 1430, fols. 26–33; 1472, fol. 178; 1494, fols. 473–78. For a version of this in verse, see Ashm. 240, fols. 33–35v. For an edition of these verses, see Robert M. Schuler, ed., *Alchemical Poetry, 1575–1700: from Previously Unpublished Manuscripts* (New York: Garland, 1995), 49–70.

84. Ashm. 1430, fol. 26. This phrase also appeared in Forman's creation tract (Ashm. 802, Item 1, fol. 8).

85. Williams, *Common Expositor,* 77.

86. Ashm. 802, Item 2, fol. 1r–v.

87. Ibid., fol. 1^v. In the text of Forman's copy of the Life, Adam was made of nine things: the slime of the earth, the sea, stones, clouds, winds, sun, light of the world, the Holy Ghost, and fire (ibid., fols. 2v–3). Cf. Owen Hannaway, *The Chemists and the Word: The Didactic Origins of Chemistry* (Baltimore: Johns Hopkins University Press, 1975), 27–28.

88. See also Ashm. 1472, fol. 64; 1494, fol. 190.

89. See also Ashm. 1472, fols. 64–66; 1494, fols. 190–200, 218.

90. Ashm. 1494, fol. 191.

91. Ibid., fols. 190–200, 218. Forman used the 1567 Paris edition of *De occulta,* and he probably owned a copy. In his manuscripts he often noted to look further in Agrippa, and gave book, chapter, and folio references.

Agrippa's previous chapter was about the creation of man, after the image of God. There were some similarities between this and the beginning of Forman's entry on "Anima & Spiritus," though here Forman was not following Agrippa directly. Forman's notes on cabala, which include extracts from Agrippa, were also written in 1610 and, perhaps coincidentally, are now bound with the notes on the motions of the heavens (Ashm. 244).

92. William Newman, "Thomas Vaughan as an interpreter of Agrippa von Nettesheim," *Aubix,* 29 (1982), 125–140.

93. Ashm. 1494, fol. 196.

94. These had been printed in German in 1573, and the second has never appeared in print in English: Ashm. 1490, Items 79 and 80. Charles Webster has identified this text as Karl Sudhoff, *Bibliographia Paracelsica,* vol. 1 (Berlin, 1894), no. 145.

95. Paracelsus, *Operum medico-chimicorum sive paradoxorum, tomus genuinus undecimus,* trans. Zacharias Palthenius, vol. 11 (Frankfurt, 1605). I am indebted to Charles Webster for this identification. Almost no printed books bearing marks of Forman's ownership have been identified.

96. Sloane 3822, Item 2, fols. 68–75. See also Forman's notes on the celestial heavens (Ashm. 244, fols. 35–60).

97. Ashm. 802, Item 2, fol. 2.

98. Ibid., fol. 3.

99. Ibid., fol. 2.

100. Paracelsus, *Operum medico-chimicorum* 11:66–110.

101. Walter Pagel, *Paracelsus: An Introduction to Philosophical Medicine in the Era of the Renaissance* (Basel, Switzerland: Karger, 1958), 83, 228.

102. Ashm. 802, Item 2, fols. 2v–3.

103. Ibid., fol. 21.

104. Ibid., fol. 21v. Some versions of the Life gave the number of diseases God had inflicted on Adam as seventy, some as seventy-two.

105. These elements appear in other Books of Adam and Eve. See for instance S. C. Malan, ed., *The Book of Adam and Eve* (London, 1882).

106. Ashm. 802, Item 2, fol. 9.

107. Ibid., fols. 9v–10.

108. Ibid., fol. 15v.

109. Ibid., fol. 18v (Life, chap. 24). Another medicinal theme in the Life, this of Christian origin, was the oil of mercy (ibid., fols. 23v, 25r–v; Life, chap. 36). Forman's text contained additions about the tree of mercy, which seem to have been drawn from the gospel of Nicodemus.

110. Gervase Babington, *Certain Plaine, Briefe, and Comfortable Notes, Upon Every Chapter of Genesis* (London, 1596), 14–15.

111. Nicholas Gibbens, *Questions and Disputations Concerning the Holy Scripture* (London, 1602), 39–42.

112. Ibid., 39.

113. Ibid., 161.

114. Ibid., 162–63.

115. Ibid., 218–19; for antediluvian vegetarianism see also 361–65.

116. Gibbens's inclusion of Jean Fernel in his sources further illustrates the overlap between theology and medicine. For recent work on Fernel, see Lawrence Brockliss and Colin Jones, *The Medical World of Early Modern France* (Oxford: Oxford University Press, 1997), 128–38; Stuart Clark, "Demons and Disease: The Disenchantment of the Sick," in Marijke Gijswijt-Hofstra, Hilary Marland, and Hans de Waardt, eds., *Illness and Healing Alternatives in Western Europe* (London: Routledge, 1997), 38–58; Linda Deer Richardson, "The Generation of Disease: Occult Causes and Diseases of the Total Substance," in A. Wear, R. K. French, and I. M. Lonie, eds., *The Medical Renaissance of the Sixteenth Century* (Cambridge: Cambridge University Press, 1985), 175–94; Siraisi, *Cardano and Renaissance Medicine*, 158–61.

117. The phrase appeared in verses entitled "Christ's Affliction" following one of the English versions of the Life in a contiguous script (Bodley 596, fols. 12v–13).

118. Roger Bacon, *Opus Majus,* ed. John H. Bridges (Oxford: Oxford University Press, 1900), 204–13; William Newman, "The Philosopher's Egg: Theory and Practice in the Alchemy of Roger Bacon," *Micologus: Nature, Sciences and Medieval Societies,* 3 (1995): 75–101. See also Michela Pereira, "*Mater Medicinarum:* English Physicians and the Alchemical Elixir in the Fifteenth Century," in Roger French, Jon Arrizabalaga, Andrew Cunningham, and Luis Garcia-Ballester, eds., *Medicine from the Black Death to the French Disease* (Aldershot: Ashgate, 1998), 26–52.

119. Pagel, *Paracelsus,* 153–58.

120. Agrippa, then Vaughan, used this term. See Newman, "Thomas Vaughan," 132, n. 68.

121. Bostocke, *Auncient Phisicke,* sigs. Biiii–[vi]v.

122. I. W., *Coppie of a Letter,* sig. [B6]v.

123. Ibid., sig. [B7]v.

124. Ashm. 1429, fol. 77. William Black, *A Descriptive, Analytical and Critical Catalogue of the Manuscripts Bequeathed unto the University of Oxford by Elias Ashmole* (Oxford: Oxford University Press, 1845), attributed this text to Forman, though his only evidence is the following, probably written by Richard Napier: "September 1611. Doctor Formans booke for the use of Clement his sonne." This appears at the front of the volume. The title page is missing, and there are no references to Forman in the text.

125. Ashm. 1419, fols. 57–82v; for another copy of this text, made by Isaac Newton, see Cambridge, King's College, Keynes MS 22. I am indebted to William Newman for identifying this text for me and for pointing me in the direction of Lawrence Principe's discussion of it in *The Aspiring Adept: Robert Boyle and his Alchemical Quest* (Princeton: Princeton University Press, 1998), p. 198. I have explored the "food of angels" further in "Reading for the philosophers' stone" in Marina Frasca-Spada and Nick Jardine, eds., *Books and the Sciences in History* (Cambridge: Cambridge University Press, 2000), 132–50.

126. Ashm. 1419, fols. 63–64v.

127. Ibid., fol. 63v.

128. Ibid., fol. 64.

129. Ashm. 1433, Item 1. I do not know when he made his initial transcript.

130. Very few manuscripts of Dunstan's alchemical tract seem to have survived. There are three copies in the Ashmole collection, made by Forman, Thomas Robson, and Ashmole. Ashmole noted that Sir Thomas Browne had a Latin copy, which was confirmed by a letter from Brown to Ashmole (Elias Ashmole, *Autobiographical and Historical Notes, Correspondence, and Other Sources,* ed. C. H. Josten [Oxford: Oxford University Press, 1966], 2:755, n. 2). Brown had sent Ashmole a list of alchemical manuscripts and was inviting Ashmole to look at them. This letter is from January 1659. Josten identifies the Dunstan text as possibly Sloane MS 1255, Item 2. See also Sloane MSS 1744 (which also contains some Forman and some Robson), 1876, 3738, and 3757. Oxford, Corpus Christi College, MS 128, is the only medieval copy identified. It may have been owned by John Dee and is listed in Julian Roberts and Andrew Watson, eds., *John Dee's Library Catalogue* (London: Bibliographical Society, 1990) as DM [129].

131. Elias Ashmole, *Theatrum Chemicum Britannicum* (London, 1652), sig. B1v. Ashmole recorded that he compared the copy of Edwardus's manuscript now bound in Ashm. 1419 with one in the hand of its author (Ashm. 1419, fol. 82v). Josten thought that Ashmole might have learned this scheme from Father Backhouse: Ashmole, *Autobiographical and Historical Notes,* 1:86.

132. Ashm. 802, Item 2, fols. 21v–23.

133. I. W., *Coppie of a Letter,* sig. B7r–v.

134. The tradition of using medical and scientific material in Hexamera was present in St. Ambrose, who probably drew on Cicero's *De natura deorum* (Nancy Siraisi, *Medieval and Early Renaissance Medicine* [Chicago: Chicago University Press, 1990], 7–8). The critical edition of Sylvester's translation of Du Bartas's text defers sources to the critical edition of the original text (Susan Snyder, ed., *The Divine Weeks and Works of Guillaume de Saluste Sieur du Bartas,* trans. Josuah Sylvester [Oxford: Oxford University Press, 1979]). In the introduction to this, the editors point to the problems of identifying sources for Hexamera, and in the section of the Furies here at issue (beginning line 305), they note that "[t]he list of diseases which Du Bartas now gives cannot be traced to any particular source. It is probable that he used the works of Hippocrates" (Urban T. Holmes, John C. Lyons, and Robert W. Linker, eds., *The Workes of Guillaume de Salluste* [Chapel Hill: University of North Carolina Press, 1935–40], 733). Ashmole described an old Latin text, owned by Dr. Barlow, Bishop of Lincoln, which he thought Du Bartas simply translated and published as his own (Ashm. 826, fol. 119).

135. "Heere sorrow stopt the doore / Of his sad voice, and almost dead for woe, / The prophetizing spirit foresooke him so" ("The Handicrafts," lines 772–74, in Snyder, *Divine Weeks*).

136. The hands and inks in each are very similar.

137. For Forman's treatments and recipes see his casebooks, Ashm. 234, 226, 195, 219, 236, and 411 *passim*. See also the notes of his assistant and illegitimate son, Josuah Walworth, Cambridge, Trinity College MS O.2.59, Item 4, fols. 121–49.

138. Although Basil Valentine used antimonial leads for amulets, there is no evidence that Forman did this (Partington, *Chemistry,* 2:198).

139. Ashm. 392, fol. 46.

140. D. P. Walker, *Spiritual and Demonic Magic from Ficino to Campanella* (London: Warburg Institute, 1958); Webster, *From Paracelsus to Newton;* Yates, *Giordano Bruno.*

141. Yates, *Giordano Bruno*. John Dee's conversations with angels have received considerable attention. The most recent study is Harkness, *Dee's Conversations.* Harkness, "The Scientific Reformation."

142. Michael MacDonald, *Witchcraft and Hysteria in Elizabethan London: Edward Jorden and the Mary Glover Case* (London: Routledge, 1991).

143. Clark, "Demons and Disease," 38–44; idem, *Thinking with Demons: The Idea of Witchcraft in Early Modern Europe* (Oxford: Oxford University Press, 1997), 195–98; Brian Copenhaver, *Symphorien Champier and the Reception of the Occultist Tradition in Renaissance France* (The Hague: Mouton, 1978, 227–

28); MacDonald, *Witchcraft and Hysteria*, xxxii; Siraisi, *Cardano and Renaissance Medicine*, chap. 7.

144. This is most evident in Forman's treatises on plague: Ashm. 208, Item 8, fols. 110–34; 1403; 1436.

145. Ashm. 1495, fols. 4v, 32; 363, fols. 272–82, [309].

146. Ashm. 363, fols. 69v–71v; see also Ashm. 431, fols. 145–46. Ashmole attributed a note to Forman to the effect that the heavens move in three ways, one of which was natural, and two against nature (Ashm. 421, fols. 149r–v).

147. Ashm. 195, fols. 210, 218.

148. Ashm. 411, fols. 95, 99v, 115, 118v.

149. In one of his commonplace books he recorded a detailed account of several methods for casting things and noted a method that one "Mullenax" used to cast things in plaster (Ashm. 1494, fol. 324). This might be Emery Molyneaux, the globe maker, whom Forman knew in the early 1590s.

150. Ashm. 392, fol. 46. See also Ashm. 390, fol. 30.

151. In his diary for 1600 Forman noted that he "copied out also the 4 boocks of steganographia and divers other bocks" (Ashm. 208, Item 1, fol. 62v). For Forman's notes on the Picatrix see, for instance, Ashm. 244, fols. 45, 97; 431, fol. 146r–v; 1491, fol. 1128. Pingree has identified Napier's copy of the Picatrix, which might be a transcript of Forman's copy (David Pingree, ed., *Picatrix: The Latin Versions of the Ghayat al-hakim*, Studies of the Warburg Institute, vol. 39 [London: Warburg Institute, 1986], xix, liii–lv).

152. Ashm. 240, fol. 106. Michael MacDonald states that Forman taught Napier to use amulets in 1611 (*Mystical Bedlam: Madness, Anxiety, and Healing in Seventeenth-Century England* [Cambridge: Cambridge University Press, 1981], 294, n. 195). He does not include this letter in his evidence, and although the manuscripts to which MacDonald refers are evidence that Napier was constructing amulets by 1611, there is no indication that he learned how to do so from Forman.

153. Ashm. 244, Item I, fols. 6v, 11v.

154. Ashm. 1494, fol. 484. Forman cites Paracelsus, "Archidoxes," fol. 15, and although this text is included in Palthenius's vol. 11 of translation of Paracelsus's *Operum medico-chimicorum*, the page citation does not match, and Forman was probably using a different edition.

155. Ashm. 1494, fol. 574–75.

156. Ashm. 1491, fol. 679.

157. Ashm. 208, Item 1, fol. 37v. For rings that might have been magical, see ibid., fol. 49.

158. Ashm. 226, fol. 249v.

159. Ashm. 205, fol. 23.

160. Ashm. 195, fols. 29v, 56v–57v, 58.

161. Ibid., fol. 223v; 219, fol. 48.

162. Sloane 3822, fol. 11.

163. Ashm. 1494, fol. 586v. For the recipe for the ink, see ibid., fol. 402.

164. Ashm. 411, fol. 58v. This might be the same Mrs. Shackelton whose coat of arms Forman described as having been made for her burial on January 7, 1608 (Ashm. 802, Item 37, fol. 207v).

165. Ashm. 208, Item 1, fol. 43v. Forman wrote the word "foiygomercy" in bold, distinct and upright letters.

166. Ibid., fol. 46v.

167. Ibid., fol. 59v.

168. Ashm. 1472, fol. 813 (August 19, 1594).

169. Ibid. (August 23, 1594).

170. Ibid. (September 8, 1594). The text of this dream is incomplete.

171. This is probably the same man as "John Good," who came to live with Forman in 1598 and who later confessed to robbing his study (Ashm. 208, Item 1, fol. 59v; 392, fol. 136).

172. Ashm. 354, fols. 236–37.

173. Ashm. 219, fols. 135, 136.

174. Ashm. 208, Item 1, fol. 62v.

175. Lynn Thorndike, *A History of Magic and Experimental Science* (New York: Columbia University Press, 1934), 2:281–83.

176. Bodleian Library, Jones MS 1; Cambridge, Trinity College MS O.9.7; Jerusalem, National Library, Yahuda MS VAR. 34. For Forman's drafts of the drawings see Ashm. 820, Item 3. Other copies of this text include Bodley MS 951; London, British Library, Harleian MS 181; and Sloane MS 1712. A translation of this text, without the illustrations, appears to be in the hand of Richard Napier (Ashm. 1515).

177. Ashm. 1515, fol. 23v.

178. In "Of Appoticarie Druges" Forman included a section entitled "Writing of Books" in which he outlined the ritual constituents that Raziel told Solomon were necessary for wills and petitions to be fulfilled. These included using pure, virgin parchment and special inks. This may have been why Forman made the effort to write some manuscripts on parchment (Ashm. 1491, 1303–9).

179. Jerusalem, National Library, Yahuda MS VAR. 34, fol. 8.

180. Ibid., fol. 21.

181. Cambridge, Trinity College MS O.9.7, fols. 107v, 115.

182. Ashm. 1515, fol. 34.

183. See Forman's guide to astrological physic, Ashm. 363, and his casebooks, Ashm. 234, 226, 195, 219, 236, and 411.

184. Cf. Yates, *Giordano Bruno,* 151.

8
Some Problems with the Historiography of Alchemy

Lawrence M. Principe and William R. Newman

The well-recognized complexity and opacity of alchemical literature has long constituted a barrier to its proper understanding. Indeed, since the eighteenth-century disappearance of its last serious practitioners within the community of chemists, alchemy has been the subject of several radically distinct schools of historical interpretation. The current understanding of alchemy among historians of science, not to mention the general public, remains strongly colored by one or more of these divergent schools of interpretation, which stem, respectively, from the Enlightenment rejection of obscurity and the later Romantic disenchantment with Newtonian science that led to a new embrace of the occult.[1] At present, when historical interest in alchemy is in the ascendant, it is appropriate to reexamine the content, origins, effects, and validity of these interpretations. We will argue in this chapter that none of these established interpretive schools is satisfactory, for none represents alchemy in a way that is consistent with the historical record, and all severely distort the content and context of the discipline. As we will show, this distortion is often the inevitable consequence of the adoption, frequently unwitting, of principles derived from nineteenth-century occultism, which have become widespread tenets in the historiography of alchemy. We hope, therefore, to clear away some of the detritus that has gradually accumulated around the topic, enabling scholars more accurately to chart and later to follow a path through what remains a partly uncharted domain and thus arrive finally at a clearer and more accurate understanding of alchemy.

The Eighteenth-Century View of Alchemy

The increasing rejection of traditional alchemy during the eighteenth century is generally well known, even though more detailed studies would be beneficial to define the exact course and means of that repudiation. Increasingly from the beginning of the century there was a tendency to sequester the "older" alchemy from the "newer" science of chemistry, and this divorce appears clearly in the etymological distinctions between "alchemy" and "chemistry," which became entrenched in the first decades of the eighteenth century. As we have shown elsewhere, the words "alchemy" and "chemistry" were used interchangeably to refer to the same body of activities throughout most of the seventeenth century, and only during the eighteenth century were distinctions similar to those in common modern usage rigidly drawn between the two. In the early eighteenth century, the domain of "alchemy" was for the first time widely restricted to gold making—or what had previously been termed "chrysopoeia" or *alchemia transmutatoria*—and many writers (such as the Lemerys, Geoffroy, and Fontenelle) focused ever more exclusively on the cheating practices of alchemical charlatans, eventually indicting the whole subject as a fraud.[2] Indeed, for most writers and thinkers of the eighteenth century, alchemy was synonymous with gold making and fraud.

By the middle of the eighteenth century, "alchemy" was in fairly universal disrepute among scientific authors, save for scattered continuing support from a few writers particularly in Germany. In the lingering German debates over the validity of alchemical transmutation (which continued until the end of the century), the opponents of chrysopoeia continued to insist upon alchemy's putatively fraudulent character. These Enlightenment writers drew heavily on metaphors of light and darkness to describe the dawning of chemistry out of the misty obscurity of the medieval delusion of alchemy.[3] This strongly negative viewpoint endured well after alchemy ceased to be a topic of actual debate, being adopted wholesale by many members of the early generation of chemical historians including Johann Friedrich Gmelin and Thomas Thomson. Indeed, judgments drawn in the eighteenth century persist to the present day, even among some historians of science.[4] In protecting the developing discipline of modern chemistry from the censures to which traditional alchemy (i.e., chrysopoeia) was liable, the Enlightenment writers produced the appearance of a radical disjunction in the history of chemistry, as if the newly redefined "alchemy" and "chemistry" were only marginally contiguous. This movement paralleled the attempts of early eighteenth-century chemical practitioners to legitimize their discipline and enhance their status by divorcing themselves sharply from the foregoing alchemical tradition, which had fallen into disrepute. The recasting of alchemy as "other" relative to chemistry was to have significant consequences in the nineteenth century and thence to the present day.

Alchemical symbolism continued to survive in some nonscientific quarters, however, even after alchemy itself was no longer reputable. For example, during the eighteenth century the Pietists both in Germany and in America propagated in a spiritual setting the alchemical imagery employed by Jakob Boehme in expounding his ecstatic visions.[5] The mystical brands of alchemical thought propounded by Heinrich Khunrath and the Rosicrucian enthusiast Robert Fludd persisted among several secret societies. The *Gold- und Rosencreutz* of the eighteenth century relied heavily on alchemical symbolism and made at least rhetorical gestures toward the importance of alchemical practice. The "Convent of the Philalethes," a Parisian masonic order sharing its name with one of the seventeenth century's most popular alchemical writers, took up alchemy along with other "occult sciences."[6] The secrecy universally connected with earlier alchemical writers and practices facilitated a juxtaposition of alchemy with different sorts of "secret knowledge" such as natural magic. Alchemical works deliberately written to be obscure and secretive in their own age sometimes became meaningless in the next. Such obscurity was a boon to those striving to display the fraudulent and "nonscientific" character of alchemy, as in the case of Johann Christoph Adelung, whose *Geschichte der menschlichen Narrheit* casts together magicians, soothsayers, and alchemists into a common bin.[7]

At the beginning of the nineteenth century, references to alchemy were to be found predominantly in association with magic, witchcraft, and the other practices commonly grouped together as "occult," topics that the eighteenth-century "triumph of reason" had ridiculed. Thus, when occult revivals began in the early nineteenth century, alchemy received new attention not so much in itself but as one of a number of "occult" sciences. The work of the occult writer Francis Barrett, whose 1801 *The Magus* involved alchemy along with natural magic, astrology, and demonology as "a complete system of occult philosophy," provides one example. This book was followed some years later by *Lives of the Alchemystical Writers*, often ascribed to Barrett as well, the very orthography of whose title signals its author's tendencies.[8] This yoking of alchemy to such disciplines as natural magic, astrology, and theurgy, although already begun in some quarters during the Renaissance (see chapter 1 of this volume), was consummated only during the final years of the *ancien régime* in France, at the time when Franz Anton Mesmer was disseminating his pneumatic "science" to the world.[9] Mesmerism, as we shall see, laid the foundations for the mid-nineteenth-century success of the "spiritual" interpretation of alchemy.

The "Spiritual" Interpretation of Alchemy

Although nineteenth-century occultism is a complex subject in its own right, it is possible to distinguish a set of characteristic features of a "spiritual" interpretation of alchemy in this period. The esoteric or occultist school, which interpreted alchemy in this spiritual way, held (and holds) that the operations recorded in alchemical texts corresponded only tangentially or not at all to physical processes.[10] Although it was in fact a commonplace of the early modern period to build extended religious conceits on alchemical processes and to draw theological parallels therefrom—an aspect of alchemical writing Luther praised in passing[11]—the occultists of the nineteenth century went much further to claim that alchemy itself was an art of internal meditation or illumination rather than an external manipulation of apparatus and chemicals. Nineteenth-century occult writers claimed that the alchemists did not aim primarily at changes of a chemical kind but rather used chemical language and terminology only to couch spiritual, moral, or mystical processes in allegorical guise. The alchemists' important goals were supramundane. The transmutation of base metals into noble ones is thus to be read as a trope or allusive instruction for the transcendental transformation of the alchemist himself, or of all mankind through him, from a base, earthly state into a more noble, more spiritual, more moral, or more divine state. The philosophers' stone that effects this transmutation/transformation may be corporeal or noncorporeal but in either case represents a mystical or spiritual power either intrinsic or extrinsic to the contemplative spiritual alchemist. This point of view, which we shall refer to as "spiritual alchemy," sees alchemical adepts as possessors of vast esoteric knowledge and spiritual enlightenment.

The nineteenth-century focus on "spiritual alchemy" succeeded in bringing about a massive transformation in the general perception of alchemy. Although the more extreme exponents of this school are rather easily relegated to places beyond the fringes of respectability, the overall impact of the esoteric school on present-day perceptions of alchemy, even among historians of science, remains significant. When Herbert Butterfield famously derided historians of alchemy as being "tinctured with the same type of lunacy they set out to describe," it is likely that he had this esoteric/occultist interpretation in mind.[12] The two most seminal figures in the history of the "spiritual" interpretation of alchemy date from the middle of the nineteenth century. Mary Anne Atwood in England and Ethan Allen Hitchcock in the United States independently produced spiritual explanations of alchemy in the 1850s, and they were rapidly joined by many others caught up in the Victorian fascination with the occult.

Mary Anne Atwood was born in 1817, daughter of Thomas South, a gentleman of Hampshire. At their home of Bury House in Gosport, father and daughter studied classical and esoteric literature together and became involved in animal magnetism, Mesmerism, and other manifestations of occultism then in vogue. In 1846, Thomas South, probably in collaboration with his daughter, published a small tract entitled *Early Magnetism, in its Higher Relations to Humanity as Veiled in the Poets and Prophets* (under the anagram Θυος Μαγος). This treatise is an early example of the Souths' reading of esoteric subtexts into historical literature, in this case the supposed allusive revelation of animal magnetism in the Homeric Hymns. Thereafter, the two read alchemical literature, and believing they had found therein great esoteric spiritual knowledge and practice in allegorical guise, settled upon writing two expositions, the father writing in verse and the daughter in prose. In 1850, Mary Anne's treatise was completed, sent to the press, and published as *A Suggestive Inquiry into the Hermetic Mystery*. Immediately upon publication, however, the author and her father recalled the book and, on the lawn of Bury House, burned all of the copies plus South's notes and unfinished poetical work.[13] Only the few copies that had already been purchased or sent out to libraries survived. These few examples do seem to have been eagerly read and cited by occultists in the latter half of the nineteenth century. It remains open to question whether this recall and destruction arose from a "moral panic" upon a sudden "realization of the sanctity of the Art" and a fear of being "betrayers of the sacred secret," as claimed by Atwood's followers, or was instead intended to excite future interest in the volume.[14]

The balance of Atwood's life was without further authorship; she married Rev. Alban Thomas Atwood, vicar of Leake, Yorkshire, in 1859, with whom she bore no children, and who died in 1883. Atwood herself died in 1910 and bequeathed her papers and few copies of the *Suggestive Inquiry* (bearing her revisions) to her friend and confidante Isabelle de Steiger. From these copies a new edition was produced in Belfast in 1918, introduced and edited by Walter Leslie Wilmshurst, an editor of esoterica.[15] It has remained in print almost continuously ever since.

The *Inquiry* begins with a cursory (and naive) history of alchemy from Egyptian antiquity to the seventeenth century, when a "mistrust, gathering from disappointment" generated the "absolute odium" under which alchemy and its practitioners had lain since the Enlightenment.[16] Atwood gives special place to Eugenius Philalethes (alias Thomas Vaughan) but reserves her highest (and lengthiest) praise for Jakob Boehme (1575–1624), the Lutheran cobbler-mystic, whom she calls "the plainest, simplest, and most confidential exponent" of all. She adds to her praise of Boehme a surviving fragment of her father's immolated versification that declares that the accumulated wisdom and esoteric knowledge of Orpheus, Zoroaster, Pythagoras, Plato, and every other "saint and sage"

In Böhme's wondrous page we view
Discovered and revealed anew.[17]

Atwood concludes that the world is "fully ignorant of the genuine doctrine" of alchemy because the adepts' recipes "though at variance with all common-sense probability, have been the means of surrounding many a literal soul with stills, coals, and furnaces, in the hope by such lifeless instruments to sublime the Spirit of nature, or by salt, sulphur and mercury, or the three combined with antimony, to extract the Form of gold."[18] What appear to be laboratory operations are but "wisdom's envelope, to guard her universal magistery from an incapable and dreaming world."[19] Such laboratory processes were doomed from the start, according to Atwood, because although the adepts could in fact transmute base metals into gold, that was the lowest form of their craft and was not accomplished by means of normal chemical operations. The physical transmutation of metals required that the alchemist first transform himself so that he could operate in a quasi-magical and pneumatic fashion. The materialist recipes of the alchemists were purely delusory: in fact, they were window dressing.

The balance of Atwood's treatise presents her thesis wrapped in a dense and often incoherent hodgepodge of decontextualized and often unattributed quotations from alchemical and classical authors, piled high with obscure assertions, enraptured exclamations, and bizarrely twisted scientific notions.[20] Atwood asserts that the alchemists' prime matter, philosophical mercury, primordial chaos, or Spirit of Life is an imponderable incorporeal ether. The alchemical vessel is man himself, and when in a trancelike state, the adept can "magnetically" draw in this ether or primordial light and condense it into the philosophers' stone, "pure Ethereality of Nature" or "Light inspissate," a noncorporeal agent of universal change and exaltation dwelling within the enlightened adept:[21] "Man is the true laboratory of the Hermetic Art; his life the subject, the grand distillatory, the thing distilling and the thing distilled, and Self-Knowledge [is] at the root of all Alchemical tradition."[22]

The alchemical process thus involved a self-purification and exaltation to a "higher plane of existence." Simultaneously, the spiritualized esoteric adept could also manipulate matter by the application of the same forces, advancing lead into gold by purely spiritual means. Atwood declares an underlying unity of creation and the ability of all its manifestations—mineral, vegetable, animal, or spiritual—to be exalted within their sphere by the same power. Clearly alchemy is totally distinct from chemistry: "No modern art or chemistry, notwithstanding all its surreptitious claims, has any thing in common with Alchemy."[23]

Atwood's exposition of the "Hermetic Mystery" is intimately bound up with her early enthusiasm for Mesmerism. The ether that forms the foundation of hermetic manipulations is the same medium by which Mesmer's planetary influences and animal magnetism were transmitted. Atwood claims that the trancelike state necessary for self-purification and the concentration or manifestation of the "matter" was achieved in ancient times by the devotees of the Eleusinian mysteries and contemporaneously by the practitioners of Mesmerism—"the first key opening to the vestibule of this Experiment."[24] She is generally dismissive of modern science, considering it too intent upon the physical, lower order of things and its practitioners thereby both unfitted and unable to ascend to the ancient and eternal mysteries.

Shortly thereafter, Ethan Allen Hitchcock (1798–1870), a general in the U.S. Army, propounded a considerably different, but still spiritual, view of alchemy. His *Remarks upon Alchymists* was a short work published in 1855, attempting to show "that the philosopher's stone is a mere symbol signifying something which could not be expressed openly without incurring the danger of an *auto da fé*." After an unfavorable review of the book, Hitchcock issued a response two years later in the form of the longer and more detailed *Remarks upon Alchemy and the Alchemists*.[25] Hitchcock's reading of alchemical texts is moral and Christian, arguing that they are allegories or veiled descriptions of the moral life. He asserts that the alchemists actually did nothing akin to chemistry and that "*Man* was the *subject* of Alchemy; and that the *object* of the Art was the perfection, or at least the improvement, of Man."[26] Philosophical Mercury is a clean and pure conscience, difficult to obtain but potentially available everywhere, and once obtained it leads easily (through "women's work and children's play") to the philosophers' stone, which is the consummate moral life. The true nature of the alchemical quest was hidden in secrecy owing to the "intolerance of the Middle Ages . . . known to every one." Hitchcock casts the alchemists as "Reformers" and asserts that an "open expression of their opinions would have brought them into conflict with the superstition of the time, and thus exposed them to the stake"; unfortunately, he never explains exactly why an exhortation to Christian morality would have been viewed as heretical.[27]

Hitchcock indulges in none of the bizarre occultism of Atwood's extravagant thesis. For Hitchcock, improvement of the human being comes not by psychic exaltation to higher planes of existence but rather through the practice of true religion and morality. He interprets the substances, theories, and operations of the alchemists allegorically after the manner of a preacher unfolding a scriptural parable. For Hitchcock, alchemy is wholly a dimension of orthodox religion. Accordingly, Hitchcock's work, although cited frequently by the close of the nineteenth century, did not create the furor that followed upon Atwood's more daring *Suggestive Inquiry*.

These midcentury interpretations fed a general revival of alchemy among spiritualists and occultists in both Europe and America in the late nineteenth century. In France, a broad esoteric movement followed the occultist Eliphas Lévi (alias Alphonse Louis Constant, 1810–75), and in England, occultist groups grew and flourished.[28] The esoterics Anna Kingsford (1846–88) and Edward Maitland (1824–97) created a Hermetic Society, published a *Corpus Hermeticum*, and prosecuted the

"Higher Alchemy," finding alchemical mysteries couched in the Old Testament and equating alchemy with religion itself.[29] William Wynn Westcott (1848–1925), the "Supreme Magus of the Rosicrucian Society in England, and Master of the Quatuor Coronati Lodge," summarized these esoteric formulations under the pseudonym *Sapere Aude* ("Dare to be wise") in 1893.[30] Westcott saw a link between Eastern mysticism and Western alchemy, undoubtedly provoked by the garbled Hinduism of his associate Helena Petrovna Blavatsky (1831–91) and enshrined in the Theosophical Society she founded in 1875, of which Westcott was a member. In 1888, Westcott helped organize the "Hermetic Order of the Golden Dawn," which flourished for fifteen years (before splintering into sects) and which had a seminal influence on W. B. Yeats. Its influence on late Victorian society is only beginning to be recognized.[31] Westcott also wrote a preface to a translation of Nicholas Flamel's *Hieroglyphics* and edited and annotated a *Collectanea Hermetica* (published by the Theosophical Publications Society) that included several alchemical and mystical tracts.[32]

Across the Channel in France, several occult alchemical societies were founded, including "L'Association Alchimique de France," which under its leader F. Jollivet-Castelot (author of *La synthèse de l'or* in 1909) published the monthly review *L'Hyperchimie* beginning in 1896. By the late nineteenth century the incorporation of alchemy into esoteric and magical subjects and the mythic history of secret societies had become a matter of course. Hargrave Jennings's popular work on the Rosicrucians presents alchemical adepts as mysterious, unaging, semi-immortal wanderers endowed with a knowledge and being far above that of mere mortals. Likewise, Albert Pike recapitulates the esoteric interpretation of alchemy in a Masonic context.[33] The revival of alchemy under esoteric/occult guise grew so vigorous that a report on the various groups and their tenets was presented before the American Chemical Society in 1897 and printed in the annual report of the Smithsonian Institution, wherein the author complained of the "company of educated charlatans" then engineering the revival.[34]

The most prolific of the esoterics was Arthur Edward Waite (1857–1942),[35] who published a multitude of books on occult topics ranging from Freemasonry and Rosicrucianism to devil worship and cabala. The majority of his publications, however, dealt with alchemy, and he exerted a significant influence on the development of twentieth-century interpretations of alchemy. His first book, *Lives of the Alchemystical Writers* (1888), was a massively rewritten version of the book by the same title issued anonymously seventy years earlier. Whereas the original *Lives* was merely a compilation of uncritical and largely fictitious biographies of about fifty alchemists, Waite's expanded (but no more critical) version used these accounts to promote his own theosophical view of alchemy. His introductory essay finds fault with both Atwood's "psychical" and Hitchcock's "moral" interpretation because neither recognizes the real physical operations that the alchemists carried out in the laboratory, where they did (according to Waite) labor successfully to produce a physical philosophers' stone able to transmute metals.[36] Waite asserts that the "attempt to enthrone [alchemists] upon the loftiest pinnacles of achievement in the psychic world, however attractive and dazzling to a romantic imagination, and however spiritually attractive, must be regretfully abandoned."[37] Nonetheless, he retains a largely spiritual view of alchemy, arguing that the true "Hermetic interpretation lies in a middle course": The alchemists labored on actual physical processes, but these were only the corporeal manifestations of "a theory of Universal Development" that "had an equal application to the triune man" as to the triune metals.[38] Alchemy, or "psycho-chemistry is a grand and sublime scheme of absolute reconstruction . . . the divinisation, or deification in the narrower sense, of man the triune by an influx from above."[39] The adepts' success in physical processes is, by the "Hermetic doctrine of correspondences, . . . analogically a substantial guarantee of the successful issue of parallel methods when applied in the psychic world with the subject man."[40]

Waite dealt with this esoteric "alchemical transformation of humanity" to provide "the perfect youth to come" at length in his *Azoth, or the Star in the East,* published in 1893. There he refers to alchemy as "physical mysticism" claiming that "all alchemists [were] Mystics, and alchemy . . . a mystic work." While maintaining his view that the exclusively spiritual interpretations of Atwood and Hitchcock are "errors of enthusiasm," Waite nevertheless asserts that "alchemical literature deals primarily at least with the conscious intelligence of man, and with the unevolved possibilities of the body and mind of humanity."[41] Thus alchemy presents the means of a "spiritual evolution" of mankind as a whole into a higher form of being. The (successful) transmutation of metals, according to Waite, is the lowest form of alchemical study. Waite urges experimental trials of psychic alchemy and provides the reader with pithy (and often bizarre) rules for activity and advancement in the spiritual realm.

Waite is perhaps best known for the alchemical treatises he translated in the early 1890s.[42] But these editions have proven to be historically pernicious, for although they appeared as translations, they are often nothing of the kind. They are almost invariably based upon corrupt editions and offer texts butchered to unrecognizability by the silent excision of large portions of material and adulterated by the addition of occultist elements and slants completely alien to the originals. The fact that there are in many cases no other modern translations with which to compare or replace them has intensified their ill effects over time. Waite's corrupt translations were used regularly by historians of science until the middle of this century, as witnessed by their frequent citation in articles in *Ambix* and *Isis*, as well as in scholarly books; some authors still continue to refer to them. Nearly all have been reprinted and are currently available in inexpensive editions.

After his string of translations in the 1890s, Waite did not publish again on alchemy for almost thirty years. In 1926, however, he produced his last book, *The Secret Tradition in Alchemy*, whose contents mark a sharp departure from his earlier interpretations. Here Waite attacks Atwood and Hitchcock much more vigorously than before, proposing to "survey alchemical literature and its history" in order to ascertain its real degree of esoteric or spiritual content. He then proceeds through biographies akin to his earlier *Lives*, but a remarkable metamorphosis is evident: a degree of skepticism and historical discretion appears, and Waite cites his "own hardened unbelief about things occult"![43] He concludes that "between the age of Byzantine records and the age of Luther there is no vestige" of a spiritual alchemy in the historical record and actually refers to the history of alchemy as a record of experimental physics.[44] At no point does Waite explicitly repudiate his esoteric theosophical theories from the *Lives*, his doctrine of the "alchemical transformation of mankind" from *Azoth*, or any of his many other clear expressions of belief in occult matters. In one place he even criticizes a reviewer for maintaining the very same esoteric view of alchemy he himself had expounded at length in *Azoth*. The occultist has marvelously transmuted himself into a positivist; whether his mind was changed by further studies or by a convenient abandonment of Victorian occultism for 1920s positivism is unclear.

Fate and Validity of the "Spiritual" Interpretation

The esoteric school remains strong to this day and continues to have an extensive impact on both the general and the learned perceptions of historical alchemy. Writers such as Julius Evola (1898–1974) and Titus Burckhardt have extended the movement through the twentieth century.[45] Outside of serious scholarship, a number of esoteric alchemical circles in the United States and Europe perpetuate the transcendental views of alchemy so popular in nineteenth-century occultism, sometimes amalgamated with heterogeneous notions from more recent occultist/spiritualist movements such as the New Age movement or radical environmentalism. Many of these groups operate presses that regularly release books on "spiritual alchemy" as well as poor editions and translations of earlier texts. Over half of the books on alchemy published since 1970 either espouse an esoteric/occultist view or are reprints originating from esoteric organizations or presses. A 1976 reprint (by Shambhala Press in Berkeley, California) of Waite's 1894 translation of Paracelsus bears a foreword by one Charles Poncé of the "Azoth Foundation," who declares alchemy to be the "product of Soul Imagining" and the "archetypal language of the soul." An even more extreme example may be found in Kenneth Rexroth's preface to a reprint of Waite's edition of the *Works of Thomas Vaughan*. Being completely ignorant of the Sendivogian alchemy that Vaughan espoused, Rexroth is free to interpret it as a veiled expression of tantric yoga. Hence Vaughan's sentimental references to his dead wife lead Rexroth to the sensational claim that "Thomas Vaughan and his wife, his *soror mystica* wrapped in entranced embrace at the Pinner of Wakefield were, it is true, blundering into a region of revelation which they little understood and which, it would seem, eventually destroyed both of them."[46]

Many books embracing the esoteric interpretation continue to be taken quite seriously. The grip of occultism was, at least at one time, particularly strong in France, where esoteric/occultist volumes have been published regularly from the last century to the present day; a reviewer of one such book referred to its contents as the "typical French approach to the subject of alchemy," and Robert Halleux has briefly catalogued such writers.[47] This unremitting flow of esoterica continues to seep into the field to the detriment of serious scholarship. Such volumes often appear on university library shelves alongside scholarly works (sometimes outnumbering them) and thus present a trap for the unwary. Within academe, scholars writing on the relationship of alchemy to religion, art, and literature in particular have frequently embraced rather uncritically part or all of the spiritual interpretation of alchemy. Additionally, those who do not utilize primary sources are particularly liable to acquiesce to the esoteric view. One attraction of this interpretation is that it allows writers with a generally sympathetic view of alchemists to attribute grand and cosmic designs to their subjects in opposition to the (equally unwarrantable) criticism that all their efforts toward the manufacture of gold were mercenary or fraudulent.

The chief problem with the esoteric view is that even laying aside the more extreme positions, the historical record (as Waite, for whatever reason, finally concluded in 1926) simply does not countenance it. Although the works of many alchemical writers contain (often extensive) expressions of period piety, imprecations to God, exhortations to morality, and even the occasional appearance of an angelic or spiritual messenger, we find no indication that the vast majority of alchemists were working on anything other than material substances toward material goals. The distinctions in tone and attitude toward spirituality that quite admittedly exist between many "alchemical" texts and more modern "chemical" texts can be explicated without recourse to the spiritual interpretation's disjunction between "alchemy" and "chemistry" and its labeling of them as esoteric and exoteric traditions, respectively. First, it must be remembered that transmutatory alchemy fell out of wide popularity at around the time of the widespread secularization of intellectual culture that occurred in the eighteenth century. Most alchemical texts originated in a culture of greater religious sensibility than our own and thus naturally exhibit more spiritual and religious expressions than do later works of "chemistry." Second, the secrecy and "initiatic style" ubiquitous in works on transmutation led quite naturally to a tone of mystery absent from the later, more "open" writings of eighteenth-century chemistry. This emphasis on secrecy led originally to the fairly common contemporaneous invocation of morality or divine agency as "gatekeepers" to secret knowledge, but in the nineteenth century to a linkage of the arcana of alchemy to the secrets of "the occult" as a whole.[48] These culturally based differences of expression and tone do not countenance the spiritual interpretation, which fails to recognize the cultural context of alchemical texts.

This is not to say that there was nothing whatsoever akin to a "spiritual alchemy" in the broad historical spectrum of alchemy. The relationship between alchemy and religion, theology, and spirituality is complex, but still does not countenance the esoteric spiritual school of interpretation. It is true that a tradition of using alchemical terms and imagery in religious and spiritual literature did in fact develop in the early modern period; indeed, such works sometimes use the term "spiritual alchemy." St. Francis de Sales, for example, writes of Christian love as the "divine powder of projection" and labels the transforming power of that love as a "holy and sacred alchemy." The alchemical emphasis on purification, transformation, and the quest for the perfection of gold naturally provided an abundant supply of similitudes to religious writers. Secular writers and poets made a similar use of alchemy and its ideas, and the Scriptures themselves occasionally invoke metallurgical images of the refiner's fire to express spiritual trials in a similar way.[49]

Alchemical theories and processes also provided theological and devotional object lessons. Sir Thomas Browne remarked that his study of material relating to the philosophers' stone "taught me a great deale of Divinity." More dramatically, the alchemist Pierre-Jean Fabre wrote a book entitled *Alchymista Christianus*, in which "most of the mysteries of the Christian faith are explained by means of chymical analogies and figures."[50] Complementary to the use of alchemy to instantiate and illustrate religious truths, the same religious truths and revelations were also employed to direct alchemical work. There exist numerous "alchemical readings" of the Bible, just as there are alchemical readings of pagan mythology.[51] It was this practice of some seekers after transmutation which Thomas Sprat criticized when he wrote dismissively that "they believe they see some footsteps of it, in every line of *Moses, Solomon*, and *Virgil*." Interpretations of holy texts also contributed occasionally to the cause of alchemical secrecy, as in the case of a description of how to make the philosophers' stone disguised under the forms and words of the Mass.[52] Of course, many fundamental terms and themes in alchemy—such as death and resurrection, exaltation and sublimation—derive from Christian theology.

But in all these interactions of alchemy with spirituality, it is clear that alchemy functions as a source of tropes and imagery for rhetorical embellishment or didactic exemplification rather than as an inherently spiritual exercise which elevates the practitioner by some esoteric illumination. It might be noted that the deployment of alchemical images to express spiritual themes metaphorically is in some sense related to Hitchcock's exposition of alchemical themes as metaphorical descriptions of Christian principles and devotions, but it has nothing in common with the "spiritual alchemy" that stems from Atwood and her followers, with its links to Mesmerism and other nineteenth-century occultist movements. When alchemical authors deploy sacred texts or spiritual terminology, this is a relatively unproblematic use of images, concepts, and terms drawn from the religious culture of the time, rather than evidence that alchemical practices were concerned primarily or essentially with the spiritual enlightenment or development of the practitioner. These linkages were made by minds more attuned to the drawing of similitudes and the reading of "meanings" (and more convinced of the epistemological value of similitudes in general) than are those of our highly literal modern world.

Perhaps the most pervasive use of alchemical language in spiritual writings appears in Jakob Boehme's mystical theology. The language of the Paracelsian *tria prima*, the divine *Salitter (sal nitrum)*, and related concepts are used to express divine qualities, powers, and activities and to expound the cobbler's ecstatic visions. But Boehme's use of alchemical language and imagery—as extensive as it is—remains clearly of a different order than, for example, the practical and theoretical antimonial exercises of Basil Valentine, Alexander von Suchten, Eirenaeus Philalethes and others, or the rigorous Scholastic alchemy of "Geber," Albert the Great, Petrus Bonus, or Gaston Duclo.[53] Even if Boehme's work were taken as evidence of the "spiritual alchemy" promoted by esoterics and occultists, it would remain to be proven by historical argument that he falls into the mainstream of early modern alchemical thought, and that extrapolations about alchemy in general could be reliably or usefully made from him.

We should also mention the little-recognized school of "supernatural alchemy" which seems to have developed in seventeenth-century England. This school held that certain alchemical products had supernatural effects either upon the external world or upon the possessor. Robert Boyle, for example, thought that the philosophers' stone might summon angels and facilitate communication with them, and Elias Ashmole and others mentioned yet greater "supernatural stones" which gave the possessor special intellectual or spiritual powers.[54] But even these notions are only superficially similar to the spiritual interpretation of alchemy as it was conceived in the nineteenth century. Now it is possible that this school may have initiated a certain tradition of alchemical interpretation which persisted at some low level through the eighteenth century, and then provided a nucleation point for nineteenth-century occultist writers, but there is presently no clear historical evidence for this conjecture. (The same conjecture might also be made regarding Behmenist thought.) The seventeenth-century supernatural school needs more study to define its origins, content, extent, and influence. Nonetheless, at present it is clear that this "school" was a small and perhaps fairly localized subset of alchemy, and so it would be wrong to extend its characteristics to "alchemy" as a whole.

Thus it goes without question that alchemy and religion (or spirituality of various kinds) interpenetrated one another in the medieval and early modern periods, and that each borrowed terms and concepts from the other. This fact is not, however, remarkable in itself, nor is such interpenetration with religion unique to alchemy. Recent work across the entire spectrum of the history of science displays clearly the ubiquity and the importance of religious and theological concerns and influences in early modern natural philosophy; alchemy should be neither an exception nor a special case. By rejecting the "spiritual interpretation" of alchemy we do not intend to imply that the discipline over the *longue durée* was unconcerned with religion any more than modern historians of science would wish to deny the religious and theological dimensions and motivations of the works of Kepler, Boyle, or Newton. We do argue that the view which sees alchemy as an essentially spiritual activity, and which maintains that the degree or character of alchemy's religious or theological content renders it distinct from other branches of contemporaneous natural philosophy (and particularly from "chemistry") is an ahistorical formulation which postdates the early modern period and was fully developed only in the context of nineteenth-century occultism.

In addition to its direct effects on the historical understanding of alchemy, the prevalence of the esoteric interpretation in the late nineteenth and early twentieth centuries seems to have had even greater indirect effects. The currency of the notion of an internal alchemy whose goal was the transformation of the soul cannot have failed to influence the construction formulated by Carl Gustav Jung, with which it shares an emphasis on psychic states and spiritual self-development. Indeed, the

Jungian view, as we shall see, seems little more than nineteenth-century occultism translated into "scientific" terminology.

The Jungian Interpretation of Alchemy

Carl Jung has probably exercised a greater influence on the common perception of alchemy than any other modern author. His psychologizing view of alchemy has been propelled into the cultural mainstream by such writers as the historian of mythology Joseph Campbell, the literary critic Northrop Frye, the philosopher Gaston Bachelard, and the historian of religion Mircea Eliade.[55] But Jung has had an even more pronounced effect on the historiography of alchemy itself. The Jungian approach to alchemy is a stock element of most popularizing texts on the subject, for example, Gareth Roberts's 1994 *The Mirror of Alchemy* and Allison Coudert's 1980 *Alchemy: The Philosopher's Stone*. Even such serious students of the subject as F. Sherwood Taylor and E. J. Holmyard felt obliged to consider the Jungian perspective in their own surveys.[56] Among recent serious historians of science, Betty Jo Teeter Dobbs promoted the Jungian approach in her famous *Foundations of Newton's Alchemy* and reaffirmed it unequivocally as recently as 1990.[57] Even the praiseworthy *Norton History of Chemistry* pays homage to the Jungian analysis of alchemy, citing it as a "traditionalist" view of the subject. This note of approbation is shared by the debunker of alchemical symbolism Marco Beretta.[58]

Jung was deeply interested in occultism from at least his adolescence. His doctoral dissertation, "On the Psychology and Pathology of So-Called Occult Phenomena," was based on the spiritualist seances of his cousin Helly Preiswerk, in which he was an active participant.[59] As early as 1913, he had adopted a "spiritualist and redemptive interpretation of alchemy," and one can be sure that this reflected his wide reading in the occult literature of the nineteenth century.[60] In addition, Jung was aware of and probably influenced by earlier work on alchemical symbolism by the Freudian psychologist Herbert Silberer.[61] But only in the 1920s did he begin writing on the subject, a pursuit that was to occupy the rest of his life.[62] By far the clearest specimen of Jung's approach can be found in his "Die Erlösungsvorstellungen in der Alchemie," published in the *Eranos-Jahrbuch* of 1936 and translated into English as "The Idea of Redemption in Alchemy."[63] Here Jung put forth his soon-to-be-famous claim that in the analysis of alchemy, "we are called upon to deal, not with chemical experimentations as such, but with something resembling psychic processes expressed in pseudo-chemical language."[64]

According to Jung, alchemists were concerned less with chemical reactions than with psychic states taking place within the practitioner. The practice of alchemy involved the use of "active imagination" on the part of the would-be adept, which led to a hallucinatory state in which he "projected" the contents of his psyche onto the matter within his alembic.[65] The Jungian alchemist literally "saw" his own unconscious expressing itself in the form of bizarre archetypal images, which were "irruptions" of the collective unconscious into his conscious mind. Because he viewed the primary role of alchemy in the light of the unconscious, Jung pointedly devalued the chemical content of alchemical texts. The alchemist's "experience had nothing to do with matter in itself," and consequently, the attempt to decipher alchemical texts from a chemical point of view was quite "hopeless."[66] In this claim one can see clear vestiges of Jung's immersion in Victorian occultism. Like Atwood and Hitchcock, Jung pointedly rejects the image of alchemy as protochemistry. Unlike the more radical upholders of spiritual alchemy, however, Jung did not altogether deny the role of the laboratory in alchemy. In this regard, Jung's view of alchemy greatly resembles A. E. Waite's notion of "alchemical transformation" expressed in *Azoth*, rather than those of inveterate spiritual alchemists such as Atwood and Hitchcock. Perhaps this should not be surprising, given that Waite's works were circulating among members of Jung's Zurich Psychological Club in the 1910s.[67] Like Waite, Jung did not completely reject the claim that alchemy involved laboratory experiment, but in effect he wrote it out of the picture, since alchemy's real concern was the transformation of the psyche, which could project its contents onto any sort of matter. The actual substances employed in a process made no difference at all to the alchemist so long as they stimulated the psyche to its act of projection. In reference to the many operations intended to produce the philosophers' stone, Jung explicitly stated that "[o]ne can make nothing of these from the standpoint of our modern chemical knowledge; if we turn to the texts and the hundreds and hundreds of processes and recipes which the Middle Ages and Antiquity have left behind, we find among them relatively few which contain any chemical sense."[68] According to Jung, therefore, the alchemical experience had more in common with the ecstasy of the illuminé in a state of *unio mystica* than with that of the laboratory technician.

Beyond his confident view that alchemy represented a process of psychic transformation, Jung also had comments to make on the subject of alchemical secrecy. Alchemists did not use obscure language to conceal chemical ingredients or to delude the uninitiated. Instead, they employed their bizarre terminology of dragons, dying kings, and copulating couples because these were the forms in which the unconscious projected itself onto matter.[69] Since the Latin Middle Ages were dominated by Christianity, the minds of medieval alchemists were especially filled with images of Christ, leading to the elaboration of a widespread "Lapis-Christus parallel," in which the philosophers' stone was symbolized by Christ.[70] The multiplicity of names for the *materia prima*, the starting point of the alchemical work, was a necessary consequence of the fact that "projection derives from the individual, and is different for each individual."[71]

Because the unconscious, according to Jung, always reveals itself in the form of hints and images, ambiguity was therefore an essential element of "genuine" alchemy. From this it follows that any alchemical text that could be clearly decoded into chemical language would be a second-rate product, and indeed, Jung went so far as to say that "there are good and bad authors in alchemical literature as elsewhere. There are productions by charlatans, simpletons, and swindlers. Such inferior writings are easily recognizable by their endless recipes, their careless and uneducated composition, their studied mystification, their excruciating dullness, and their shameless insistence on the making of gold. Good books can always be recognized by the industry, care, and visible mental struggles of the author."[72]

This claim incorporates, of course, an important self-validating principle. If a historian, by successfully decoding a given allegorical recipe into chemical language, were to challenge Jung's assertion that alchemical processes are expressions of the psyche, Jung (or a Jungian) could simply reply that the very fact that such a recipe could be translated into modern chemistry means that it is not a specimen of "good" or "genuine" alchemy. Genuine alchemy, by definition, cannot be decoded.

Jung's belief in a "good" and a "bad" alchemy led him to make numerous historiographical statements, and these have had a marked effect on subsequent historians. Jung believed that the history of alchemy could be divided into two chronological periods. The first, which he called the "classical period," fell between late antiquity and the end of the sixteenth century.[73] The second, which Jung called the period of alchemy's "decay," began with Paracelsus and Jakob Boehme, who split the field into two divergent realms. Paracelsus was responsible for putting alchemy on the path that would convert it into a "natural science" by emphasizing its medical aspect, whereas Boehme transformed it into a purely speculative mystical theology.[74] Jung put this metamorphosis of alchemy into typically graphic terms: "As we have seen, the Gnostic vision of the nous entangling itself in physical nature flashes out again from these late-comers to alchemy. But the philosopher who, in earlier days, descended like a Hercules into the darkness of Acheron to fulfill a divine opus has now become a laboratory worker given to speculation."[75]

In other words, the Paracelsians of the seventeenth century were no longer engaged in the "integration of the personality" by means of alchemy: They were mere technicians and scientists with a side interest in symbolism. The followers of Boehme, on the other hand, forsook the laboratory altogether and converted alchemy into a devotional literature. They too abandoned the path of ancient and medieval alchemy, which had relied on the integration of laboratory practice and visionary projection for its success as a path to "individuation" (the healing of the psyche). Jung believed, then, that alchemy began to divide into something like modern chemistry on the one hand, and Hermetic theosophy on the other, around the beginning of the seventeenth century. He added that this "degeneration" of alchemy continued into the late eighteenth century and that the classical tradition received its coup de grâce with the parallel occurrence of the chemical revolution of Lavoisier and Goethe's fully conscious use of alchemical metaphor.

Criticism of the Jungian Interpretation
Despite the widespread acceptance of Jung's approach to alchemy, a number of his key suppositions have been challenged recently. First, Richard Noll's fundamental study of Jung as a cult figure has cast considerable doubt on the validity of Jung's "collective unconscious," supposedly the deepest level of the mind, from which profound "archetypal images" can irrupt into the conscious.[76] Part of Jung's fascination with alchemy lay in his observation that "alchemical" visions and dreams spontaneously oc-

curred in the minds of his own psychotic patients, along with other archaic, esoteric images.[77] As Noll has argued, the showpiece in Jung's theory was a patient known subsequently as "the Solar Phallus Man." The patient had visions of the sun endowed with a swaying penis from which the winds issued. Jung was struck by the similarity of this image to a Mithraic liturgy stemming from late antiquity. On numerous subsequent occasions, Jung claimed that the patient could not have had any knowledge of the ancient mysteries and that the particular Mithraic text had not even been published at the time of his vision. Thus the Solar Phallus Man assumed an important role for Jung and his followers, providing evidence of the universal, transchronological character of archetypes drawn from a collective unconscious. As Noll has convincingly shown, however, Jung knew perfectly well that the Mithras liturgy had been translated and published both in German and English before the patient's remarkable vision and that the symbol was available in other sources as well. Jung engaged in a deliberate pattern of deception to buttress the scientific validity of his collective unconscious, itself an outgrowth of nineteenth-century occultism.[78]

Although it is not our purpose here to engage in an ad hominem attack on Carl Jung, it is important to emphasize the weakness of his theory of the collective unconscious, for without it, his interpretation of alchemy cannot stand. Jung explicitly argued that in the "classical period" of alchemy, the collective unconscious provided the stock of alchemical imagery: "The alchemistic process of the classical period (from antiquity to the end of the sixteenth century) was a chemical research into which there entered an admixture of unconscious psychic material by the way of projection. For this reason the alchemistic texts frequently emphasize the psychological prerequisites of the work. The contents that come into consideration are those that suit themselves to projection upon the unknown chemical substance. Because of the impersonal nature of matter, it was the collective archetypes that were projected."[79]

Since Jung's theory of the chronological development of alchemy depends on his belief in the complex mechanism of "projection," according to which the contents of the collective unconscious acquire an external, hallucinatory reality, it follows that the fortune of his interpretation is yoked to the fate of the collective unconscious itself. The evidence of fraud that Noll uncovered therefore acquires a significance beyond the unsavory light in which it places Jung, for by exposing the deeply problematic character of the collective unconscious, it undercuts the foundation of his interpretation and "historiography" of alchemy.

In addition to Noll's probing analysis of Jung himself, the historians Barbara Obrist and Robert Halleux have presented detailed arguments against Jung's interpretation based upon their extensive reading of late medieval and Renaissance alchemical texts, indeed, some of the very same figurative texts that Jung found most attractive.[80] Obrist in particular has shown that the *Aurora consurgens,* a highly emblematic alchemical text of the late Middle Ages ascribed to Thomas Aquinas, can be interpreted more simply and easily without the elaborate apparatus of analytical psychology. Nor do we need to invoke the gratuitous hypothesis made by its Jungian commentator Marie Luise von Franz that St. Thomas wrote the *Aurora consurgens* in a state of visionary ecstasy on his deathbed. The work is manifestly pseudonymous, having little relation either to the philosophical or theological work of Thomas Aquinas.[81]

Finally, we have elsewhere argued extensively against the validity and utility of the Jungian model. Briefly, if the images used in alchemical texts are in fact irruptions of the unconscious, then there would be no possibility of "working backward" from them to decipher such images into actual, valid laboratory practice. Nonetheless, we have presented comprehensive decodings of alchemical symbolism into modern, replicable chemical terminology.[82] Even some of the most allegorical writings—even when describing operations intimately linked with the making of the philosophers' stone—can be sensitively "decoded" and the chemical effects reproduced in a modern laboratory. Since these decoded authors include Eirenaeus Philalethes and Basilius Valentinus, who were among Jung's favorite examples for use in his psychological interpretation, Jungians cannot dismiss them as "bad alchemists."

Furthermore, we have shown clearly how extravagant alchemical imagery was consciously constructed to hide actual laboratory operations and how the very same alchemists who penned bizarre allegorical descriptions in print were able routinely to express their knowledge in clear, unambiguous "chemical" terms in private communications. A clear example of this is the preparation of the "sophic mercury" darkly veiled in the allegorical writings of Eirenaeus Philalethes but lucidly expressed in operational terms in the private 1651 letter of George Starkey (the real author of the Philalethes tracts) to his friend Robert Boyle. Indeed, contemporaneous alchemical readers were highly eager to "decode" allegorical texts such as Philalethes' *Introitus apertus ad occlusum regis palatium*, and many were successful in doing so as well as in reproducing the results described.[83]

Additionally, some alchemists were quite specific in cogently describing and designing apparatus for particular operations. Were alchemy a psychic phenomenon involving the collective unconscious and an external, hallucinatory reality, there would be no requirement for intelligent apparatus design as a prerequisite for success. Finally, chemical replications of results described in alchemical texts under the guise of extravagant imagery, including some of those that Jung himself used to argue for the non-chemical basis of alchemy, have demonstrated that the images that Jung claims originate in the collective unconscious actually have much more reasonable origins, namely, in the sometimes evocative physical appearances of chemicals reacting in flasks.[84] Clearly, then, alchemical texts, even highly emblematic and chrysopoetic ones, are not mere irruptions of the unconscious; they are descriptions of laboratory operations consciously and purposefully outfitted in sometimes outlandish guise. The alchemists' images are not unconscious productions, but rather expressive metaphors developed under the guidance of actual observation of chemical reactions coupled with the need to maintain secrecy and the outlook fostered by the "emblematic world-view" characteristic of the premodern period.[85]

Despite the fact that Jung's interpretation of alchemy is riddled with problems, many continue to accept his historiographical model. Surprisingly, few have challenged his claim that an alchemy previously unified by its single-minded projection of psychic contents onto matter began splitting into a natural science and a devotional art as a result of the early modern iatrochemical movement. To the contrary, Dobbs explicitly adopted this model in her *Foundations of Newton's Alchemy*, and it has striking resonances with her later attempts to distinguish between seventeenth-century "chemistry" and "alchemy."[86] According to her, the former was characterized by its straightforward practicality, whereas the latter had such themes as the redemption of matter and the salvation of the soul.[87] We have argued elsewhere against similar anachronistic attempts to distinguish "alchemy" from "chemistry" in seventeenth-century contexts, such as those made by Dietlinde Goltz and Marco Beretta.[88] In fact, the overarching discipline of late antique and medieval alchemy already contained a multitude of practical laboratory pursuits along with religious motifs; sometimes the two tendencies are combined in one text, but often they are not. The same is true of premodern alchemy's successor, the "chymistry" of the sixteenth and seventeenth centuries. There is no compelling reason to adopt nor any clear evidence to support the Jungian theory that a previously homogeneous discipline disintegrated into the two realms of devotional and practical alchemy upon the advent of Paracelsianism and that these immediately came to be identified as "alchemy" and "chemistry" respectively.[89]

Although Jung was an important contributor to the notion that an essential difference of philosophical or spiritual outlook divided alchemy from chemistry, the idea is not restricted to Jungians. Many scholars characterize alchemy as an essentially vitalistic, organic view of nature as well as a spiritual illumination granted to its adepts. Hence alchemy supposedly incorporates a "mentality" radically different from that of modern chemistry. Traces of nineteenth-century occultism are clearly evident in this viewpoint, though its twentieth-century proponents do not always perceive these vestiges. In the following section, we shall refer to this view of alchemy, in which illuminism and vitalism are viewed as atemporal, essential characteristics of the discipline, as the "panpsychic" model.

The Panpsychic Interpretation of Alchemy

The view that alchemy was a historical constant, suffused with ecstatic illuminism and vitalism over its entire period, and that the loss of this outlook signified the death of the discipline had already been promoted as early as 1938 by the comparative religionist Mircea Eliade, whose essay "Metallurgy, Magic and Alchemy" formed the nucleus around which he built his immensely popular *The Forge and the Crucible*.[90] Eliade, like Jung, had early ties with the popular occultism of the late nineteenth and nascent twentieth centuries. In his student years, he was a devotee of Rudolph Steiner's "Anthroposophy" and developed a keen interest in alchemy.[91] This juvenile interest bore fruit later in "Metallurgy, Magic, and Alchemy," in which Eliade, like Jung, portrayed alchemy as a discipline concerned primarily with soteriology. Although the alchemist might work tangentially with chemicals and metals, his real quest concerned the soul:

"The alchemist, while pursuing the 'perfecting' of the metal, its 'transformation' into gold, pursued in fact his own perfection."[92]

Eliade was not a psychoanalyst, and he therefore eschewed the language of analytical psychology, yet this part of his message is identical with Jung's.[93] Like the Swiss psychologist, Eliade thought that alchemists experienced an initiatic experience leading to "certain states of consciousness inaccessible to the uninitiated."[94] And like Jung, Eliade stressed that the chemical side of alchemy became pronounced only when the discipline "decayed" or "degenerated" from its primeval simplicity.[95] As the "sacred" side of alchemy declined, the ecstatic experiences of the adept abated, making it possible for the newly "profane" science of chemistry to emerge and for precise laboratory observations to be made. This division of sacred alchemy from profane chemistry also recalls the spiritual interpretation of alchemy. Eliade differs from Jung, however, in his extreme emphasis on the vitalism of alchemy, which he claimed to be a defining characteristic of the discipline. Drawing numerous parallels with African smith traditions, Eliade focused on alchemical texts that spoke of the growth of metals within the earth and of their "love" and "marriage" with one another.[96] His alchemists described the furnace as a surrogate for the "great tellurian matrix" of the earth, in which the philosophers' stone as an "embryo" would be incubated.[97]

Eliade further claimed that alchemy represented a sort of "organic" worldview in contradistinction to the mechanism of modern science.[98] Only with the birth of the scientific worldview during the early modern period did alchemy lose its cosmic vision: "After the mental revolution accomplished by the Renaissance, the physico-chemical operations and cosmic events achieve their autonomy from the laws of universal life, enclosing themselves, though, into a system of 'dead' mechanical laws."[99]

For Eliade, the development of mechanism killed alchemy and inaugurated modern science. Indeed, in a remarkably prescient turn of phrase, Eliade made it clear that for him, the death of alchemy was synonymous with the death of nature. It is only through the persistence of a few alchemical topoi in Western civilization, he said, that "the Cosmos 'dies' very late in [the] European imagination."[100]

Eliade's panpsychic view of alchemy has a surprising counterpart in the writings of a contemporaneous and influential historian of chemistry, Hélène Metzger. Although Metzger seems to have influenced Eliade little if at all, she had already proposed important parts of his interpretation in a fundamental article of 1922, which later resurfaced as an integral part of her well-known *Les doctrines chimiques en France du début du XVIIe à la fin du XVIIIe siècle* of 1923.[101] Metzger maintained, like Eliade, that alchemists took a hylozoic approach to matter and that they believed that metals and minerals grew like vegetables or animals within the earth.[102] This very vitalism, according to Metzger, made alchemy possible, for the "Hermetic philosophers" based their belief that base metals could be perfected on an organic model: "Now in order to justify their research, the Hermetic philosophers said that gold is a metal that has attained the final limit of its perfection. The imperfect metals, like green fruits exposed to the sun, ripen spontaneously and transform themselves naturally into gold."[103]

Hence the very notion of metallic transmutation was based, according to Metzger, on a transference from the plant and animal realm to that of minerals and metals. Metzger went so far as to argue that such specious analogical reasoning was at the heart of alchemy. Just as alchemists made an invalid comparison between the inanimate mineral realm and that of the living, so they assumed that there were operative correspondences working between stars and metals or between minerals and the parts of the body.[104] She believed such "primitive," "illogical" thought to be responsible for alchemy's origin and longevity.[105] As Jan Golinski has argued, Metzger was probably influenced here by the famous philosopher-anthropologist Lucien Lévy-Bruhl, who happened to be her uncle.[106] Lévy-Bruhl too had spoken of a "primitive mentality" characterized by the reification of analogies, and Metzger explicitly used this concept to explain alchemy in several of her later articles.[107]

Metzger shared with Eliade not only the belief that alchemy was by its very nature vitalistic, but also the conviction that the mechanical worldview of modern science killed it off. According to Metzger, the emergent corpuscular philosophy of the seventeenth century was fundamentally opposed to alchemical transmutation, for the matter theory of the Cartesians left no room for "perfectability" in the mineral realm. Metzger was unequivocal about this point, which resurfaces at various places in her oeuvre: "What did manage to ruin [alchemy] was the Cartesian theory, which did not even attack it directly; indeed, to admit, as Descartes had done, that 'all varieties which are found in matter depend on the movement of its parts,' was in effect to admit that matter is similar to itself everywhere; it is therefore to render absurd the idea of the perfection of chemical substances or even of Nature, which remains always as created."[108]

According to Metzger, there was an inviolable schism between corpuscular theory and alchemy. With the accession of the former, the latter had to fall. One result of this observation is that whenever Metzger found early modern writers on chemical subjects invoking the language of "corpuscles" or "minimal parts," they automatically became "chemists" rather than "alchemists." This was one tool that she used in defining the emergent discipline of chemistry, as opposed to alchemy. "Alchemy" could not coexist with mechanism, whereas "chemistry," as in the writing of the important iatrochemist Nicolas Lemery, could and did.[109]

Metzger and Eliade both asserted, then, that alchemy was fundamentally stamped by its insistence that matter was alive, indeed, ensouled, and when this view came into question during the Scientific Revolution, alchemy had to pass into oblivion. Hylozoism was therefore an essential characteristic of alchemy, according to these two writers. Jung too upheld that all alchemists viewed matter as alive and ensouled. Only this notion of the soul of matter allowed alchemists to develop their myth of redemption, according to which the philosophers' stone became an analogue of Christ. All three authors were therefore forced to see the development of early modern chemistry as a divorce from a radically different mentality represented by alchemy. In the case of Eliade and Jung this schism appeared in the nostalgic terminology of "decay" or "degeneration," whereas Metzger, driven by other sentiments, employed a different language. But she too, no less than they, believed the seventeenth century to have witnessed the birth of our modern science of chemistry, to the immediate discomfiture of the alchemists.

This panpsychic view of alchemy has received a more modern expression in some of the less overtly Jungian writings of Dobbs, and indeed, she employed both Eliade and Metzger directly.[110] It is interesting to note that Dobbs makes no reference at all to Jung in two of her last publications, *Alchemical Death and Resurrection* and *The Janus Faces of Genius,* despite her simultaneous endorsement of his views elsewhere.[111] In these two works Dobbs stresses that alchemy has always been characterized by a desire to view nature in biological and vitalistic terms.[112] Not only is vitalism a fundamental trait of alchemy, however; so too is the quest for religious illumination.[113] This fusion of vitalism and illuminism that characterizes alchemy over its entire history is in turn the product of a primitive understanding of life, death, and resurrection that imposes these categories on the animal, vegetable, and mineral realms.[114] What is instructive about these comments is the ease with which Dobbs manages to pass from an overtly Jungian point of view to one that is bereft of the apparatus of analytical psychology while maintaining Jung's essential point that alchemy was above all a quest for religious revelation.

A romantically colored rendition of the panpsychic view also forms a central thesis of Carolyn Merchant's popular *The Death of Nature* (1980), whose very title could be a restatement of Eliade. Merchant uses Eliade and Jung to argue that the alchemists—again considered as a homogeneous body—held a sacred view of nature in which the earth was revered as female.[115] She sees the (supposed) triumph of the mechanical philosophy over alchemy in the seventeenth century as a central example of "the transition from the organism to the machine."[116] Evelyn Fox Keller has also expressed these themes from the panpsychic interpretation in her *Reflections on Gender and Science* of 1985.[117] According to Keller, "the hermetic tradition" (here uncomplicatedly equated with alchemy) and "mechanism" provided the two poles available to natural scientists in mid-seventeenth-century Britain.[118] Unlike the upholders of a mechanical worldview, the alchemists employed a highly gendered language whose "basic images" were "the hermaphrodite and the marital couple."[119] Hence, although she asserts that the alchemists were not actually feminists themselves, Keller sees them as having championed the "view of nature and woman as Godly," a position that she claims the mechanical philosophers defeated.[120] The traces of Eliade's interpretation lie just below the surface of such claims, whether explicitly invoked or transmitted by authors such as Merchant and Dobbs.[121]

Criticism of the Panpsychic Interpretation
One can see, then, how the panpsychic model of alchemy, based primarily on the work of Eliade and a diluted version of Jung's conceptions, but also abetted by Metzger's focus on vitalism, has colored the current view of alchemy. It seems to us that there are three fundamental problems with the panpsychic interpretation. From a historian's point of view, one of its most obvious weaknesses lies in its failure to acknowledge the development of

alchemical theory and practice over the *longue durée* as well as its internal diversity during a particular time. It is an essentialist picture of alchemy. Its goal has often been the segregation of the field from other scientific disciplines for one or another polemic or historiographical purpose. As we and others have argued elsewhere, alchemy as a historical phenomenon is too diverse to permit such overly reductionistic views.

This diversity comes to the fore when we treat the second problem: vitalism. The panpsychic notion that alchemy is inherently and necessarily vitalistic draws some support from alchemical sources, which often speak in a language that attaches images of vegetable and animal life to the mineral realm. Images from agriculture and reproduction are particularly common. But here we encounter a problem analogous to that found in the spiritual interpretation, namely, how much of such imagery is to be understood in a literal sense? We argue that among many alchemists these images are merely metaphorical or heuristic. Although we cannot declare that this is true for all alchemical writers, because of the inherent diversity of the subject, there are sufficient numbers of nonconforming alchemists to subvert any attempt to characterize alchemy on the basis of its vitalistic content. For example, the famous medieval *Testamentum* ascribed falsely to Ramon Lull uses the term *menstruum* to describe a corrosive, drawing an elaborate parallel between the Galenic theory of human generation, with its male and female sperms and menstrual blood, and the mineral realm. But pseudo-Lull explicitly says that this analogy between the mineral and the animal is only a metaphor: "the mineral genus is added only figuratively, for the sake of similarity."[122] What may have appeared at first to be a naive case of hylozoism turns out to be a deliberate choice of metaphor.

Unlike post-Enlightenment writers on alchemy, early modern scholars were sensitive to this distinction between analogy and identity in alchemical thought. When the polymath Daniel Georg Morhof wrote his 1672 overview of alchemical thought, for example, he explicitly considered the question "an metalla vivant" (whether metals live). Morhof's response is enlightening, for he excuses himself from answering the question at length and lists only Giordano Bruno and the little-known Berigardus as proponents of such vitalistic or hylozoic notions before he passes on to other topics. Clearly, the widely read Morhof did not see vitalism as ubiquitous or even widespread in alchemy as demanded by the panpsychic model.[123]

Additionally, the popular late sixteenth-century chrysopoetic writer Gaston Duclo, or "Claveus," explicitly discusses the use of vitalistic imagery. In his 1590 *Apologia chrysopoeiae* (written against the attack of Thomas Erastus on transmutation), Duclo states clearly that the laws governing the animate world of animals and vegetables are different from those governing the inanimate world of minerals and metals. The anti-Paracelsian Erastus had in fact used cross-realm analogies in his attack on alchemy, and Duclo unequivocally asserts their faulty nature. Nevertheless, Duclo himself uses terms like *anima* and *semen* and the image of the alchemist imitating the farmer growing crops, but he firmly denies any literal-minded interpretation of these expressions, stating that they are used only *metaphorice*.[124] Such terminology is only superficially vitalistic: it does not imply the hylozoism claimed by the panpsychic interpretation.

But there is a third flaw imbedded here as well. Adherents of the panpsychic model insist on a thoroughgoing break between alchemy and chemistry, which they claim to have been implicit throughout much of the seventeenth century. This notion reflects Jung's assertion that alchemy ceases to be alchemy when it becomes clear enough to be understood in chemical terms, and this seems also to be the case for Eliade, with the added criterion that such "degeneration" from the older tradition of alchemy marks a passage from the sacred to the profane. This break also appears in Metzger's work, which contains a crucial erroneous element beyond the mere assertion that much of the seventeenth century saw alchemy and chemistry as widely divergent practices. As we have already stated, the definitive break between alchemy and chemistry occurs for Metzger when the iatrochemists of the seventeenth century adopt an explicit corpuscular theory as in the case of Nicolas Lemery. This distinction displays the flip side of the supposed vitalist-mechanist dichotomy: alchemists are vitalists, chemists are mechanists. But this distinction will not hold. It is now a demonstrated fact that an important corpuscularian tradition was associated with alchemy from the thirteenth century onward. Indeed, this alchemical corpuscular theory influenced the conceptions of no less a mechanical philosopher than Robert Boyle himself.[125] Thus not only is the assertion that alchemy was necessarily vitalistic flawed, so is the complementary assertion that it was nonmechanical and noncorpuscularian. In fact, the very notion of a clean distinction between vitalism and mechanism in the seventeenth century, regardless of their putative attachment to "alchemy" and "chemistry" respectively, is open to question.[126]

Metzger, Jung, and Eliade all employ a supposedly nonarbitrary, historically founded criterion of demarcation for alchemy that they can then use to repel all countervailing criticisms. If one points out an alchemist who employed clear language, an expressly nonvitalist system, or the language of atomism, then these authors can simply reply that by virtue of their definition, he was not really an alchemist, or at least not a "good" one. But it does not require tremendous acumen to see that such a response is little more than a species of begging the question.

Positivist and Presentist Treatments

The reader will note that a common theme in all of the three foregoing interpretations of alchemy is the tendency to downplay or eliminate any natural philosophical or "scientific" content in alchemy. Although we argue that the artificial segregation of alchemy from the scientific tradition is an error, we wish equally to steer away from a "positivist" position that there is no real distinction between alchemy and later science. It is pertinent then, briefly to mention the positivist view of alchemy, whose tutelage we must likewise decline.

We are ill at ease with the label "positivist," however, because of the diffuseness of its common use. The case of the nineteenth-century organic chemist Justus Liebig provides a good example of the problems inherent in the term. In the third letter of his *Chemische Briefe,* Liebig treats the history of chemistry and states that "alchemy has never been anything other than chemistry."[127] Although Liebig's position here has often been characterized as positivist because he dwells upon the positive contributions of alchemy without due consideration of its historical context, at the end of the same letter Liebig explicitly refutes the threefold development thesis of Auguste Comte on which classical positivism is based.[128] The kind of "positivism" that is generally alluded to by historians of science and that we mean to critique in its application to alchemy shares much with "presentist" or "Whig" historiography, which assigns relative importance to historical ideas based upon their level of connection with or similarity to current scientific notions and shows insufficient interest in the historical and cultural context of those ideas. For the sake of simplicity, then, we will henceforth refer to this type of historical writing as "presentist."

This presentist historiography has had two quite opposite effects on the scholarly study of alchemy. Shortly before the appearance of Atwood's *Suggestive Inquiry*, Hermann Kopp published the first of his several historical studies on the history of chemistry.[129] Kopp began by listing the specific contributions of alchemists to chemistry—for example, methods of separation and purification, apparatus, and various chemical products—denominating alchemy as a developmental phase of chemistry. Yet he also dismissed as mere error whatever did not make a positive, experimental contribution to later chemistry. As one historian has remarked, Kopp "merely takes the seeds out of the fruit, which to him are the only things of value. He has no interest in the uniqueness of the fruit as a whole, its shape, color, and smell."[130] But in spite of this criticism, it should be remembered that before Kopp, alchemy was seen, following Enlightenment judgments, primarily as a fraud without redeeming qualities; thus Kopp's reading offers not only a more accurate representation of the historical account, but also a rehabilitation of alchemy. Kopp's scholarship much exceeds that of his predecessors Johann Friedrich Gmelin and Karl Christoph Schmieder, and in his later contributions he develops a somewhat broader view of alchemy. Yet the approach of picking and choosing nuggets of positive contributions to chemistry out of their alchemical context continued to characterize much of the serious historical literature of the nineteenth and early twentieth centuries, particularly general histories of chemistry.[131] Such well-intentioned rehabilitation of alchemists and their art lies behind the early twentieth-century denomination of the reality of radioactive decay as a "vindication" of the alchemical belief in transmutation; this spurious connection is still encountered in popular texts on alchemy and serves as a point of confusion rather of clarification.

At the same time, presentist attitudes have led certain historians to dismiss alchemy from serious scholarly consideration. The spiritual and nonscientific interpretations outlined earlier in this chapter made the subject appear even more unpalatable to those with a positivist bent. For example, George Sarton expressed a notorious revulsion for alchemy, which led him to declare as late as 1950 that alchemists were all "fools or knaves, or more often a combination of both in various proportions." His presentist outlook is fossilized in the *Isis* classification system he devised, which continues to this day to classify alchemy with witchcraft and divination under the opprobrious rubric of "pseudo-science."[132] Alchemy seemed insufficiently "scientific" to merit serious consideration in the history of science. Such views stand behind the resistance to revelations that respected figures of early modern science, such as Newton and Boyle, were devotees of traditional alchemy.[133] Similar sentiments undergird the continued casual use of alchemy as a convenient foil against which to set off modern science. On the flip side, the presentist extractions of "scientific germs" from the totality of alchemy left the overall context of the field largely unexplored, allowing esoteric and psychological notions to fill that vacuum with little resistance, thus further removing from our reach an accurate understanding of the sum of the discipline. The unsatisfactory nature of a presentist approach need not be insisted upon to modern historians.

Summary

The reader may already have noticed that of the three most influential interpretations of alchemy presented and critiqued here—spiritual, psychological, and panpsychic—none (excepting Metzger's role in the panpsychic model) were devised by historians. Jung and Eliade, moreover, were directly influenced by late nineteenth-century occultism, and Metzger by the anthropological musings of her uncle Lévy-Bruhl. Yet in spite of their origins outside of properly historical studies, a fact sufficiently manifested by their tendency to view alchemy as a chronological constant, these interpretations have all permeated the historiography of alchemy to such an extent that many historians have adopted them without being aware of either their origins or their unsuitability.

A factor common to these interpretations is their tendency to separate alchemy from "science" or natural philosophy; all insist upon psychological, ecstatic, or irrational elements as fundamental to alchemy. Much of this view arises from the often rather alien nature of alchemical writings, whether we consider the highly metaphorical style, the commonplace religious sensibilities and sentiments, or the outlandish emblems and figures. Clearly these modes of expression are far removed from those encountered in the writings of more recent and more well-established scientific figures. But differences of expression need not translate directly into differences of intent or content. We have mentioned above the readings of even ostensibly bizarre texts recently offered by us and others that "decode" alchemical language into a language of the laboratory and of natural philosophy. We do not deny that alchemical thought often embodied cultural and intellectual presuppositions and intents far different from those typical of the modern age; we do, however, deny the validity of interpretations that artificially, unwarrantably, and most of all, ahistorically introduce a chasm between "alchemy" and "chemistry." We argue that this putative divide is largely an artifact of the interpretations critiqued in this chapter.

We have shown moreover that not only the esoteric spiritual interpretation but also, to varying degrees, both the Jungian and the panpsychic interpretations draw their inspiration from nineteenth-century occultism. The similarity of Jung's psychologizing view to the "spiritual evolution" system of A. E. Waite's *Azoth* is clear, and what we now know of Jung's juvenile interest in the occult and the currency of Victorian esoterica in Jung's early circles supports this observable similarity. Likewise, the concept in Eliade and others that "genuine" alchemy had spiritual illumination as its goal and that early modern alchemy was a product of a degeneration from a more "sacred," purer, or more "organic" time are developments of ideas that can be found in the work of Atwood and other occultists. Since Jung's and Eliade's views have been widely accepted by subsequent historians of alchemy, we therefore come to the rather surprising conclusion that the residues of Victorian occultism have deeply colored the historical study of the discipline. It seems unlikely that many historians would continue to engage in the blithe generalizations criticized in this chapter if they realized their dubious origins.

Future of Alchemical Studies

Thus far we have devoted this chapter to a study of various influential interpretations and approaches to alchemy, all of which we find unsatisfactory. We hope that this exercise will succeed in clearing away these skewed historiographies and their claims and allow for more accurate and more penetrating future studies. Although it would be premature to sketch out in any detail what a fresh view of alchemy would look like, it is not amiss to conclude with suggested directions for the historical study of the subject, some of which are currently being explored.

A fundamental difficulty in the study of alchemy has been the lack of reliable historical data relating to both authors and their texts. This twin problem, which inhibits the accurate situation of authors and texts into their historical and cultural contexts, presents a major hurdle to be overcome before thorough and substantive advances in the understanding of alchemy can be made. Some notable advances have been effected in this arena, and further advances are to be expected from the work now in the hands of prominent scholars.[134] Nonetheless, rigorous historical attention to issues of textual purity and authorial biography should be one important focus for alchemical studies over the next decades. Critical editions of important individual works are needed and more comprehensively, editions of the complete *opera* of important figures, containing careful discrimination between the strata of authentic, interpolated, and spurious works. Even the writings of so important a character as Paracelsus, in spite of the labors of Karl Sudhoff decades ago, are still quite problematic.[135] The present corpora of many important figures are heterogeneous masses of texts; some are validly attributable to the real author, whereas others were composed by students or imitators who in some instances lived centuries later.[136] Clearly, the possession of emended texts localizable to a time and a place, coupled with information regarding the intellectual, temporal, religious, social, and political situation of their authors, are prerequisites to solid and contextual historical inquiry.

A common failing of the interpretations critiqued in this chapter is the depiction of alchemy as a uniform and constant monolith; consequently, future studies should pay attention to mapping out the development and fine structure of the discipline. One potentially fruitful method of approaching this problem is by executing a variety of focused case studies of specific alchemists or their schools; broad surveys of alchemy in the style of Taylor and Holmyard are no longer of value for advancing the field. Such precise studies could then be drawn upon for making comparisons and contrasts between styles and contents among different schools and epochs.[137] For example, we already know a great deal about Paracelsians, and it would now be useful to contrast this group with non-Paracelsian alchemical workers. Such a segregation of alchemical schools would sort out the conflicting works of rival alchemical practitioners, possibly showing, for example, that Paracelsus, somewhat too casually marked as a chief doyen of alchemy, was as much an outsider and iconoclast to the alchemical tradition as he was to classical Galenism. Likewise, the bracketing of Van Helmont and the Helmontians would resolve some of the paradoxes presented by apparent self-contradictions in seventeenth-century alchemy.

Since there is strong evidence that there is now a new interest in alchemy among historians of science, we hope that the kind of studies advocated here will be carried out with renewed vigor and liberated from the misconceived interpretations that this chapter has undertaken to criticize. Now that the importance of alchemy to the origins of early modern science is a more or less established fact, future studies will serve to define more rigorously what the precise lines of influence were. These future developments should also serve to elucidate the spectrum of notions, attitudes, and pursuits generally grouped under the wide umbrella of "alchemy" and to portray it as a vastly more dynamic field than has hitherto been presumed.

Acknowledgments

We wish to thank the National Science Foundation (SBR-9510135) and the National Endowment for the Humanities (RH-21301–95) for their support of our collaboration, part of the results of which are presented in this chapter.

Notes

1. For nineteenth-century occultism and the Romantic movement, see the still indispensable August Viatte, *Les sources occultes du romantisme,* 2 vols. (Paris: Librairie ancienne Honoré Champion, 1928). The Romantic rejection of Newtonianism is particularly strong in the works of William Blake. For his occult connections, see Kathleen Raine, *Blake and Tradition,* 2 vols. (Princeton: Princeton University Press, 1968), esp. 1:99–125, and 2:189–213.

2. William R. Newman and Lawrence M. Principe, "Alchemy vs. Chemistry: The Etymological Origins of a Historiographic Mistake," *Early Science and Medicine,* 3 (1998): 32–65.

3. W. Ganzenmüller, "Wandlungen in der geschichtlichen Betrachtung der Alchemie," *Chymia,* 3 (1950): 143–54; Dietlinde Goltz, "Alchemie und Aufklärung, Ein Beitrag zur Naturwissenschaftgeschichtsschreibung der Aufklärung," *Medizinhistorisches Journal,* 7 (1972): 31–48. For examples of the late debate against gold making, see J. C. Wiegleb, *Historich-kritische Untersuchung der Alchemie*

(Weimar, 1777), and Siegmund Heinrich Güldenfalk, *Sammlung von mehr als hundert wahrhaftigen Transmutationsgeschichten* (Frankfurt and Leipzig, 1784).

4. Johann Friedrich Gmelin, *Geschichte der Chemie*, 3 vols. (Göttingen, 1797–99); Thomas Thompson, *History of Chemistry*, 2 vols. (London, 1830). The exception that proves the rule is Karl Christoph Schmieder, *Geschichte der Alchemie* (Halle, 1832). Schmieder not only provides a positive view of alchemy but seems actually to have been a believer in alchemical transmutation. An example of a contemporary author who continues the eighteenth-century equation of alchemy with fraud (copious recently published evidence to the contrary notwithstanding) is Maurice Crosland; see, for example, his "Chemical Revolution of the Eighteenth Century and the Eclipse of Alchemy in the 'Age of Enlightenment,'" in Z. R. W. M. von Martels, ed., *Alchemy Revisited* (Leiden, the Netherlands: E. J. Brill, 1990), 67–77; compare his article with the balance of the volume.

5. Julius Sachse, *The German Pietists of Provincial Pennsylvania, 1694–1708* (Philadelphia: P. C. Stockhausen, 1895), passim.

6. Christopher McIntosh, *The Rose-Cross and the Age of Reason* (Leiden, the Netherlands: Brill, 1992); Charles Porset, "Les enjeux 'alchimiques' du Convent des Philalèthes," in Didier Kahn and Sylvain Matton, eds., *Alchimie: Art, histoire et mythes: Actes du 1er colloque international de la Société d'Étude de l'Histoire de l'Alchimie* (Paris: Société d'Étude de l'Histoire de l'Alchimie, 1995), 757–800.

7. Johann Christoph Adelung, *Geschichte der menschlichen Narrheit* (Leipzig, 1785).

8. Francis Barrett, *The Magus, or Celestial Intelligencer* (London, 1801); [Barrett?], *Lives of the Alchemystical Writers* (London, 1815).

9. See the treatments of Mesmerism in Robert Darnton, *Mesmerism and the End of the Enlightenment in France* (Cambridge: Harvard University Press, 1968); Henri Ellenberger, *The Discovery of the Unconscious* (New York: Basic Books, 1970); Alan Gauld, *A History of Hypnotism* (Cambridge: Cambridge University Press, 1992); and George Bloch, *Mesmerism* (Los Altos: W. Kaufmann, 1980).

10. If we restrict the meaning of esoteric to "accessible to only a small group or elite" and occult to "hidden," then alchemy is both esoteric (its meaning being consciously restricted) and occult (hidden from the common reader) during much of its history; however, the meanings attached to these words in common parlance since the nineteenth century and by the school now under consideration do not fall within such narrow denotative limits. The modern construction "occultist" appears to have been coined by Eliphas Lévi in the nineteenth century and was used first in English in 1881, connoting something more than a mere believer in one or more of the occult sciences. It refers primarily to a fusion of Eastern and Western magical and mystical beliefs that reached its apotheosis in the works of Madame Blavatsky and has undergone a revival in the "New Age" movements. See Joscelyn Godwin, *The Theosophical Enlightenment* (Albany: State University of New York Press, 1994), esp. 49; Mircea Eliade, *Occultism, Witchcraft, and Cultural Fashions* (Chicago: University of Chicago Press, 1976), esp. 47–54; and Brian P. Copenhaver, "Natural Magic, Hermetism, and Occultism in Early Modern Science," in

David C. Lindberg and Robert S. Westman, eds., *Reappraisals of the Scientific Revolution* (Cambridge: Cambridge University Press, 1990), 261–301, esp. 289.

11. Martin Luther, *Colloquia Mensalia* [*Tischreden*], trans. Henry Bell (London, 1652), 480; here Luther praises alchemy "for the sake of the Allegorie and secret signification . . . touching the Resurrection of the dead."

12. Herbert Butterfield, *The Origins of Modern Science 1300–1800* (New York: Macmillan, 1952), 98.

13. A fragment of the Thomas South's poem, entitled "Enigma of Alchemy," was found in 1918 as proof sheets folded into a secondhand book in a London bookshop. This fragment was published by William Leslie Wilmshurst (the editor of the reprinted version of *Suggestive Inquiry*) in *The Quest*, 10 (January 1919): 213–25 (reprinted, Edmonds, WA: Alchemical Press, 1984).

14. Mary Anne Atwood, *A Suggestive Inquiry into the Hermetic Mystery* (Belfast: William Tait, 1918), from the introduction by Walter Leslie Wilmshurst, 6–9.

15. A revised edition appeared in 1920 and a reprint of it in 1960 (New York: Julian Press); another reprint (1918 edition) is still in print (Yogi Publication Society).

16. Atwood, *Inquiry*, 52–53.

17. Ibid., 57; Philalethes on 61–62.

18. Ibid., 70–71.

19. Ibid., 26.

20. As an example: "chemical affinity, called Elective Attraction, is ruled by the same laws; and it is found that when two matters unite, one is attractive and the other repulsive; when either attraction or repulsion predominates in a matter, the circulation is in ellipse; but when they are in equilibrium, a circle is produced" (ibid., 155). A generous reader might compare this to Atwood's adored Boehme admixed with a sort of Newtonianism and Berzelian dualism; alternatively, more skeptical minds may see a resemblance to the superficial examination answers cobbled together by ingenious but unlearned students.

21. Ibid., 78–85, 96–98, 162, 454–55.

22. Ibid., 162.

23. Ibid., 143.

24. Ibid., 543; 527–58: "Mesmerism, as it is mechanically practiced in the present day, is a first step indeed, and this only before the entrance of that glorious temple of Divine Wisdom which a more scientific Handicraft enabled the ancients experimentally to enter, and from its foundation build up, as it were, a crystalline edifice of Light and Truth."

25. Ethan Allen Hitchcock, *Remarks upon Alchymists* (Carlisle, Penn., 1855); *Remarks upon Alchemy and the Alchemists* (Boston, 1857). The review appeared in the *Westminster Review*, 66 (October 1856): 153–62. See I. B. Cohen, "Ethan Allen Hitchcock: Soldier-Humanitarian-Scholar, Discoverer of the 'True Subject of the Hermetic Art,'" *Proceedings of the American Antiquarian Society*, 61 (1951): 29–136.

26. Hitchcock, *Remarks upon Alchemy*, iv.

27. Ibid., viii and 30. The fact that he implicitly extends the medieval "midnight of darkness" into the eighteenth century can be passed over with only a smirk. Not surprisingly, there are glimpses of period anti-Catholicism even though Hitchcock seems to speak well of the Catholic Church at one juncture (64–65).

28. See, for example, Ellic Howe, ed., *The Alchemist of the Golden Dawn: The Letters of the Reverend W. A. Ayton to F. L. Gardner and Others 1886–1905* (Wellingborough, U.K.: Aquarian Press, 1985). Eliphas Lévi, *Dogme et rituel de la Haute Magie*, 2 vols. (Paris, 1854–56). Lévi's volume was translated into all the important Western European languages; the English-language version was by A. E. Waite, *Transcendental Magic: Its Doctrine and Ritual* (London, 1897). See also Thomas A. Williams, *Eliphas Lévi: Master of Occultism* (University of Alabama Press, 1975), and Christopher McIntosh, *Eliphas Lévi and the French Occult Revival* (New York: Samuel Weiser, 1975).

29. See *The Story of Anna Kingsford and Edward Maitland, and of the New Gospel of Interpretation*, 3rd ed. (Birmingham, Ala.: 1905); Godwin, *Theosophical Enlightenment*, 333–37. Only two volumes of the *Corpus* appeared: a reissue of John Everard's 1649 translation of *Pymander* (London, 1884), and *The Virgin of the World of Hermes Trismegestus* (London, 1885; reprinted, Madras, 1900). One of the characteristics of Kingsford and Maitland—reproduced in their followers to this day—is a predilection for ludicrous etymologies; e.g., the "Guardin" of Eden, and Noah as a form of *nous*.

30. S. A. [Sapere Aude, alias William Wynn Westcott], *The Science of Alchymy* (London: Theosophical Publishing Society, 1893), 17–20. Westcott thought Atwood to be a clergyman of the Church of England, 17. Besides these esoteric notions, Westcott was known in a wholly different context as the coroner of northeast London and the coeditor of *The Extra Pharmacopaeia of Unofficial Drugs* (1883) and its fifteen supplements dating up to 1920; he is briefly mentioned (with a photograph) in *The Chemist and Druggist* for September 2, 1922, 339.

31. Ellic Howe, *The Magicians of the Golden Dawn* (New York: Samuel Weiser, 1978); R. A. Gilbert, *The Golden Dawn: Twilight of the Magicians* (San Bernardino, Calif.: 1988); see also the issues of *Cauda Pavonis* dedicated to the Golden Dawn, new series 8 (Spring and Fall 1989).

32. The Flamel text was a modernized version of the 1624 English translation by "Irenaeus Orandus" and was published at London in 1870. Seven volumes were issued in the *Collectanea Hermetica* between 1893 and 1896: Jean d'Espagnet's *Hermetic Arcanum* (Westcott's translation of the 1623 Latin edition); *Pymander; Lover of Philalethes, Hermetic Art* (1714); *Aesch Mezareph*, collected from *Cabala Denudata* of Rosenroth; *Somnium Scipionis; Chaldean Oracles of Zoroaster;* and Eugenius Philalethes, *Euphrates* (1655 edition).

33. Hargrave Jennings, *The Rosicrucians* (London, 1870); see esp. 20–39; Albert Pike, *Morals and Dogma of the Ancient and Accepted Scottish Rite* (London, 1871).

34. H. Carrington Bolton, "The Revival of Alchemy," *Science*, 6 (1897): 853–63, and *Annual Report of the Smithsonian Institution 1897*, 207–17.

35. See R. A. Gilbert, *A. E. Waite: Magician of Many Parts* (Wellingborough, U.K.: Crucible, 1987).

36. Arthur Edward Waite, *Lives of the Alchemystical Philosophers* (London, 1888), 9–27. The work was reissued as *Alchemists through the Ages* (New York: Rudolf Steiner Publications, 1970).

37. Waite, *Lives of the Alchemystical Philosophers*, 273.

38. Meaning, of course, the spirit, soul, and body of man and the mercury, sulfur, and salt of metals.

39. Ibid., 30–36 and 273.

40. Ibid., 274.

41. A. E. Waite, *Azoth, or the Star in the East, embracing the first matter of the Magnum Opus, the evolution of the Aprodite-Urania, the supernatural generation of the son of the sun, and the alchemical transfiguration of humanity* (London, 1893), 54, 58, and 60.

42. Waite was the editor and translator of each of the following: *Alchemical Writings of Edward Kelly* (1893); Basil Valentine, *Triumphal Chariot of Antimony* (1893); *Musaeum hermeticum*, 2 vols. (1893); Benedictus Figulus, *Golden Casket* (1893); *Hermetic and Alchemical Writings of Paracelsus*, 2 vols. (1894); Petrus Bonus, *Pearl of Great Price* (1894); *Turba Philosophorum* (1896). On Waite's translating activity, its context, and effects in terms of Paracelsus, see Andrew Cunningham, "Paracelsus Fat and Thin: Thoughts on Reputation and Realities," in Ole Peter Grell, ed., *Paracelsus: The Man and His Reputation, His Ideas, and Their Transformations* (Leiden, the Netherlands: Brill, 1998), 64–68.

43. A. E. Waite, *The Secret Tradition in Alchemy* (New York: Alfred Knopf, 1926), 165.

44. Ibid., 366.

45. Julius Evola, *La tradizione ermetica* (Bari: Guis, 1931); recently reprinted in English translation as *The Hermetic Tradition* (Rochester, Vt.: Inner Traditions International, 1995); Titus Burckhardt, *Alchimie, Sinn und Weltbild* (Olten: Walter Verlag, 1960), translated as *Alchemy, Science of the Cosmos, Science of the Soul* (London: Stuart and Watkins, 1967).

46. A. E. Waite, ed., *The Works of Thomas Vaughan*, foreword by Kenneth Rexroth (New York: University Books, 1968), 10.

47. G. Heym's review of René Alleul, *Aspects de l'alchimie traditionelle* (Paris: 1953), in *Ambix*, 5 (1956): 129–30; Robert Halleux, *Les textes alchimiques* (Turnhout, Belgium: Brepols, 1979), 56–58; "L'unique mérite de leur travaux est de montrer qu'une approche érudite de l'alchimie est plus qu'une nécessité scientifique: c'est une exigence de santé mentale," p. 57.

48. For a consideration of the "initiatic style" and its sources, see William R. Newman, *The Summa perfectionis of pseudo-Geber* (Leiden, the Netherlands: Brill, 1991), 85–99. On divine agency in alchemical laboratories see our forthcoming study on chymical laboratory practice.

49. Sylvain Matton, "Thématique alchimique et littérature religieuse dans la France du XVIIe siècle," *Chrysopoeia*, 2 (1988): 129–208 (De Sales quotation on pp. 199–200); Sylvia Fabrizio-Costa, "De quelques emplois des thèmes alchimiques dans l'art

oratoire italien du XVIIe siècle," *Chrysopoeia*, 3 (1989): 135–162; Stanton J. Linden, *Darke Hierogliphicks: Alchemy in English Literature from Chaucer to the Restoration*, (Lexington, KY: University of Kentucky Press, 1996). See also Daniel Merkur, "The Study of Spiritual Alchemy: Mysticism, Gold-Making, and Esoteric Hermeneutics," *Ambix* 37 (1990): 35–45. Refining metaphors occur in the Bible at, for example, I Peter 1:7, Proverbs 17:3 and 27:21, Wisdom 3:6, Job 23:10.

50. Thomas Browne, *Religio medici* in *Works of Sir Thomas Browne*, ed. Geoffroy Keynes, 4 vols. (Chicago: University of Chicago Press, 1964) 1:50; on Fabre, see Bernard Joly, *Rationalité de l'alchimie au XVIIe siècle* (Paris: Vrin, 1992), esp. pp. 42–45; Pierre Jean-Fabre, *Alchymista Christianus, in quo Deus rerum author et quam plurima fidei christianae mysteria per analogias chymicas et figuras explicantur, christianorumque orthodoxa doctrina, vita et probitas non oscitanter ex chymica arte demonstantur* (Toulouse, 1632).

51. Sylvain Matton, "Une lecture alchimique de la Bible: les 'Paradoxes chimiques' de François Thybourel," *Chrysopoeia*, 2 (1988): 402–22. Although the best-known exposition of Classical mythology as alchemy is Antoine-Joseph Pernety, *Les fables égyptiennes et grecques devoilées* (Paris, 1758), the genre was well-established much earlier; see the fifteenth-century Vincenzo Percolla, *Auriloquio*, ed. Carlo Alberto Anzuini, in *Textes et Travaux de Chrysopoeia* 2, 1996, and for other examples, Michael Maier, *Arcana arcanissima* (s.l., s.d. [London, 1613 or 1614]), and Pierre-Jean Fabre, *Hercules piochymicus* (Toulouse, 1634).

52. Thomas Sprat, *History of the Royal Society* (London, 1667), p. 37; Nicholas Melchior, *Processus sub forma missae*, in *Theatrum chemicum*, 6 vols. (Strasbourg, 1659–61), 3:758–61.

53. Yet even in the case of Boehme, it is not clear that practical alchemy was entirely excluded. See Will-Erich Peuckert, *Das Leben Jakob Boehmes* (Jena, Germany: Eugen Diederichs, 1924), 167.

54. Lawrence M. Principe, *The Aspiring Adept: Robert Boyle and His Alchemical Quest* (Princeton: Princeton University Press, 1998), 188–90, 197–201; Robert M. Schuler, "Some Spiritual Alchemies of Seventeenth-Century England," *Journal of the History of Ideas*, 41 (1980): 293–318.

55. Joseph Campbell, *The Flight of the Wild Gander: Explorations in the Mythological Dimension* (New York: Harper Perennial, 1990; first ed., 1951), 86–87, 218–19. Idem, *The Masks of God: Primitive Mythology* (Harmondsworth, U.K.: Penguin, 1982; first ed., 1959), 72. Northrop Frye, *Anatomy of Criticism* (Princeton: Princeton University Press, 1957). For Bachelard, see Barbara Obrist, *Les débuts de l'imagerie alchimique (XIVe–XVe siècles)* (Paris: Le Sycomore, 1982), 22–23. Mircea Eliade, *The Forge and the Crucible* (Chicago: University of Chicago Press, 1978; first ed., 1962; first French ed., 1956), 221–26.

56. Gareth Roberts, *The Mirror of Alchemy* (London: British Library, 1994), 7, 66. Allison Coudert, *Alchemy: The Philosopher's Stone* (London: Wildwood House, 1980), 148–60. F. Sherwood Taylor, *The Alchemists: Founders of Modern Chemistry* (New York: Henry Schuman, 1949), 159, 228. E. J. Holmyard, *Alchemy* (New York: Dover, 1990; first ed., 1957), 163–64, 176.

57. Betty Jo Teeter Dobbs, *The Foundations of Newton's Alchemy* (Cambridge: Cambridge University Press, 1975), 26–35; "From the Secrecy of Alchemy to the

Openness of Chemistry," in Tore Frängsmyr, ed., *Solomon's House Revisited* (Canton, Mass.: Science History Publications, 1990), 75–94, esp. 76. Dobbs's endorsement of the Jungian historiographical model has served not only to further its acceptance among historians of science but to justify the model to other Jungians. Hence Nathan Schwartz-Salant's *Encountering Jung on Alchemy* (Princeton: Princeton University Press, 1995), 10, explicitly cites Dobbs in support of the claim that historians of science "have recognized the value of Jung's approach."

58. William H. Brock, *The Norton History of Chemistry* (New York: Norton, 1993), 17 and 678. Marco Beretta, *The Enlightenment of Matter* (Canton, Mass.: Science History Publications, 1993), 77, n. 6, and 331.

59. Richard Noll, *The Jung Cult* (Princeton: Princeton University Press, 1994), 144. Idem, *The Aryan Christ* (New York: Random House, 1997), 25–30, 37–41.

60. Noll, *Aryan Christ*, 171.

61. Luther H. Martin, "A History of the Psychological Interpretation of Alchemy," *Ambix*, 22 (1975): 10–20, esp. 12–16.

62. Jung's earliest writing on alchemy is found in *The Secret of the Golden Flower*, published with Richard Wilhelm in 1929: see Martin, "A History," 16. Sustained treatments are found in *The Collected Works of Carl Gustav Jung*, 20 vols. (London: Routledge, 1953–79), vol. 9, pt. 2: *Aion*; vol. 12: *Psychology and Alchemy*; vol. 13: *Alchemical Studies*; vol. 14: *Mysterium Coniunctionis*.

63. Jung, "Die Erlösungsvorstellungen in der Alchemie," in *Eranos-Jahrbuch 1936* (Zurich: Rhein-Verlag, 1937), 13–111; in English, "The Idea of Redemption in Alchemy," in Stanley Dell, ed., *The Integration of the Personality* (New York: Farrar & Rinehart, 1939), 205–80. A retranslated and much expanded version of the original Eranos lecture appears in Jung, *Works*, vol. 12: *Psychology and Alchemy*, 227–471.

64. Jung, "Redemption in Alchemy," 210.

65. Ibid., 215. The German text is quite unequivocal in its reference to "hallucinations" (Jung, "Die Erlösungsvorstellungen," 23–24): "Wie die beiden vorhergehenden Texts, so beweisen auch Hoghelande's Ausführungen, dass während der praktischen Arbeit halluzinatorische oder visionäre Wahrnehmungen erfolgten, die nichts anderes sein können als Projektionen unbewusster Inhalte."

66. Jung, "Redemption in Alchemy," 213, 206.

67. Noll, *Aryan Christ*, 229–30.

68. Jung, "Die Erlösungsvorstellungen," 16: "Man kann sich vom Standpunkt unseres modernen chemischen Wissens darunter nichts vorstellen; und greifen wir zu den Texten und zu den Hunderten und Aberhunderten von Verfahren und Rezepten, die uns das Mittelalter und die Antike hinterlassen haben, so finden wir darunter relativ wenige, welche einen erkennbaren chemischen Sinn erhalten."

69. Jung, "Idea of Redemption," 239–47.

70. Ibid., 250–67.

71. Ibid., 239.

72. Jung, *Works*, vol. 12: *Psychology and Alchemy*, 316. Here the *Collected Works* version gives a better rendition of the German ("Die Erlösungsvorstellungen," 59) than does Dell's translation, "Redemption in Alchemy," 239.

73. Jung, "Redemption in Alchemy," 269–71.

74. Ibid., 268. For an elaboration of Jung's view of Paracelsus as a reformer of alchemy, see his early "Paracelsus," in Jung, *Works*, vol. 15: *The Spirit in Man, Art, and Literature*, 3–12. The essay is a translation of an address given in 1929.

75. Jung, "Redemption in Alchemy," 267.

76. Noll, *The Jung Cult*, 181–84.

77. Jung, *Works*, vol. 13: *Alchemical Studies*, 253–349.

78. Noll, *Aryan Christ*, 22–52, 98–119.

79. Jung, "Redemption in Alchemy," 269–70. "Die Erlösungsvorstellungen," 103–4: "Die alchemische Prozess der klassischen Zeit (von der Antike bis zum Ende des 16. Jahrhunderts) war eine an sich chemische Untersuchung, in welche sich auf Wege der Projektion unbewusstes psychisches Material mischte. Die psychologische Bedingung des Werkes wird daher in den Texten vielfach betont. Die in Betracht kommenden Inhalte sind solche, welche sich zur Projektion in den unbekannten chemischen Stoff eignen. Wegen der unpersönlichen, rein dinglichen Natur des Stoffes finden Projektionen von unpersönlichen, sogenannten kollektiven Archetypen statt."

80. Obrist, *Les débuts de l'imagerie alchimique*, 15–21, 183–245; Halleux, *Les textes alchimiques*, 55–58.

81. Marie-Luise von Franz, *Aurora consurgens: A document attributed to Thomas Aquinas* (New York: Pantheon Books, 1966), 407–31.

82. William R. Newman, *Gehennical Fire: The Lives of George Starkey, an American Alchemist in the Scientific Revolution* (Cambridge: Harvard University Press, 1994), 125–33; Lawrence M. Principe, "'Chemical Translation' and the Role of Impurities in Alchemy: Examples from Basil Valentine's Triumph-Wagen," *Ambix*, 34 (1987): 21–30.

83. William R. Newman, "The Authorship of the 'Introitus Apertus ad Occlusum Regis Palatium,'" in von Martels, *Alchemy Revisited*, 139–44; "'Decknamen or pseudochemical language?' Eirenaeus Philalethes and Carl Jung," *Revue d'histoire des sciences*, 49 (1996): 159–88; and *Gehennical Fire*, 128, 132. In our forthcoming study of the alchemical collaboration of Starkey and Boyle we will detail the methods of analysis used by Starkey to "decode" the writings of earlier alchemical writers such as Bernard of Trier.

84. Lawrence M. Principe, "Apparatus and Reproducibility in Alchemy," in Trevor Levere and Frederick L. Holmes, eds., *Instruments and Experimentation in the History of Chemistry* (Cambridge: MIT Press, 2000), 57–74.

85. On the emblematic worldview, see William B. Ashworth, "Natural History and the Emblematic World-View," in Lindberg and Westman, *Reappraisals of the Scientific Revolution*, 303–31.

86. Dobbs, *Foundations of Newton's Alchemy*, 26–29.

87. Betty Jo Teeter Dobbs, "Conceptual Problems in Newton's Early Chemistry: A Preliminary Study," in Margaret J. Osler and Paul Lawrence Farber, eds., *Religion, Science, and Worldview: Essays in Honor of Richard Westfall* (Cambridge: Cambridge University Press, 1985), 3–32, esp. 4. Dobbs claims to base this on a distinction made by Newton himself between "vulgar" and "vegetable" chemistry. See the critique of Dobbs's position in William R. Newman's review of *The Janus Faces of Genius*, *Isis* 84, no. 3 (1993): 578–79.

88. Newman and Principe, "Alchemy vs. Chemistry," passim.

89. Ibid.

90. Mircea Eliade, "Metallurgy, Magic and Alchemy," in *Cahiers de Zalmoxis, publiés par Mircea Eliade*, vol. 1 (Paris: Librairie Orientaliste Paul Geuthner, [1938]).

91. Mac Linscott Ricketts, *Mircea Eliade: The Romanian Roots, 1907–1945* (Boulder, Colo.: East European Monographs, 1988), 141–53, 313–25, 804–8, 835–42.

92. Eliade, "Metallurgy, Magic and Alchemy," 44. See also his "Alchemy as a Spiritual Technique" in *Yoga: Immortality and Freedom*, second ed., Bollingen Series 55 (Princeton: Princeton University Press, 1969), 290–92.

93. Indeed, *The Forge and the Crucible* is heavily influenced by Jung; see 52, 158, 161, 163, and 221–26.

94. Ibid., 162.

95. Eliade, "Metallurgy, Magic and Alchemy," 44.

96. Ibid., 23–25.

97. Ibid., 27 and 38.

98. Ibid., 23.

99. Ibid.

100. Ibid., 38.

101. Hélène Metzger, "L'évolution du règne métallique d'après les alchimistes du XVIIe siècle," *Isis*, 4 (1922): 466–82; *Les doctrines chimiques en France du début du XVIIe à la fin du XVIIIe siècle* (Paris: Les Presses Universitaires de France, 1923), 99–142.

102. Metzger speaks of "la théorie alchimique" as a homogeneous whole. According to "les philosophes hermetiques," "les transmutations sont non seulement possibles, mais certaines, qu'elles se font dans un sens determinée toujours la même, et qu'elles aboutissent forcement à donner aux metaux imparfaits l'admirable forme de l'or." See "L'évolution du règne métallique," 472.

103. Metzger, "L'évolution du règne métallique," 473.

104. Hélène Metzger, *Chemistry*, trans. Colette V. Michael (West Cornwall, Conn.: Locust Hill Press, 1991; first French ed., 1930), 13–18.

105. Metzger, *Les doctrines chimiques*, 162–63.

106. Jan Golinski, "Hélène Metzger and the Interpretation of Seventeenth Century Chemistry," *History of Science*, 25 (1987): 85–97.

107. Hélène Metzger, "L'a priori dans la doctrine scientifique et l'histoire des sciences" (1936) and "La philosophie de Lucien Lévy-Bruhl et l'histoire des sciences" (1930), in Gad Freudenthal, ed., *La méthode philosophique en histoire des sciences: textes 1914–1939* (Paris: Fayard, 1987), 41–56, 113–28 (esp. 50–51, 119–20). Golinski has already drawn attention to these passages in "Metzger and Chemistry," 96, n. 10.

108. In order for alchemy to collapse, "il fallait tout d'abord qu'une autre notion de la perfection succedât au mysticisme des alchimistes: or, à l'époque du triomphe de la philosophie cartésienne, les savants, qu'ils donnent ou refusent leur adhésion à l'ensemble de la nouvelle doctrine, les savants refusaient d'admettre qu'une substance quelconque qui occupe de l'étendue soit théoretiquement plus parfaite que toutes les autres substances analogues. Les métaux ont été créés par Dieu pour demeurer ce qu'ils sont; et le monde entier reste constamment semblable à lui-même, semblable à ce qu'il était au moment de la création; si, comme le prétend Descartes, 'toutes les variétés qui sont en la matière dependent du mouvement de ses parties,' si d'autre part 'Dieu qui est la première cause du mouvement en conserve toujours une égale quantité dans l'univers,' la nature qui est parfaite n'a donc aucune tendance au perfectionnement! L'idée même du perfectionnement paraîtra fantaisiste et inintelligible" (Metzger, *Les doctrines chimiques,* 138).

109. Metzger, *Les doctrines chimiques,* 27, 93, 133, 247, 289–91, 423–24. Dobbs has correctly criticized Metzger on this point: *Foundations of Newton's Alchemy,* 44–47. But Dobbs herself, though recognizing that corpuscularian views need not be at odds with transmutation, did not recognize that alchemical texts themselves often embody a corpuscular theory of matter having medieval roots.

110. Dobbs, *Foundations of Newton's Alchemy,* 26, n. 2; 43, n. 46 (for Eliade); 44–45, 45, n. 54 (for Metzger).

111. Betty Jo Teeter Dobbs, *Alchemical Death and Resurrection* (Washington, D.C.: Smithsonian Institution Libraries, 1990), and *The Janus Faces of Genius* (Cambridge: Cambridge University Press, 1991). For her 1990 endorsement of Jung, see "From the Secrecy of Alchemy to the Openness of Chemistry," 76.

112. Dobbs, *Alchemical Death,* 4; *Janus Faces,* 46.

113. Dobbs, *Alchemical Death,* 13, 25.

114. Ibid., 24.

115. Carolyn Merchant, *The Death of Nature* (San Francisco: Harper & Row, 1983; first ed., 1980), 17–20, 25–27. Merchant's use of Eliade is documented on 296, n. 1, her use of Jung on 298, n. 21.

116. Ibid., xxii.

117. Evelyn Fox Keller, *Reflections on Gender and Science* (New Haven: Yale University Press, 1985), 43–65. See the critique of Merchant and Keller in William Newman, "Alchemy, Domination, and Gender," in Noretta Koertge, ed., *A House Built on Sand* (Oxford: Oxford University Press, 1998), 216–26.

118. Keller, *Reflections,* 44.

119. Ibid., 48.

120. Ibid., 53–54.

121. Keller cites Merchant and Dobbs explicitly in *Reflections*. See 54–55, nn. 11–12.

122. See Newman, *Gehennical Fire*, 101–6, for a discussion of this issue. For the passage from pseudo-Lull, see *Il Testamentum alchemico attribuito a Raimondo Lullo*, ed. Michela Pereira and Barbara Spaggiari (Florence: Sismel, 1999), pp. 28–29.

123. Daniel Georg Morhof, *Epistola de metallorum transmutatione ad Joelum Langelottum*, in Manget, *Bibliotheca chemica curiosa*, 1:168–92, at 179; Lawrence M. Principe, "Daniel Georg Morhof's Analysis and Defence of Transmutational Alchemy," in *Mapping the World of Learning: The Polyhistor of Daniel Georg Morhof*, ed. Françoise Waquet (Wiesbaden: Harrassowtiz, 2000), 139–153.

124. Lawrence M. Principe, "Diversity in Alchemy: The Case of Gaston 'Claveus' DuClo, a Scholastic Mercurialist Chrysopoeian," in Allen G. Debus and Michael Walton, eds., *Reading the Book of Nature: The Other Side of the Scientific Revolution* (Kirksville, Mo.: Sixteenth Century Press, 1998), 181–200; Gaston DuClo, *Apologia chrysopoeiae et argyropoeiae*, in Lazarus Zetzner, ed., *Theatrum chemicum*, 6 vols. (Torino: Bottega d'Erasmo, 1981; reprint of 1659–61 Strasburg ed.), 1:4–80, at 36–37, 39, 43–44, and 65.

125. Newman, *Gehennical Fire*, 92–114; 141–69; "The Corpuscular Transmutational Theory of Eirenaeus Philalethes," in Piyo Rattansi and Antonio Clericuzio, eds., *Alchemy and Chemistry in the XVI and XVII Centuries* (Dordrecht, the Netherlands: Kluwer, 1994), 161–82; "The Corpuscular Theory of J. B. van Helmont and its Medieval Sources," *Vivarium*, 31 (1993): 161–91; "Boyle's Debt to Corpuscular Alchemy," in Michael Hunter, ed., *Robert Boyle Reconsidered* (Cambridge: Cambridge University Press, 1994), 107–18; and "The Alchemical Sources of Robert Boyle's Corpuscular Philosophy," *Annals of Science*, 53 (1996): 567–85.

126. Antonio Clericuzio, "A Redefinition of Boyle's Chemistry and Corpuscular Philosophy," *Annals of Science*, 47 (1990): 561–89; see 583–87, and John Henry, "Occult Qualities and the Experimental Philosophy: Active Principles in Pre-Newtonian Matter Theory," *History of Science*, 24 (1986): 335–81.

127. Justus Liebig, *Chemische Briefe* (Heidelberg, 1851), 59.

128. For an instance where Liebig is used to exemplify the positivist view of alchemy, see Beretta, *Enlightenment of Matter*, 330. Liebig dissents from Comte in *Briefe*, 68–70.

129. Hermann Kopp, *Geschichte der Chemie* (Brunswick, 1843–47); *Alchemie in Älterer und Neuerer Zeit* (Heidelberg, 1886).

130. Ganzenmüller, "Wandlungen," 69.

131. Jost Weyer, "The Image of Alchemy and Chemistry in Nineteenth and Twentieth Century Histories of Chemistry," *Ambix*, 23 (1976): 65–70. There are some notable exceptions, such as E. O. von Lippman, *Die Entstehung und Ausbreitung der Alchemie* (Berlin: Springer, 1919).

132. George Sarton, "Boyle and Bayle, the Sceptical Chemist and the Sceptical Historian," *Chymia*, 3 (1950): 155–89, at 160.

133. Dobbs, *Foundations of Newton's Alchemy,* 6–12, 16–18; Principe, *Aspiring Adept,* 18–23.

134. Only a few examples of recent and ongoing work are Barbara Obrist's translation and edition of Constantine of Pisa's *The Book of the Secrets of Alchemy* (Leiden: Brill, 1990); Robert Halleux, *Les alchemistes grecs* (Paris: Les Belles Lettres, 1981); William Newman, *The Summa perfectionis of the pseudo-Geber* (Leiden: E. J. Brill, 1991); the publications of *Chrysopoeia* under the direction of Sylvain Matton; Michela Pereira and Barbara Spaggiari, *Il Testamentum alchemico attribuito a Raimondo Lullo,* (Florence: Sismel, 1999); and Fabre's *Manuscriptum* in Bernard Joly, *Rationalité de L'alchimie au XVIIe siècle,* (Paris: Vrin, 1992).

135. Andrew Weeks, *Paracelsus: Speculative Theory and the Crisis of the Reformation* (Albany: State University of New York Press, 1997), 36–44. Problems of disputed authenticity occur with the *Paracelsian De natura rerum,* among others.

136. One example of a work untangling knotty attributions is Michela Pereira, *The Alchemical Corpus Attributed to Raymond Lull,* Warburg Institute Surveys and Texts no. 18 (London: Warburg Institute, 1989).

137. For a few of many possible examples of such case studies, see the collections Rattansi and Clericuzio, *Alchemy and Chemistry in the XVI and XVII Centuries;* Jean-Claude Margolin and Sylvain Matton, eds., *Alchimie et Philosophie à la Renaissance* (Paris: Vrin, 1993); and von Martels, *Alchemy Revisited;* as well as Karin Figala, "Zwei Londoner Alchemisten um 1700: Sir Isaac Newton und Cleidophorus Mystagogus," *Physis,* 18 (1976): 245–73; Principe, *Aspiring Adept* and "Diversity in Alchemy"; Newman, *Gehennical Fire;* and Robert M. Schuler, "William Blomfild, Elizabethan Alchemist," *Ambix,* 20 (1973): 75–87.

Contributors

Nicholas H. Clulee is Professor of History at Frostburg State University in Maryland. He received a B.A. from Hobart College and a Ph.D. from the University of Chicago. He published *John Dee's Natural Philosophy* (Routledge, 1988) and is currently working on Roger Bacon and an edition of John Dee's works.

Germana Ernst is professor of philosophy at the University of Rome, III. She is the author of *Religione, ragione e natura* (Franco Angeli, 1991) and the editor of many sixteenth- and seventeenth-century texts, notably Tommaso Campanella's *Articuli prophetales* (1977) and *Compendio di filosofia della* natura (1999) and Ortensio Landi's *Paradossi* (1999).

Anthony Grafton teaches European history and history of science at Princeton University. His publications include *Joseph Scaliger* (Oxford, 1983), *Defenders of the Text* (Harvard, 1991), and *The Footnote: A Curious History* (Harvard, 1997); he coedited, with Nancy Siraisi, *Natural Particulars* (MIT, 2000).

Didier Kahn currently occupies the position of "chercheur" at the CNRS. He is the author of a doctoral thesis, "Paracelsisme et alchimie en France à la fin de la Renaissance" (Université de Paris IV, 1998, to be published by Droz). His work concerns the relationship between literature, science, and religion in the Europe of the alchemists (1567–1666), as well as the editing of Diderot's correspondence.

Lauren Kassell is an assistant lecturer in history and philosophy of science at Cambridge University and a fellow of Pembroke College. She is completing a book on Simon Forman, medicine, and the occult in early modern England.

William R. Newman is a member of the Department of History and Philosophy of Science at Indiana University. He has published *Gehennical Fire: The Lives of George Starkey, An American Alchemist in the Scientific Revolution* (Harvard, 1994) and *The Summa perfectionis of pseudo-Geber* (Brill, 1991).

Lawrence M. Principe is an associate professor at Johns Hopkins University in the Departments of History of Science and of Chemistry. He is author of *The Aspiring Adept: Robert Boyle and his Alchemical Quest* (Princeton, 1998).

H. Darrel Rutkin is a doctoral candidate in the Department of History and Philosophy of Science at Indiana University. His dissertation will study Giovanni Pico della Mirandola's *Disputationes adversus astrologiam divinatricem,* its context and influence. A Rome Prize winner, he is currently conducting dissertation research at the American Academy in Rome.

Nancy G. Siraisi is professor of history at Hunter College and the Graduate Center of the City University of New York. Her field of research is history of medieval and Renaissance medicine and life sciences in intellectual and social context. Her most recent book is *The Clock and the Mirror: Girolamo Cardano and Renaissance Medicine* (Princeton: Princeton University Press, 1997).

Index

Adam, Charles, 296
Adelung, Johann Christoph, 387
Admiranda Methodus, 252
Adorno, Theodor Wiesengrund, 2
Advantages of Life in the Papal Curia, The, 7
Against the Astrologers, 45
Agrippa, Henry Cornelius, 96, 357–358, 369
Airs Waters Places, 80–82
Alberti, Leon Battista, 5–9
Alberti, Piero, 5–9
Alchemical Death and Resurrection, 411
Alchemy
  arguments against, 21–24
  and Aristotle, 214
  and Arthur Edward Waite, 393–395
  and astrology, 14–27
  books, 19, 198–199
  and Carl Gustav Jung, 401–408
  and *Consideratio*, 213–224
  eighteenth century view of, 385–388
  Ethan Allen Hitchcock on, 391–392
  future of, 418–420
  Johannes Trithemius on, 191–194, 196
  John Dee on, 174–178, 182–194, 195
  and Lullists, 251–252
  and magic, 25–26, 367–373
  Mary Anne Atwood on, 389–391
  and metals, 16–17, 214–217
  and Mircea Eliade, 408–415
  panpsychic interpretation of, 408–415
  and Philipp à Gabella, 201
  and planets, 18–22, 184–185
  positivist view of, 415–417
  presentist view of, 415–417
  and psychology, 401–404
  and religion, 352–360
  and Rosicrucianism, 301–302, 309–310
  and Simon Forman, 345–352, 367–373
  spiritual interpretation of, 388–396
  supernatural, 399–400
  twentieth century view of, 396–401
  and vitalism, 411–412
Alchemy: The Philosopher's Stone, 401
Alchymista Christianus, 398
Alciato, Andrea, 45
Al-Kindi, 209
Aloysia Sigea, 245
Alumbrados of Seville, the, 264–268
Andreae, Johann Valentin, 238, 240, 262
Aphorisms, 82–83, 86
Archinto, Filippo, 43, 56
Aristotle, 214
Arnold, Paul, 295
Ars alchemie, 19
Ashmole, Elias, 15–16, 19–21, 24–26, 364, 399

436   Index

Astrologers. *See also* Cardano, Girolamo; Dee, John; Galilei, Galileo; Kepler, Johann
  accuracy of predictions made by, 4, 82–84, 103–104
  dignity of, 46–49
  German, 9–14
  of Italy, 5–9
  physicians as, 77-98, 107–110, 367–373
  practices of Renaissance, 3–4, 12–13, 106
  and Ptolemy, 98–107
Astrologicorum aphorismorum segmenta septem, 45
Astrology
  and alchemy, 14–27
  arguments against, 21–24, 60, 88–89
  as art, 46–49, 57, 105–106
  and astronomy, 9, 18, 59–60, 154, 177–178, 203–213
  basis of, 59–62
  books, 19, 39–40, 44–46, 74–76, 95–96, 198–199
  and Christianity, 50–56
  and the course of illness, 71–73
  and the critical days doctrine, 86–89
  critiques of, 74–76, 91–92
  dedicatory letters, 135–159
  genitures, 3–4, 10–12, 40–43, 56–62, 96–98, 104–105
  and intellectuals, 7–8
  and language, 178–182
  medical, 77–92
  and medical interrogations, 69–74
  and memory, 143
  and metals, 16–17, 184–185
  and numbers, 182
  and occult sciences, 17, 26, 45, 149–150, 391–392, 396–401
  and physicians, 77–92
  and plagues, 85
  and poetry, 139–140
  and Ptolemy, 98–107
  and the Pythagorean tetractys, 180
  and religion, 50–56, 61, 201, 237–238, 263, 281–283, 287–289, 352–360
  and the Roman Empire, 99–100
  and royalty, 3, 5–9, 40–43, 96, 142–155, 255
  and self-scrutiny, 11–12
  skepticism of, 4–5
  and social relations, 8–9
  and spirituality, 43–44
  and the stock market, 2
  textbooks, 9–10
Astronomiae Instauratae Mechanica, 134, 142–143
Astronomia inferior, 173–174
Astronomia Nova, 133–134, 147–159
Astronomy and astrology, 9, 18–22, 59–60, 154, 177–178, 184–185, 190f, 203–213
Atheism, 277
Atwood, Mary Anne, 14, 30, 389–391
Azoth, 394–395

Bachelard, Gaston, 401
Bacon, Roger, 191, 209, 362
Baduel, Claude, 96–97
Baillet, Adrien, 245, 264, 295–300
Baillif, Roch Le, 275
Barbavara, Francesco, 6–9
Beck, Hans-Georg, 10
Belot, Jean, 251
Benzi, Francesco, 89
Beretta, Marco, 407
Bernard of Trier, 23
Besold, Christoph, 238
Béthune, Philippe de, 254
Biagioli, Mario, 133–134
Bianchi, Lorenzo, 245, 283
Blavatsky, Helena Petrovna, 393
Bodier, Thomas, 92
Boehme, Jakob, 387, 390
Boissat, Pierre de, 245–247
Boivin de Villeneuve, Jean, 302
Books on astrology, 19, 39–40, 44–46, 74–76, 95–96, 198–199
Bostocke, Richard, 345, 362–363

Boucher, Jean, 277, 279, 293
Boyle, Robert, 407, 414
Brahe, Tycho, 134
  dedicatory letter to *Astronomiae Instauratae Mechanica* of, 142–143
Brasavola, Antonio Musa, 83
Brind'Amour, Pierre, 11
Browne, Thomas, 398
Bruno, Giordano, 413
Burckhardt, Jacob, 3–4, 7–8
Burckhardt, Titus, 396
Burggraf, Johann Ernst, 277
Butterfield, Herbert, 389

Calvinists, 241–242
Camerarius, Joachim, 9, 13
Campanella, Tommaso, 5
Campbell, Joseph, 401
Campioni, Antonio de, 69–70
Cardano, Battista, 70–71
Cardano, Girolamo, 12–13, 27–28
  and astrology as art, 105–106
  books by, 39-40, 44–46, 95–96
  on Christianity, 50–56
  on the course of illness, 71–73
  on critical days, 86–89
  critiques by, 74–76, 91–92, 109
  early work of, 43–46
  genitures by, 40–43, 44, 56–62, 92, 96–98, 104–105
  on God, 85
  on large-scale events, 50–52
  and medical interrogations, 69–74
  medical work of, 92–98
  on plagues, 85
  on Ptolemy, 98–107
  use of astrology as a physician, 92–98, 107-110
  on value of medical astrology, 77–92
  on weather, 84–85
Casanate, Guglielmo, 42
Casato, Pietro, 84
Cellini, Benvenuto, 109
Centiloquium, 102, 106
Charpentier, Marc-Antoine, 308
Charron, Pierre, 287–288

Chaume, Étienne, 29, 245–247, 255–256, 259, 264, 294
Cheke, John, 41
Chesne, Joseph Du, 275
Chevallier, Pierre, 261
Choricr, Nicolas, 245, 247–248, 258–259
Christianity and astrology, 50–56, 61, 85, 201
  and alchemy, 396–401
  and Libertinism, 287–289
  and *Life of Adam and Eve*, 352–360
  and origin of disease, 360–367
  and the Rosicrucian movement, 237–238, 241, 263–268, 277–278, 281–283
Chymische Hochzeit, 238
City of the Sun, 5
Cohen, Gustave, 296, 300
Colletet, Guillaume, 284–287
Confessio Fraternitatis, 238, 240–241
Conradus, Rodolphus, 249–250
Consideratio, 199–201
  and alchemy, 213–224
  and astronomy, 202–204
  on creation, 204–207
  vs. *Emerald Tablet*, 201–213
  and *Monas hieroglyphica*, 222
  and *Novum lumen chemicum*, 223
  and *Propaedeumata aphoristica*, 202–207
  and *Scholia*, 218–220
Constantine of Pisa, 19
Continuation de l'histoire du regne de Louys le Juste, 267
Coppie of a Letter . . . by a Learned Physician, A, 345, 363, 365–366
Corneille, Thomas, 307–308
Corpus Hermeticum, 392
Cosmology, 201. *See also* Christianity and astrology
Coudert, Allison, 401
Critical days doctrine, the, 86–89

D'Abano, Pietro, 79, 88–89
Da Castiglionchio, Lapo, 7

Da Montegeltro, Federigo, 7
De Angelis, Alessandro, 60–61
Death of Nature, The, 412
Dedicatory letters
  to Brahe's *Astronomiae Instauratae Mechanica,* 142–143
  chronological considerations of, 155–159
  features of, 135–136
  to Galileo's *Sidereus Nuncius,* 136–142, 143–147
  and historiography, 148–149
  to Kepler's *Astronomia Nova,* 147–155
Dee, John, 14–15, 28–29, 173–174, 224–226
  and alchemy, 174–178, 182–194, 195
  and astronomy, 177–178, 186–188f, 190f, 204–213
  and the Emerald Tablet, 182–194
  and Johannes Trithemius, 194–197
  on language, 178–182
  and Philipp à Gabella, 197–213
  and *Propaedeumata aphoristica,* 202–207
  and Rosicrucianism, 197–198, 236–237
De La Brosse, Guy, 306
De Lavalle, Claude, 41
Della Mirandola, Pico, 4, 80, 87–88
Del Rio, Martin, 282
De occulta philosophia, 109
De Petri Boessatii vita, 247
De Pizan, Christine, 23
De Ranconnet, Aimar, 41, 96
Descartes, René, 252, 295–300
Des Freres de la Roze Croix, 302
D'Espagnet, Jean, 254
Des Réaux, Gédéon Tallemant, 248
D'Este, Leonello, 7
De stella nova, 158
De Visé, Donneau, 307–308
Die Morgenland Fahrt, 235
Difference Betwene the Auncient Phisicke, The, 345, 362–363
Dillon, John, 15

Disease and astronomy
  and the critical days doctrine, 86–89
  and the food of angels, 360–367
  Girolamo Cardano on, 74–92
  and medical interrogations, 69–74
  and plagues, 85
  and uses of medical astrology, 107–110
  and weather, 84–85
Divine Weeks and Works, 366
Dobbs, Betty Jo Teeter, 401, 407, 411–412
Donville, Louise Goddard de, 268
Dorn, Gerhard, 202–203
Du Bartas, Guillaume Saluste, 366
Duclo, Gaston, 414
Dupèbe, Jean, 11
Dupleix, Scipion, 266–268
Dupuy, Jacques, 274, 301

Economist, The, 2
Edighoffer, Roland, 237
Effroyables Pactions, 243, 248, 253, 259, 261, 270–275
Eglinus, Raphael, 197
Eliade, Mircea, 401, 408–415
Emblems, 45
Emerald Tablet, 25, 173–174, 179, 182–194
  vs. *Consideratio,* 201–213
England, 11, 40–41, 96, 237, 388–393
Epidemics, 80–83, 86, 103
Epistola ad Thomam de Bononia, 23
Epistolae medicinales, 89
Erastus, Thomas, 281
Esteve, Pedro Jaime, 82
Evola, Julius, 396
Examen sur l'Inconnue et Nouvelle Caballe des Freres de la Croix Rosee, 261, 272–273

Fabre, Pierre-Jean, 398
Fama Fraternitatis, 238–240, 258
Farnese, Pier Luigi, 42, 58
Ficino, Marsilio, 14, 24, 79, 109
Figulus, Benedictus, 240, 278

Fludd, Robert, 387
Forge and the Crucible, The, 408
Forman, Simon
 and alchemy, 345–346, 367–373
 on Alexander von Suchten, 346–352
 and *Of Cako*, 346–352
 and the food of angels, 360–367
 medical practice of, 367–373
 on origins of disease, 360–367
 on Paracelsus, 357–360
 transcription of *Life of Adam and Eve* by, 352–360
Foul, Robert Le, 251
Foundations of Newton's Alchemy, 401, 407
France. *See* Rosicrucian movement
Frisius, Gemma, 47
Frye, Northrop, 401
Fubini, Riccardo, 9

Gabella, Philipp à, 174, 224–226
 and alchemy, 201, 213–224
 and astronomy, 203–213
 and *Consideratio*, 201–213
 and cosmology, 201
 on creation, 204–206
 vs. John Dee, 201–213
 and *Propaedeumata aphoristica*, 202–207
 and *Secretioris Philosophiae Consideratio Brevis*, 197–201, 207f, 210–212f, 221–222f
 and truth, 201
Galen, 99–105
Galilei, Galileo, 133–135
 chronological considerations and, 155–159
 dedicatory letter to *Sidereus Nuncius* of, 136–142, 143–147
 vs. Johann Kepler, 154–155, 156–159
 and poetry, 139–140
Ganay, Germanus de, 191–193
Garasse, François, 268, 278, 280–283
Gaspard, M., 273
Gaulmin, Gilbert, 274
Gaultier, Jacques, 263, 266, 278, 293

Gaurico, Luca, 57–58, 98, 105
Gellius, Aulus, 99
Genitures, astrology, 3–4, 10–12, 40–43, 44, 56–62, 92, 96–98, 104–105
 of Jesus Christ, 52
Germany, 236–242, 262–263
Geschichte der menschlichen Narrheit, 387
Gilly, Carlos, 198, 235, 237, 245
Giordano Bruno, 237
Gmelin, Johann Friedrich, 416
Goclenius, Rudolph, 276
Goclenius Heautontimorumenos, 276–277
Gogava, Antonio, 39
Gohory, Jacques, 218–221
Goltz, Dietlinde, 407
Gonzaga, Ludovico, 7
Gonzago, Alvise, 84–85
Goodage, John, 370
Gouhier, Henri, 295, 296, 299
Gras, NIcolas Le, 306
Great Work, the, 218–219
Grisogono, Federico, 88–89
Grosseteste, Robert, 209
Guarinoni, Hippolytus, 240
Guicciardini, Francesco, 11
Gutman, Aegidius, 239

Halleux, Robert, 406
Hamilton, John, 39, 96
Harris, John, 369
Hartmann, Johann, 238
Haslmayr, Adam, 238, 239–240, 242, 262, 278, 293
Henry II (England), 74
Hess, Tobias, 238, 242
Hesse, Hermann, 235
Hillgarth, J. N., 251
Histoire generale du progrez et decadence de l'Heresie moderne, 261
Hitchcock, Ethan Allen, 30, 389, 391–392
Holland, 300
Holmyard, E. J., 401
Horky, Martin, 156
Horoscopes. *See* Genitures, astrology

Ingegno, Alfonso, 58
Instruction à la France sur la verité de l'histoire des Freres de la Roze-Croix, 261, 283–287, 291–294
Intellectuals and astrology, 7–8
Italy, astrology practice in, 5–9. *See also* Cardano, Girolamo

Janus Faces of Genius, The, 411
Jesus Christ, 40
Jollivet-Castelot, F., 393
Jung, Carl Gustav
  criticism of, 404–408
  interpretation of alchemy by, 401–404
  and occult sciences, 401–404

Keller, Evelyn Fox, 412
Kepler, Johann, 28, 133–134
  and astronomy, 154
  chronological considerations and, 155–159
  dedicatory letter to *Astronomia Nova*, 147–155
  vs. Galileo Galilei, 154–155, 156–159
  and historiography, 148–149
  and nativity of Rudolf II, 152–154
  and occult sciences, 149–150
Khunrath, Heinrich, 254, 387
Kingsford, Anna, 3é2
Kircher, Athanasius, 1–2
Kirsop, Wallace, 261
Kopp, Hermann, 416

La Doctrine curieuse des beaux Esprits de ce temps, 268–275, 284–287
L'Alemant, Adrien, 82, 91–92
Lancre, Pierre de, 287–288
Language and astrology, 178–182
Lautensack, Paul, 239
Lemery, Nicolas, 14, 411, 414
Lenoble, Robert, 296, 299
Leonello, Marquis, 3
Leowitz, Cyprian, 12, 98
Lerner, Michel-Pierre, 278

Les Enffans de la Croix Roze, 268
Lévi, Eliphas, 14, 392
L'Hyperchimie, 393
Liber de propria vita, 107
Liber de providentia ex anni constitutione, 83
Liber secretorum alchimie, 19
Libertinism, 287–289
Liebig, Justus, 415
Life of Adam and Eve, 352–360
Lives of the Alchemystical Writers, 387
Locke, John, 248
Los Angeles Times, 2
Lull, Ramon, 251–252, 413
Luther, Martin and Lutheranism, 4, 290–291, 388

MacDonald, Michael, 12
Magic and alchemy, 25–26, 367–373
Magini, Giovanni Antonio, 69–74, 156–157
Magnus, Albert, 53
Maier, Michael, 263
Mainardi, Giovanni, 89
Maitland, Edward, 392
Malingre, Claude, 261–262, 265–266, 286
Matton, Sylvain, 295
Mayerne, Théodore Turquet de, 254–259, 275
Medical astrology
  and alchemy, 214–215, 367–373
  and the critical days doctrine, 86–89
  and Galen, 99–105
  Girolamo Cardano on, 74–92, 107–110
  and medical interrogations, 69–74
  and origin of disease, 360–367
  Philipp à Gabella on, 214–215
  and plagues, 85
  and Ptolemy, 98–107
  and Rosicrucianism, 289–291, 303–304
  uses of, 107–110
  and weather, 84

Medical interrogations, 69–74
Médici, Catherine de, 74
Médici, Cosimo II de, 133–134, 136–142, 143–147
Médici, Giovan Giacomo, 41–42
Meijer, Willem, 300
Melanchthon, Philipp, 9, 13
Memory and astrology, 143
Merchant, Carolyn, 412
Mercure François, 243–244, 266
Mersenne, Marin, 277, 280–283
Mesmer, Franz Anton, 388
Metals and astrology, 16–17, 184–185, 214–217
Metzger, Hélène, 409–411
Mirror of Alchemy, The, 401
Monas hieroglyphica, 14, 173–174, 176–177f, 186–188f
  and alchemy, 174–178, 182–194
  and astronomy, 202–204
  vs. *Consideratio*, 201–213
  on creation, 204–206
  and language, 178–182
Moran, Bruce, 197–198
Moreau, René, 289–291
Morestel, Pierre, 251
Morhof, Daniel Georg, 413
Moritz, Landgraf, 197–198
Mydorge, Claude, 296–297
Mysterium Cosmographicum, 157

Napier, Richard, 369
Naudé, Gabriel, 243, 247, 254, 261
  on *La Doctrine curieuse*, 284–287
  on Libertinism, 287–289
  on Paracelsianism, 283, 289–291
  on Rosicrucianism, 291–294
Neuhus, Heinrich, 260–261
Newman, William, 191, 248
Niccoli, Ottavia, 4–5
Nifo, Agostino, 47
Noll, Richard, 404–406
Norton, Thomas, 19–20
Norton History of Alchemy, 401
Nostradamus, 11
Novum lumen chemicum, 200, 217, 223

Numbers and astrology, 182
Nutton, Vivian, 102

Obrist, Barbara, 406
Occult sciences, 17, 26, 45, 391–392, 396–401
  and Carl Gustav Jung, 401–404
  and royalty, 149–150
Offenbahrung Göttlicher Majestät, 239
On Painting, 9
On the Art of Building, 9
On the Family, 5–9
On the Seven-Month Child, 80
Optica, 158
Ordinall of Alchemy, The, 19, 185
Oresme, Nicole, 23
Ottato, Cesare, 88

Pailleur, Jacques Le, 248–252
Panpsychic interpretation of alchemy, 408–415
Paracelsus, 14, 218
  and France, 275
  and François Garasse, 280–283
  Gabriel Naudé on, 283, 289–291
  and Rosicrucians, 275–294
  Simon Forman on, 357–360
Pascal, Blaise, 248
Pasquier, Étienne, 275
Paterculus, Velleius, 100
Penot, Bernard G., 291
Pereyra, Benito, 61
Peucer, Caspar, 13
Philalethes, Eugenius, 390
Philosophers' Stone, The, 307
Physicians and astrology. *See* Medical astrology
Pighius, Albertus, 47
Pike, Albert, 393
Plagues, 85
Poetry and astrology, 139–140
Pomata, Gianna, 108
Poncé, Charles, 396
Pope Paul III, 42–43, 57–58
Positivist view of alchemy, 415–417
Presentist view of alchemy, 415–417

Prognostic, 80–83
Propaedeumata aphoristica, 175f, 202–207
Propertius, 139–140
Protestants and the Rosicrucian movement, 263, 278
Prutenic Tables, 83
Psychology and alchemy, 401–404
Ptolemy, Claudius, 10, 15, 39, 46–47
  and Cardano, Girolamo, 98–107
  on large-scale events, 50–51
Pythagorean tetractys, 180

Quadripartitum, 39, 47
Quaestiones celeberrimae in Genesim, 277
Quigley, Joan, 2

Raemond, Florimond de, 261
Reagan, Nancy, 2
Reagan, Ronald, 2
Recherches sur les Rose-Croix, 274
Reflections on Gender and Science, 412
Reiske, Jacob, 10
Religion and the Decline of Magic, 17
Rhenanus, Johannes, 198
Rheticus, Georg Joachim, 12, 57
Roberti, Jean, 276–277
Roberts, Gareth, 401
Rodis-Lewis, Geneviéve, 296
Roman empire, 99–100
Rosarium philosophorum, 199
Rosicrucian Enlightenment, The, 236, 237, 245, 262
Rosicrucian movement, the, 28–29, 218, 393–394
  and alchemy, 301–302, 309–310
  and the Alumbrados of Seville, 264–268
  authors of, 244–252
  books against, 278–279
  books on, 238–239
  and Claude Malingre, 261–262
  content of, 253–259
  denunciation of, 291–294
  and Father Mersenne, 280–283, 295–300

  and François Garasse, 280–283
  and Gabriel Naudé, 291–294
  and Heinrich Neuhus, 260–261
  influence of, 306–311
  and John Dee, 197–198
  and *La Doctrine curieuse,* 284–287
  and Libertinism, 287–289
  origins of, 236–242
  and Paracelsianism, 275–294, 289–291
  placards of 1623, 235–236, 242–244
  and politics, 242
  and Protestants, 263, 278
  reaction to, 260–275
  and religion, 237–238, 241, 277–278
  and René Descartes, 295–300
  results of, 300–311
  and royalty, 239–240
  and Scipion Dupleix, 266–268
  and Théophile de Viau, 268–275
  versions of, 254–255, 259
  and Wallace Kirsop, 261
Röslin, Helisaeus, 239
Royalty and astrology, 3, 5–9, 40–43, 96, 142–155, 255
  and the Rosicrucian movement, 239–240
Rudolf II, Emperor, 142–143, 147–155

Sanches, Francisco, 52
Sarton, George, 416
Schmieder, Karl Christoph, 416
Scholem, Gershom, 255
Scholia, 218–220
Schweighart, Theophilus, 301
Scot, Michael, 19
Scotland, 39–40
Scribani, Charles, 250
Secret, François, 235, 245
Secretioris philosophie consideratio brevis, 174, 197–201, 207f, 210–212f, 221–222f
Secret Tradition of Alchemy, The, 395
Sendivogius, Michael, 200–201, 215–216
Sfondrato, Francesco, 42

Sforza, Paolo, 94–95
Sidereus Nuncius, 133–135, 143–147, 156–159
  dedicatory letter to, 136–142
Silberer, Herbert, 401
Skepticism of astrology, 4–5
Smith, Thomas, 11–12
Social relations and astrology, 8–9
Solomon, Howard, 305
Speculum Rhodostauroticum, 301
Spirituality and astrology, 43–44
Spon, Charles, 84
Starkey, George, 406–407
Steinmoel, Johann, 192
Suggestive Inquiry into the Hermetic Mystery, A, 389
Supernatural alchemy, 399–400

Table chronographique, 266
Tabula smaragdina. *See* Emerald Tablet
Taylor, F. Sherwood, 401
Tetrabiblos, 10, 73, 101–107
Theatrum chemicum britannicum, 15–16, 19, 25–26, 240, 364
Theatrum Honoris, 249–250
Themis aurea, 263
Thomas, Keith, 12
Thomas of Bologna, 23
Thompson, E. P., 2
Thoynard, Nicolas, 248
Thresor de l'histoire generale de nostre temps, 273
Toscanelli, Paolo, 9
Trismegistus, Hermes, 192–193, 199–200, 345
Trithemius, Johannes, 25, 173–174, 224–226
  and alchemy, 191–194, 196
  and John Dee, 194–197
Turnèbe, Adrien, 52

Uffel, Bruno Carolus, 198

Valla, Giorgio, 99
Van Helmont, Jean Baptiste, 276, 278, 295, 300

Van Wassenaer, Nicolaes Janszoon, 300
Vaughan, Thomas, 16–17, 390
Viau, Théophile de, 268–275
Vickers, Brian, 17
Vie de Monsieur Des-Cartes, 248, 264
Visconti, Giangaleazzo, 5–9
Vitalism, 411–412
Voet, Gisbert, 251–252
Von Anhalt, August, 238–239, 241
Von Bezold, Friedrich, 3
Von der Pfalz, Friedrich, 241
Von Franz, Marie Luise, 406
Von Nettesheim, Agrippa, 14, 16–17, 24
Von Suchten, Alexander, 346–352
Von Uffel, Bruno Carlus, 197

Waard, Cornelis de, 296
Waite, Arthur Edward, 272, 393–395
Warburg, Aby, 3–4
Weather and disease, 84–85
Wessel, Wilhelm, 197
Westcott, William Wynn, 393
Wilmshurst, Walter Leslie, 390
Wolf, Hieronymus, 9–14
Worm, Ole, 238

Yates, Frances, 17, 28, 236–241, 262

Zetzner, Lazarus, 251